This highly illustrated monograph provides a comprehensive treatment of the study of the structure and function of proteins, nucleic acids and viruses using synchrotron radiation and crystallography.

Synchrotron radiation is intense, polychromatic and finely collimated, and is highly effective for probing the structure of macromolecules. This is a fast-expanding field, and this timely monograph gives a complete introduction to the technique and its uses. Beginning with chapters on the fundamentals of macromolecular crystallography and macromolecular structure, the book goes on to review the sources and properties of synchrotron radiation, instrumentation and data collection. There are chapters on the Laue method, on diffuse X-ray scattering and on variable wavelength dispersion methods. The book concludes with a description and survey of applications including studies at high resolution, the use of small crystals, the study of large unit cells, and time-resolved crystallography (particularly of enzymes). Appendices are provided which present essential information for the synchrotron user as well as information about synchrotron facilities currently available and planned. A detailed bibliography and reference section completes the volume. Many tables, diagrams and photographs are included.

This book is aimed at crystallographers, physicists, chemists and biochemists in universities, research institutes and in the pharmaceutical industry.

MACROMOLECULAR CRYSTALLOGRAPHY WITH SYNCHROTRON RADIATION

Macromolecular crystallography with synchrotron radiation

JOHN R. HELLIWELL

Professor of Structural Chemistry
at the
University of Manchester
and
Joint Appointee
at
SERC, Daresbury Laboratory

CAMBRIDGE
UNIVERSITY PRESS

Published by the Press Syndicate of the University of Cambridge
The Pitt Building, Trumpington Street, Cambridge CB2 1RP
40 West 20th Street, New York NY 10011–4211, USA
10 Stamford Road, Oakleigh, Victoria 3166, Australia

First published 1992

Printed in Great Britain
at the University Press, Cambridge

A catalogue record of this book is available from the British Library

Library of Congress cataloguing in publication data

Helliwell, John R.
Macromolecular crystallography with synchrotron radiation / John
R. Helliwell.
p. cm.
Includes bibliographical references and index.
ISBN 0–521–33467–5 (hardcover)
1. X-ray crystallography. 2. Synchrotron radiation. 3. Proteins—
Analysis. 4. Nucleic acids—Analysis. 5. Viruses. I. Title.
QP519.9.X72H45 1992
547.7'046—dc20 91–10541 CIP

ISBN 0 521 33467 5 hardback

To
my Mother and the memory of my Father
and
Madeleine, James, Nicholas and Katherine

Contents

Preface

The scope of this book covers the use of synchrotron radiation in the X-ray analysis of single crystals of proteins, nucleic acids and viruses. The impact of this new X-ray source with its polychromatic nature and associated high intensity and fine collimation has brought important advances in the field of macromolecular crystallography. It has extended structure determinations to higher resolution, allowed use of smaller samples and larger, more complex, unit cells. Several new methods have come to the fore and some old methods have been revived. Firstly, the Laue method is being developed and used now for quantitative, time resolved analysis of structure. Secondly, variable wavelength methods are being developed and used for phase determination for metallo-proteins or derivatised proteins. Thirdly, the diffuse scattering is being measured more easily and procedures for analysing it are being developed in order to study molecular flexibility; hopefully its use will be increasingly widespread but at present it is the least developed of these three methods. The availability of the synchrotron is a very modern development but it has reopened fundamental questions of which crystallographic method to use. It is interesting to wonder what von Laue, W. H. and W. L. Bragg and the other early pioneers would have made of the synchrotron instead of starting with the X-ray emission tube. Certainly the Braggs were advocates of the monochromatic rotating crystal method. Wyckoff and Pauling used the Laue method although the weakness of the Bremsstrahlung continuum argued against it. With these difficulties it was fine then to be dismissive, as the Braggs were, of this method. The Braggs raised other fundamental objections to Laue geometry. These were the 'multiplicity problem', the 'wavelength normalisation problem' and the problem of determining absolute cell parameters from Laue data. Only the last of these three is limiting although progress is being made even there.

Variable wavelength approaches to phase determination using anomalous dispersion were discussed in the late 1950s by Okaya and Pepinsky as well as Mitchell, in the 1960s by Herzenberg and Lau and Karle and pioneered by Hoppe and Jakubowski. Technically these methods and others have only become really feasible with the synchrotron.

Diffuse scattering was pioneered as a technique by Lonsdale and others. One imagines that these investigators would have been delighted to see the diversity of the diffuse scattering in the single crystal diffraction patterns recorded from macromolecular crystals at the synchrotron. As station master for the two instruments for protein crystallography at the SERC Synchrotron Radiation Source in Daresbury, England from 1980 to 1985 I saw hundreds of samples and their diffraction patterns during routine data collection runs. The uniqueness and diversity of the diffuse scattering was most striking.

This book will, I hope, serve as a source of information on the properties of synchrotron radiation from storage rings, diffraction instrumentation (such as optics and detectors) as well as diffraction methods and applications. To help the newcomer to the field or to assist experts from other disciplines, there are two chapters covering the fundamentals. The basics of macromolecular crystallography are described. Also, the principles of macromolecular structure are covered and various aspects of biological functions are discussed including oxygen transport and storage, enzyme catalysis, the ribosome and protein synthesis and virus structure. Applications in biotechnology such as protein engineering and drug design are mentioned. There is a wealth of results where synchrotron radiation has been critical to a given structure determination or definition of molecular function. The role of synchrotron radiation in furthering particular crystallographic analyses is therefore addressed and tabulated in detail.

The underlying importance of structure in defining function has meant that a wide range of structure determining methods have been developed in the last few decades. These other methods, such as extended X-ray absorption fine structure (EXAFS) spectroscopy (see Appendix 4), fibre diffraction and nuclear magnetic resonance (NMR), have been brought to bear in various studies as complementary tools to X-ray crystallography. In the Bibliography therefore I have given references to texts on these other techniques. However, macromolecular crystallography is the main method for precisely determining three-dimensional molecular structure over a huge range of molecular weights (up to several million in the case of viruses).

Synchrotron radiation is used when a conventional, home, X-ray source with electronic or image plate area detector proves inadequate. This is not to belittle the X-ray tube or rotating anode. On the contrary, I have also endeavoured to develop conventional X-ray source data collection capabilities based on area detectors. It is fair to say that a synchrotron radiation source is a national and international resource for tackling the most technically demanding problems and an investigator may turn to it when necessary. The usefulness of synchrotron radiation in this research has been recognised by pharmaceutical companies who are now participating actively at synchrotron radiation facilities in the building and use of instruments dedicated to macromolecular crystallography.

Manchester, UK
January, 1991   John R. Helliwell

Acknowledgements

A variety of people have kindly offered comments and criticisms on drafts of sections of the manuscript of this book. I am grateful therefore to D. W. J. Cruickshank, M. Helliwell, W. N. Hunter, J. Raftery, R. Beddoes, I. Haneef, A. W. Thompson, S. Harrop, M. M. Harding, K. Moffat, S. S. Hasnain, N. M. Allinson, D. J. Thompson, R. P. Walker, S. A. Rule, T. Higashi, I. D. Glover, A. Liljas, R. Liddington , S. Popov and D. S. Moss for their help which has undoubtedly improved it. I am particularly grateful to Dr C. Nave for his critical reading of the final copy. Any errors that there may be are, of course, my own responsibility.

Dr S. A. Rule and Mr S. Harrop assisted in the preparation of the tables in chapters 6, 9 and 10 as well as appendix A2. Professor J. Drenth kindly gave permission for use of figures 2.2–2.7. Dr R. P. Walker kindly provided table 4.2 and figure 4.18(b). Dr S. Sasaki kindly provided data included in appendices A3.1 and A3.2. Professor M. Hart was very helpful with appendix A3.4. Dr J. Raftery helped with the preparation of figures 3.8, 3.9, 3.10 and 3.12 using coordinates deposited in the Protein Data Bank; specific acknowledgements are given in the caption to each figure.

I must also thank Rufus Neal of Cambridge University Press who took me through the mechanics of publishing a book and also Mrs M. Storey of CUP who, as sub-editor, considerably improved the submitted draft.

Mrs Yvonne Cook was tireless in the typing of the manuscript and great thanks must go to her. Claire Murphy helped in the proof-reading. Miss Julie Holt typed some of the tables and provided secretarial help which was greatly appreciated. John Rowcroft and Alan Gebbie are thanked for their draughtsmanship.

I would like to express my gratitude to Ossett School and the Universities of York, Oxford, Keele, Manchester and Alabama (in Birmingham, USA) as well as the Science and Engineering Research Council's (SERC) Daresbury Laboratory for providing such splendid environments in which to study and work.

Finally, I would like to thank my wife, Madeleine Helliwell, and family for their support.

A note on units

Currently accepted units of measurement are used in this book. This generally means the SI system. An exception is the use of ångstrom (Å) rather than nanometre (nm). Although the latter is the SI standard the Å is the unit in common use. Inevitably therefore the Å is adopted here also. Indeed this is a reasonable unit of length because a carbon–hydrogen bond length is of the order of 1 Å (10^{-10} m).

CHAPTER 1

Introduction

Macromolecular crystallography is a very powerful method used to study complex biological systems. The structures of a wide variety of proteins, nucleic acids and their assemblies have been determined at atomic or near-atomic resolution. As a result, a detailed understanding has been gained of various living processes such as enzyme catalysis, the immune response, the encoding of hereditary information, viral infection and photosynthesis.

The first X-ray diffraction photograph ever taken was from copper sulphate by Friedrich and Knipping at von Laue's suggestion in 1912. In the following year W. L. Bragg deduced the crystal structure of sodium chloride from Laue photographs. A variety of relatively small molecular structures were then solved at an increasing rate.

The first X-ray diffraction pictures of a protein crystal were taken in 1934 by Bernal in Cambridge, but in those days the data quality was crude and the techniques for deriving a crystal structure of a macromolecule from the X-ray data were not sufficiently developed. The advent of the computer has been a critical development.

The first protein structures to be determined were myoglobin and haemoglobin in the late 1950s by Kendrew *et al* (1958) and Perutz *et al* (1960). From then on a steadily increasing number of protein structures have become known. Nowadays, once a suitable crystal is available, a new structure of a protein or even a virus can, in favourable circumstances, be determined in a year or less. In the case where there is a closely related structure available then a new crystal structure may be obtained in as little as 1–2 weeks.

Crystallographic techniques are facilitated to a considerable extent by the degree of sophistication of the technology used at the various stages

1

leading to a macromolecular crystal structure determination. These
stages are:

crystallisation of the pure sample;
determination of crystal unit cell parameters;
X-ray data collection to a given resolution;
solution of the crystallographic phase problem;
interpretation of the electron density map;
refinement of the molecular model against the observed
  data.

Detailing the function of the molecule then additionally involves
substrate or inhibitor (drug) binding in the cases of enzymes or viruses
and/or site directed mutagenesis of a protein and subsequent X-ray
analysis. Sophisticated technologies are used which include:

genetic engineering to improve protein preparation and
  purification;
robotic machines to automate crystallisation procedures;
synchrotrons to provide intense, collimated and tunable X-
  rays to deal with problems associated with small samples,
  dense diffraction patterns (large unit cells), weak
  scattering, radiation damage and the phase problem, and
  to allow time resolved studies and the measurement of
  diffuse scattering;
electronic detectors and novel X-ray sensitive materials
  such as image plates and storage phosphors coupled to
  powerful computer workstations to improve the efficiency
  with which data are recorded and processed;
advanced computer graphics for molecular modelling and
  interpretation;
parallel and vector processor computers for refinement of
  the large numbers of parameters against the available
  data;
computer data bases and expert systems to study the
  relationships between protein structures.

This huge investment of skills and technologies is indicative of the
paramount importance of X-ray crystallography in unravelling the
detailed mechanisms of life.

1.1 WHY DO WE WANT TO KNOW A MACROMOLECULAR STRUCTURE?

The processes of life depend fundamentally on the atomic and geometrical structure and interactions of the molecules in the living cell. Thus molecular biology is fascinating because knowledge of the structure and action of these molecules gives a clue to the understanding of life. Enzymes, for example, are nature's catalysts. Nearly all reactions in the living cell occur at the right moment and with sufficient speed because of the catalytic action of an enzyme. They are extremely efficient, highly selective and orders of magnitude better than inorganic and organo-metallic catalysts. Most enzymes are proteins and composed of one or more polypeptide chains constructed from the 20 naturally occurring amino acids. The chains are folded into a globular shape and this three-dimensional structure determines the chemical and physical behaviour and the resulting function of the protein.

To determine these complicated structures the only general method available is X-ray diffraction of the single crystals of these materials. Although the structure of small proteins (molecular weight (MW) less than about 10000 daltons (D)) can be determined in solution with nuclear magnetic resonance (NMR) spectroscopy and the assembly of proteins in a complex can be studied with electron microscopy, only X-ray diffraction can give the three-dimensional structure of small as well as large proteins with a precision of about 0.1–0.2 Å.

Because most macromolecules exert their biological action in an aqueous solution, one may ask whether the molecular structure of a macromolecule in the crystal is a fair representation of the protein structure in solution. Since enzymes can be fully active in the crystalline state the answer is obviously that it is.

A macromolecular crystal usually consists of between 30% and 80% solvent of crystallisation. Hence, the enzyme active site, for example, can be accessible to these solvent channels and able therefore to catalyse conversion of a reactant to product. Because of this observation alone one is able to say that the crystal structure is directly relevant in helping to determine the macromolecule's functional state. Of course, the results themselves, defining the structure, do make chemical sense.

The greatest variety of biological structures is within proteins and much of the structural work in biology has been devoted to proteins. Also, the structures of a number of t-RNAs have been solved; a specific t-RNA transports an amino acid to the protein synthetic machinery for translation into a growing polypeptide chain. More recently the structures of a number of DNA-binding proteins have been determined and from this work general principles for the interaction between DNA and proteins have been derived (for a review, see Steitz (1990)).

X-ray crystallography has also been used to unravel the structure of various enormous virus particles. Viruses have a protein coat which envelops the nucleic acid. The function of this coat is to protect the nucleic acid. In higher organisms antibodies are generated against this coating as a defence mechanism. From its three-dimensional structure one can see which parts of the protein coat are at the surface of the virus particle and which therefore are the potential sites for the generation of antibodies. These form the basis of ideas for the design of new vaccines.

Synchrotron radiation (SR) has been used to determine the structure of a variety of virus structures. The determination of the rhinovirus and the mengo virus structures has revealed an interesting surface morphology comprising a 'canyon'. The canyon has been shown to accommodate certain drug molecules, the effect of which is to prevent the virus from injecting its nucleic acid into the host. It may prove feasible therefore to treat certain viral infections directly with drugs, providing a major breakthrough in the treatment of active viral infections. Subsequently the structure of the foot and mouth disease virus has been determined. Although the canyon appears not to be present in this structure, it is expected that new forms of treatment for this disease will be forthcoming as a result of the knowledge of the structure.

The need for the three-dimensional structure of an enzyme to be known in order to understand its catalytic action has been emphasised. However, the structure is also indispensable for two other purposes:

> The rational design of drugs for inhibiting the action of enzymes related to illnesses (e.g. penicillin destroying enzymes in bacteria): this can replace the more conventional, time-consuming screening of enormous numbers of compounds as the usual method for drug design.
>
> Protein engineering: a development of great technical importance is the application of genetic engineering for the production of modified proteins with improved properties. Modification should be of the right kind in the right position (i.e. site specific) and this can only be

derived from the three-dimensional structure of the
protein.

Hence, we can say that molecular structures of biological materials are
needed for several purposes:

a complete understanding of the processes in the living cell,
e.g. the action of enzymes, or the transfer of genetic
information;
the rational design of drugs;
the modification of proteins via genetic engineering;
vaccine design.

X-ray crystallography is the most versatile method for determining these
structures in detail, provided that suitable crystals can be grown.

1.2 THE IMPORTANCE OF SR IN MACROMOLECULAR CRYSTALLOGRAPHY

Particle accelerators were originally developed for high energy physics
research into the subatomic structure of matter. The SR, which was
produced by accelerators in that context, was a nuisance by-product – an
energy loss process. The early stages of the utilisation of SR were therefore parasitic on the high energy physics machines whose parameters
were, of course, not optimised for SR. However, SR has become well
recognised in its own right as a major research tool in biology, chemistry
and physics. Particle accelerators began to be designed as dedicated to SR
production with parameters optimised solely for this work, e.g. with long
lifetime beams and stable source positions; the Daresbury Synchrotron
Radiation Source (SRS) was the first dedicated high energy source, which
came on-line in 1981 (figure 1.1).

The technology of particle accelerators designed for SR production has
advanced considerably so that it is now possible to induce sophisticated
particle trajectories in special magnets known as wigglers and undulators (table 1.1). These new types of magnets can extend both the available spectral range of emitted photons as well as the brightness and
brilliance compared with radiation from charged particles in simple
circular trajectories in ordinary bending magnets. These enhancements
from undulator and multipole wigglers at low emittance sources, compared with a bending magnet, cover five orders of magnitude. The emission from a bending magnet is, in itself, already two orders of magnitude
higher than a laboratory X-ray, emission line source.

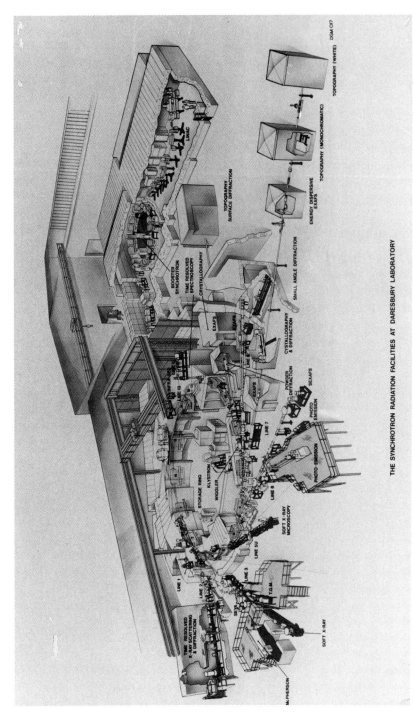

Figure 1.1 A layout of the SRS at Daresbury and its beam lines and workstations. The two labelled crystallography are for macromolecular crystal X-ray data collection. Obviously many other techniques are also housed at SR sources as is evident from this figure. Reproduced with the permission of SERC, Daresbury Laboratory.

Table 1.1. *The first eighteen beam lines planned at the ESRF Grenoble (from Miller (1990) reproduced with permission). The ESRF overall layout is shown in Figure 4.7 (in chapter 4).*

Beam line	Scientific areas	Source
1. Microfocus	Microdiffraction Small angle scattering High pressure	Undulator 0.8–3.0 Å
2. Multipole wiggler/ materials science	Small molecule crystallography Magnetic scattering	Wiggler 4–60 keV
3. Multipole wiggler/ biology	Macromolecular crystallography Laue and monochromatic modes High pressure energy dispersive studies	Wiggler 4–60 keV
4. High flux beam line	Real time small angle scattering Monochromatic protein crystallography	Undulator, λ tunable around 1.0 Å
5. High energy X-ray scattering	Gamma-ray diffraction Small angle scattering Compton scattering	Multipole wiggler or wavelength shifter
6. Circular polarization	Dichroism in EXAFS, SEXAFS Spin-dependent photoemission Microscopy at 2.5 keV	Helical undulator $E \leqslant 4$ keV
7. Surface diffraction	Surface structural studies Phase transitions Growth mechanisms Liquid surface diffraction	Undulator $K_{max} = 1.85$
8. Dispersive EXAFS	Time resolved structural studies	Wiggler, or tapered undulator
9. Undulator, open beam line		
10. Bending magnet open beam line		
11. Mössbauer/ High resolution Inelastic	Nuclear Bragg scattering High resolution (5–100 meV) Inelastic scattering at 0–5 eV energy transfer; electronic and vibrational excitations	Undulator ~ 14 keV
12. Asymmetric wiggler beam line	Magnetic scattering	Asymmetric wiggler
13. Surface science	SEXAFS and standing waves techniques	Undulator
14. High energy wiggler	Microtomography; possibly angiography	Wiggler
15. Powder diffraction	Powder diffraction for structure determination	Bending magnet later undulator

Table 1.1. (*cont.*)

Beam line	Scientific areas	Source
16. Wiggler Long beam line (75 m)	Topography (possibly second Laue station)	Multipole wiggler
17. Anomalous scattering beam line	Anomalous scattering	Undulator
18. EXAFS	Two EXAFS stations	Bending magnet

The properties of SR can be summarised as:

(a) high flux;
(b) high brightness (well collimated);
(c) high brilliance (small source size and well collimated);
(d) tunable over a wide wavelength range (X-rays–infra-red);
(e) polarised;
(f) defined time structure;
(g) calculable spectra.

The special combination of properties of SR finds widespread application in macromolecular crystallography.

Weak scattering is a feature of these crystals. The atomic numbers of the elements constituting the crystal (mainly carbon, nitrogen, oxygen, hydrogen and sulphur) means that the scattering of X-rays is relatively weak. In addition, the large fraction of solvent further reduces the scattering into the Bragg reflections. The resolution limit is rather poor, indicative of extensive flexibility and/or disorder in the crystal. In order to maximise the information that is derived from these crystals it is necessary, on conventional X-ray sources, to measure the high resolution, high angle data with long counting times. Unfortunately, radiation damage to the sample degrades the high resolution data first. The high intensity and collimation of the SR beam reduces the exposure time required by several orders of magnitude. As a result it is observed that radiation damage to the sample is reduced, i.e. more reflections can be measured from a crystal with a short exposure and a high intensity than with a long exposure and a low intensity. Moreover, short wavelength SR beams reduce the fraction of absorbed X-rays and again the damage. The main alternative for beating radiation damage is to use sample freezing. If successful the lifetime of such a sample is very long. Unfortunately, it is not always possible to do this and when successful it is almost always accompanied by an increase in mosaic spread (e.g. 0.02° increases up to 0.5°). This causes the smearing out of diffraction spots. In some instances of large unit cell studies, e.g. crystallography of ribosomes, freezing is essential to preserve the resolution limit (to ≈ 5 Å) and the SR beam is then used to reduce the time and therefore burden of data collection to an acceptable level.

Virus and ribosome crystallography is one of the main uses of SR and one which takes advantage of a combination of SR properties such as the high flux, the fine collimation and short wavelengths.

Often large crystals can be elusive and so the intense SR beam is used to compensate for small sample volumes. The high brightness of SR allows a tiny incident beam to be brought onto a small crystal cross section area thus minimising any extraneous matter in the beam and therefore producing a reasonable signal to noise in the diffraction pattern. Small crystals are also vulnerable to radiation damage in that there are fewer unit cells available to yield diffraction intensity data of the required statistical precision. Freezing of the sample is required to preserve the lifetime of a small sample in the beam.

Rapid data collection has opened up the possibility of doing time resolved macromolecular crystallography of, for example, enzymatic processes. That this is feasible at all relies on the large solvent content of these crystals which permits function in this solid state. It is possible now to measure accurate diffraction data with an exposure time of a second or so, capable of yielding molecular structure detail of the necessary precision and sensitivity (e.g. to see a single water molecule binding to a protein). Exposure times have also been realised as short as 120 ps using an undulator as source.

The solution of a new macromolecular structure by crystallography requires that the phases of individual reflections are determined to allow a Fourier synthesis to be calculated. The method of multiple isomorphous replacement (usually with anomalous scattering) is used extensively whereby intensity changes are induced. As a result the positions of the heavy atoms are determined either by Patterson or direct methods and then the phases calculated. Usually, but not always, it is possible to interpret the electron density map. The anomalous scattering phasing power of a heavy atom derivative can now be rather easily optimised by appropriate choice of wavelength of the SR beam. The absorption edges of these high atomic number derivative atoms occur in the range

$0.6\,\text{Å} < \lambda < 1.1\,\text{Å}$. In this range the protein crystal absorption is virtually eliminated for a typical crystal <1 mm in dimension. Sometimes it is not possible to make heavy atom derivatives owing to the chemical nature of a specific protein or the particular crystal form. Many proteins contain an essential metal atom or alternatively selenium can be incorporated into a protein. Similarly bromine can be incorporated into a nucleotide. In all these cases data can be collected at multiple wavelengths using SR and this allows phases to be determined. Protein structures have now been solved by several variants of these methods. This is an important technical capability because it either reduces the number of heavy atom derivatives that need to be found for isomorphous replacement or allows phase determination from a single crystal.

Future applications of SR in macromolecular crystallography involve extrapolations from the use of existing SR sources to the next generation of sources such as the European Synchrotron Radiation Facility (ESRF) in Grenoble or the Advanced Photon Source (APS) at Argonne or the Super Photon Ring (SPRING) in Japan. Shorter exposure times in time resolved work will be realised and smaller sample volumes and larger unit cells studied. The sophistication of our technical understanding is also improving. For example, the polarisation properties of the beam have yet to be exploited in the use of the dichroism of the scattering of heavier atoms in phasing reflections. It may also become routine for the diffuse scattering, which is rather weak, to be measured and interpreted. There are also other possible uses of SR, e.g. magnetic scattering. Moreover, the use of ultra-short wavelength ($0.33\,\text{Å}$) beams from X-ray undulators could well provide a leap forward in data accuracy and molecular structure determination.

SR has revolutionised experimental X-ray crystallography and there are other fascinating developments in store.

CHAPTER 2

Fundamentals of macromolecular crystallography

X-rays are used to probe the atomic or molecular structure of matter because the wavelength of the radiation is of approximately the same dimension as an atom. Similarly longer wavelength visible light is appropriate for studying larger structures, e.g. cell organelles. However, since there is no known X-ray lens the equivalent function of a glass lens for visible light in a conventional microscope has to be performed by computational transformation of X-ray diffraction patterns.

The basic steps in a macromolecular crystal structure analysis involve:

(*i*) crystallisation;
(*ii*) space group and cell parameter determination;
(*iii*) data collection;
(*iv*) phase determination;
(*v*) electron density map interpretation;
(*vi*) refinement of the molecular model.

Figure 2.1(*a*)–(*f*) illustrates some of these steps showing, as an example, the structure determination of human erythrocyte purine nucleoside phosphorylase (PNP) (Ealick *et al* (1990)). A list of general texts on crystallography is given in the bibliography, section 1.

2.1 CRYSTALLISATION, CRYSTALS AND CRYSTAL PERFECTION, SYMMETRY

Crystallisation is a process involving precipitation of the dissolved protein from solution. This is achieved by decreasing the protein solubility, decreasing any repulsive forces between individual protein molecules and/or increasing the attractive forces. The crystals that might be produced need to be of 'X-ray diffraction quality'. This means

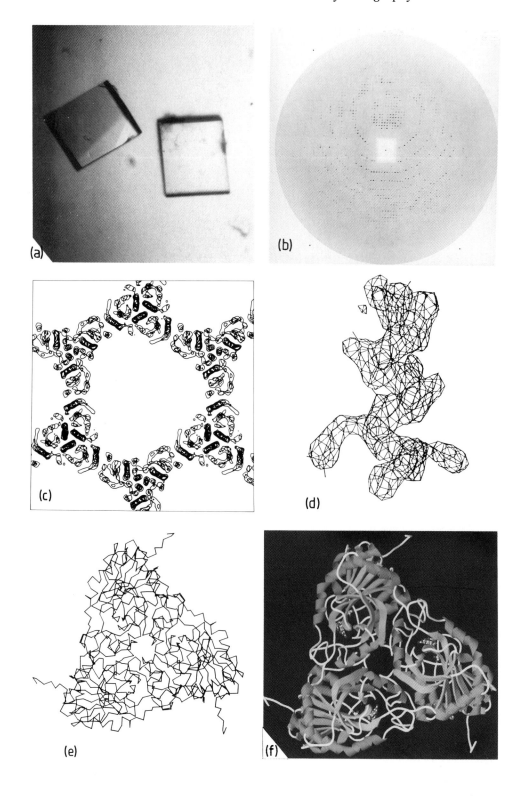

that they have to be of a reasonable size and perfection. In some ways synchrotron radiation allows use of poorer quality crystals in that, for example, a smaller crystal volume can be used. However, there is no substitute for a good quality crystal of a reasonable size. A variety of macromolecular crystals are shown in plate I. The principles of crystallisation will be illustrated with respect to proteins primarily.

2.1.1 The principles of protein crystallisation

Protein solubility can be decreased by adding a salt, usually ammonium sulphate, or polyethylene glycol (PEG). The ions of the salt or the PEG are hydrated, withdrawing water from the solution, thus causing the protein to become more concentrated. Also, the ions screen any charges at the surface of individual protein molecules which may lead to coulombic repulsion. A change of pH also leads to a change of charge on the protein and therefore varies the ionic forces between them. Attractive forces between molecules are mediated by hydrogen bonds and/or van der Waals or hydrophobic interactions. The latter vary with temperature.

There are a large number of variables available in a protein crystallisation trial, namely:

the purity of the preparation;
the concentration of the protein;
the concentration and type of the precipitant;
pH;

Figure 2.1 The basic steps of macromolecular crystal structure solving are illustrated with respect to the enzyme, PNP MW 30000×3 D. (a) A crystal of human PNP; space group R32. (b) Monochromatic oscillation diffraction photograph recorded at the Daresbury SRS; resolution limit of outermost diffraction spots ≈ 3 Å. (c) Electron density map, calculated at 6 Å resolution, viewed down the hexagonal c axis. The diameter of the central solvent channel is ≈ 130 Å. Six trimers are visible. (d) A portion of the 3 Å electron density map with fitted molecular model. (e) The PNP trimer molecular model. (f) The PNP trimer with bound inhibitor; the protein here is represented in ribbon format for α-helix and β sheet (see chapter 3 for details of macromolecular structure). Based on Ealick et al (1990). These figures kindly supplied by Dr S. Ealick and reproduced with permission.

the temperature;
the kind of buffer;
the presence of additives (ions or organic components) in
   low concentration.

Obviously, to find the correct crystallisation condition requires the controlled variation of a great number of variables. There are texts devoted to this subject (e.g. McPherson (1982); Giegé *et al* (1988)).

2.1.2 Practical aspects

Crystallisation experiments are probably best carried out using as small amounts of protein as possible. This is because the total amount of protein may be very limited and yet to find the ideal crystallisation conditions may require a large number of experiments. Hence, the amount of protein per experiment should be as small as possible (typical volume of a solution: $2-4\,\mu\ell$). Note that a typical, good sized crystal of volume $0.3\times0.3\times0.3\,mm^3$ contains $\approx 15\,\mu g$ of protein. For 50 experiments, $0.75\,mg$ of protein is therefore the minimum used, but usually at least $5\,mg$ is required overall.

The different methods of crystallisation include use of one of the following:

drops, hanging or sitting (figure 2.2);
capillary tubes (figures 2.3 and 2.6);
dialysis (figures 2.5 and 2.6).

Equilibration is achieved by

vapour diffusion (figure 2.4);
liquid/liquid diffusion (figure 2.3);
concentration gradient (ultra-centrifuge method).

There is also the so-called 'batch method', which involves the simple mixing of some precipitant to the protein solution and leaving the mixture to stand (see section 2.1.3 for an example).

It is advisable to follow a two-step process of, firstly, finding the conditions for just precipitating the protein and then, secondly, trying variations around these conditions. This has been facilitated in recent years by use of robotic pipetting of solutions (see, e.g. Cox and Weber (1987)).

Plate I Examples of macromolecular crystals

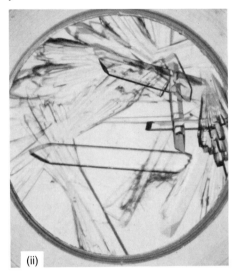

0·1 mm

(iii)

(i)   Monoclinic crystals of human carbonic anhydrase II. Space group P2₁, cell parameters $a=42.7$ Å, $b=41.7$ Å, $c=73.0$ Å, $\beta=104.6°$. Typical largest dimensions ≈ 1 mm. Photograph kindly provided by Dr J. Vidgren and Professor A. Liljas with permission.

(ii)   Orthorhombic crystals of concanavalin-A (complexed with α-methyl mannoside). Space group P2₁2₁2₁, cell parameters $a=123.9$ Å, $b=129.1$ Å, $c=67.5$ Å, $\alpha=\beta=\gamma=90°$. Photograph kindly provided by Dr J. Yariv with permission.

(iii) Orthorhombic crystals of concanavalin-A (without saccharide). Space group I222, cell parameters $a=89.9$ Å, $b=87.2$ Å, $c=63.1$ Å, $\alpha=\beta=\gamma=90°$. Typical size ≈ 1 mm longest dimension. Photograph kindly provided by Dr J. Yariv with permission.

(iv)

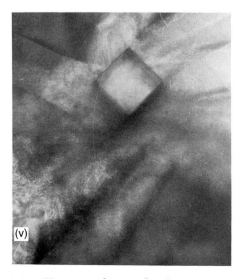

(v)

(iv)   Orthorhombic crystals of 6-
        phosphogluconate
        dehydrogenase. Space group
        C222$_1$, cell parameters a=72.7 Å,
        b=148.2 Å, c=102.9 Å,
        $\alpha=\beta=\gamma=90°$. Typical largest
        dimension $\approx$1 mm. Photograph
        kindly provided by Dr M. J.
        Adams and Dr D. Somers with
        permission.

(v)   Tetragonal crystals of
        trypanothione reductase. Space
        group P4$_1$, cell parameters
        a=b=128.6 Å, c=92.5 Å,
        $\alpha=\beta=\gamma=90°$. Typical largest
        dimension $\approx$1 mm. Photograph
        kindly provided by W. N.
        Hunter with permission.

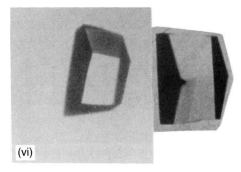

(vi)

(vi)   Trigonal crystals of p21^{1}c
        protein complexed with caged
        GTP, used in a time-resolved Laue
        experiment (section 10.5). Space
        group P3$_2$21, cell parameters
        a=b=41.0 Å, c=164.8 Å,
        $\alpha=\beta=90°$, $\gamma=120°$. From
        Schlichting et al (1989) with
        the permission of the authors.

(vii) Rhombohedral crystal of human PNP. Space group R32, cell parameters $a=b=c=99.2$ Å, $\alpha=\beta=\gamma=92.3°$. The corresponding hexagonal cell parameters are $a=b=142.9(1)$ Å, $c=165.2(1)$ Å, $\alpha=\beta=90°$, $\gamma=120°$. Typical size ≈0.7 mm. Photograph kindly provided by Dr S. Ealick with permission.

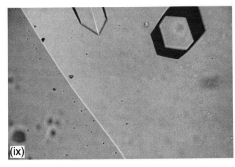

(ix) As example (viii) but with one of the hexagonal needles viewed on axis. Photograph kindly provided by Dr S. Ealick with permission.

(viii) Hexagonal crystals of *Escherichia coli* PNP. Space group $P6_122$ or $P6_522$, cell parameters $a=b=106.5$ Å, $c=241.3$ Å, $\alpha=\beta=90°$, $\gamma=120°$. Typical longest dimension ≈0.6 mm. Photograph kindly provided by Dr S. Ealick with permission.

(x) Cubic crystals of concanavalin-A (complexed with α-methyl glucoside). Space group $I2_13$, cell parameters $a=b=c=167.8$ Å, $\alpha=\beta=\gamma=90°$. Crystals can grow up to 2 mm in size. Photograph kindly provided by Dr J. Yariv with permission.

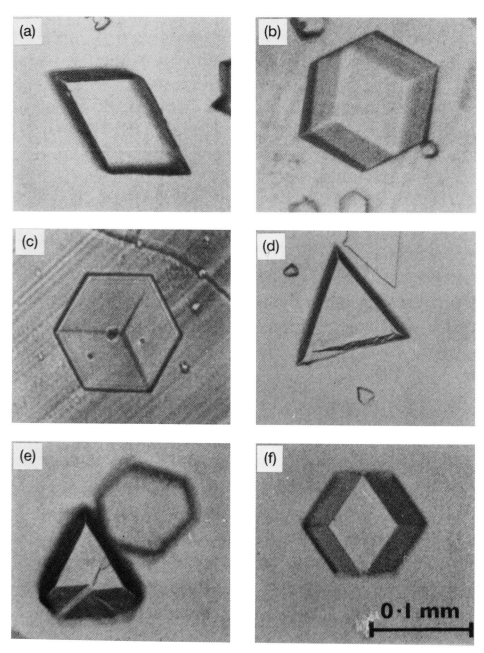

(xi)   (a)–(f) A selection of crystals of different strains of
foot and mouth disease virus. Cubic crystals type (b)
were used by Acharya *et al* (1989) to solve the
structure (see chapter 10); space group I23, *a*=345 Å.
From Fox *et al* (1987) with the permission of the
authors.

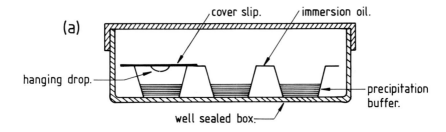

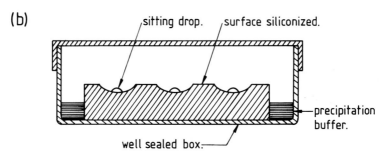

Figure 2.2 Vapour diffusion method of macromolecular crystallisation. The hanging drop method is shown in (*a*). It can be carried out in different ways but a typical procedure is the following:

> Preparatory work: cut plastic trays in the required size and siliconise microscope glass cover slips of 22×22 mm. Keep the cover slips in an oven at 110°C during the night.
>
> Dissolve the protein in a buffer solution to a concentration of approximately 10 mg ml^{-1}.
>
> Prepare a concentration range of the precipitant in the same buffer. Precipitant: ammonium sulphate, PEG 2000, 4000, 6000 or 10 000, potassium phosphate or sodium chloride, etc.
>
> Add to each of the depressions in the plastic tray 1 mℓ of the precipitant solutions.
>
> 'Paint' the upper surface of the trays with immersion oil.
>
> Put the drop on the cover slip and use, e.g. 5 $\mu\ell$ protein solution+5 $\mu\ell$ of the liquid in the tray depression.

If the surface tension of the protein solution is rather low one should use the sitting drop method (*b*) which can be carried out in single drop or multiple drop vessels. Figure kindly supplied by J. Drenth, University of Groningen and reproduced with permission.

If these methods fail then one can try the crystallisation of homologous proteins from a different source whereby slight changes in the amino acid sequence might be at the protein surface and influence nucleation. In those cases where only small crystals grow initially or where a very big crystal is required then seeding can be used (figure 2.7).

Microgravity is being explored as a means for improving crystal size, habit and perfection (DeLucas *et al* 1989). The major motivation behind these experiments is to eliminate density-driven convective flow that

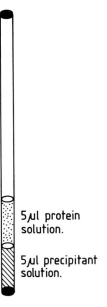

5 μl protein solution.

5 μl precipitant solution.

Figure 2.3 In the liquid/liquid diffusion method a column of the protein solution and of the precipitant touch each other in a capillary tube. The solution with the higher density should be the lower one.

Materials: melting point capillaries, diameter=1.4 mm, or somewhat wider tubes; syringe with injection needle; a gas flame for closing the tubes; this can also be done with plastic modelling clay; styrofoam for storing the capillaries.

Working procedure: the capillary is closed by melting one end. $5\mu\ell$ of the precipitant solution (e.g. ammonium sulphate in a buffer) is injected into the capillary and spun down in a table centrifuge. Next $5\mu\ell$ of the protein solution in the same buffer follows. This is also centrifuged down and a sharp boundary with the precipitant solution is formed. Now the open end of the tube is closed. Figure kindly supplied by J. Drenth, University of Groningen and reproduced with permission.

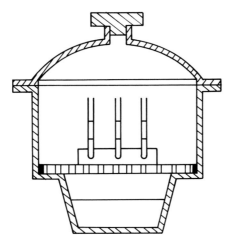

Figure 2.4 In a scaled-up version of the vapour diffusion method the protein solution is kept in an atmosphere which is slowly changed by the diffusion of vapour into it. This can be the vapour of an organic solvent or of a volatile acid or base. The vessels or small tubes with the protein solution can be placed, e.g. into a desiccator which has at its bottom the evaporating solution. This is a method more commonly applied to small molecule crystal growth. Figure kindly supplied by J. Drenth, University of Groningen and reproduced with permission.

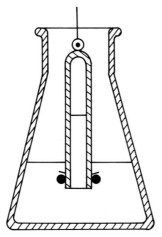

Figure 2.5 As with all the other methods a great many varieties exist for executing the dialysis method. For not too small amounts of protein, dialysis tubes as shown in the figure can be used. The dialysis membrane is attached to the tube by a rubber ring. Figure kindly supplied by J. Drenth, University of Groningen and reproduced with permission.

accompanies crystal growth in gravitational fields (Kroes and Reiss 1984; Pusey, Witherow and Naumann 1988). In addition, sedimentation of growing crystals, which can interfere with the formation of single crystals, is eliminated in the absence of gravity. So far, it has been shown that more highly ordered crystals have been grown in microgravity compared with the same method on Earth (e.g. 1.65 Å resolution on Earth versus 1.3 Å in microgravity (DeLucas *et al* 1989)). It is thought that because protein crystals are relatively weakly bonded, with water bridges playing predominant roles, molecular-packing patterns may be more regular in the absence of convective turbulence. In addition, both crystal habit and crystal volume have been affected.

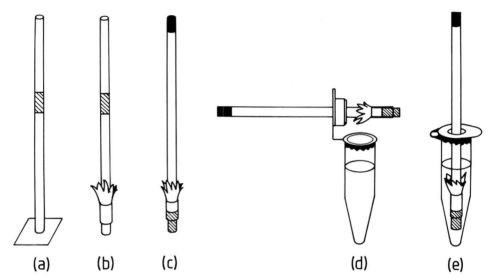

(a)   (b)   (c)   (d)   (e)

Figure 2.6 The microdialysis crystallisation procedure. (*a*) $5 \mu \ell$ of a protein solution is injected into a capillary and covered by a dialysis membrane of $\pm 1 \times 1 \, cm^2$. (*b*) The membrane is fastened with a piece of teflon tubing. (*c*) The protein solution is spun down in a table centrifuge and the capillary is closed by a piece of modelling clay. (*d*) Next, the capillary is placed in an Eppendorf reservoir containing the dialysis solution and capped, (*e*).

Crystal growth can be conveniently viewed by holding the capillary horizontally under a microscope. The advantages of a dialysis method are, of course, that the capillary can be placed easily into reservoirs with different solutions. Figure kindly supplied by J. Drenth, University of Groningen and reproduced with permission.

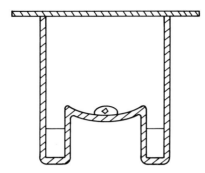

Figure 2.7 The seeding method: A small crystal is washed in a series of solutions in which it slowly dissolves, e.g. solutions with a decreasing precipitant concentration. In this way the surface of the crystal is etched and cleaned. It is then transferred to a fresh drop of protein solution with so high a concentration of precipitating agent that the crystal does not dissolve. The drop is then equilibrated with a more concentrated precipitating solution in either the hanging drop or sitting drop mode. Figure kindly supplied by J. Drenth, University of Groningen and reproduced with permission.

2.1.3 Example crystallisation recipe by simple batch method: lysozyme

One of the easiest proteins to crystallise, even in inexperienced hands, is lysozyme; all, except one, of many undergraduate and graduate students have been successful with the following recipe and got lysozyme crystals at least up to 0.3 mm in size. Moreover, lysozyme of high quality is available from various chemical companies (e.g. Sigma) on three-day delivery at a small cost (~£10 per gram). So, for the curious from fields other than protein crystallography and the novice, try it!

Dissolve 80 mg of lysozyme (Sigma grade III recrystallised three times; Sigma Chemical Company), in 1 mℓ of 0.04 M acetate buffer at pH 4.7. Gently stir the solution for at least 5 min, avoiding frothing, until all the protein is dissolved. Add 1 mℓ 10% (weight/volume) sodium chloride solution over a period of 5 min and continue stirring for another 5 min. Centrifuge the solution. Place it in a plastic container in a quiet corner at room temperature. Lysozyme crystals appear after a day or so and after two days are of a reasonable size. (This recipe is based on Blundell and Johnson (1976).)

2.1.4 Crystal quality and perfection

The factors which are commonly used to specify the quality of a crystal for X-ray crystal structure analysis include:

(*a*) size;
(*b*) mosaicity;
(*c*) resolution limit (see figure 2.1(*b*) for example);
(*d*) radiation lifetime;
(*e*) crystal habit or shape.

The size is important because the energy in a given diffraction spot (i.e. the integrated intensity of a spot on a photographic film) is directly proportional to the crystal volume. A large crystal allows a better (in terms of statistical accuracy) measurement of intensity or a shorter exposure time. There is a limit to the usable size because X-ray absorption will increase; up to a certain size (dependent on wavelength) an absorption correction to the measured intensities can be applied (see chapter 6). The crystal habit may be quite irregular (e.g. thin plates), which also can necessitate a large absorption correction factor. A symmetric crystal shape can be advantageous for measuring anomalous differences (see section 2.4.2 and chapter 9). However, the shape is not critical at shorter wavelengths ($<1.0\,\text{Å}$) where sample absorption is small. In any case, macromolecular crystals are usually mounted in closed capillaries along with a plug of mother liquor so that they do not lose their solvent (Bernal and Crowfoot 1934). The capillary, normally cylindrical in shape, adds an extra component to the sample absorption correction.

A protein crystal usually consists of between 30% and 80% solvent of crystallisation and as such is an unusual state of crystalline matter. In figure 2.1(*c*) we saw a view down the three-fold axis of the electron density distribution of a crystal of the enzyme PNP which contains nearly 80% solvent. In this view it is clear that relatively few intermolecular contacts contribute to making up the crystal lattice. The bulk of the protein molecular surface projects into the solvent channels of the crystal.

The perfection of a crystal refers to the precision with which each and every unit cell associates. In a perfect crystal there are no dislocations in the laying down of one unit cell next to another as the crystal grows. In such a case the angular range over which a crystal is rocked, and that a given reflection occurs, is approximately given by the angle defined by the length of the unit cell divided by the length of the crystal (the reciprocal of the number of unit cells) – figure 2.8(a). Many protein crystals appear to have very small angular rocking widths (figure 2.8(c) and Helliwell (1988)), so small that actually the angular divergence and spectral bandwidth in the incident, monochromatised, synchrotron beam is usually larger than the sample effect. Specialised instrumentation is needed to test this feature of macromolecular crystals at the synchrotron; equipment which was originally developed to assess the perfection of crystals such as silicon or germanium. However, some protein crystals have genuinely broad angular rocking widths (e.g. 0.5–1°). These crystals fall into what is termed 'ideally imperfect' or mosaic crystals (figure 2.8(b)). The mosaic concept assumes that a crystal is made up of a large number of blocks of perfectly ordered unit cells with each block angularly misaligned. A small mosaic spread (rocking width) is important in Laue diffraction where the patterns are dense with spots (see chapter 7).

The resolution limit of the sample is not necessarily correlated with the mosaic spread. As is evident from figure 2.1(c) with the PNP crystal relatively few intermolecular contacts make up the crystal periodicity. The unit cells could be perfectly aligned one with another but the protein surface projecting into solvent is able to flop around in the solvent channels. Hence, the angular rocking width of the crystal can be small indicating perfection, but the resolution limit could be relatively poor. In the case of PNP the crystal diffracts only to a resolution limit of ≈ 2.8 Å.

Irradiation of the crystal sample leads to radiation damage. The mosaicity increases, which is a manifestation of the disruption of the intermolecular contacts which hold the crystal together. In addition the resolution limit decreases indicative of increased disorder of the molecules in the crystal, e.g. fusion of side chains, disruption of the protein molecule. These effects have been found to be reduced with the SR beam in the following ways. Use of short wavelengths decreases the absorption of the beam and leads to less damage. Also, faster exposure times mean that the diffraction data can be collected before the radiation damage takes effect. These benefits and their application to specific cases is dealt with in chapters 6 and 10.

2.1.5 Symmetry

Just what is a crystal? A crystal is built up by a periodic repetition, in three dimensions, of an object or motif. A *unit cell* can be chosen to

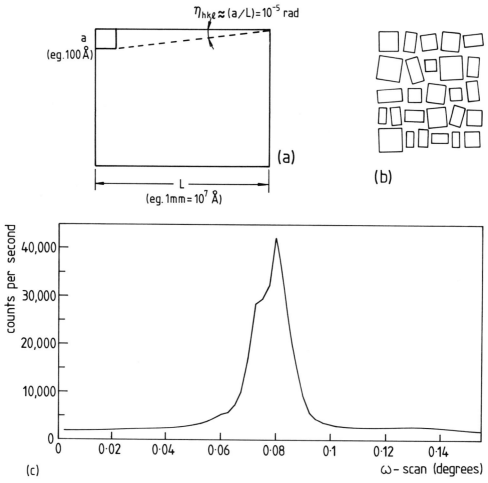

Figure 2.8 Crystal perfection: (a) For a perfect crystal the angular range over which a reflection occurs is $\eta_{hk\ell} \approx a/L$ where a is the cell parameter and L the length of the crystal. This simple explanation assumes that the crystal is X-ray transparent, i.e. the extinction length is greater than the crystal size. (b) An imperfect crystal is referred to as mosaic, i.e. a collection of somewhat misaligned blocks but with each block on its own being a perfectly aligned set of unit cells. The angular misalignments of the mosaic blocks is usually referred to as the mosaic spread, η ($\eta \geqslant \eta_{hk\ell}$). (c) The angular rocking widths ((ϕ_R) see equation (6.5)) of protein crystal reflections usually lie in the range ≈ 0.02–$0.75°$. Example: 2-Zn-insulin measured on the ADONE diffractometer in Frascati with $0.0025° \omega$ step (Colapietro, Helliwell, Spagna and Thompson unpublished; sample kindly supplied by G. Dodson).10^{-5} rad$=0.0006°$.

define the volume containing one of these motifs. The edges of this unit cell, vectors **a**, **b** and **c**, define the *basis* of a lattice of points, known as the *direct lattice*. It is often convenient to think of the points on this lattice being the origins of each and every unit cell in a crystal. The environment of each point is identical in a true crystal, i.e. the structure of the crystal is the same at $(p\mathbf{a}+q\mathbf{b}+r\mathbf{c})$ for all integer values of p, q and r (the only exception being at the surface of the crystal).

The choice of unit cell shape and volume is arbitrary but there are preferred conventions. A unit cell containing one motif and its associated lattice is called *primitive*. Sometimes it is convenient, in order to realise orthogonal basis vectors, to choose a unit cell containing more than one motif, which is then the non-primitive or centred case. In both cases the motif itself can be built up of several identical component parts, known as *asymmetric units*, related by crystallographic symmetry internal to the unit cell. The asymmetric unit therefore represents the smallest volume in a crystal upon which the crystal's symmetry elements operate to generate the crystal.

The total number of *crystal systems* is seven encompassing the possible values of $|\mathbf{a}|$, $|\mathbf{b}|$ and $|\mathbf{c}|$ and the angles between them and essential symmetry (table 2.1). The total number of lattices (Bravais lattices (Bravais 1849)) is 14 representing the number of possible combinations of crystal systems and primitive and centred arrangements (figure 2.9).

A molecule may be symmetric. This symmetry, through its centre point, is referred to as its point symmetry. This symmetry may be invariant to rotation axes (2,3,4,5,6,7 etc fold), mirror planes, inversion centres or a combination of these.

For a given molecule the combination of symmetry elements is known as its *point group*. There are 32 crystallographic point groups that are possible. The asymmetric unit volume can contain an object which may or may not have symmetry. It is quite common in protein crystallography that this object has symmetry of its own, known as *non-crystallographic symmetry* (examples include two-fold, three-fold, four-fold, five-fold etc, i.e. any point symmetry is possible in the asymmetric unit because it need not obey the translational symmetry of the crystal).

A crystal can have additional symmetry based on rotation plus translation elements. The combination of a rotation axis and a translation is a screw axis (e.g. a 2_1 screw axis involves a 180° rotation and translation by half a unit cell dimension). The combination of a mirror plane and a translation is a glide plane. These symmetry elements lead to *systematic absences* in a diffraction pattern; e.g. for a 2_1 axis in the **c**

Table 2.1. *The seven crystal systems.*

	Constraints on unit cell parameters	Essential symmetry
Triclinic	None i.e. $a \neq b \neq c$ $\alpha \neq \beta \neq \gamma$	No axes of symmetry
Monoclinic	$a \neq b \neq c$ (see Note (1)) $\alpha = \gamma = 90°$	Diad axis or mirror plane (inverse diad axis)
Orthorhombic	$a \neq b \neq c$ $\alpha = \beta = \gamma = 90°$	Three orthogonal diads or inverse diad axes
Tetragonal	$a = b \neq c$ and $\alpha = \beta = \gamma = 90°$	One tetrad or inverse tetrad axis
Trigonal (rhombohedral) (or as hexagonal)	$a = b = c$; $\alpha = \beta = \gamma \neq 90°$	One triad or inverse triad axis
Hexagonal	$a = b \neq c$ and $\alpha = \beta = 90°$, $\gamma = 120°$ (see Note (2))	One hexad or inverse hexad
Cubic	$a = b = c$ and $\alpha = \beta = \gamma = 90°$	Four triad axes

Notes:
(1) By convention the **b** axis is chosen as the unique axis.
(2) By convention the **c** axis is chosen as the unique axis.
(3) Since protein molecules are handed, protein crystal space groups actually cannot contain the following symmetry elements: inversion centre, mirror or glide planes.
(4) The non-equal signs for the unit cell parameters mean no necessity to be equal i.e. logically a triclinic cell with $a = b = c$ (within experimental error) is possible.

direction 00ℓ reflections are observed only for $\ell=2n$ (n integer), a 3_1 axis requires $\ell=3n$ etc. Systematic absences also occur for centred lattices (e.g. for a C centred lattice $(h+k)$ must be even for all $hk\ell$ for a reflection to be observed). Some rotational symmetry elements are forbidden in a crystal because they do not satisfy the periodic, translational symmetry of the crystal. Those that are allowed are two-, three-, four- and six-fold axes. To prove to yourself that a five-fold, for example, is forbidden, cut out a set of pentagons and try tiling them together. There is no arrangement possible to cover a surface fully. The symmetry symbols used in macromolecular crystallography are depicted in table 2.2.

The group of symmetry elements observed in a given crystal is known as its *space group*. There are 230 possible space groups. These are tabulated in the *International Tables for Crystallography* (Volume 1 or, more recently updated, Volume A).

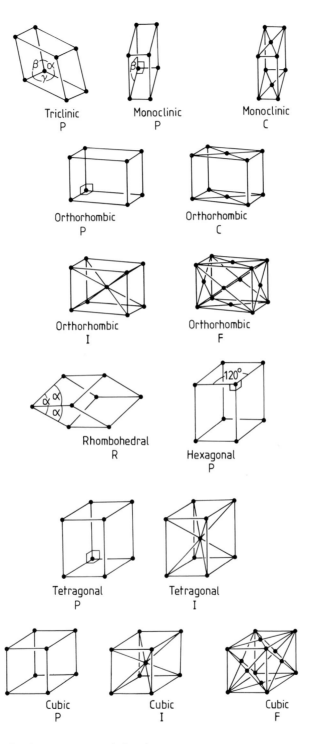

Figure 2.9 The fourteen Bravais lattices.

Table 2.2. *Symmetry symbols relevant to macromolecular crystallography.*

Type	Written	Graphical
Centre of symmetry	$\bar{1}$	○
Mirror plane (shown perpendicular to plane of this paper)	m	———
Rotation axes	2	
	3	
	4	
	6	
Screw axes	2_1	
	$3_1, 3_2$	
	$4_1, 4_2, 4_3$	
	$6_1, 6_2, 6_3, 6_4, 6_5$	

Notes:

(1) As remarked in note (3) of Table 2.1 crystals of macromolecules cannot contain an inversion centre, a mirror or a glide plane. The diffraction pattern from a protein crystal can, however, contain an inversion centre (otherwise known as a centre of symmetry and a mirror plane. The symmetry symbols given here are for those symmetry elements seen therefore for macromolecular crystals and their diffraction patterns.

(2) There are additional symbols for the cases of a two-fold rotation axis ($\rightarrow$) and a two-fold screw axis ($\rightarrow$) when viewed in the plane of the paper. There are no agreed symbols for the other rotation or screw axes when viewed in plane.

(3) The additional symmetry elements which are necessary for the 230 space groups to define the symmetry of *all* crystals (i.e. enantiomorphic and non-enantiomorphic) are: glide planes (i.e. mirror reflection + translation) and 'improper' rotation axes (rotation axis + inversion).

(4) This table is derived from Woolfson (1970) with permission.

Proteins and nucleic acids are optically active (i.e. chiral or handed), e.g. naturally occurring amino acids are all left-handed. This restricts the number of crystal space group types (that macromolecules can crystallise in) to 65 and which contain, as symmetry elements, rotation or screw rotation axes. The ones excluded are all those which contain mirror or glide planes or an inversion (x,y,z to $\bar{x},\bar{y},\bar{z}$), i.e. the centrosymmetric ones. Proteins crystallise therefore in non-centrosymmetric space groups (table 2.3). Four examples are given in figure 2.10, space groups $P2_1$, $P2_12_12_1$, C2 and $C222_1$; these are the most frequently occurring in macromolecular crystallography (table 2.4).

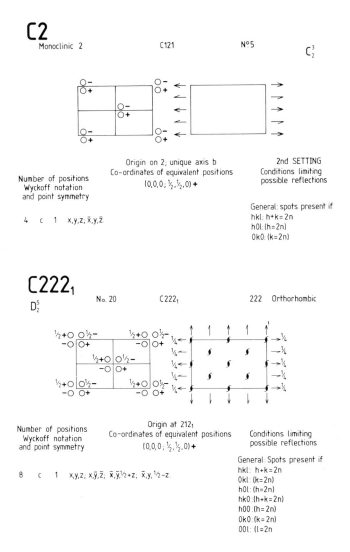

C2

Monoclinic 2   C121   N°5 C_2^3

Origin on 2; unique axis b
Co-ordinates of equivalent positions
$(0,0,0; \frac{1}{2},\frac{1}{2},0)+$

2nd SETTING
Conditions limiting
possible reflections

Number of positions
Wyckoff notation
and point symmetry

General: spots present if
hkl: h+k=2n
h0l:(h=2n)
0k0:(k=2n)

4   c   1   x,y,z; x̄,y,z̄.

C222₁

D_2^5 No. 20   C222₁   222   Orthorhombic

Origin at 212₁
Co-ordinates of equivalent positions
$(0,0,0; \frac{1}{2},\frac{1}{2},0)+$

Number of positions
Wyckoff notation
and point symmetry

Conditions limiting
possible reflections

General: Spots present if
hkl: h+k=2n
0kl: (k=2n)
h0l: (h=2n)
hk0:(h+k=2n)
h00:(h=2n)
0k0:(k=2n)
00l: (l=2n

8   c   1   x,y,z; x,ȳ,z̄; x̄,ȳ,½+z; x̄,y,½−z.

Figure 2.10 The most frequently occurring space groups for macromolecules C2 (7.7%), C222₁ (5.9%), P2₁ (12.4%), and P2₁2₁2₁ (27.7%) are shown. The nomenclature of this figure is based on *International Tables for Crystallography* (for P2₁ and P2₁2₁2₁ from Volume A (1987) and for C2 and C222₁ from Volume 1 (1959), to illustrate the newer and older types of nomenclature in these tables of space groups). Reproduced with the permission of the International Union of Crystallography.

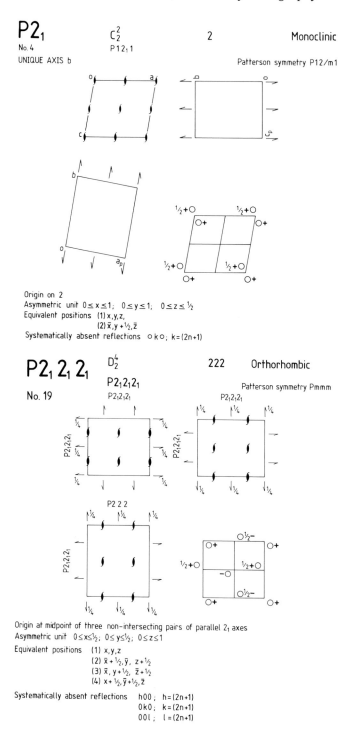

P2₁

No. 4

UNIQUE AXIS b

C_2^2

P12₁1

2   Monoclinic

Patterson symmetry P12/m1

Origin on 2

Asymmetric unit $0 \leq x \leq 1$; $0 \leq y \leq 1$; $0 \leq z \leq \frac{1}{2}$

Equivalent positions (1) x,y,z,

(2) $\bar{x}, y + \frac{1}{2}, \bar{z}$

Systematically absent reflections o k o; k = (2n+1)

P2₁2₁2₁

No. 19

D_2^4

P2₁2₁2₁

222   Orthorhombic

Patterson symmetry Pmmm

P2₁2₁2₁

P2₁2₁2₁

P2 2 2

P2₁2₁2₁

Origin at midpoint of three non–intersecting pairs of parallel 2₁ axes

Asymmetric unit $0 \leq x \leq \frac{1}{2}$; $0 \leq y \leq \frac{1}{2}$; $0 \leq z \leq 1$

Equivalent positions (1) x,y,z

(2) $\bar{x} + \frac{1}{2}, \bar{y}, z + \frac{1}{2}$

(3) $\bar{x}, y + \frac{1}{2}, \bar{z} + \frac{1}{2}$

(4) $x + \frac{1}{2}, \bar{y} + \frac{1}{2}, \bar{z}$

Systematically absent reflections h00 ; h = (2n+1)

0k0 ; k = (2n+1)

00l ; l = (2n+1)

Figure 2.10 (cont.)

Table 2.3. *The non-centrosymmetric space groups accessible to macro-molecular crystals and the symmetry of their diffraction patterns (and Patterson maps).*

System	Crystal space group	No. of unique occurrences in Tables 6.3, 9.6, 10.1, 10.2, 10.5, 10.7(A) and A2.2 (see note (1))	Symmetry of the diffraction pattern or Patterson map
Triclinic	P1	2	P$\bar{1}$
Monoclinic	P2	0	P2/m
	P2$_1$	21	P2/m
	C2	13	C2/m
Orthorhombic	P222	0	Pmmm
	P222$_1$	0	Pmmm
	P2$_1$2$_1$2	1	Pmmm
	P2$_1$2$_1$2$_1$	47	Pmmm
	C222$_1$	10	Cmmm
	C222	0	Cmmm
	F222	0	Fmmm
	I222	4	Immm
	I2$_1$2$_1$2$_1$	0	Immm
Tetragonal	P4	0	P4/m
	P4$_1$	0	P4/m
	P4$_2$	0	P4/m
	P4$_3$	4	P4/m
	I4	2	I4/m
	I4$_1$	0	I4/m
	P422	1	P4/mmm
	P42$_1$2	3	P4/mmm
	p4$_1$22	1	P4/mmm
	P4$_1$2$_1$2	7	P4/mmm
	P4$_2$22	1	P4/mmm
	P4$_2$2$_1$2	0	P4/mmm
	P4$_3$22	0	P4/mmm
	P4$_3$2$_1$2	8	P4/mmm
	I422	3	I4/mmm
	I4$_1$22	0	I4/mmm
Trigonal (rhombohedral)	P3	0	P$\bar{3}$
	P3$_1$	0	P$\bar{3}$
	P3$_2$	1	P$\bar{3}$
	R3	2	R$\bar{3}$
	P312	0	P$\bar{3}$1m
	P321	3	P$\bar{3}$m1
	P3$_1$21	9	P$\bar{3}$1m
	P3$_1$12	0	P$\bar{3}$1m
	P3$_2$12	0	P$\bar{3}$1m
	P3$_2$21	5	P$\bar{3}$m1
	R32	2	R$\bar{3}$m1

Table 2.3. (*cont.*)

System	Crystal space group	No. of unique occurrences in Tables 6.3, 9.6, 10.1, 10.2, 10.5, 10.7(A) and A2.2 (see note (1))	Symmetry of the diffraction pattern or Patterson map
Hexagonal	P6	0	P6/m
	P6$_5$	1	P6/m
	P6$_4$	0	P6/m
	P6$_3$	2	P6/m
	P6$_2$	0	P6/m
	P6$_1$	3	P6/m
	P622	0	P6/mmm
	P6$_1$22	4	P6/mmm
	P6$_2$22	0	P6/mmm
	P6$_3$22	3	P6/mmm
	P6$_4$22	0	P6/mmm
	P6$_5$22	1	P6/mmm
Cubic	P23	0	Pm3
	P2$_1$3	2	Pm3
	F23	0	Fm3
	I23	3	Im3
	I2$_1$3	0	Im3
	P432	0	Pm3m
	P4$_1$32	0	Pm3m
	P4$_2$32	1	Pm3m
	P4$_3$32	0	Pm3m
	F432	0	Fm3m
	F4$_1$32	0	Fm3m
	I432	0	Im3m
	I4$_1$32	0	Im3m
		170	

Notes:

(1) Table 10.6 excluded from this count because the choice of enantiomeric related space groups is not made at the preliminary assessment stage of crystal quality and diffraction resolution.

(2) The symmetry of the diffraction pattern is known as the Laue class of which there are 11 types (i.e. excluding the nature of the lattice centering).

The symmetry of the diffraction pattern can be derived from the symmetry of the crystal. The inversion centre ($hk\ell \rightarrow \overline{hk\ell}$) is introduced as an additional symmetry element because, in the absence of anomalous scattering, $|F_{hk\ell}|^2 = |F_{\overline{hk\ell}}|^2$; this is known as Friedel's law. In the presence of anomalous scattering this law is broken (see equations (2.17) and (2.18)). The Laue class is that set of possible diffraction pattern symmetries. There are 11 possible Laue classes, see table 2.3.

Table 2.4. *Percentage of common space groups in macromolecular crystallography (for 170 unique examples quoted in Tables 6.3, 9.6, 10.1, 10.2,10.5, 10.7(A), and A2.2) in ranking order of occurrence.*

$P2_12_12_1$	27.7%
$P2_1$	12.4
C2	7.7
$C222_1$	5.9
$P3_121$	5.3
$P4_32_12$	4.7
$P4_12_12$	4.1
$P3_221$	2.9
$P6_122$	2.4
I222	2.4
$P4_3$	2.4
$P42_12$	1.8
I422	1.8
P321	1.8
$P6_1$	1.8
$P6_322$	1.8
I23	1.8
Others (i.e. with only one or two examples)	11.3
	100%

Number of space groups with no example to date is 34 (out of 65 possible space groups), i.e. 52.3% not used.

Notes:

(1) These can be contrasted with the percentage of common space groups found for 29 059 organic compounds quoted by Stout and Jensen (1989) as $P2_1/c$ (36.0%), $P\bar{1}$ (13.7%), $P2_12_12_1$ (11.6%), $P2_1$ (6.7%), C2/c (6.6%).

(2) Padmaja, Ramakumar and Viswamitra (1990) have also analysed the frequency of space groups. For all 51 611 structures in the Cambridge Structural Database they report $P2_1/c$ (36.6%), $P\bar{1}$ (16.9%), $P2_12_12_1$ (11.0%), C2/c (7.0%) and $P2_1$ (6.4%). For chiral small molecules the most popular space groups were $P2_12_12_1$ (44.7%) and $P2_1$ (33.4%). For proteins (these authors choose to exclude nucleic acids, viral structures and oligopeptides) they report $P2_12_12_1$ (26.9%), C2 (12.9%) and $P2_1$ (10.6%) as the three most popular space groups for 208 proteins. Also, they quote 36 unoccupied space groups out of the 65 non-centrosymmetric ones.

Symmetry is a fascinating topic, which can be treated in considerable depth in its own right. There is a wide selection of texts available which are listed in the bibliography.

2.2 GEOMETRY: BRAGG'S LAW, THE LAUE EQUATIONS THE RECIPROCAL LATTICE AND THE EWALD SPHERE CONSTRUCTION

Bragg's law predicts the angle of reflection of any diffracted ray from specific atomic planes whereby

$$n\lambda = 2d \sin \theta \tag{2.1}$$

where d is the interplanar spacing of that set of planes, λ is the wavelength of the X-rays and n is an integer. For a fixed λ, the closer the separation of the planes then the larger the value of θ, the diffraction angle; this corresponds to higher resolution data.

Bragg's law is a special case of the Laue equations which define the condition for diffraction (constructive interference, figure 2.11(a)) to occur:

$$
\left.
\begin{aligned}
\mathbf{a} \cdot \mathbf{S} &= h \\
\mathbf{b} \cdot \mathbf{S} &= k \\
\mathbf{c} \cdot \mathbf{S} &= \ell
\end{aligned}
\right\} \tag{2.2}
$$

where h, k, ℓ are the Miller indices defining a unique plane of reflection (or diffraction spot), $\mathbf{a}$, $\mathbf{b}$ and $\mathbf{c}$ are the vectors which define the unit cell of the crystal and $\mathbf{S}$ is the vector path difference of the incident and reflected rays for the $hk\ell$ plane and is called the scattering vector (figure 2.11(b)).

The Laue equations can be recast by using the concept of the reciprocal lattice. New vectors $\mathbf{a}^*$, $\mathbf{b}^*$ and $\mathbf{c}^*$ can be defined by the following relationships:

$$
\left.
\begin{aligned}
\mathbf{a} \cdot \mathbf{a}^* &= 1 \quad \mathbf{b} \cdot \mathbf{a}^* = 0 \quad \mathbf{c} \cdot \mathbf{a}^* = 0 \\
\mathbf{a} \cdot \mathbf{b}^* &= 0 \quad \mathbf{b} \cdot \mathbf{b}^* = 1 \quad \mathbf{c} \cdot \mathbf{b}^* = 0 \\
\mathbf{a} \cdot \mathbf{c}^* &= 0 \quad \mathbf{b} \cdot \mathbf{c}^* = 0 \quad \mathbf{c} \cdot \mathbf{c}^* = 1
\end{aligned}
\right\} \tag{2.3}
$$

The vectors $\mathbf{a}^*$, $\mathbf{b}^*$ and $\mathbf{c}^*$ define a reciprocal lattice whereby the magnitude of $\mathbf{a}^*$, for example, is inversely related to $\mathbf{a}$ and likewise $\mathbf{b}^*$ to $\mathbf{b}$ and $\mathbf{c}^*$ to $\mathbf{c}$. The vector $\mathbf{a}^*$ is perpendicular to $\mathbf{b}$ and $\mathbf{c}$, $\mathbf{b}^*$ is perpendicular to $\mathbf{a}$ and $\mathbf{c}$ and $\mathbf{c}^*$ is perpendicular to $\mathbf{a}$ and $\mathbf{b}$.

We see

$$
\left.
\begin{aligned}
(h\mathbf{a}^* + k\mathbf{b}^* + \ell\mathbf{c}^*) \cdot \mathbf{a} &= h \\
(h\mathbf{a}^* + k\mathbf{b}^* + \ell\mathbf{c}^*) \cdot \mathbf{b} &= k \\
(h\mathbf{a}^* + k\mathbf{b}^* + \ell\mathbf{c}^*) \cdot \mathbf{c} &= \ell
\end{aligned}
\right\} \tag{2.4}
$$

Comparing equations (2.2) and (2.3) we can say

$$\mathbf{S} = h\mathbf{a}* + k\mathbf{b}* + \ell\mathbf{c}* \tag{2.5}$$

A very useful geometric construction that describes the diffraction condition is the Ewald sphere construction. A sphere is drawn of unit radius,

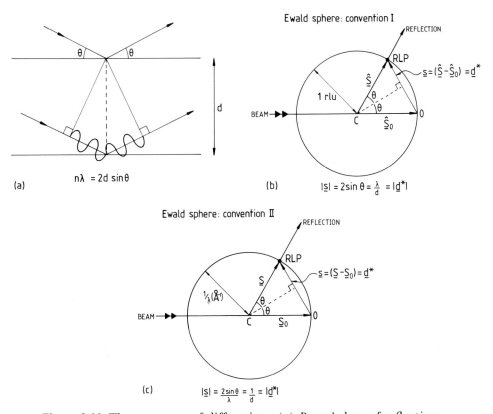

Figure 2.11 The geometry of diffraction: (*a*) Bragg's law of reflection; (*b*) the Ewald sphere construction: convention I with the radius set at unity (known as 1 reciprocal lattice unit, rlu) (RLP=reciprocal lattice point) and $|d*|=\lambda/d$; (*c*) the Ewald sphere construction: convention II with the radius set at $1/\lambda$ and $|d*|=1/d$.

centred in the beam direction and passing through the origin of the reciprocal lattice (i.e. $h=0$, $k=0$, $\ell=0$). The beam direction is a diameter of this sphere. A line drawn from the centre of the sphere and out through any reciprocal lattice point which sits on the sphere describes the direction of a reflection (figure 2.11(*b*)).

Appendix 1 describes the geometries of data collection in detail.

2.3 STRUCTURE FACTOR AND ELECTRON DENSITY EQUATIONS

The intensity measured for a given reflection with Miller indices $(hk\ell)$ is proportional to $|\mathbf{F}(hk\ell)|^2$ where

$$\mathbf{F}(hk\ell) = \sum_{j=1}^{N} f_j \, \exp[2\pi i(hx_j + ky_j + \ell z_j)] \qquad (2.6)$$

and f_j is the atomic scattering factor for X-rays for the jth atom of coordinate (x_j, y_j, z_j) expressed as fractions of the cell $\mathbf{a}$, $\mathbf{b}$, $\mathbf{c}$. This is the structure factor equation in the absence of thermal motion or disorder. The Fourier inverse of this equation is the electron density (ϱ) equation which relates the contents of the unit cell to the set of structure factors $\mathbf{F}(hk\ell)$

$$\rho(x, y, z) = \frac{1}{V} \sum_{hk\ell} |\mathbf{F}(hk\ell)| \exp(i\alpha_{hk\ell}) \exp[-2\pi i \, (hx + ky + \ell z)] \qquad (2.7)$$

whereby if the amplitude $|\mathbf{F}(hk\ell)|$ and phase $\alpha_{hk\ell}$ of the structure factor are known for 'all' $hk\ell$ planes or reflections, then the electron density can be calculated for all points (x,y,z) in the cell and so the crystal structure is then solved. Of course it is 'impossible' to measure 'all' h, k, ℓ reflections so the summation is usually terminated with a finite number of terms at a certain Bragg resolution limit $(d_{min} = \lambda/2 \sin \theta_{max})$. The problem of phase determination, however, is the fundamental one in any crystal structure determination since $|\mathbf{F}(hk\ell)|^2$ is the measured quantity and not $\alpha_{hk\ell}$.

2.4 PHASE DETERMINATION

2.4.1 Multiple isomorphous replacement

Heavy atom derivatives of a macromolecular crystal can be prepared (Green, Ingram and Perutz 1954) which for a minimum of two derivatives (and the native crystal) and in the absence of errors, leads to a unique determination of the phase $\alpha_{hk\ell}$ in equation (2.7) (figure 2.13(a)). This requires the site and occupancy of the heavy atom to be known for the calculation of the vector $\mathbf{F}_H$ (the heavy atom structure factor). In the absence of any starting phase information the heavy atom is located using an isomorphous difference Patterson synthesis $P(u,v,w)$ where the isomorphous difference is given by

$$\Delta_{ISO}(hk\ell) = (|\mathbf{F}_{PH}| - |\mathbf{F}_P|) \qquad (2.8)$$

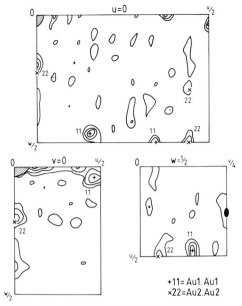

+11= Au1. Au1
×22=Au2.Au2

Figure 2.12 Example of an isomorphous difference Patterson map, extracted from Adams, Helliwell and Bugg (1977). A gold (KAu(CN)$_2$) derivative of the enzyme 6-phosphogluconate dehydrogenase, crystal cell parameters a=72.72 Å, b=148.15 Å, c=102.91 Å. The space group of the crystal is C222$_1$; see below for a list of the equivalent positions and Harker vectors between these equivalent positions.

The three Harker sections are shown. The largest non-origin peaks yield the coordinates of two independent gold sites, Au1 and Au2, at (x,y,z) values of (0.2293, 0.0792, 0.2291) and (0.4969, 0.2228, 0.1428). Vectors between Au1 and Au2 occur on general sections (not shown here). The origin peak is 1000 and contours start at 25 and increase in intervals of 25. The equivalent positions in the space group C222$_1$ are:

$(0,0,0)$; $(\tfrac{1}{2},\tfrac{1}{2},0)$

$(x,$	$y,$	$z)$
$(x,$	$\bar{y},$	$\bar{z})$
$(\bar{x},$	$\bar{y},$	$\tfrac{1}{2}+z)$
$(\bar{x},$	$y,$	$\tfrac{1}{2}-z)$

The Harker vectors (u,v,w) between these equivalents are:

$(0,0,0)$; $(\tfrac{1}{2},\tfrac{1}{2},0)$

$(0,$	$2y,$	$2z)$
$(2x,$	$2y,$	$\tfrac{1}{2})$
$(2x,$	$0,$	$\tfrac{1}{2}-2z)$

The Harker sections are, therefore, u=0, v=0, and w=$\tfrac{1}{2}$.

and the subscript P represents protein and PH represents protein plus heavy atom for the structure amplitudes for an $hk\ell$ reflection. The Patterson function is given by

$$P(u,v,w) = \frac{2}{V}\sum_{hk\ell} \Delta^2_{\mathrm{ISO}}(hk\ell)\cos[2\pi(hu + kv + \ell w)] \qquad (2.9)$$

The largest non-origin peaks on the Patterson map (figure 2.12) should be due to the heavy atom vectors provided the derivative protein crystal is isomorphous with the unmodified or 'native' protein crystal.

2.4.2 Use of anomalous dispersion data

The 'strength' with which atoms scatter X-rays is described by the atomic scattering factor f which is given by

$$f = f_0 + f'(\lambda) + \mathrm{i}f''(\lambda) \qquad (2.10)$$

The atomic scattering factor f was used in equation (2.6) (for the jth atom). For wavelengths not close to absorption edges, f' and f'' are small, though significant, quantities compared with f_0. At wavelengths close to absorption edges, X-rays cause the ejection of photoelectrons. These effects considerably change the amplitude and phase of the elastic Bragg scattering (related to f_0). To take account of this, f' and f'' represent the wavelength-dependent changes needed to correct for the change in amplitude and phase of the 'normal', f_0, scattering (see tables in appendix 3.2). For a *point* atom of Z (the atomic number) electrons, f_0 would have a value Z for all scattering angles but, because of the finite atom size, f_0 decreases with $(\sin\theta)/\lambda$ quite considerably. In contrast, because f' and f'' arise due to the inner shell electrons, the size of which is much smaller than the size of an atom, they are essentially constant, with $(\sin\theta)/\lambda$. Thermal vibration and molecular disorder in the crystal reduce the magnitudes of f_0, f' and f'' with $(\sin\theta)/\lambda$.

The values of f' and f'' for a given λ or frequency, ω, are related by a Kronig–Kramers (1928) transformation

$$f'(\omega) = \sum_i \int_0^\infty \frac{\omega^2(\mathrm{d}g/\mathrm{d}\omega)_i\,\mathrm{d}\omega'}{\omega^2 - \omega'^2} \qquad (2.11)$$

$$f''(\omega) = \tfrac{1}{2}\pi\omega\sum_i(\mathrm{d}g/\mathrm{d}\omega)_i \qquad (2.12)$$

where dg/dω is the oscillator density for oscillators of any given type and the summation is over all possible absorption edges. In fact

$$(dg \,/\, d\omega)_i \propto \mu(\omega)_i \qquad (2.13)$$

$$f'' \propto \mu(\omega)_i \, \omega_i \qquad (2.14)$$

or in terms of the photon energy, E,

$$f'' \propto \mu(E)_i \, E_i \qquad (2.15)$$

and μ is the absorption coefficient of the element.

If only one isomorphous heavy atom derivative can be prepared then the anomalous scattering signal of the heavy atom can be used to resolve the phase ambiguity. At a wavelength somewhat removed from an absorption edge this signal is small, though significant. We can write for the derivative structure factor

$$|\mathbf{F}_{PH}(hk\ell)| = \sum_{j=1}^{N} f_j \exp[2\pi i \, (hx_j + ky_j + \ell z_j)]$$
$$+(f_{N+1} + f'_{N+1} + if''_{N+1}) \exp[2\pi i \, (hx_{N+1} + ky_{N+1} + \ell z_{N+1})] \qquad (2.16)$$

for the case of one anomalous scatterer in the unit cell with coordinates $(x_{N+1}, y_{N+1}, z_{N+1})$.

Because of the $\pi/2$ phase advance of f''_{N+1} relative to f_{N+1} then, as illustrated in figure 2.14,

$$|\mathbf{F}_{PH}(hk\ell)| \neq |\mathbf{F}_{PH}(\bar{h}\bar{k}\bar{\ell})| \qquad (2.17)$$

It is usual to define the quantity known as an anomalous difference, Δ_{ANO} given by

$$\Delta_{ANO}(hk\ell) \;=\; |\mathbf{F}_{PH}(hk\ell)| \;-\; |\mathbf{F}_{PH}(\bar{h}\bar{k}\bar{\ell})| \qquad (2.18)$$

The anomalous difference for each reflection, $\Delta_{ANO}(hk\ell)$, can be used, along with the isomorphous difference to determine its phase, α_P (figure 2.13(b)).

This approach is known as single isomorphous replacement with anomalous scattering (SIRAS). Trigonometric equations can be derived

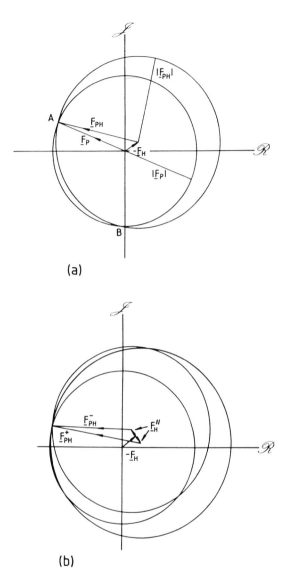

Figure 2.13 (a) The Harker plot. For acentric reflections the use of a single isomorphous derivative (SIR – single isomorphous replacement) leaves a phase ambiguity at A and B. The use of a second derivative at a different binding site, in the absence of errors would decide uniquely between A and B for the protein phase, α_p.

(b) The single isomorphous ambiguity at A and B can be resolved by using the anomalous scattering information from the heavy atom derivative (SIRAS).

These are Argand diagrams in the complex plane where $\mathcal{I}$ is the imaginary and $\mathcal{R}$ is the real axis.

relating the measured amplitudes to the phase angles of interest (Kartha 1975). From figure 2.14, dropping the $(hk\ell)$ suffix for clarity and writing $|\mathbf{F}_{\text{PH}}|$ etc as F_{PH} etc.

$$\alpha_{\text{PH}} = \alpha_{\text{H}} \pm \cos^{-1}\left(\frac{F_{\text{PH}}^2 - F_{\text{P}}^2 + F_{\text{H}}^2}{2F_{\text{PH}}F_{\text{H}}}\right) \tag{2.19}$$

and

$$\alpha_{\text{PH}} = \frac{\pi}{2} + \alpha_{\text{H}} \pm \cos^{-1}\left(\frac{F_{\text{PH}}^{+2} - F_{\text{PH}}^{-2}}{4F_{\text{PH}}F_{\text{H}}''}\right) \tag{2.20}$$

where α_{H} is calculated based on the heavy atom coordinate derived from the difference Patterson (equation (2.9)).

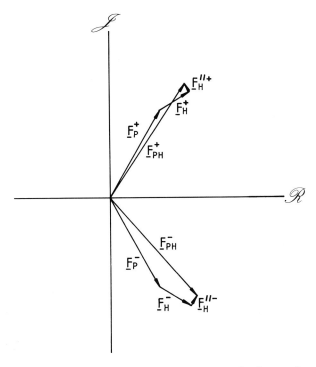

Figure 2.14 For the SIRAS case a vector diagram can be drawn from $hk\ell$ and $\overline{hk\ell}$ labelled '+' and '−' respectively for protein 'P' and heavy atom 'H' contributions to the structure factor. From this, equations (2.19), (2.20) and (2.21) can be derived. This is an Argand diagram in the complex plane where $\mathcal{I}$ is the imaginary and $\mathcal{R}$ is the real axis.

Equations (2.19) and (2.20) give a unique value of α_{PH} and the phase α_p of the protein reflection is given by

$$\alpha_p = \tan^{-1} \frac{(F_{PH} \sin \alpha_{PH} - F_H \sin \alpha_H)}{(F_{PH} \cos \alpha_{PH} - F_H \cos \alpha_H)} \tag{2.21}$$

2.4.3 Accuracy of amplitude measurements required for phase determination

Since the phase is determined directly from the small isomorphous and anomalous differences in structure amplitudes for a given reflection $(hk\ell)$, how well do these small differences need to be measured? In a typical protein of, say, 50 000 molecular weight comprised of mainly carbon, nitrogen and oxygen atoms, there are going to be approximately 3500 atoms (equivalent to 25 000e$^-$) distributed at various positions in an asymmetric unit scattering the X-rays to produce a given F_P^2 measurement. The addition of a single platinum atom, for example, of 78e$^-$ is a small fraction of 25 000e$^-$. It is, however, concentrated at one site. Clearly, Δ_{ISO} as the small difference between two large numbers, is very susceptible to errors of measurement. Similarly, since the f'' for platinum at Cu Kα wavelength is only 7e$^-$, the value of Δ_{ANO} is even more vulnerable to errors of measurement than Δ_{ISO}.

An expression for the average Friedel amplitude difference that can be expected is given by

$$\frac{\langle F_{PH}^+ - F_{PH}^- \rangle}{\langle F_P \rangle} = 2 \left(\frac{N_A}{N} \right)^{\frac{1}{2}} \frac{F_H''}{f_0} \tag{2.22}$$

where N is the number of non-anomalously scattering atoms in the structure and N_A is the number of anomalously scattering atoms (based on Crick and Magdoff (1956)). For the case quoted above, ($f''=7e^-$ in 3500 nitrogen atoms ($f_0=7e^-$ for $(\sin \theta)/\lambda=0$) then from equation (2.22) the average fractional difference expected is 3.4%. The maximum possible anomalous difference is $2N_A f''$. Rewriting equation (2.22), the expected isomorphous difference (N_{ISO} is the number of isomorphously substituted atoms) is given by (Crick and Magdoff 1956)

$$\frac{\langle F_{PH} - F_P \rangle}{\langle F_P \rangle} = \left(\frac{N_{ISO}}{N} \right)^{\frac{1}{2}} \frac{F_H}{f_0} \tag{2.23}$$

The expected isomorphous amplitude difference is 18.8%. Hence an accuracy of intensity measurement of 5% for F_P^2 or F_{PH}^2 is adequate for isomorphous-based phasing, whereas 1% or better is needed, ideally, for non-optimised anomalous dispersion-based phasing for the example quoted.

SR offers the possibility of (a) optimising f'' as a function of λ to maximise signal to noise in the diffraction pattern, and (b) simulating isomorphous replacement through $f'(\lambda)$ to avoid the need for preparing isomorphous heavy atom derivatives. In the case of a metallo-protein, in principle, no derivatives are needed at all if multiple wavelength data sets can be collected. Each of these data sets will be exactly isomorphous with each and every other data set. This is an advantage over the heavy atom replacement method since not all derivatives that are prepared are isomorphous with the native and almost certainly not with respect to each other. On the other hand, heavy atom derivatives have L absorption edges in the short wavelength range $0.6\,\text{Å}<\lambda<1.1\,\text{Å}$. In this range protein crystal absorption is greatly reduced thus removing a serious systematic error in the measurement. For a metallo-protein with the natural cofactor being, say, manganese, the K edge at $1.896\,\text{Å}$ is a wavelength for which systematic absorption errors of the sample will be large. These matters are discussed in detail in chapter 9.

2.4.4 Calculation of the phase set incorporating a treatment of errors

In the absence of errors the phase of the native structure factor can be known exactly for the centric reflections using a single heavy atom derivative, the acentric reflections with a single derivative and anomalous scattering information (figure 2.13(b)) or with two derivatives in the absence of the anomalous differences (figure 2.13(a)). In practice, errors are a serious consideration and all possible sources of information need to be incorporated for an accurate evaluation of the phase. The effect of the errors is to prevent the structure amplitude circles (figures 2.13(a) and (b)) intersecting at unique points.

Blow and Crick (1959) made the assumption that for acentric reflections all the errors (experimental and those from non-isomorphism) lie in the magnitude of $\mathbf{F_{PH}}$.

Then for any phase angle α, the probability of that angle being correct is:

$$P(\alpha) = N \ \exp(-\varepsilon^2 / 2E^2) \tag{2.24}$$

where

$$\varepsilon(\alpha) = |\mathbf{F}_{PH}| - |\mathbf{F}_P + \mathbf{F}_H| \tag{2.25}$$

is the lack of closure error and where

$$\langle E^2 \rangle = \langle (|\mathbf{F}_{PH} \pm \mathbf{F}_P| - |\mathbf{F}_H|)^2 \rangle \tag{2.26}$$

is the average error.

Anomalous scattering information can be incorporated similarly (North 1965; Matthews 1966).

$$P_{ano}(\alpha) = \exp(-\varepsilon''^2 / 2E''^2) \tag{2.27}$$

$$\varepsilon''(\alpha) = \Delta_{ANO} - \Delta_{CALC} \tag{2.28}$$

where $\Delta_{CALC} = (2F_P F_H'' / F_{PH})\sin(\alpha_P - \alpha_H)$ and $\Delta_{ANO} = F_{PH}^+ - F_{PH}^-$.
For several derivatives the probabilities may be multiplied.

Blow and Crick (1959) showed that if the electron density map is calculated using phases corresponding to the centroid of the probability of the phase distribution the map has the smallest rms error.

The centroid coordinates (x, y) are calculated as:

$$x = \frac{\int_0^{2\pi} p(\alpha)\cos\alpha \, d\alpha}{\int_0^{2\pi} p(\alpha) \, d\alpha} \quad ; \quad y = \frac{\int_0^{2\pi} p(\alpha)\sin\alpha \, d\alpha}{\int_0^{2\pi} p(\alpha) \, d\alpha} \tag{2.29}$$

$$\tan\alpha_{best} = y / x \tag{2.30}$$

and

$$m = (x^2 + y^2)^{\frac{1}{2}} \tag{2.31}$$

where m is the magnitude of the vector from the origin. It is indicative of the quality of the phase determination for that reflection and is known as the figure of merit (see figure 2.15).

The best calculated electron density map for the native protein is:

$$\varrho(x,y,z) = \frac{1}{V} \sum_{hk\ell} F_P \, m \exp(i\alpha_{best}) \exp[-2\pi i(hx + ky + \ell z)] \tag{2.32}$$

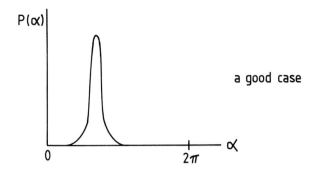

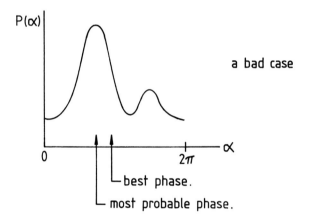

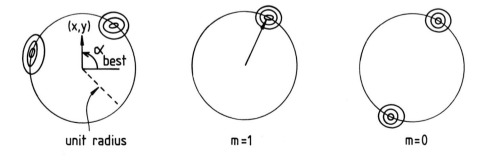

Figure 2.15 Some schematic examples of phasing and the meaning of figure of merit, m (following the treatment of Blow and Crick (1959)).

2.5 THE DIFFERENCE FOURIER TECHNIQUE IN PROTEIN CRYSTALLOGRAPHY

Once the structure of the protein has been established, even at low resolution, it is of interest to study the binding of small molecules to that protein crystal form. In practice, crystals are soaked in a solution containing a concentration of the small molecule that is higher than its equilibrium binding constant. There is no guarantee that the small molecule will bind at the same site as in the dissolved protein, in fact, it may not bind at all if the binding site is inaccessible to the solvent channels in the crystal or the ligand might not be soluble in the crystallisation precipitant. In the case of enzymes where the mechanism of enzyme catalysis is of primary interest, these studies can involve the binding of substrates, coenzymes and inhibitors. The latter may be required in studies aimed at designing drugs and/or antibacterial agents. The difference Fourier technique is also used in the solving of new heavy atom derivatives when the α_p are available.

The difference Fourier $\Delta\varrho$ is defined by:

$$\Delta\varrho_{\mathrm{OBS}}(x,y,z) = \frac{1}{V}\sum_{hk\ell} m(F_{\mathrm{PI}} - F_{\mathrm{P}})\exp(i\alpha_{\mathrm{p}})\exp[-2\pi i(hx + ky + \ell z)] \qquad (2.33)$$

(where I≡the bound molecule) which is an approximation to the true structure of the bound molecule alone:

$$\Delta\varrho_{\mathrm{true}}(x,y,z) = \frac{1}{V}\sum_{hk\ell} F_{\mathrm{I}}\exp(i\alpha_{\mathrm{I}})\exp[-2\pi i(hx + ky + \ell z)] \qquad (2.34)$$

It is important that the measurements are as precise as possible so that small changes in structure factor amplitude are greater than the standard deviations of their measurements.

Because the structures of the active sites of enzymes are extremely responsive to the binding of substrates and substrate analogues, conformational changes may take place which give rise to cracking or disorder of the crystal. It is important, however, that the crystals of protein plus inhibitor be closely isomorphous with the native protein crystals, at least with respect to the cell dimensions so that the molecular transform is sampled at the same reciprocal lattice points. In those instances where large conformational changes occur, *de novo* structure determination has to be attempted.

2.6 REFINEMENT OF THE STRUCTURES OF BIOLOGICAL MACROMOLECULES

The interpretation of the experimentally determined electron density map involves the fitting of a molecular model using a graphic display device.

The ease with which a molecular model can be fitted to the observed density depends on the resolution at which the map is calculated (compare figures 2.1(c), 2.1(d), 10.4 and 10.5) and its quality. The resolution limit of the calculation is set by the phase determination method. For the method of isomorphous replacement, phasing is successful usually to ≈ 2.5–$3.0\,\text{Å}$. In the case where a related structure is already known the method of molecular replacement (Rossmann (1972)) can be used whereby rotation and translation matrices are determined and then calculated phases used. Clearly, these two procedures are both approximate methods. The model is usually improved by using least-squares methods of refinement (for a collection of papers see Machin, Campbell and Elder (1980)) and higher resolution data (better than $2\,\text{Å}$ or so). Refinement methods involve the determination of shifts to the atomic parameters (coordinates and thermal parameters) so as to agree better with the observed diffraction data whilst preserving the known stereochemical features of proteins and nucleic acids. This is achieved by minimising a composite observational function:

$$\Phi = \Phi_{\text{X-ray diffraction}} + \Phi_{\text{bonds}} + \Phi_{\text{bond angles}} + \Phi_{\text{torsion angles}} + \Phi_{\text{non-bonded interactions}} \tag{2.35}$$

where

$$\Phi_{\text{X-ray diffraction}} = \sum_{hk\ell} \frac{1}{\sigma_F^2} (F_{\text{obs}} - F_{\text{calc}})^2 \tag{2.35a}$$

$$\Phi_{\text{bonds}} = \sum_{\text{distances}} \frac{1}{\sigma_D^2} (d_{\text{ideal}} - d_{\text{model}})^2 \tag{2.35b}$$

$$\Phi_{\text{bond angles}} = \sum \frac{1}{\sigma^2} (\tau_{\text{ideal}} - \tau_{\text{model}})^2 \tag{2.35c}$$

$$\Phi_{\text{torsion angles}} = \sum \frac{1}{\sigma^2} [1 + \cos(m\theta_i + \delta)] \tag{2.35d}$$

$$\Phi_{\text{non-bonded interactions}} = \sum \left(\frac{A}{r_i^{12}} + \frac{B}{r_i^6} \right) \tag{2.35e}$$

where d_{ideal} refers to the ideal bond distances and d_{model} those of the current structure, τ_{ideal} are ideal bond angles and τ_{model} those for the current structure, m is the periodicity of a torsion energy barrier and δ the phase. A and B are repulsive and long-range non-bonded parameters respectively. The σ's provide weighting of the terms. (From Steigemann (1980) based on Levitt (1974).)

The terms (2.35b)–(2.35e), i.e. excluding the X-ray diffraction term, are collectively known as the energy terms. The relative weight of the X-ray and the energy terms can be varied. The number of parameters to be refined is large. Because of the limited resolution of the X-ray diffraction pattern of a protein crystal the energy terms are exceedingly useful effectively to increase the data-to-parameter ratio. SR is widely used to help either increase the diffraction resolution limit of the crystal or improve the data quality at high resolution. Table 2.5 illustrates the relationship of resolution to the number of X-ray diffraction data. High resolution refinement has allowed determination of thermal parameters used to study the mobility of protein structures (Artymiuk *et al* 1979; Frauenfelder, Petsko and Tsernoglou 1979).

The stereochemical observations restrain the model to be compatible with prior knowledge regarding the distributions about 'ideal' values for particular features. Restraints are included relating to bonding distances, planarity of groups, chirality at asymmetric centres, non-bonded contacts, restricted torsion angles, non-crystallographic symmetry and thermal parameters. Many of these yield observational functions that are equivalent to terms in typical potential energy descriptions (Hendrickson 1980). It has been essential to do alternating rounds of refinement and model building using difference Fourier electron density maps either of the form $(F_{obs}-F_{calc})\exp(i\alpha_{calc})$ or $(2F_{obs}-F_{calc})\exp(i\alpha_{calc})$. It is also customary to use F_{calc}'s based on a model with a fragment of the model deleted. The fragment deleted (e.g. ten residue lengths) can be systematically varied through the whole protein and so the model is rebuilt to a whole series of these difference maps.

Computational refinement techniques have been extended to include the use of molecular dynamics and simulated annealing (Brünger, Kuriyan and Karplus 1987). This is proving useful in the early stages of refinement to reduce the amount of graphics model rebuilding work that is needed. In contrast to conventional minimisation, which allows only energetically 'downhill' steps, simulated annealing has a greater radius of convergence because it can cross barriers between minima by taking 'uphill' steps with probability $\exp(-E/kT)$, using an effective

Table 2.5. *Number of reflections accessible to various* d_{spacing}*'s for a* $100 \times 100 \times 100$ *Å unit cell for one octant of the resolution sphere.*

d_{spacing} (Å)	Number	Relative number
7	1527	1.0
6	2424	1.6
5	4189	2.7
4	8181	5.4
3	19393	12.7
2	65450	42.9
1.5	155140	101.6
1.2	303009	198.4

Notes:
(1) The volume of one octant of the resolution sphere is $\frac{1}{8}(1/d^3)\frac{4}{3}\pi$. This is appropriate for the approximate number of independent reflections for an orthorhombic crystal.
(2) The volume between adjacent reciprocal lattice points is 10^{-6} Å^{-3}.
(3) The improvement of the resolution of diffraction for crystals of the nucleosome using SR was 7–5 Å (Richmond *et al* 1984; Richmond, Searles and Simpson 1988), for phosphorylase b $\approx$ 2.8–2 Å (Wilson *et al* 1983), for hen egg white lysozyme $\approx$ 2–1.4 Å (Beavis, Sowerby and Helliwell, unpublished). This table illustrates therefore the improvement of the number of unique data via improvements in resolution limit.

temperature T as a control parameter (k is Boltzmann's constant here). In some ways therefore it is equivalent to manual model rebuilding but can be done automatically. The goal of a fully refined structure can be realised more quickly therefore. For a collection of papers see Goodfellow, Henrick and Hubbard (1989).

The basis of accurate refinement, and therefore stereochemically correct structures, is good high resolution X-ray diffraction observations (2 Å or better). These data are relatively weak owing to the nature of the atomic scattering factor, thermal vibration, static disorder and radiation damage effects. The high intensity of SR compensates for the intrinsically weak scattering of biological material. Also, the rapid data collection and use of short wavelengths reduces radiation damage. The short wavelength improves the signal to noise by improving the geometry of the diffraction measurement and reduces the systematic errors of absorption which leads to better atomic temperature factors. These matters are dealt with in chapter 6.

CHAPTER 3

Fundamentals of macromolecular structure

The understanding of biology has been transformed from being at a gross anatomical level to a molecular level. A major contribution to this has come from the application of physical techniques, especially X-ray diffraction, for the determination of structures which have provided, therefore, explanations for many key functions of organisms. The molecular basis of heredity followed from the discovery of the double-helix structure of deoxyribonucleic acid (DNA). The mechanism of action of many different protein molecules can be explained on the basis of their three-dimensional structures, for example, oxygen transport and storage, enzymes, membrane proteins and the immune response. Also the means by which viral infection takes place is currently being unravelled. The application of all this structural information has started with the engineering of new proteins with enhanced or modified functions and also the rational design of drugs. On the horizon is the detailed structure determination of the ribosome, which is the cell organelle involved in protein synthesis. A list of books dealing with the structure and function of macromolecules in detail is given in the bibliography, section 5.

The scale of atoms, molecules and macromolecules is illustrated in figure 3.1. The dimensions of bond lengths are of the order of 10^{-10} m or 1 Å and determine, therefore, the resolving power and the wavelength required in the technique of X-ray diffraction applied to determining molecular structure. A globular protein may typically be 50 Å in diameter, a virus is typically 300 Å whereas a cell is of the order of a few μm. The resolution of the optical microscope is only good enough to provide a gross image of cells and other subcellular organelles. However, all the functions of the cell are determined at the molecular level via the chemical reactivity of the valence electrons of individual atoms. The

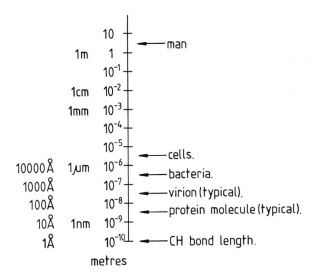

Figure 3.1 The scale of molecules, macromolecules, viruses and organelles.

function of a macromolecule involves motion and dynamics associated with specific conformational changes. The timescale of various biological processes is illustrated in figure 3.2.

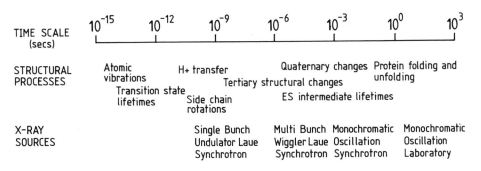

Figure 3.2 The timescale of various biological processes and approximate time resolutions of diffraction data recording, afforded with different synchrotron radiation sources. From Moffat (1989a) reproduced with permission of the author and from the *Annual Review of Biophysics and Biophysical Chemistry* **18**, © 1989 Annual Reviews Inc.

3.1 PRINCIPLES OF PROTEIN STRUCTURE

Proteins are linear polymers (polypeptides) with the amino acids as the monomer units. The molecular weight of a protein is typically in the range of 10 000–100 000 for a single chain. Higher molecular weights occur by the association of several protein chains. There are 20 amino acids commonly associated in proteins. The general formula of an amino acid is as follows:

$$NH_2 - \underset{\underset{R}{|}}{\overset{\overset{H}{|}}{C_\alpha}} - COOH$$

The R group refers to the side chain of which there are 20 different kinds varying in size, shape, charge, hydrogen bonding capacity and chemical reactivity. The properties of the side chains can be classed in terms of their acidic or basic nature and whether they are hydrophobic (water hating) or hydrophilic (water loving). Table 3.1 gives the chemical formulae of all the 20 commonly found amino acids, along with their three-letter and one-letter codes.

The arrangement of the four different groups about the α-carbon atom can be either left- (laevo, L) or right- (dextro, D) handed. Only L-amino acids are constituents of natural proteins.

In proteins, the α-carboxyl group of one amino acid is joined to the α-amino group of another amino acid by a peptide bond. The formation of a dipeptide from two amino acids follows the reaction:

$$^+H_3N - \underset{\underset{R_1}{|}}{\overset{\overset{H}{|}}{C}} - C\overset{\nearrow O}{\underset{\searrow O^-}{}} \quad + \quad ^+H_3N - \underset{\underset{R_2}{|}}{\overset{\overset{H}{|}}{C}} - C\overset{\nearrow O}{\underset{\searrow O^-}{}}$$

$$\downarrow$$

$$^+H_3N - \underset{\underset{R_1}{|}}{\overset{\overset{H}{|}}{C_\alpha}} - \overset{\overset{O}{\|}}{C} - \underset{\underset{H}{|}}{N} - \underset{\underset{R_2}{|}}{\overset{\overset{H}{|}}{C_\alpha}} - C\overset{\nearrow O}{\underset{\searrow O^-}{}} \quad + H_2O$$

peptide unit

Table 3.1. *The twenty common, naturally occurring, amino acids, their abbreviated names and one letter codes.*

amino acid (general formula)
–L isomer shown.

Name	R	Abbreviation	One letter code
glycine	—H	gly	G
alanine	—CH₃	ala	A
serine	—CH₂OH	ser	S
threonine	—CH(OH)CH₃	thr	T
valine	—CH(CH₃)CH₃	val	V
leucine	—CH₂—CH(CH₃)CH₃	leu	L
isoleucine	—CH(CH₃)—CH₂—CH₃	ile	I
aspartate	—CH₂—COO⁻	asp	D
asparagine	—CH₂—CONH₂	asn	N
glutamate	—CH₂—CH₂—COO⁻	glu	E
glutamine	—CH₂—CH₂—CONH₂	gln	Q
arginine	—CH₂—CH₂—CH₂—NH—C(NH₂)=NH	arg	R
lysine	—CH₂—CH₂—CH₂—CH₂—NH₃+	lys	K
cysteine	—CH₂—SH	cys	C
methionine	—CH₂—CH₂—S—CH₃	met	M
phenylalanine	—CH₂—(phenyl)	phe	F
tyrosine	—CH₂—(phenyl)—OH	tyr	Y
histidine	—CH₂—C=CH / H⁺N—NH / C—H (imidazole)	his	H
tryptophan	—CH₂—C (indole)	trp	W

Table 3.1 (*cont.*)

Name	R	Abbreviation	One letter code

proline is different from the above and has the full formula

$$\begin{array}{c} \text{COO} \\ | \\ {}^{+}\text{H}_2\text{N} \longrightarrow \text{C} - \text{H} \\ | \\ \text{H}_2\text{C} \diagdown \quad \diagup \text{CH}_2 \\ \text{CH}_2 \end{array}$$

pro   p

Notes:
(1) Non-polar amino acid side chains are
    gly, ala, val, leu, ile, pro, phe, trp, met.
(2) Polar side chains
    ser, thr, cys, tyr, asn, gln.
(3) Acidic side chains
    asp, glu.
(4) Basic side chains
    lys, arg, his.

The successive reaction of the carboxyl with the amino ends of amino acids leads to the building up of the linear polypeptide chain. Every α-carbon to which a side chain R is attached represents one residue. The start of the chain is by convention, the amino end. The peptide linkage between CO and NH undergoes hybridisation to form a planar bond, referred to as the peptide plane. The definition of the conformation of a protein chain can be simplified considerably as a result by defining only two parameters per residue, the conformational angles ϕ and ψ (see figure 3.3). The precise structures of amino acids and the peptide unit were determined by X-ray crystallographic studies undertaken by Pauling and Corey and their collaborators in the late 1940s and early 1950s.

The particular sequence of side chains or amino acids along the polypeptide backbone is known as the primary structure. This structure is determined by a particular sequence of nucleic acids in the gene; the relationship between a nucleic acid sequence and an amino acid sequence is known as the genetic code (see section 3.2).

Proteins have well-defined three-dimensional structures in their functional form. It is possible, under extreme conditions of acid or heat, to denature a protein, i.e. for the molecule to lose its specific shape and also then its function. Conversely by returning to mild conditions the linear chain refolds to a specific three-dimensional structure. Generally speaking, therefore, it is apparent that a particular primary sequence contains the information to specify a particular folded protein. There are then perturbations of this folded structure as the protein undertakes its function, involving the motion of side chains or extended lengths of chain or of domains within a protein. In principle, however, it should be possible to predict the basic three-dimensional fold from the primary structure. This has not yet been achieved. Some progress has been made in predicting particular substructures within the overall protein structure. We will shortly discuss these so-called secondary structures. It has been found in recent years that there are chaperone proteins involved in assisting the folding process and also disulphide isomerases to avoid 'incorrect' disulphide bridge formation. The mechanics of protein folding are not as simple, therefore, as was once believed.

There are two frequently occurring types of structure found as parts of a globular protein. These are the α-helix and the β pleated sheet. Both these structures are stabilised by hydrogen bonding, although much weaker than a covalent bond (i.e. the relative bond strengths are $10\,\text{kcal mole}^{-1}$ or less versus 100–$800\,\text{kcal mole}^{-1}$ respectively) many hydrogen bonds contribute to the stability of these units.

In the α-helix (figure 3.4) a hydrogen bond occurs between the CO group of an amino acid and the NH group of the amino acid that is four residues ahead in the linear sequence. This hydrogen bonding pattern can be achieved for each and every residue to form a helix. Hence, all the main chain CO and NH groups are hydrogen bonded, forming a stable

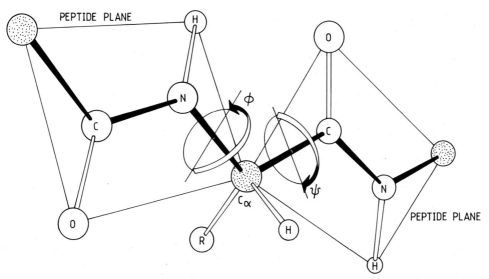

Figure 3.3 The conformational angles ϕ and ψ at each α-carbon in a polypeptide chain.

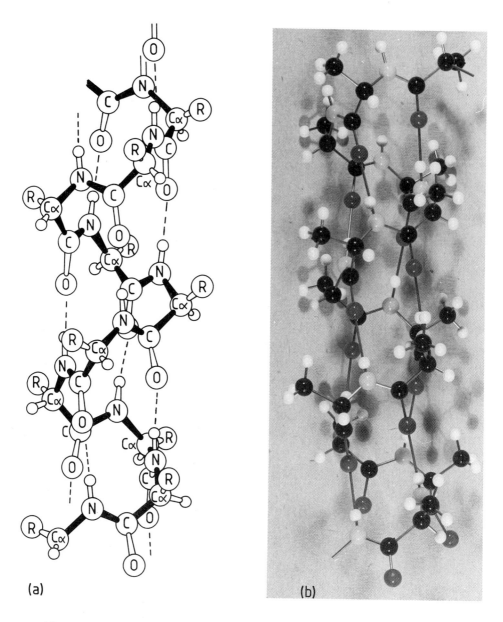

(a)　　　　　　　　　　　　　　　　　(b)

Figure 3.4　The right-handed α-helix formed by the polypeptide chain of
a protein with 3.6 amino acid residues per turn of helix. (a)
Schematic (hydrogen bonds shown as ––––). (b) Beevers
molecular model (purchased from Beevers Molecular
Models, Department of Chemistry, University of Edinburgh,
with permission to be reproduced here).

structure via many, weak hydrogen bonds. The pitch of the α-helix is 5.4 Å with 3.6 residues per turn. The screw sense of a helix can be right- or left-handed; the α-helices found in natural proteins are right-handed.

In the β sheet several, extended, polypeptide chains run approximately parallel or antiparallel to each other allowing hydrogen bonds to form between the NH and CO in adjacent chains. Both antiparallel and parallel β sheet types can occur in a globular protein (figures 3.5 and 3.6). The formation of many hydrogen bonds in the sheet therefore produces a stable structure.

Clearly hydrogen bonding is critical in stabilising a globular protein structure. Also of importance are electrostatic and van der Waals interactions, of respective bond strengths $5\,\mathrm{kcal\,mole^{-1}}$ and $1\,\mathrm{kcal\,mole^{-1}}$.

The electrostatic bond strength depends strongly on the value assumed for the dielectric constant, and therefore on the separation of the charges and the proximity of any water molecules. One reason why the theoretical prediction of protein three-dimensional structure from the amino acid sequence is so difficult is because of the prevalence of water *in vivo* and the large value of the dielectric constant for water (80); as a result the force between charges varies dramatically depending on the number of water molecules interposed between the charges.

The average thermal energy of atoms at room temperature is $0.6\,\mathrm{kcal\,mole^{-1}}$. It is clear that a protein is relatively loosely held together as a globular fold. It is susceptible to vibration and conformational change, which is, of course, necessary for its function. Vibrations and flexibility of protein molecules in the crystal give rise to significant diffuse X-ray scattering features (see chapter 8). Conformational flexibility occurs in a protein crystal because of the large solvent content thus permitting enzyme activity, for example, to be studied via time resolved X-ray crystallography (see chapter 10).

All the above bonding interactions are important in stabilising the globular fold. The prevalence of α-helix or β sheet varies from one protein to another. In some proteins such as myoglobin and haemoglobin (figure 3.7) the α-helix predominates. In other proteins such as concanavalin-A (figure 3.8) the β sheet predominates. Usually, there exists a mixture of α and β secondary structures.

There are families of proteins of related function which have similar structures. We have already mentioned the similarity between the two oxygen-binding proteins, myoglobin and haemoglobin, which contain haem units and are nearly all α-helical. Concanavalin-A is a member of a family of proteins known as lectins, involved in saccharide binding and

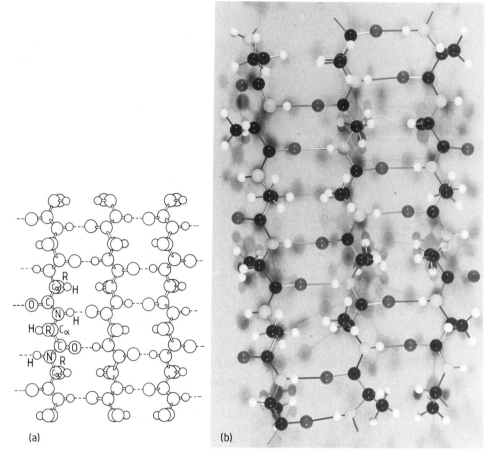

Figure 3.5 The antiparallel β pleated sheet formed by the polypeptide chain of a protein. (a) Schematic (hydrogen bonds shown as ————). (b) Beevers molecular model (with permission to be reproduced here). (c) As (b) but edge on view.

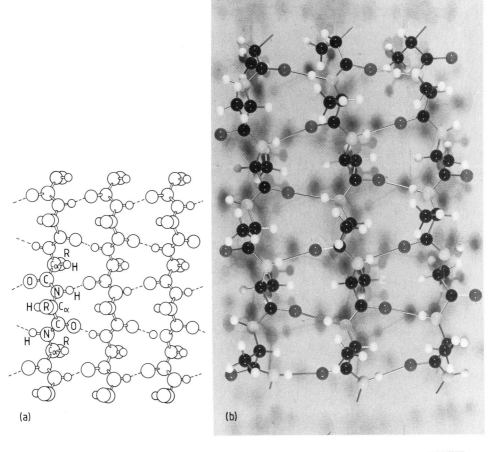

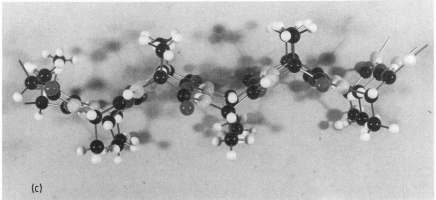

Figure 3.6 The parallel β pleated sheet formed by the polypeptide chain of a protein. (a) Schematic (hydrogen bonds – – – –). (b) Beevers molecular model (with permission to be reproduced here). (c) As (b) but edge on view.

cell recognition processes. Lectins possess very similar, nearly all β, structures. Another family of structures is the dehydrogenase family of enzymes which bind a small molecule, NAD (nicotinamide adenine dinucleotide) as a cofactor to aid the catalytic conversion of a reacting substrate molecule to a product molecule. Wherever the NAD is bound a very similar structure of a part of the globular protein is found, forming a

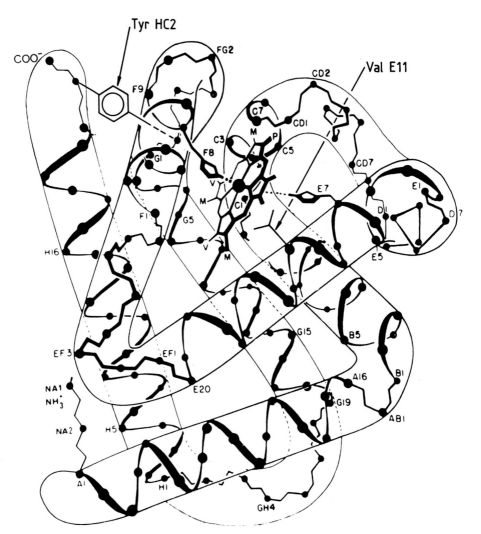

Figure 3.7 The β subunit of the haemoglobin (Hb) tetramer. The iron porphyrin embedded in the molecule is flanked on each side by the proximal and distal histidines. The Hb structure is dominated by α-helices. From Perutz (1970) reprinted with permission from *Nature* **228**, 726–37 fig 1, copyright © 1970 Macmillan Magazines Ltd.

Figure 3.8 The structure of the protein
concanavalin-A, a saccharide
binding protein, is predomi-
nantly β sheet. (a) Beevers
molecular model (with permission to be reproduced
here). (b) Ribbon diagram
(kindly prepared by Dr J.
Raftery). Based originally on
the coordinates of Reeke,
Becker and Edelman (1975).
The ribbon representation was
introduced by Richardson
(1985). The computer pro-
gram RIBBON was authored
by Priestle (1988).

(a)

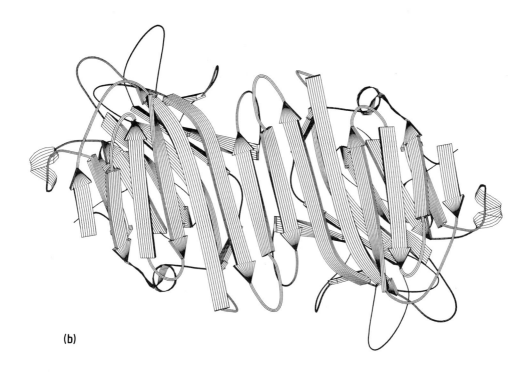

(b)

domain. This domain consists of a six-stranded parallel β sheet with four α-helices (figure 3.9(a)). An increasingly common structural motif found in globular proteins is the so-called β barrel (figure 3.9(b)), first identified in triosephosphate isomerase by Phillips, Wilson and coworkers. This barrel is made up of a central core of eight parallel β strands (i.e. the sheet is wrapped into a cylinder with the first and eighth strands interacting). Outside the β barrel are eight α-helices. This 'supersecond-

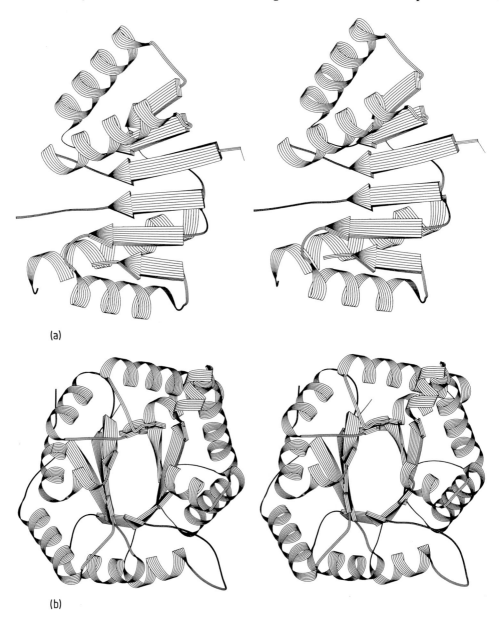

(a)

(b)

ary' structural motif has been found in 16 proteins so far. The β barrel obviously represents a stable structural motif.

The prediction of the three-dimensional structure from the amino acid sequence is a difficult problem. It is possible to predict the occurrence of an α-helix or β sheet with reasonable certainty and the supersecondary structural folds referred to above. This is not the same as the structure determination, however, that is available from experimental X-ray crystallography. The nature of biological function depends on precise conformations. These are not going to be predicted from theory since only one conformation will be calculated and the functional process involves *conformational change*. It is possible that the confident prediction of structural units as large as the β barrel will be an aid to X-ray structure determination in future years.

Homologous proteins of different sequence can be reasonably successfully predicted provided one of the family has been determined by X-ray crystallography. Computer graphics modelling involves a 'cut and paste' procedure of removing one sequence and adding on the new sequence.

Figure 3.9 Supersecondary structural motifs found within a globular protein. (*a*) Stereo drawing of the NAD binding domain found in dehydrogenase enzymes which consists of six strands of β sheet flanked on each side by two α-helices. Based on the dogfish lactate dehydrogenase coordinates of Abad-Zapatero *et al* (1987).

(*b*) Stereo drawing of the 'β barrel' found in at least 16 globular proteins; it consists of a β sheet joined round on itself to form a barrel and then flanked on the outside by α-helices. Based on the chicken triosephosphate isomerase coordinates of Banner *et al* (1976).

The ideal way to view stereo diagrams is with a stereo viewer (reflecting stereoscope) which can be purchased as follows: Order Ref number Z 15,675-2 cost £24.80 p. 1650 in the Aldrich Chemical Company Ltd (Catalogue 1990–1991); address The Old Brickyard, New Road, Gillingham, Dorset SP8 4JL, UK. Some people can achieve stereopsis with a small card placed between the individual drawings. Others find that focussing the eyes on an object at infinity (thus making the two lines of sight parallel and then dropping the lines of sight to the stereo view allows stereo vision without any tools whatsoever. Drawings kindly prepared by Dr J. Raftery; this type of representation was introduced by Richardson (1985). The computer program used was RIBBON (Priestle 1988).

Experimental methods of structure determination will always be a necessity but will, of course, be combined with theoretical modelling methods to elucidate biological functions.

We now consider the functional behaviour of some proteins in detail.

3.1.1 Globins

Myoglobin (Mb) and haemoglobin (Hb) are important in the history of macromolecular crystallography because they were the first protein crystal structures elucidated. This work came to fruition in the late 1950s and earned Kendrew and Perutz the Nobel Prize for Chemistry in 1962. Mb and Hb are very important molecules. Together they allow the storage and transport of oxygen in muscle and blood respectively. As a result of an organism being able to use oxygen, 18 times more energy is extracted from glucose than without oxygen.

Mb is a single globular protein with 151 amino acid residues of molecular weight 17 000, consisting of 1260 non-hydrogen atoms. Hb consists of four polypeptide chains (two of one type and two of another type) (figure 3.10) each of which makes a compact globular fold similar to Mb. The four folded chains associate together to form a larger compact entity in what is referred to as a quaternary structure. The association of the four molecules is also a precisely defined arrangement and is critical in dictating the somewhat different modes of oxygen binding in Mb and Hb.

The capacity of Mb and Hb to bind oxygen depends on the presence of a non-polypeptide unit. This is the haem group or porphyrin. The iron atom in the centre of the haem can form two additional bonding interactions, one on either side of the haem plane (see figure 3.7). Within the haem plane the iron atom binds to four, pyrrole, nitrogen atoms. The haem group is located in a crevice in the Mb molecule. Mb itself folds up into mainly α-helices which are by convention labelled A to H (i.e. eight in all). The iron atom of the haem is directly bonded to a histidine residue from the F-helix (residue F8, i.e. the eighth residue along the F-helix), and is known as the proximal histidine. The iron atom in Mb is about 0.3 Å out of the plane of the porphyrin, disposed towards F8. The oxygen-binding site is on the other side of the haem plane at the sixth coordinated position. Another histidine (E7), known as the distal histidine, is near to the oxygen site. Histidine E7 is critical to prevent binding of carbon monoxide in preference to oxygen. A free haem would bind carbon monoxide 25 000× more strongly than oxygen whereas in Mb, carbon monoxide binding to a porphyrin is 100× less effective

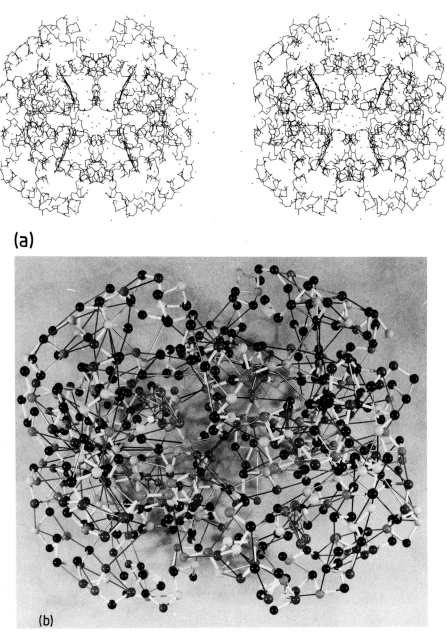

(a)

(b)

Figure 3.10 (a) A stereo drawing of the Hb tetramer, extracted from the Protein Data Bank based on the work of Perutz and coworkers (Fermi *et al* 1984). The dots are ordered water molecules. Drawing prepared by Dr J. Raftery. (b) A Beevers molecular model of the tetramer (with permission to be reproduced here).

because of the E7 histidine. However, carbon monoxide remains a dangerous poison.

In Hb, the four subunits are similar to Mb in structure. However, there are major differences in the oxygen-binding behaviour of the Hb which results from the association of subunits. The binding of one oxygen molecule to one haem in one subunit of Hb *affects* the binding of an oxygen to another subunit. There is a cooperative interaction between subunits. Since the four haems in Hb are never closer than $\approx 25\,\text{Å}$ a structural signal of oxygen binding at one subunit has to be sent. Perutz (1970) explained the structural basis of this from his X-ray crystal structure analyses. In deoxyhaemoglobin the iron atom is about $0.4\,\text{Å}$ displaced out of the haem plane towards the F8 histidine. On oxygenation the iron atom moves into the haem plane. In doing so the iron atom pulls F8 with it thus shifting the whole of the F-helix, and the corners of the E- and F-helices and the F- and G-helices. These conformational changes lead to the breaking of electrostatic interactions at the subunit interfaces (figure 3.11). Hence, a structural signal is sent from one subunit to another. Fundamental to this process is the relatively 'loose' association of the globular fold itself thus allowing the conformational change to take place. Life itself seems a precariously balanced affair! Evolution amplifies success, however, and the globins are found in many species serving this function of oxygen binding and transport. There are other mechanisms that some organisms have but these are not dealt with here.

This structural explanation of the cooperative binding of oxygen to Hb in 1970 by Perutz was called into question by EXAFS (extended X-ray absorption fine structure) spectroscopy results. The EXAFS technique is explained in appendix 4. It is a method, largely synchrotron-based, capable of providing more precise metal atom nearest neighbour distances than macromolecular crystallography (for reviews see Hasnain (1988) and Garner (1990)). EXAFS studies of Hb in the 1980s resulted in a fierce controversy between the crystallographers and the EXAFS spectroscopists (Perutz *et al* 1982; Galloway 1985; Fermi, Perutz and Shulman 1987).

The EXAFS of the various states of Hb was interpreted as showing that the iron atom did not move on oxygenation. Hence, the structural

Figure 3.11 Balsa wood models of the Hb tetramer in the (a) oxygenated and (b) deoxygenated forms. Note the increased separation of the β subunits on deoxygenation. Reproduced with the permission of M. F. Perutz.

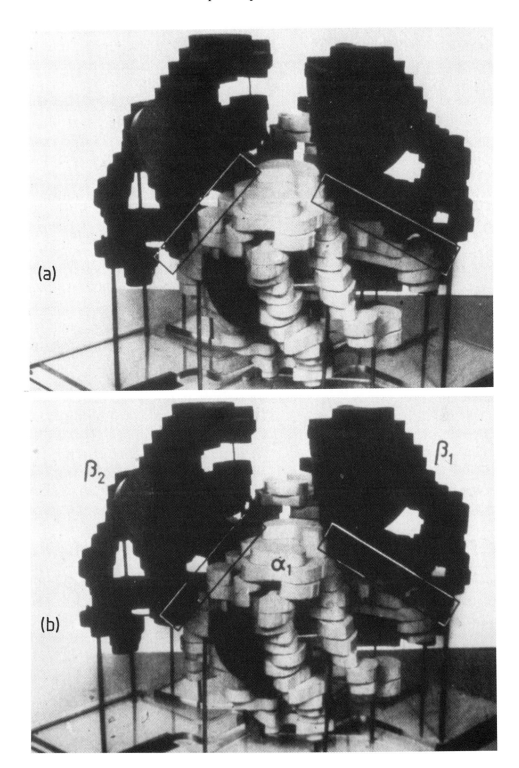

explanation of the cooperativity of Hb was challenged. The matter was concluded in favour of the original Perutz (1970) hypothesis. It was found that the EXAFS estimation of the Fe–N$_{pyrrole}$ distances had been converted by a dubious triangulation procedure to give the change in the Fe–haem plane distance. However, the assumption that the haem remained planar throughout is incorrect thus invalidating the basis of the triangulation (Perutz *et al* 1982). Although the basic EXAFS result was correct, the data were over-interpreted. More recently, a multiple scattering analysis of the EXAFS and XANES (X-ray absorption near edge structure) data has been done (Hasnain and Strange 1990). The multiple scattering of the photoelectron waves from the pyrrole rings has been included in the analysis and, by imposing constraints on the pyrrole ring geometry, the position of the iron atom relative to the porphyrin plane was obtained. For deoxyhaemoglobin the iron atom was shown to lie 0.5 ± 0.1 Å out of the porphyrin plane, in good agreement with the crystallography.

The original X-ray analysis of Hb was at a resolution of 2.8 Å. In order to define the X-ray crystal structure better, SR was used to collect the data to 1.7 Å at LURE (Fermi, Perutz, Shaanan and Fourme 1984) and then to 1.5 Å at the SRS Daresbury (Waller and Liddington (1990); see section 10.1 case study C). The extra precision of the X-ray analyses and model refinements (section 2.6) of Hb (Liddington 1985) confirmed the original interpretation of Perutz (1970) and brought the Fe–N$_{pyrrole}$ distance in excellent agreement with the EXAFS value (Perutz *et al* 1982).

This illustrates nicely one of the major roles that SR plays in macromolecular crystallography. For the cases of Mb and Hb the protein is easily available and readily crystallises into large crystals. This was very important for the success of the X-ray analyses. However, the measurement of the highest resolution data, necessary to settle the controversy referred to above, required the high intensity of the SR beam. The intensity compensated for the decrease in the atomic scattering factor for X-rays at high angle $((\sin\theta)/\lambda)$. A major application of SR in macromolecular crystallography, in general, is in the measurement of high resolution data (see table 10.1) to provide the most accurate structure possible.

3.1.2 Enzymes

The chemical reactions needed for living organisms to exist require a multitude of catalysts to make these reactions occur at reasonable rates and follow pathways useful for the organism. The catalysts in biological systems are enzymes and nearly, but not all, of them are proteins. They enhance reaction rates by factors of up to 10^6. Whole pathways of reactions exist involving the handing of one molecule on from one enzyme to another, the product of one being the reactant for the next. These are known as metabolic pathways. The 'throughput' of these pathways is often regulated by the binding of a final product molecule to a binding site on an enzyme at the head of a pathway. Hence, an enzyme may well have, as well as its catalytic site, a regulatory binding site. The process of conformational change on binding at the regulatory site of an inhibitor is transmitted to the catalytic site to inhibit binding of the true substrate.

Enzyme–substrate interactions are very specific, i.e. the substrate for one enzyme will not bind to the active site of another enzyme and vice versa. The specificity of a substrate for its enzyme arises because of the complementary shapes of the substrate and the hole or cleft into which it fits. In addition, specific hydrogen bonds or electrostatic interactions are available to entice its natural substrate to bind or to reject other molecules.

The recognition of the substrate by an enzyme usually involves a conformational change that may involve the straining of a bond in a substrate thereby making it amenable to rupture. In any case the first key step in catalysis is formation of the enzyme–substrate complex.

The first enzyme structure to be determined was that of lysozyme (see Phillips (1966)). Lysozyme is present in nasal mucus and tears, for example, and is capable of dissolving certain bacteria by cleaving the polysaccharide in their cell walls. Lysozyme was discovered in 1922 by Fleming. The precise details of its molecular function were revealed with the X-ray analysis of lysozyme crystals by Phillips and coworkers in the 1960s. Generally, enzymes' names end in 'ase'. The name lysozyme is an exception to this rule (which came from Fleming) – lyso because of lysing bacteria and zyme from enzyme. The three-dimensional structure of lysozyme is shown in figure 3.12. There are 129 amino acid residues (see table 3.2 for the sequence) of total molecular weight approximately 14 600. There is a mixture of α-helix and β sheet regions in the protein. The active site of the enzyme was studied by binding an inhibitor to the enzyme in the crystal. The inhibitor was a trimer of N-acetylglucosamine (tri-NAG). The X-ray analysis of the tri-NAG complex with lysozyme revealed the position of the active site. A variety of hydrogen bonds stabilised the complex as well as van der Waals interactions. The tri-NAG filled about half of a cleft on the surface of the enzyme molecule. Three additional sugar residues could be fitted into the cleft. It

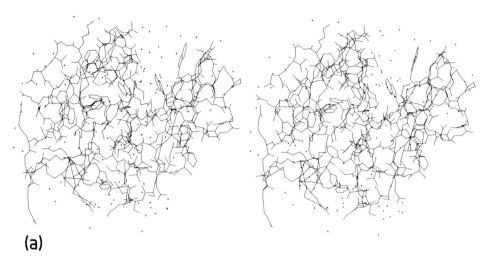

(a)

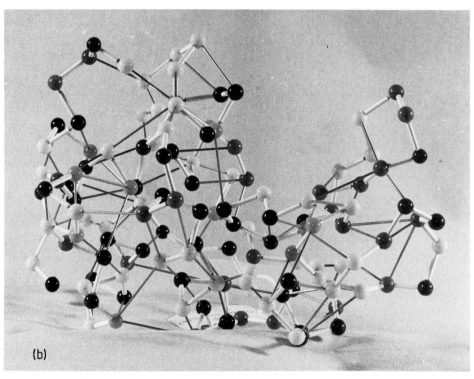

(b)

glycosidic linkage attacked by residues
asp 52 and glu 35 of the enzyme lysozyme

R = H₃C–C–H

(c)

Figure 3.12 The enzyme lysozyme viewed to illustrate the active site
cleft. (*a*) Skeletal diagrams in stereo (kindly prepared by Dr
J. Raftery). (*b*) Beevers molecular model (with permission
to be reproduced here). (*c*) Schematic of the alternating
NAG–NAM substrate and the bond cleaved, catalysed by
the enzymatic residues asp 52 and glu 35. Based on the
coordinates of Diamond (1974).

was inferred that the hydrolysis of hexa-NAG was facilitated by a distortion of the fourth sugar ring and that the bond between the fourth and
fifth sugar rings was the one broken. The detailed catalytic mechanism
proposed by Phillips and his colleagues involved the amino acid residues
glutamic acid 35 (glu 35) and aspartic acid 52 (asp 52) which are to be
found on either side of the hexa-NAG model built into the cleft.

The COOH group of glu 35 donates a proton to the glycosidic oxygen
atom between the fourth and fifth sugar ring thereby cleaving the bond.
This creates a positively charged C-1 (carbonium) ion on the fourth ring
which is stabilised by the carboxy group of asp 52. The dimer of sugar
rings then diffuses off the enzyme. The carbonium ion then reacts with
OH⁻ from the solvent. Glu 35 becomes protonated and then the tetra-
NAG diffuses off the enzyme. This particular enzyme molecule is then
ready for the next hexa-NAG molecule.

The experimental study of the tri-NAG inhibitor rather than hexa-
NAG was done for two reasons. One was that the full hexa-NAG would
be converted within one second or so (the natural turnover rate of a
lysozyme molecule in solution). The second reason was that the full
active site of the enzyme in the crystals of hen egg white lysozyme, the
form studied by Phillips and coworkers, is actually blocked by adjacent
molecules in the crystal lattice.

Table 3.2. *The amino acid sequence of (hen egg white) lysosyme, extracted from the Protein Data Bank. (In standard format.)*

1	lys	val	phe	gly	arg	cys	glu	leu	ala	ala	ala	met	lys
14	arg	his	gly	leu	asp	asn	tyr	arg	gly	tyr	ser	leu	gly
27	asn	trp	val	cys	ala	ala	lys	phe	glu	ser	asn	phe	asn
40	thr	gln	ala	thr	asn	arg	asn	thr	asp	gly	ser	thr	asp
53	tyr	gly	ile	leu	gln	ile	asn	ser	arg	trp	trp	cys	asn
66	asp	gly	arg	thr	pro	gly	ser	arg	asn	leu	cys	asn	ile
79	pro	cys	ser	ala	leu	leu	ser	ser	asp	ile	thr	ala	ser
92	val	asn	cys	ala	lys	lys	ile	val	ser	asp	gly	asp	gly
105	met	asn	ala	trp	val	ala	trp	arg	asn	arg	cys	lys	gly
118	thr	asp	val	gln	ala	trp	ile	arg	gly	cys	arg	leu	

With the advent of SR and the revival of the Laue method of data collection (chapter 7) the time required to measure a near-complete data set, with a broad wavelength bandpass, has been reduced to the millisecond range and there are immediate prospects of reducing this to microsecond data collection times. In one feasibility study exposure times have been realised of 120 ps from a lysozyme crystal using an undulator, narrow bandpass, X-ray source at the CHESS synchrotron. In order to study the lysozyme reaction further a different crystal form is being looked at, that of turkey lysozyme whereby the full active site cleft is accessible to the solvent channels running through the crystal; the specific intention being to study by X-ray crystallography the transition state complex (Howell *et al* to be published).

A large number of enzyme structures has now been determined by X-ray crystallography. Lysozyme was presented here because it was the first enzyme to be studied in this way. It also illustrates the essential features of catalysis, of what experimental information is available and how much has to be arrived at by model building and guesswork. A good electron density map of the enzyme without substrate was determined and some details of the interaction with the substrate were revealed by a binding study of a portion of the substrate. The direct imaging of the enzyme–substrate complex should be feasible at room temperature using rapid synchrotron data collection techniques. The technical problem to be solved is to coordinate the reaction of each and every enzyme molecule in the crystal ($\approx 10^{15}$ of them typically) with its substrate. In some cases the reaction can be slowed down sufficiently by using a substrate analogue such that simple diffusion of the substrate into the crystal can be used.

An alternative method is to cool the crystal (e.g. to $-50\,°C$ to $-100\,°C$) to slow the natural catalytic process. This is known as cryoenzymology (Douzou and Petsko 1984) and has been applied to elastase, a digestive enzyme. The problem with such low temperature studies is that quite often the crystal disorders. It is much more convenient to use room temperature and the capabilities of the synchrotron. Time resolved studies of enzyme structures are dealt with in chapter 10.

3.1.3 Membrane proteins

Biological membranes contain or separate a cell, a cell organelle or an organism from its environment. A membrane is a very active 'wall' involving various essential processes. These include molecular and ionic transport systems regulating the intracellular contents, receptors for external stimuli and signal transducers. Photosynthesis is the process whereby light is converted into chemical bond energy and is mediated by proteins embedded in a membrane.

X-ray crystallographic studies of such proteins obviously require that crystals of the pure protein are obtained. As a first step large quantities of a single protein need to be obtained. Some membrane proteins can be solubilised by mild treatment (e.g. with salt). Other membrane proteins are much more strongly bound and can be solubilised only by treatment with, for example, detergent which competes with the hydrocarbon chains of the membrane lipid for the non-polar interactions of the protein (figure 3.13). Crystallisation of membrane bound proteins has

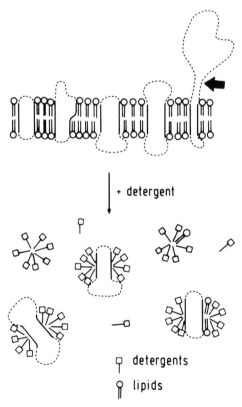

+ detergent

detergents

lipids

Figure 3.13 Schematic drawing of a biological membrane (top) consisting of a lipid bilayer and membrane proteins embedded into it, and its solubilisation by detergents (bottom). The polar part of the membrane protein surface is indicated by broken lines. From Deisenhofer and Michel (1989) with the permission of the authors, *EMBO J*, Oxford University Press and copyright © The Nobel Foundation (1989).

proved to be exceedingly difficult because of the unusual distribution of hydrophobic amino acids on a large part of the protein surface which inhibits association of the protein into a three-dimensional crystal. Breakthroughs in this research involved crystallisation of such proteins with detergent, e.g. the photosynthetic reaction centre from the purple bacterium *Rhodopseudomonas viridis* by Michel (1982) and the *E. coli* matrix protein by Garavito and Rosenbusch (1980) (figure 3.14).

There are several examples of parts of membrane proteins determined crystallographically. Here, a protein anchored in the membrane is 'clipped' to provide just the hydrophilic domain for crystallisation. These studies were of cytochrome b_5 (Mathews, Argos and Levine 1972), haemagglutinin (Wilson *et al* 1981) and neuraminidase (Varghese, Laver and Colman 1983) from influenza virus and the human class I histocompatibility antigen, HLA-A2 (Bjorkman *et al* 1987).

The X-ray crystal structure of the photosynthetic reaction centre was the first membrane bound protein to be determined (Deisenhofer *et al* 1984, 1985). X-ray data were measured to 3 Å for native and heavy atom derivatives with a conventional X-ray source and extended to 2.3 Å using SR.

An overall view of the photosynthetic reaction centre from *R. viridis* is shown in figure 3.15. It is a complex of four protein subunits and of fourteen cofactors. The total length of the reaction centre is ~130 Å. The

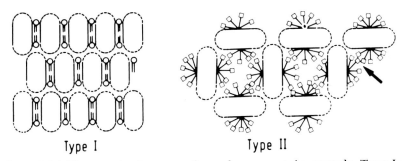

Type I   Type II

Figure 3.14 The two basic types of membrane protein crystals. Type I: stacks of membranes containing two-dimensionally crystalline membrane proteins, which are then ordered in the third dimension. Type II: a membrane protein crystallised with detergents bound to its hydrophobic surface. The polar surface part of the membrane proteins is indicated by broken lines. The symbols for liquids and detergents are the same as in figure 3.13. From Deisenhofer and Michel (1989) with the permission of the authors, *EMBO J*, Oxford University Press and copyright © The Nobel Foundation (1989).

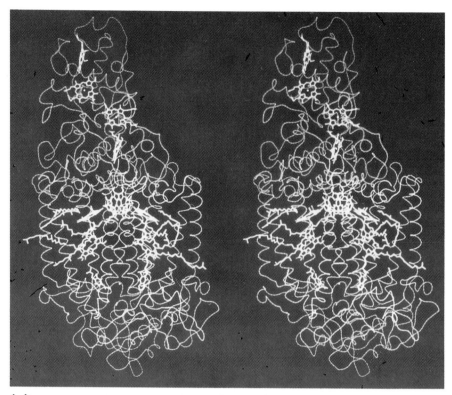

(a)

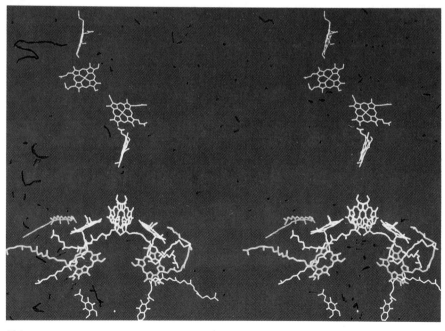

(b)

core of the reaction centre has an elliptical cross section with axes of 70 and 30 Å. Absorption of a photon generates an excited state and an electron is transferred via the protein cofactors across the membrane with time constants of 2.8 ps, 200 ps and 100 μs. The whole process can be described as a light-driven cyclic electron flow, the net effect of which is the generation of a proton gradient across the membrane that is used to synthesise the energy-rich molecule, ATP (adenosine triphosphate). The whole structural basis of the process was therefore revealed from the crystal structure. This work earned Deisenhofer, Huber and Michel the Nobel Prize for Chemistry in 1989. There are still unanswered questions concerning the nature of the electron transfer, its speed and temperature dependence. One of the major surprises from the structural work was the symmetry of the core structure whereby, of the two possible branches for electron transfer, only one branch is apparently used.

3.1.4 The proteins of the immune system

The immune system is essential to survival because without it death from infection would result. The critical event in mounting an immune response is the recognition of chemical markers that distinguish self from non-self. The diversity of the proteins of the immune response is based on structural variability whereby there are many millions of slightly different amino acid sequences enabling each molecule to recognise a specific target pattern. For a review see Tonegawa (1985).

The most familiar of the recognition proteins are antibodies or immunoglobulins. Another class of recognition molecules consists of the proteins called T-cell receptors but less is known of their properties and structure. A third class of proteins with a vital role in immune recognition is those in the major histocompatibility complex (MHC). The MHC proteins were discovered in tissue-grafting experiments. The structure of the major histocompatibility antigen has been determined by Bjorkman *et al* (1987).

Antibody molecules mark a foreign organism for destruction by other parts of the defensive system, e.g. the complement system, which has the

Figure 3.15 (*a*) Stereo view of the photosynthetic reaction centre. (*b*) Stereo view of the cofactors (including haem groups, bacteriochlorophyll-bs, bacteriophaeophytin-bs, carotenoid, quinones and a single non-haem iron). From Deisenhofer and Michel (1989) with the permission of the authors, *EMBO J*, Oxford University Press and copyright © The Nobel Foundation (1989).

ultimate effect of perforating the cell membrane, or macrophages which engulf and digest foreign particles. The key structural question is then how does an antibody recognise the foreign antigen?

An antibody molecule consists of four polypeptide chains. There are two identical 'light' chains of about 220 amino acids and two identical 'heavy' chains of either 330 or 440 amino acids. The four chains are held together by disulphide bonds and non-covalent interactions as shown in figure 3.16. In parts of the antibody, variations in amino acid sequence are found and there are three short segments which are especially variable (known as 'hypervariable' regions). These segments come together to form a cleft that acts as the antigen-binding site. The specificity of the molecule depends on the shape of the cleft and on the properties of the chemical groups that line its walls. Hence, the nature of the antigen–antibody recognition process is determined primarily by the amino acids in the hypervariable regions.

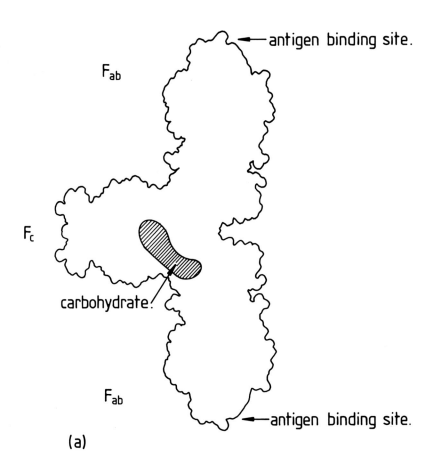
(a)

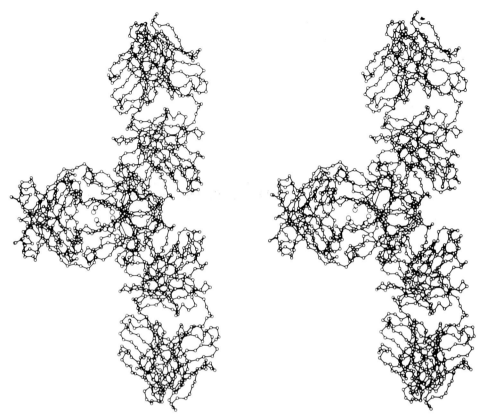

Figure 3.16 (*a*) Schematic diagram and (*b*) stereo view of the structure
of an antibody (F_{ab} is the antigen bonding fragment and F_c a
fragment which crystallises readily). From Silverton, Navia
and Davies (1977) with permission.

A wide variety of crystal structure studies have been done on parts of
an antibody and these have led to an overall detailed three-dimensional
structure. In addition, the structures of an antibody–antigen complex
(Amit *et al* 1986) and a neuraminidase–antibody complex (Colman *et al*
1987) have been determined.

3.2 PRINCIPLES OF NUCLEIC ACID STRUCTURE

The genetic information is contained within the chemical structure of
DNA, deoxyribonucleic acid, namely the sequence of bases along the
polynucleotide chain. RNA, ribonucleic acid, is used by an organism to
act as a messenger between the genes and the site of protein synthesis,
the ribosome. The genes of all cells and many viruses are made of DNA.

Some viruses, however, use RNA as their genetic material. A virus is a complex composed of protein and nucleic acid. Vast quantities of information have to be stored in the DNA sequences in complex organisms. The chromosomes have a distinct structural character that allows a condensation of the volume of DNA which is necessary to encode the information required to specify all the proteins in a given organism. A fundamental unit within the chromosome, for packing the DNA, is a nucleosome, a complex of proteins and DNA. The genetic information is converted into a specific protein at the ribosome, the site of protein synthesis within the cell. The ribosome is also a complex of proteins and RNA molecules. Crystals of viruses, the ribosome and nucleosomes require SR, in whole or in part, for their structural elucidation. To understand these structures and the role of the synchrotron in their X-ray analyses we must first explain the principles of nucleic acid structure.

3.2.1 DNA and RNA

DNA is a polymer of deoxyribonucleotide units. A nucleotide consists of a nitrogenous base, a sugar and one or more phosphate groups. The sugar in a deoxyribonucleotide is deoxyribose. The base is a purine or pyrimidine. The purines in DNA are adenine (A) and guanine (G) and the pyrimidines are thymine (T) and cytosine (C) (figure 3.17).

The three-dimensional structure of DNA was deduced by Watson and Crick (1953) by analysing X-ray fibre diffraction photographs of DNA fibres recorded by Franklin and Wilkins (Franklin and Gosling 1953). This structure for DNA was arrived at by building a model which could explain the basic features of the fibre pattern.

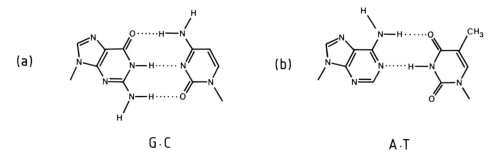

Figure 3.17 The structure of DNA is based on specific hydrogen bonding pairs (known as Watson–Crick base pairs), adenine (A) with thymine (T) and guanine (G) with cytosine (C). In RNA thymine is replaced by uracil. Figure kindly provided by Dr W. N. Hunter with permission.

The important features of the Watson and Crick model are as follows. The DNA consists of a double helix whereby two polynucleotide chains are coiled around a common axis (figure 3.18). The bases are on the inside of the helix whereby a base on one chain hydrogen bonds with a base on the other chain. There is a very specific pairing of bases (figure 3.17); adenine (A) must pair with thymine (T) whereas guanine (G) must pair with cytosine (C). These pairs fit perfectly into the space available on the inside of this helix whose dimensions are consistent with the X-ray fibre diffraction pattern.

The precise sequence of bases along the DNA carries the genetic information. An individual amino acid is coded by a triplet of bases. The genetic code for all the amino acids is shown in table 3.3.

The double helical model for DNA immediately suggested a mechanism for the replication of DNA. Because A must always pair with T and G with C one polynucleotide chain is the complement of the other. Hence, a daughter DNA molecule can be made by taking only one polynucleotide strand and then synthesising a new one that is complementary to it. As mentioned above, the Watson and Crick model of DNA was deduced from fibre diffraction patterns of natural DNA fibres. Each individual fibre had a variable sequence and so this, plus the fact that the fibre diffraction pattern is cylindrically averaged over several fibres, meant that the fine detail of the DNA structure could not be obtained. More information can be derived from fibres of synthetic nucleotides with defined base sequence (Mahendrasingam *et al* 1986; Fuller, Forsyth and Mahendrasingam 1990) although the cylindrical averaging is a fundamental problem.

Short stretches of DNA (oligonucleotides) can be synthesised and crystallised and therefore studied at near-atomic resolution. These studies reveal that there is a great deal of structural variability and flexibility in the DNA structure (Kennard and Hunter 1989). Most of these studies have used conventional X-ray sources. However, as will be mentioned in chapter 10, SR has been used in a few cases where the crystals have been small or radiation sensitive. Also, use has been made of a brominated form of an oligonucleotide to use multiple wavelength techniques for phase determination.

RNA is also used to store genetic information in some viruses (such as HIV, human immunodeficiency virus). RNA differs from DNA in two respects. Firstly, the sugar units in RNA are riboses rather than deoxyriboses. The second difference is that thymine is replaced by uracil; the latter does not possess a methyl group but is otherwise identical. The structure of transfer RNA (t-RNA) is shown in figure 3.19.

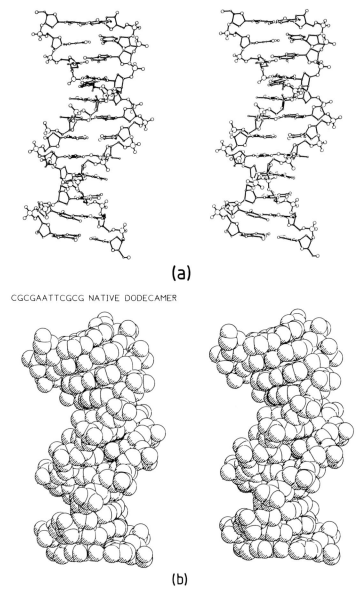

(a)

CGCGAATTCGCG NATIVE DODECAMER

(b)

Figure 3.18 The 'Watson–Crick' double helical structure of DNA illustrated by the crystal structure of an oligonucleotide. (*a*) Skeletal model representation, in stereo (note the tilting of the base pairs in certain cases; this is responsible for causing the DNA double helix to coil up, for example, into the nucleosome, a key component of the chromosome). (*b*) Space filling representation, in stereo. (*c*) Beevers molecular model. Figures kindly provided by Dr W. N. Hunter with permission.

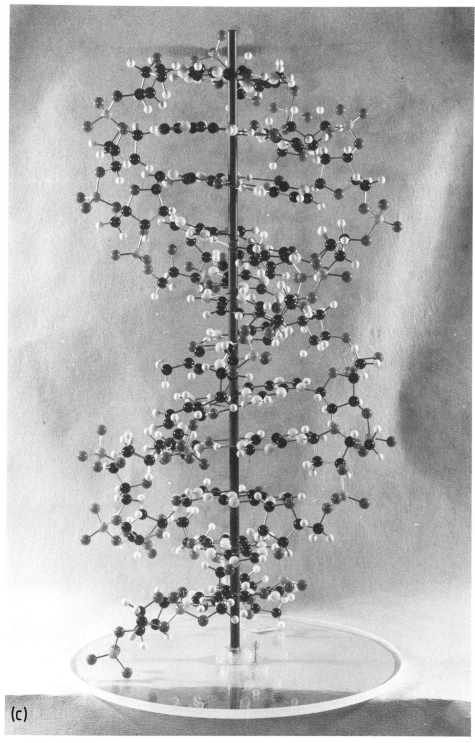

Figure 3.18 (*cont.*)

Table 3.3. *The genetic code relating the possible sequence of a triplet of nucleotides to individual amino acids.*

Position 1	Position 2				Position 3
	U	C	A	G	
U	phe	ser	tyr	cys	U
	phe	ser	tyr	cys	C
	leu	ser	stop	stop	A
	leu	ser	stop	trp	G
C	leu	pro	his	arg	U
	leu	pro	his	arg	C
	leu	pro	gln	arg	A
	leu	pro	gln	arg	G
A	ile	thr	asn	ser	U
	ile	thr	asn	ser	C
	ile	thr	lys	arg	A
	met	thr	lys	arg	G
G	val	ala	asp	gly	U
	val	ala	asp	gly	C
	val	ala	glu	gly	A
	val	ala	glu	gly	G

Note:

Uracil replaces thymine in RNA compared to DNA. The chemical difference is that the methyl group for thymine is replaced by a hydrogen in uracil.

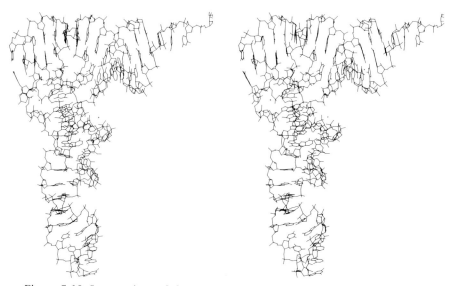

Figure 3.19 Stereo view of the structure of t-RNA. Based on the coordinates of Hingerty, Brown and Jack (1978). Figures kindly prepared by Dr J. Raftery.

3.3 MULTIMACROMOLECULAR COMPLEXES

3.3.1 Chromosomes, chromatin, nucleosomes

The DNA of an organism, if it is stretched out as a double helix, is a very long molecule. The DNA of the bacterium *Escherichia coli* consists of 4×10^6 base pairs and the extended DNA would be 1.4 mm long; in humans the DNA consists of 3.9×10^9 base pairs and the DNA would be 990 mm long. These figures can be arrived at by taking the 3.4 Å separation of successive base pairs in the helix and multiplying by the total number of base pairs. The huge size of the DNA in humans, referred to as the human genome, and its perceived importance in aiding the understanding of a wide variety of genetic disorders is leading to the attempt to sequence it. This huge project, administered by HUGO (Human Genome Organisation) is expected to take several decades of effort by many laboratories participating in the project. A fundamental structural question is how is the DNA packaged?

A group of proteins known as histones bind to the DNA to form what is called a nucleosome. Many nucleosomes constitute a chromatin fibre. This fibre resembles beads on a string, each bead being one nucleosome.

Nucleosome cores have been crystallised and studied by X-ray diffraction and electron microscopy. The X-ray diffraction model to 7 Å resolution has been published (Richmond *et al* 1984). The resolution of the X-ray diffraction data has been extended to $\approx$5 Å using SR (Richmond, Searles and Simpson (1988); see table 10.3).

3.3.2 The ribosome

Protein synthesis requires the conversion of the genetic information stored in the DNA into the step by step addition of one amino acid to a growing polypeptide chain (figure 3.20(*a*)). In the language of the molecular biologist the genetic information is *translated* in the three stages of initiation of the process, elongation of the polypeptide chain and termination of the process. Firstly, however, the genetic information stored in the DNA has to reach the site, in the cell, of protein synthesis. The information is carried by a messenger known as messenger RNA (m-RNA), which is a single-stranded RNA molecule.

The ribosome is at the centre of the protein synthesis machinery (figure 3.20). It consists of a large and a small subunit, each consisting of about two-thirds RNA and one-third protein by weight. In *E. coli* the molecular weight of the ribosome is 2.7 million. In addition to the ribo-

(a)

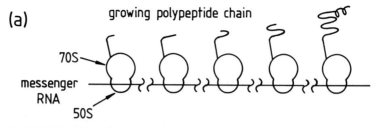

(b)

Figure 3.20 The ribosome is the site of protein synthesis. (a) Schematic of several ribosomes in the process of manufacturing a particular protein. (b) Model derived from electron microscopy image reconstruction. From Bartels *et al* (1988) with permission.

some there are another hundred or so macromolecules involved in a coordinated way to synthesise a protein.

The architecture of the ribosome is being studied by a variety of physical techniques such as electron microscopy (figure 3.20(b)) and neutron, as well as X-ray, diffraction and NMR. Some progress has been made in the X-ray crystallographic analysis of some of the individual proteins in the ribosomal subunits (for a review see Liljas (1982)). These proteins are, however, very difficult to crystallise. The structures of some of the smaller ones are now also being determined by two-dimensional NMR methods. (For a review of two-dimensional NMR see Wuthrich (1986).)

In recent years crystallisation of whole ribosome particles has been possible. The crystals have very large unit cell parameters and are very radiation sensitive. SR is being used to compensate for the weak diffraction of the crystals and freezing of the crystal preserves its lifetime in the beam. This is a most exciting project and also a very challenging one, not least because the structure to be determined does not possess any degree of internal symmetry. Internal symmetry is a very useful aid to improving the quality of electron density maps for virus crystals. These advantages will not be present in the ribosome electron density map when it is determined. However, some of the individual proteins are available at high resolution, as referred to before, and this prior information can be useful to the overall interpretation. The role of SR in the ribosome project is covered in chapter 10.

3.3.3 Viruses

The function of a virus is to invade an organism and to use the organism to make multiple copies of itself. The consequences for the host organism in supporting this mass production of the invader include death. Death can be as a result of a cancer caused by the virus. Alternatively, in the case of AIDS, the virus destroys the immune system.

A simple virus particle consists of an outer coat made of protein, perhaps itself covered by a membrane, and an inner core of DNA or RNA. The two common structures of the coat proteins are either a cylindrical (helical) shell or a spherical shell with icosohedral symmetry. The proteins on the surface of the virus can undergo relatively rapid mutation as a result of the large number of generations of virus produced. These changes on the surface of the virus are not recognised by the host's antibodies and therefore the virus avoids rejection as an invader.

High symmetry of the spherical shell of protein is very important. The number of nucleotides in the viral DNA or RNA is actually quite small because of the limited storage space available and it is therefore necessary for the proteins making up the coat to occur in multiple copies. X-ray crystallography is able to determine the structure of the protein shell. The structure of the nucleic acid is not revealed except for some partial details in a few cases. This is either because it is randomly coiled, or because the symmetry of the nucleic acid structure is just different from that of the protein shell which satisfies the crystal lattice and symmetry.

Several plant viruses have been studied using X-ray crystallography and conventional X-ray sources. These are, in no particular order, TBSV (tomato bushy stunt virus), SBMV (southern bean mosaic virus) and STNV (satellite tobacco necrosis virus) – all spherical viruses – and TMV (tobacco mosaic virus) – a cylindrical virus. These virus crystals diffract relatively well and are reasonably stable to radiation.

Several mammalian and insect viruses have recently been solved by X-ray crystallography (e.g. see figure 3.21). These are poliovirus, rhinovirus, mengovirus and FMDV (foot and mouth disease virus) and black beetle virus. All except poliovirus were very radiation sensitive and could only be studied by collecting the data at the synchrotron. Even so a total of ≈600 crystals was required for each structure analysis. In addition, in the case of FMDV, handling a highly pathogenic sample on the synchrotron beam line is not trivial, nor is its transport to the facility!

However, the speed with which virus crystal data can be collected at the synchrotron has transformed the subject so that it is even possible rapidly to survey the binding of a large number of drugs to the virus shell by X-ray crystallography. This makes virus crystallography one of the most spectacular projects to be undertaken at the synchrotron. Its intrinsic technical difficulties have been well served by the high intensity and fine collimation of the SR beam for these dense diffraction patterns and the use of short wavelengths has been important in reducing radiation damage, although hundreds of crystals are used in the initial structure determination.

One of the most pressing problems of the age is, of course, AIDS which is caused by HIV. The HIV virion is about 1000 Å in diameter. It has not been possible so far to crystallise this virus. Instead, the individual proteins are being crystallised for detailed study. There is no doubt, however, that the architecture of the plant and mammalian viruses referred to earlier, determined by X-ray crystallography, are very important to the general understanding of HIV and thereby the AIDS disease.

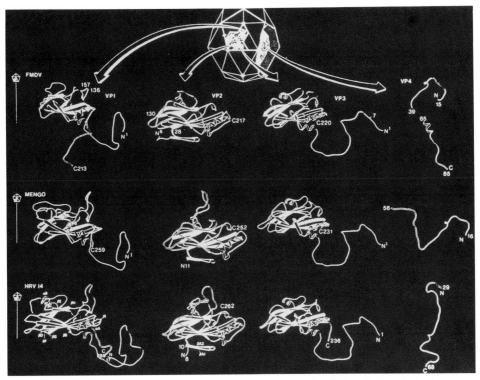

Figure 3.21 A schematic diagram of the structure of the class of virus particles known as picornaviruses; shown here are rhinovirus, mengovirus and FMDV. See section 10.5 for further details. VP=viral protein (types 1–4). From Acharya *et al* (1989), reprinted with permission from the authors and *Nature* **337**, 709–16, copyright © 1989 Macmillan Magazines Ltd.

3.4 APPLICATIONS TO MEDICINE AND INDUSTRY: DRUG DESIGN AND PROTEIN ENGINEERING

The detailed knowledge of macromolecular structure is providing the impetus for a new revolution whereby medicine is developed at the molecular level in a rational way and enzymes can be optimised for industrial rather than biological tasks.

3.4.1 Drug design

Drugs and antibiotics have, in the past, been discovered either by chance or by the system of screening thousands of compounds. The knowledge of protein and virus structure at near-atomic resolution allows the rational design of inhibitor molecules.

Penicillin is perhaps the most famous example of chance discovery. Penicillin was discovered in 1928 by Fleming, who observed that bacterial growth was inhibited by a contaminating mould. Moreover, this mould (*Penicillum*) was not toxic to animals. The penicillin molecule (figure 3.22) was extracted and characterised by Florey. Penicillin exhibits the necessary qualities of a drug; it is specific to the condition without being harmful to the host; the 'condition' here being a variety of bacterial infections. The mode of action of the penicillin molecule is to block the final step of the synthesis of a bacterial cell wall, thus preventing reproduction of the bacteria from one generation to the next. It does this by inhibiting an enzyme of the bacterium, known as a transpeptidase, which is critical to the cell wall synthesis. Penicillin is able to bind to the transpeptidase because it looks similar to part of the natural substrate (D-ala–D-ala) (figure 3.22). The bound penicillin then forms a covalent bond with a serine residue at the active site of the

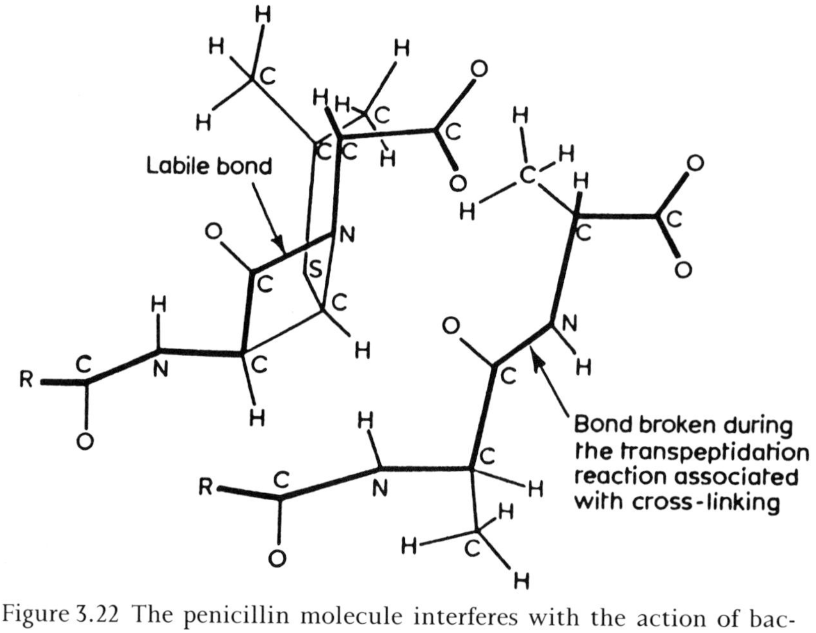

Figure 3.22 The penicillin molecule interferes with the action of bacteria (and therefore of infections) by mimicking the structure of D-ala-D-ala involved in the production of bacterial cell walls in cell division. This is a rationalisation after the event, namely the discovery of the action of penicillin by chance by Alexander Fleming. By contrast modern rational drug design strategies involve using the knowledge of the structure of receptor sites. From Helliwell (1977b) reproduced with permission of *New Scientist*.

enzyme and is therefore permanently attached to it. The enzyme molecule is then permanently inhibited.

However, bacteria have now evolved which contain a type of enzymes known as β-lactamase which are very effective at destroying penicillin (another name for the enzyme is penicillinase). Crystallographic studies of this group of enzymes are under way in various laboratories. The ultimate aim is to try and use the structure to design analogues of penicillin which have substituents that prevent binding to β-lactamase whilst still being able to attach the transpeptidase. One of the β-lactamase crystallographic analyses (*Bacillus cereus* β-lactamase I) was hampered by having only small crystals available. However, the high intensity of SR allowed high resolution data to be collected and provide important structural information on this enzyme (chapter 10 and Samraoui *et al* (1986)).

This illustrates the conceptual ideas behind the rational design of new drugs and antibacterial agents. There is a wide range of other crystallographic studies in academic and pharmaceutical industry laboratories around the world. These studies encompass attempts to deal with cancer, viral infection, sleeping sickness, hypertension, inflammation and asthma. A review of this topic has been given by Hol (1986).

3.4.2 Protein engineering

Mutated proteins can be readily produced by using genetic engineering methods. It is possible to construct new genes for proteins, with designed properties, by deletion, insertion, transposition or substitution. The chemical procedures for doing this will not be described here.

These methods allow very specific changes to be made to a protein structure. Critical residues in a protein structure can be changed, the new gene 'expressed' and a large quantity of mutated protein is produced with altered properties. This procedure is called site-directed mutagenesis and the design of a new protein is referred to as protein engineering (for a survey of the area see the book edited by Oxender and Fox (1987)). This is a tremendously important development both academically and industrially. In academic research it is assisting the understanding of how proteins fold, for example.

Industrial companies are linking up with synchrotron facilities to allow access to the powerful data collection facilities for X-ray crystallography that are established there, examples in the UK being Glaxo and Wellcome.

CHAPTER 4

Sources and properties of SR

Particle accelerators were originally developed for high energy physics research into the subatomic structure of matter. The SR, which was produced in circular electron accelerators ('synchrotrons') was a nuisance by-produce – an energy loss process. The early stages of the utilisation of SR were therefore parasitic on the high energy physics machines whose parameters were, of course, not optimised for SR. However, SR became well recognised in its own right as a major tool in research in biology, chemistry and physics. Particle accelerators began to be designed specifically for SR production with parameters optimised solely for this work, e.g. continuous beams with long lifetimes, stable source positions and magnetic insertion devices to produce radiation of specific properties; the Daresbury Synchrotron Radiation Source (SRS) which came on-line in 1981 was the first dedicated, high energy source. Table 4.1 gives a list of storage ring X-ray sources. All modern SR sources are storage rings rather than synchrotrons. The particles used may be electrons or positrons.

Studies of the properties of the radiation from accelerated charges extend over the last 100 years. Extensive theoretical work on the radiation effects in circular electron accelerators has been done by Schwinger (1949) and Sokolov and Ternov (1968) and the theory is reviewed in Jackson (1975).

The technology of storage rings used for SR production has advanced considerably so that it is now possible to induce sophisticated particle trajectories in special magnets known as wigglers and undulators. These new types of magnet can extend both the available spectral range of emitted photons as well as the brightness compared with radiation from charged particles in simple circular orbits in ordinary bending magnets. In this section the properties of SR will be discussed with reference to these three types of magnetic configuration as source.

Table 4.1. *Storage ring synchrotron X-radiation sources with machine energies, E > 1 GeV (from Winick (1989) and pers. comm. (1990) and reproduced with permission of the author and Gordon and Breach).*

Location	Ring (lab)	Electron energy (GeV)	Notes
Brazil			
Campinas	LNLS–1	1.0	Dedicated[a]
	LNLS–2	2.0	Proposed/dedicated[a]
China (PRC)			
Beijing	BEPC (IHEP)	2.2–2.8	Partly dedicated
China (ROC-Taiwan)			
Hsinchu	SRRC	1.3	Dedicated[a]
England			
Daresbury	SRS (Daresbury)	2.0	Dedicated
Europe			
Grenoble	ESRF	6.0	Dedicated[a]
France			
Orsay	DCI (LURE)	1.8	Dedicated
Germany			
Bonn	ELSA (Bonn Univ)	3.5	Partly dedicated
Hamburg	Doris II (HASYLAB)	3.5–5.5	Partly dedicated
West Berlin	Bessy II	1.5–2.0	Proposed/dedicated
India			
Indore	INDUS-II (CAT)	2.0	Proposed/dedicated[a]
Italy			
Frascati	ADONE (LNF)	1.5	Partly dedicated
Trieste	ELETTRA	1.5–2.0	Dedicated[a]
Japan			
Hiroshima	HISOR (Hiroshima Univ)	1.5	Proposed/dedicated
Nishi Harima	SPRING–8 (Sci Tech Agency)	8.0	Dedicated[a]
Kyushu	SOR (Kyushu Univ)	1.5	Proposed/dedicated
Sendai	TSSR (Tohoku Univ)	1.5	Proposed/part dedicated
Tsukuba	Photon Factory (KEK)	2.5	Dedicated
	Accumulator Ring (KEK)	6.0	Partly dedicated
	Tristan Main Ring (KEK)	8–32	Planned/dedicated
Korea			
Pohang	Pohang Light Source	2.0	Design/dedicated[a]
Sweden			
Lund	Max II	1.5	Dedicated[a]
USA			
Argonne, IL	APS (ANL)	7.0	Dedicated[a]
Berkeley, CA	ALS (LBL)	1.5	Dedicated[a]
Ithaca, NY	CESR (CHESS)	5.5	Partly dedicated
Stanford, CA	SPEAR (SSRL)	3.0–3.5	Dedicated
	PEP (SSRL)	5.0–15.0	Partly dedicated

Table 4.1. (*cont.*)

Location	Ring (lab)	Electron energy (GeV)	Notes
Upton, NY	NSLS II (BNL)	2.5	Dedicated
USSR			
Moscow	Siberia II (Kurchatov)	2.5	Dedicated[a]
Novosibirsk	VEPP-3 (INP)	2.2	Partly dedicated
	VEPP-4 (INP)	5.0–7.0	Partly dedicated[a]
Zelenograd	TNK	2.0	Dedicated[a]

[a] In construction as of 11/90.

4.1 RADIATED POWER (BENDING MAGNET)

An electron charge moving on a circular trajectory undergoes a centripetal acceleration; if viewed on edge the charge appears to undergo oscillatory motion akin to a dipole.

In the non-relativistic case the radiated power, Q, is given by the Larmor (1897) formula

$$Q = \frac{2}{3} \frac{e^2}{c^3} \left| \frac{d\mathbf{v}}{dt} \right|^2 \tag{4.1}$$

where e is the electronic charge, c is the velocity of light and $|d\mathbf{v}/dt|$ the acceleration of the electron.

In a SR source the electron moves at a speed close to c so that for a radius of curvature, ϱ, the relativistic generalisation of the Larmor formula is

$$Q = \frac{2}{3} \frac{e^2 c}{\varrho^2} \left(\frac{v}{c} \right)^4 \left(\frac{E}{mc^2} \right)^4 \tag{4.2}$$

where E is the total energy and mc^2 is the rest energy of the particle. If we define

$$\beta = v/c \quad \text{and} \quad \gamma = E/mc^2$$

then

$$Q = \frac{2}{3} \frac{e^2 c}{\varrho^2} (\beta\gamma)^4 \tag{4.3}$$

The emission of radiation by the particle is an 'energy loss process', i.e. if the orbit is to be maintained at the same energy E then the lost energy must be replenished. This is done by an oscillating radio frequency (rf) electric field. The simple circular orbit described here is the situation that exists in a bending magnet of a storage ring.

In an actual case there are many particles which then constitute a total current I. In practical units (E in GeV, ϱ in metres, I in amperes), then, the total radiated power is (in the case of electrons or positrons)

$$Q(kW) = 88.47 \frac{E^4 I}{\varrho} \tag{4.4}$$

The fourth power of E means that a dramatic rise in Q is achieved for a given change in E which, in turn, places a heavy demand on the replenishing rf field.

If we consider the case of the Daresbury SRS where $E=2\,\mathrm{GeV}$, $\varrho=5.5\,\mathrm{m}$ and the typical operating current is 300 mA, then $Q=77\,\mathrm{kW}$ in total for 2π radians or $12.3\,\mathrm{W\,mrad^{-1}}$. These estimates are very important in assessing the power loading on optical elements (sections 5.2.6 and 5.3.9).

4.2 ANGULAR DISTRIBUTION

The angular distribution of the radiation is given by the dipole radiation formula, which, in the non-relativistic case, is

$$\frac{dQ}{d\Omega} = \frac{e^2}{4\pi c^3} \left|\frac{d\mathbf{v}}{dt}\right|^2 \sin^2\theta \tag{4.5}$$

where θ is measured relative to the direction of acceleration. Integration of equation (4.5) over all solid angles, Ω, gives the Larmor formula (equation (4.1)). Equation (4.5) also describes the distribution as seen in the instantaneous frame of reference of a relativistic electron (figure 4.1).

The distribution for the relativistic case in the laboratory frame can be found by using the relativistic transformation

$$\tan\theta' = \frac{\sin\theta}{\gamma(\beta + \cos\theta)} \tag{4.6}$$

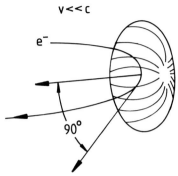

Figure 4.1 Radiation pattern for an electron in a simple circular orbit with $v \ll c$. It is of the typical dipole radiation type. This is also the pattern seen in the instantaneous frame of an electron when $v \lesssim c$.

At $\theta = 90°$, $\cos \theta = 0$ and $\sin \theta = 1$ and since $\beta = 1$ we have

$$\theta' \approx \gamma^{-1} = \frac{mc^2}{E} \tag{4.7}$$

Hence, γ^{-1} is a typical opening half-angle of the radiation in the laboratory system. It has a very small value; for an electron of rest energy 511 KeV at 2 GeV, then, $\gamma^{-1} = 0.28$ mrad or 0.016°. The angular distribution is concentrated into a very small, forward angular cone (figure 4.2). As the electron sweeps along its circular trajectory (as in the

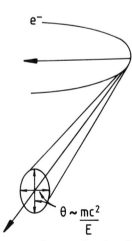

Figure 4.2 Radiation pattern, when $v \lesssim c$, in the laboratory frame of reference. It is tightly collimated in the forward direction with an opening angle $\gamma^{-1} = mc^2/E$.

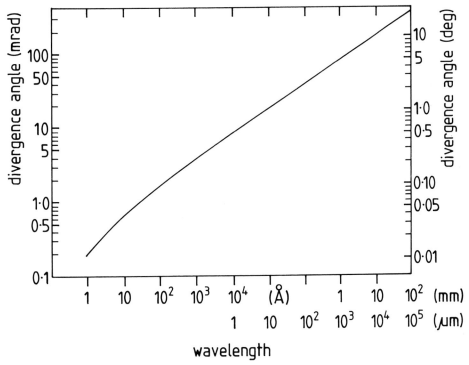

Figure 4.3 The variation of opening angle, γ^{-1}, with wavelength.
(Example: Daresbury SRS bending magnet, $\lambda_c=3.88\,\text{Å}$.)

bending magnets) the instantaneous cone is swept out in the orbital plane producing a fan of radiation. The radiation remains highly collimated in the plane perpendicular to the orbit.

The Schwinger (1949) equations (see next section) give a more detailed description of the angular distribution. In particular, the opening angle is a function of the emitted wavelength, increasing at longer wavelengths (figure 4.3).

4.3 SPECTRAL DISTRIBUTION

The spectral distribution is properly described by Schwinger (1949). The instantaneous power radiated by a monoenergetic electron in circular motion expressed as energy per unit time per unit angle per unit wavelength is given by

$$I(\psi,\lambda) = \frac{8\pi e^2 c^3}{3\omega_0\lambda^4\gamma}\,(1+\chi^2)\left[K_{2/3}^2(\xi)+\left(\frac{\chi^2}{1+\chi^2}\right)K_{1/3}^2(\xi)\right] \qquad (4.8)$$

where

$$\chi = \gamma \psi$$

and

$$\xi = \left(\frac{2\pi\varrho}{3\lambda} \right) \frac{1}{\gamma^3} (1 + \chi^2)^{3/2}$$

ψ is the angle between the line of observation and its projection on the orbital plane and is often expressed as a multiple of γ^{-1}; ω_0 is the orbital frequency and $K_{1/3}$, $K_{2/3}$ are modified Bessel functions of the second kind. Equation (4.8) offers a complete description of the spectral distribution. Figure 4.4 shows the spectral distribution in terms of the emitted number of photons per second per A per GeV per mrad horizontal per 1% bandwidth ($\delta\lambda/\lambda$). This curve is also known as the 'universal curve' because it is applicable to all SR sources (except for undulators which by definition are devices where interference effects affect the spatial and spectral distribution).

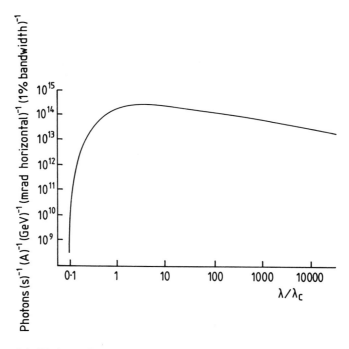

Figure 4.4 'Universal' spectral distribution of photon flux in SR.

An important wavelength describing the distribution is the characteristic (critical) wavelength, λ_c, given by

$$\lambda_c = 4\pi\varrho / 3\gamma^3 \tag{4.9}$$

or in practical units

$$\lambda_c(\text{Å}) = 5.59\varrho / E^3 = 18.64 / BE^2 \tag{4.10}$$

with ϱ in metres, E in GeV and B, the magnetic field, in tesla. The peak
photon flux occurs close to λ_c, the useful flux extends to about $\lambda_c/10$ and
half of the total power radiated is above the characteristic wavelength
and half is below.

The critical wavelength is a very important parameter. To illustrate
the relationship between λ_c, B and E in equation (4.10) several examples
will be cited. The Daresbury SRS is a 2 GeV machine with 1.2 T bending
magnets and a 5 T wiggler with, therefore, λ_c's of 3.88 Å and 0.93 Å
respectively. The ESRF is a 6 GeV specification machine with two types
of bending magnet of 0.4 T and 0.8 T and λ_c's of 1.29 Å and 0.65 Å respectively. Finally, the 1.5 GeV X-ray synchrotron light source under construction at the University of Lund (MAX-II) has in the design scheme a
7.5 T wiggler with a λ_c of 2.49 Å.

Although it is true that on the one hand, a trend is towards higher
energy machines of very large diameter to provide long straight sections
for multipole insertion devices (sections 4.10.1.2 and 4.10.2), it is also
the case that a relatively low energy in conjunction with high field
magnets produces a spectrum well into the X-ray region and suitable for
diffraction experiments. Hence, it is increasingly likely that in future
there will, as well as the very large central facility, be more and more
smaller (compact) machines built, each with a high field magnet.

4.4 POLARISATION PROPERTIES

In the plane of the orbit, the emitted SR is 100% linearly polarised with
the electromagnetic **E** vector parallel to the electron acceleration. This
is as one would expect if we simply visualised the orbit edge-on as a
dipole source. Away from the plane of the orbit there is a significant
perpendicular component.

In figure 4.5 we see a plot of the parallel and perpendicular components of polarisation as a function of ψ (expressed in multiples of the

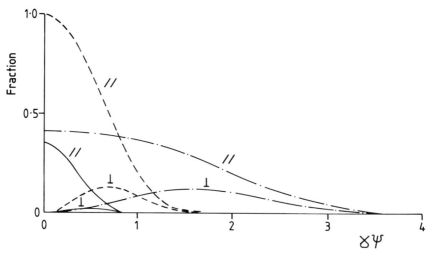

Figure 4.5 Polarisation components of SR plotted as a function of $\gamma\psi$ for three values of λ/λ_c. Full line, $\lambda/\lambda_c=0.33$; dashed line, $\lambda/\lambda_c=1$; chain line, $\lambda/\lambda_c=10$.

natural opening angle (γ^{-1})) and as a function of λ. For $\psi=0$ (i.e. in the orbit plane) $I_\perp$ is zero and the polarisation is 100% $I_\parallel$. At a certain distance from the tangent point an aperture which subtends an angle less than γ^{-1} is a simple but effective way to control the polarisation of the X-ray beam. This aperture may, in practice, be that due to a vertically focussing mirror or, in the absence of that, the camera or diffractometer collimator itself.

4.5 THE MACHINE, BEAM LINE FRONT ENDS, BEAM POSITION MONITORING AND STABILITY

The basic components of an electron storage ring radiation source are shown in figure 4.6. The ring consists of a number of dipole (bending) magnets to bend the electron beam along a circular trajectory to achieve a closed orbit. The bending magnets act as the basic sources of radiation. They are separated by straight sections in which are located numerous components including insertion devices; because of their paramount importance as radiation sources, insertion devices are dealt with separately in section 4.10. The straight sections also house quadrupole magnets to focus the electron beam. The beam properties are determined to a large extent by the arrangement of the dipole and quadrupole magnets, often referred to as the magnet lattice, which can vary greatly from one ring to another. Rf cavities are used to accelerate the beam to

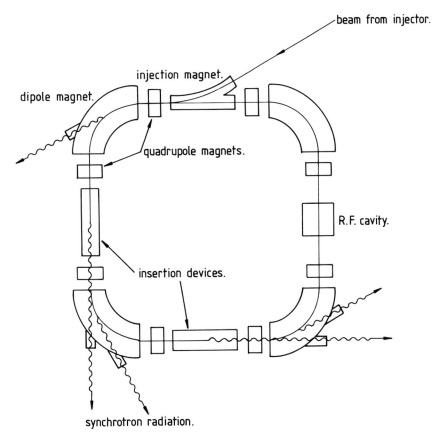

beam from injector.

injection magnet.

dipole magnet.

quadrupole magnets.

R.F. cavity.

insertion devices.

synchrotron radiation.

Figure 4.6 The basic components of an electron storage ring. From
Walker (1986) with permission.

the required energy (if necessary) and to make up the energy lost due to
the emission of SR; the rf cavities generate an electric field parallel to the
beam orbit alternating in polarity, usually sinusoidally, at high
frequency (50–500 MHz). As an example of the overall layout of a SR
machine, figure 4.7 shows the situation for the ESRF.

4.5.1 Beam line front ends

A schematic diagram of the front end of the beam line is shown in
figure 4.8. The machine vacuum has to be protected. Hence, the beam
lines emanating from the ring are 'separated' from the main ring vacuum
chamber. For an X-ray beam line this is by using a beryllium window at
the front end of the beam line. For a soft X-ray or vacuum ultra-violet
(VUV) line such a window would completely absorb the radiation of

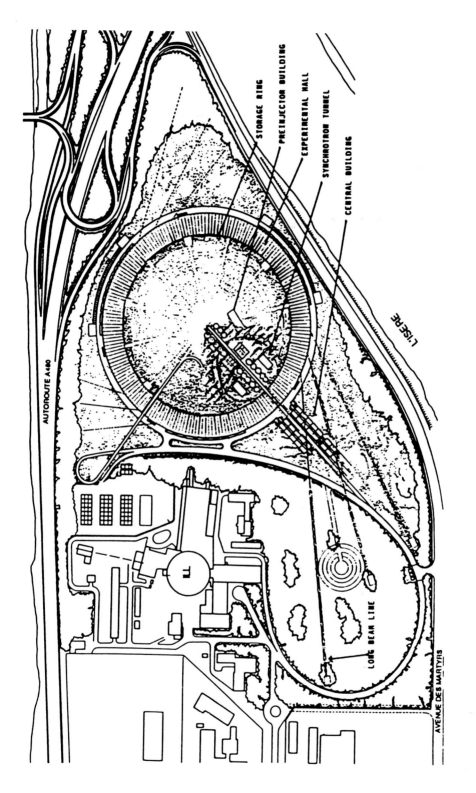

STORAGE RING

PREINJECTOR BUILDING

EXPERIMENTAL HALL

SYNCHROTRON TUNNEL

CENTRAL BUILDING

AUTOROUTE A480

L'ISÈRE

ILL

LONG BEAM LINE

AVENUE DES MARTYRS

Figure 4.7 The overall layout of the ESRF. This ring is one of the largest to be constructed specifically for SR research (the circumference of the ring is 1 km). ILL=Institut Laue Langevin neutron source. Courtesy ESRF, Grenoble, with permission.

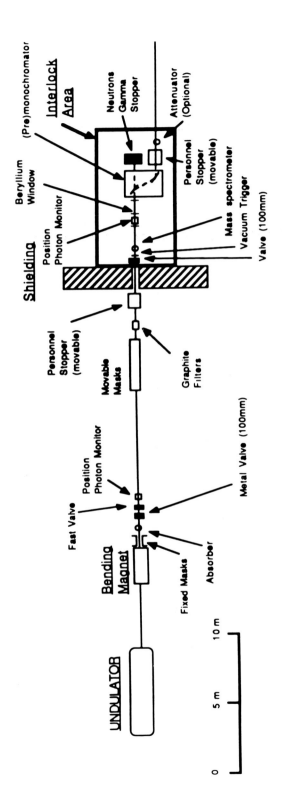

Figure 4.8 A scheme proposed for an undulator beam line front end. From the ESRF Red Book (1987) with permission.

that wavelength so instead a fast valve is placed in the beam line so as to
close quickly in the event of a vacuum failure in the beam line.

Water-cooled absorbers and masks are placed in the beam line front
end. There are fixed and movable masks. A fixed mask is located next to
the bending magnet vacuum chamber; it constrains the opening fan of
bending magnet radiation. The movable masks are the personnel safety

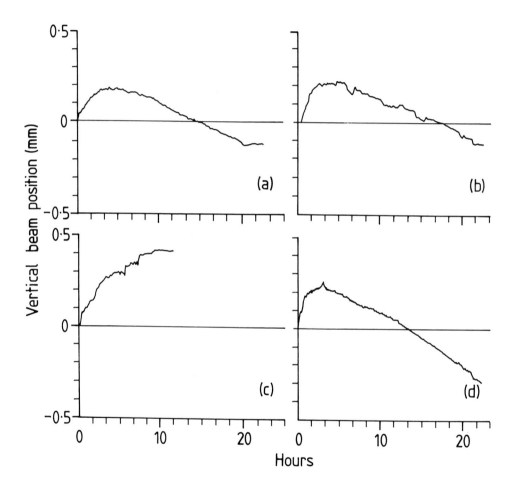

Figure 4.9 Beam position stability: the plots show the recorded movement of the vertical position of the SRS measured at 21 m
from the source in different situations: (a) the slow drift
during a typical fill; (b) sudden jumps occur sometimes due
to power supply instabilities; (c) movements in the first fill
after a two-day shutdown; and (d) abnormally high drift
with a beam current above 300 mA. From Quinn *et al* (1990)
with permission.

shutters used to close off the beam when access to an experimental hutch is required.

Insertion devices (section 4.10) present additional problems associated with the extremely high heat loads. These are capable of melting the metal components in valves in the event of a malfunction.

4.5.2 Beam position monitoring and stability

Photon beam position monitors are essential to ensure that after an injection the electron beam position is adjusted to allow the SR to strike the beam line optical components in a constant way. The wavelength output from a double crystal monochromator is especially sensitive to the vertical beam position. Also, the quality of the focus, from a toroid mirror, is especially sensitive to the horizontal beam position (figures 5.18(c) and (e)). On existing machines it is necessary to recalibrate the wavelength and the focussing of a beam line optical system after each injection.

Within a fill there are then beam position drifts and jumps as well as intermittent oscillations of magnitude $\pm100\,\mu$m and $\pm25\,\mu$rad (vertically) over a period of 12 hours. Slow position drifts are associated with temperature variations of the magnet lattice. Irregular jumps are usually due to steering magnet power supply instabilities; these jumps can be as large as $50\,\mu$m (see figure 4.9 and Quinn, Corlett, Poole and Thomson (1990)). Obviously for a focal spot of $250\,\mu$m, a $50\,\mu$m source position jump results in a corresponding shift in the focal spot position. Hence, the beam flux transmitted through a pinhole collimator will jump dramatically and, if not corrected for, will introduce errors into the experimental data. Ideally, the position and angle stabilities required are $\approx10\,\mu$m and a few μrad respectively, over short and long term.

4.6 TIME STRUCTURE

The temporal structure of the X-ray beam reproduces exactly that of the electrons (or positrons) in the storage ring (figure 4.10). The electrons travel in the ring in bunches and thus the radiation is emitted in pulses. The length of the bunch is determined partly by the frequency and amplitude of the accelerating field in the storage ring cavities and partly by the details of the magnet lattice. There are other processes which can cause the bunch to be longer than the natural length. One of these processes is the interaction on the bunch by fields induced by the bunch itself in the walls of the vacuum chamber. This has a character-

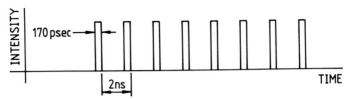

Figure 4.10 The time structure of the X-ray beam derives from that of
the electrons (or positrons) in the storage ring. The
electrons travel in the ring in bunches and thus the radiation is emitted in pulses. The values shown are for the
Daresbury SRS in multibunch mode. In single bunch mode
the light pulse occurs each orbital period of 0.321 μs.

istic signature in which the bunch length increases with the one-third
power of the bunch current, above a threshold current. The natural
bunch length is therefore only likely to be experienced at small beam
currents (Suller 1989). The bunch length normally has a value of one-
tenth or less of the bunch separation.

The bunch separation (typically 2–20 ns) is synchronised to the rf
accelerating field frequency in the main storage ring (and booster if
used) because the rf has to replenish the energy lost to SR. The circumference length of the machine can accommodate a certain maximum
number of bunches called the harmonic number; when all these bunches
are filled this is known as multibunch mode. It is also possible to operate
the machine with only one bunch filled (i.e. single bunch mode). The
time between light flashes is then determined by the orbital period,
which is useful for time resolved experiments. Moffat (1987) suggested
that if the number of photons provided by a single bunch of electrons was
sufficient to measure a diffraction pattern then the time resolution of the
experiment would be the bunch length. The feasibility of this was
demonstrated subsequently on CHESS using an undulator as the source
(section 4.10.2.1 and Szebenyi et al (1989)).

Typical values of the bunch width, separation and orbital period can
be illustrated with respect to the Daresbury SRS and the ESRF in
Grenoble. For the SRS the bunch width is 170 ps, the bunch separation
2 ns and the orbital period 321 ns. For the ESRF these values are respectively 65–140 ps, 3 ns and 2.84 μs.

4.7 BEAM CURRENT AND LIFETIME

Provided that various instabilities have thresholds at higher current
values, the main limitation on the beam current, I_b, that can be sup-

ported in a storage ring, is set by the rf power available. This has to replenish the SR power emitted which increases linearly with the beam current. The beam current in a storage ring decays with time, t, from the value at injection, I_0, with an exponential time constant, τ:

$$I_b = I_0 \, \exp\left(-t / \tau\right)$$

The lifetime of the beam is influenced by many factors. The dominant beam loss mechanism results from collisions of the electrons with residual gas molecules in the machine vacuum. Both inelastic and elastic scattering can take place off the nuclei and orbital electrons of the gas molecules. The beam lifetime is inversely proportional to the vacuum pressure that can be achieved. After the start-up of a new storage ring, or one for which the vacuum has been let up to atmosphere and then the ring pumped down and baked, the lifetime will be poor. However, it will improve with operation. This is because the gas molecules adsorbed to the vacuum vessel surfaces are desorbed by the SR itself.

Another process which can be of importance in limiting the lifetime is the scattering of one electron off another in the same bunch, known as the Touschek effect. The scattering rate depends on the electron energy and density in the bunch and so is important usually for low emittance or low energy machines, which have a high current in a short, small cross section, electron bunch.

The advent of undulators (see section 4.10) requires the pole pieces of these insertion devices to be brought close together for short wavelength emission. There is a limit to how small the gap can be made because small apertures limit the lifetime primarily due to elastic Coulomb scattering of electrons off the residual gas molecules.

4.8 SOURCE EMITTANCE

For a collection of emitting charged particles (electrons or positrons) circulating repeatedly in a storage ring there is not just a single unique trajectory; there is a finite transverse distribution and angular spread of the beam controlled by the entire magnet lattice. This lattice is the periodic configuration of quadrupoles, dipoles and straight sections which make up the ring.

The position and angular trajectory of an electron, and hence of the emitted photons, are correlated parameters. The phase space plot relates the position (x or y) of an electron and its angular trajectory (x' or y')

with respect to the mean orbit; separate plots are needed for the vertical direction (y,y') perpendicular to the mean orbit plane and for the radial direction (x,x') in the orbit plane. Figure 4.11 (a) and (b) shows typical examples. The precise shape and orientation of these phase space ellipses changes for different points on the electron orbit but their area, the electron emittance, is invariant for a given machine lattice and electron energy.

The emittance of the SR is calculated by convoluting the distribution of emitted photons with the electron emittance. Treating the angles first

$$\sigma'(\text{total}) = [\sigma'(\text{electron})^2 + \sigma'(\text{radiation})^2]^{\frac{1}{2}} \qquad (4.11)$$

We add these quantities in quadrature because to a good approximation the distributions are Gaussian. As before, the 'prime' denotes angular parameters.

The position is given by

$$\sigma(\text{total}) = \sigma(\text{electron}) \qquad (4.12)$$

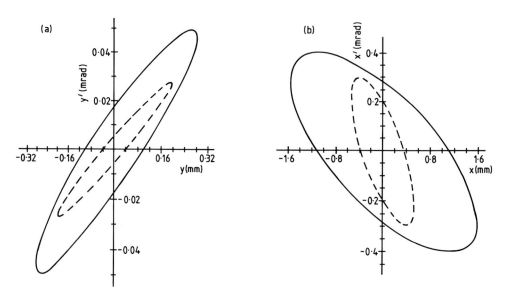

Figure 4.11 Idealised electron beam emittance phase space plots: (a) vertical; (b) horizontal (radial direction). The dotted line is a lower emittance mode than the full line. From Winick (1980) with permission.

In practice, σ'(electron) is usually a good deal less than σ'(radiation) $(=\gamma^{-1})$ so that σ'(total) is determined almost completely by γ^{-1}. The machine emittance is then, in units of metre radians for example,

$$\varepsilon \approx \sigma\sigma' \tag{4.13}$$

with values given in the horizontal and vertical plane.

Other quantities of interest are the so-called beta-functions in the horizontal and vertical where

$$\beta \approx \sigma/\sigma' \tag{4.14}$$

β varies consistently around the ring (see section 4.10.3), and so, therefore, do σ and σ'.

4.9 FLUX, INTENSITY, BRIGHTNESS AND BRILLIANCE

The principles of SR and the practicalities of producing it have now been covered. The radiation emanates with defined spectral character-istics from a finite source size and with a finite angular divergence. Macromolecular crystallography exploits some or all of these properties depending on the size of the crystal sample, the unit cell and the crystal sample mosaicity. The terms *flux*, *intensity*, *brightness* and *brilliance* are used to describe the various aspects of SR and its use in experiments.

The spectral flux, Φ, is the number of photons, emitted per unit time into a relative bandwidth $\delta\lambda/\lambda$ into an angle element in the plane of the electron orbit and integrated in the vertical plane. The units of flux Φ are usually

$[\Phi] =$ number of photons per second per mrad horizontal per 0.1% relative bandwidth

Note that the universal distribution of Schwinger described in section 4.3 is per mA per GeV whereas the flux here is a practical value for a given machine.

The intensity is used to define the flux per unit area of the wavefront some distance from the source. This is important when the sample cross section is smaller than the beam cross section. A diverging beam (1 mrad horizontal×0.2 mrad vertical), will, in the absence of focussing, be $20\times4\,\text{mm}^2$ at a distance of 20 m from the source. For a $(0.3\,\text{mm})^3$

macromolecular crystal a high flux spread over the large area of 80 mm^2 has a low intensity or, put another way, the sample intercepts only a small portion of the available flux. Focussing instruments provide a higher intensity by collecting the available flux and focussing it at the specimen; the small reflectivity losses being more than compensated for by the large aperture of collection of the optics and the small focal spot. To produce a fine focal spot at the specimen needs a small source size. Hence, another figure of merit for the source is source intensity. The source intensity is the flux divided by the source size. The intensity at the specimen is the flux, brought by the optics to the sample, divided by the focal spot area.

In unfocussed experiments as small a divergence angle as possible is beneficial. The brightness, Φ_Ω, of the source is then a useful figure of merit which has the units

$[\Phi_\Omega]$ = number of photons per second per mrad^2 per 0.1%
      relative bandwidth

In some focussing experiments where a high flux is brought to a small focal spot a small convergence angle is also required. One example here would be virus crystallography. It is easier to produce a small focal spot with small convergence angle if the beam divergence angle is small *and* the source size is small. Hence, the source *brilliance* should be large here:

[Brilliance] = number of photons per second per mrad^2 per mm^2 per
      0.1% relative bandwidth

An undulator has a very high source brilliance (see section 4.10.3).

4.10 INSERTION DEVICES (WIGGLERS AND UNDULATORS) AND RADIATION PROPERTIES

We have so far referred to the simple circular orbital motion of electrons in bending magnets as sources of SR. Magnetic structures have also been developed known as wigglers and undulators, which can be purposely designed to enhance specific characteristics of SR, namely:

    (*a*) Extend the spectral range to shorter wavelengths.
    (*b*) Increase the available intensity.
    (*c*) Provide an intense quasi-monochromatic beam.
    (*d*) Provide a different polarisation.

In these new devices the electron beam oscillates as it experiences a periodic magnetic field (figure 4.12). There is no net deflection nor displacement and so such magnets can be inserted into straight sections between ordinary bending magnets. Indeed they are referred to as insertion devices. To extend the spectrum to shorter wavelengths the radius of curvature within a wiggler needs to be less than in a bending magnet for a given machine; i.e. λ_c is reduced by using a higher magnetic field as can be seen from equations (4.9) and (4.10). Only a short straight section is needed for a three-pole wiggler, the simplest insertion device (figure 4.13(a)). To obtain light multiplication many magnetic poles are needed and this is provided in either a multipole wiggler or an undulator; the former also usually acts as a wavelength shifter. Long straight sections are needed for these devices. If the plane of the electron oscillations is vertical then the plane of polarisation of the emitted light is perpendicular to that from a bending magnet. Other polarisations (e.g. circular) can be obtained from more exotic devices (Ellaume 1989a).

The classification of a periodic magnet insertion device as a wiggler or an undulator is generally based on whether the magnitude of the angular deflection δ of the electron beam is small enough to allow interference effects to be significant. In a wiggler $\delta \gg \gamma^{-1}$ so the interference is smoothed out (figure 4.13(b)). In an undulator δ is much less and the interference effects are large (figure 4.13(c)). A parameter $K=\gamma\delta$ is used to classify the magnet so that for a multipole wiggler $K \gg 1$ and for an

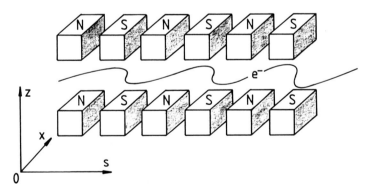

Figure 4.12 Schematic of a conventional insertion device. The magnetic field is parallel to the vertical direction ($0z$). It is a sinusoidal function of the longitudinal coordinates. The trajectory of an electron submitted to this field is a sinusoid in the horizontal plane ($0s,0x$). From Ellaume (1989b) with permission of the author and Gordon and Breach Science Publishers SA.

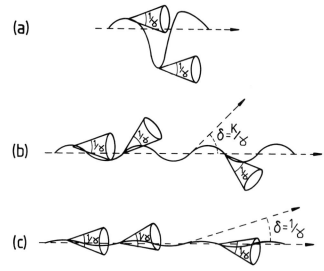

Figure 4.13 Schematic showing the electron motion and synchrotron
cone emission for the three insertion devices: (*a*)
wavelength shifter; (*b*) multipole wiggler; and (*c*)
undulator.

undulator $K \approx 1$. K may also be written as $eB_0\lambda_0/2\pi mc = 93.4\,B_0\lambda_0$ (with B_0
in tesla and λ_0 in metres).

Detailed reviews of wiggler and undulator magnets and their applications in general have been given by Winick and Knight (1977), Spencer
and Winick (1979), Winick, Brown, Halbach and Harris (1981), Brown,
Halbach, Harris and Winick (1983) and Ellaume (1989a,b). The
parameters of existing insertion devices used for the production of X-rays
(critical energy $\geqslant 1$ keV) are shown in table 4.2.

4.10.1 Wigglers

4.10.1.1 Wavelength shifter only

This is the simplest wiggler magnet and consists of only three poles
producing an oscillation of the electron beam shown schematically in
figure 4.13(*a*).

We can take as an example of this type the first Daresbury SRS
superconducting wiggler (Greaves *et al* 1983a; Marks *et al* 1983) (figures
4.14 and 4.15). The centre pole has a magnetic field of 5 T and the weaker
end poles are 2.5 T. These fields can be compared with the bending
magnet fields of 1.2 T at the SRS and the critical wavelengths of the
emitted SR for 5 T, 2.5 T and 1.2 T at 2 GeV are 0.93 Å, 1.86 Å and 3.88 Å
respectively (equation (4.10)). The side poles contribute significantly to

Table 4.2. *Parameters of existing insertion devices used for the production of X-rays (critical energy, $\varepsilon_c \geq 1$ keV). This table kindly provided by R. P. Walker and reproduced with permission.*

Ring, country	Energy (GeV)	Field (T)	Period (cm)	No. of poles[1]	ε_c (keV)	Type
SRS, England	2.0	5.0	—	1	13.3	SC W
DCI, France	1.85	4.8	26.0	3	10.9	SC W
DORIS II, Germany	5.3	0.6	13.2	31	11.2	PM U/W
	5.3	0.94	24.0	19	17.6	PM W
	5.3	0.31	12.0	7	5.8	PM U/W
	5.3	0.82	24.0	10	15.3	PM W[2]
ADONE, Italy	1.5	1.85	62.1	5	2.8	EM W
Photon Factory, Japan	2.5	5.0	—	1	20.8	SC W
	2.5	1.46	12.0	53	6.1	PM U/W
	2.5	1.5	18.0	27	6.2	PM U/W
Tristan Accumulator Ring, Japan	6.0	0.97	16.0	41	23.2	PM W[3]
UVSOR, Japan	0.6	4.0	—	1	1.0	SC W
CESR, USA	5.5	1.5	34.3	5	30.0	EM W
	5.5	0.5	3.3	123	$\varepsilon_1=4.7\text{–}8$	PM U
	5.5	1.2	19.6	25	24.1	PM W
SPEAR, USA	3.5	1.9	45.0	7	15.5	EM W
	3.5	1.2	7.0	55	9.8	PM U/W
	3.5	1.4	12.9	31	11.4	PM W
PEP, USA	14.5	0.22	7.7	51	$\varepsilon_1=12\text{–}24$	PM U
NSLS X-ray, USA	2.5	1.16	12.0	31	4.8	PM W
	2.5	5.2	17.4	5	21.7	SC W
VEPP-2M, USSR	0.65	7.5	25.0	3	2.1	SC W
VEPP-3, USSR	2.1	3.4	9.0	19	10.0	SC W

Notes:
(1) Neglecting end-poles.
(2) Asymmetric wiggler, producing circularly polarised X-rays.
(3) Elliptical multipole wiggler, producing circularly polarised X-rays.
(4) SC = superconducting, EM = electromagnet, PM = permanent magnet, U = undulator, W = wiggler, U/W = can be operated in either a wiggler or undulator mode. For undulators the tuning range of the fundamental is given.

the flux seen by an experiment, off centre line, using wavelengths in the 1–2 Å region but not for very short wavelengths (figure 4.15). The spatial extent of the source depends on the line of sight into the magnet (figure 4.16) which determines the spatial separation of the source 'points'. The

Figure 4.14 Photograph of the Daresbury SRS 5 T superconducting three-pole wiggler magnet, with permission of SERC, Daresbury.

brightness is degraded by this effect, but overall a considerable improvement in both flux and brightness for $\lambda < 2$ Å occurs for such a wiggler over a bending magnet. In protein crystallography the source is focussed at the sample and the intensity as well as the flux at the sample position is enhanced by the side pole contribution, provided that the focus does not consist of separately resolved peaks.

A major advantage of this type of insertion device is the large number of mrad being extracted (60 mrad). At the SRS this is shared between seven experiments. We can compare this with the typical 28 mrad from bending magnet 7 of the SRS which is split into three 4 mrad portions. The extra mrad of beam available for the wiggler workstations gives each experiment enough space to be 'end of line' for some camera and diffractometer configurations. Hence, a wiggler station optics can include both horizontally and vertically dispersing monochromators (sections 5.2.3 and 5.2.4) feeding monochromatic radiation to the sample.

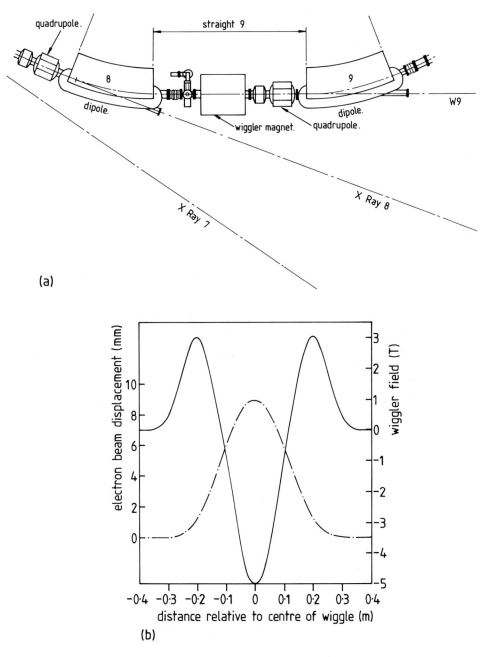

(a)

(b)

Figure 4.15 (*a*) The layout of the Daresbury SRS at straight 9 which is
occupied by the three-pole wiggler magnet. (*b*) The wiggler
field (solid line) and the electron beam trajectory (chain
line) relative to the centre of the magnet. (*c*) Contributions
to the power emitted into the wiggler beam line 9 and the
adjacent dipole magnets 8 and 9 at the SRS. (*d*) Brilliance
for the SRS wiggler magnet distributed across the horizontal
aperture. The brilliance of dipole magnet 7 is shown for

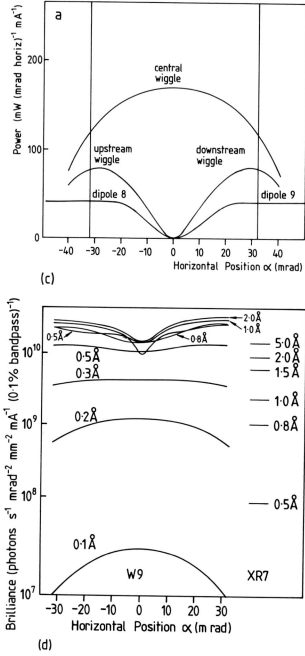

Figure 4.15 (cont.)
comparison. From Greaves et al (1983a) with permission.
The protein crystallography experiment (station 9.6)
accepts the 20–25 mrad portion. Note that in 1985/6 the SRS
lattice was upgraded to a High Brightness Lattice (HBL).
Hence, the brilliance values given in (d) and in figure 4.16
need to be increased by a factor of ten due to reduced source
size.

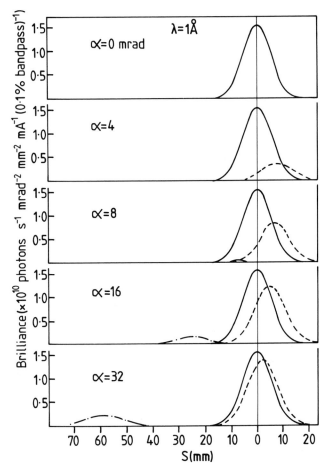

Figure 4.16 The intensities of the individual sources at 1 Å for different
viewing angles, α. The central pole of the wiggler as source
is shown by the solid line, the upstream/downstream source
(for α negative or positive respectively) by the dashed line
and the adjacent bending magnet by the chain line. S is the
displacement of each source projected onto the vertical
plane through the centre of the magnet. From Greaves *et al*
(1983a) with permission. See caption 4.15.

The wiggler at the Japanese Photon Factory (a 2.5 GeV machine, table
4.1) is also a three-pole device with a centre field of 6 T and $\lambda_c = 0.5$ Å.
This wiggler oscillates the electrons vertically and so therefore the radiation emitted has a vertical plane of polarisation. This is a useful
attribute for diffraction measurements since in this case high angle
reflections can be measured in the horizontal plane without significant
polarisation losses (chapter 5), which simplifies the mechanical mounting of diffractometers.

4.10.1.2 Light multiplication and wavelength shifting: multipole wigglers

The addition of more magnet poles to a wiggler results in a multiplication of the flux generated. One of the first multipole wigglers to be developed was on SPEAR (Winick and Spencer 1980, table 4.1) and had five full poles; this was later replaced by a seven full pole device which is detailed in table 4.2. The flux multiplication factor is this number of poles. However, because of the extra magnet length, the source size as viewed off centre line is increased and the intensity enhancement of a focussed beam is somewhat less than the number of poles.

Other multipole wigglers available, for example, include a 54-pole permanent magnet wiggler on SPEAR (Hoyer 1983) and a 6-pole wiggler on CHESS. There are others at SSRL, Photon Factory and the NSLS and wigglers are planned at ESRF and APS.

4.10.1.3 Radiated power

The combination of high flux and high photon energies produced from multipole wigglers places severe thermal loads on any component which sees the beam, e.g. beryllium windows, mirror and monochromator materials. From equation (4.4) we see that the smaller radius of curvature ϱ results in an increase in the radiated power. For the 5 T single bump wiggler at the SRS, for instance, ϱ is reduced from 5.5 m to 1.1 m resulting in an emitted power of 61 W mrad^{-1} at 300 mA and 2 GeV from the central pole. Immediately we can see the enormous power radiated from a multipole wiggler. The SSRL 54-pole wiggler with SPEAR operating at 3 GeV, 100 mA gives 0.8 kW mrad^{-1} linear power density. This wiggler has a period of 70 mm and peak operating field of 1.2 T which corresponds to a K value of 7.8. On the ESRF (table 4.1) the multipole wiggler will deliver up to 1 kW mrad^{-1} of power in the SR beam. The power density per mrad^2 is also an important parameter. This is usually given in terms of kW cm^{-2} at, say, 10 m from the source, e.g. at the ESRF a 1.5 T ten-pole device used at 6 GeV, 100 mA yields 10.5 kW cm^{-2}. Undulators, to be described in detail in the next section, also have a very high peak power density, e.g. a 2 m long, 55 mm period undulator at 6 GeV, 100 mA gives 25 kW mrad^{-2} although the total power is (only) 1 kW.

4.10.1.4 General comments

The effect of wiggler operations on the rest of a storage ring should be minimal though in less sophisticated magnet lattices the source emittance may be altered. The maximum beam current may not be as high with the wiggler as without, if the limit is set by the rf power available rather than by other effects.

4.10.2 Undulators

The unique characteristic of undulators is that the electron motion through the periodic magnet allows the radiation to undergo constructive interference causing large deviations of the emitted spectrum from the universal SR spectral curve. Extremely high fluxes of radiation are emitted at specific wavelengths. The interference conditions are met if the source emittance and the angular deflections of the electrons are very small. The undulator radiation is, as a result, tightly confined to the magnet axial direction and so can only serve one experiment at a time. The wavelengths of the emitted peaks depend on the machine energy and the magnetic field of the undulator.

There is a gradual transition, as the magnetic field decreases, from multipole wigglers to undulators; at the same time the spectrum shifts to longer wavelengths and the electron deflection decreases. That is, it is possible for a single device to be operated in either mode. For example, the Stanford 54-pole insertion device, referred to earlier, is primarily a multipole wiggler with critical wavelength 1.7 Å with a fundamental at 319 Å. However, it has been operated at lower field strengths in an undulator mode with a field of 0.15 T giving $K=0.95$ and a third harmonic at 4.9 Å. An undulator is ideally conceived as a device to produce separated very brilliant spectral lines. The possibility of operating an undulator for 1.5 Å radiation or shorter (either as the fundamental or harmonic) requires a high machine energy to compensate for the low magnetic field; we will explore this later.

The small electron deflection allows significant interference to occur between the radiation emitted by a single electron at two successive crests of its sinusoidal path through the periodic magnet.

If λ_u is the undulator period, i.e. the magnet pole lattice spacing, then the time for an electron to traverse this length is, for $\beta = v/c$,

$$\approx \lambda_u / \beta c \tag{4.15}$$

and should be compared with the time for a photon of

$$\lambda_u / c \tag{4.16}$$

Interference occurs with the condition that

$$\lambda/c = \lambda_u/\beta c - \lambda_u/c \tag{4.17}$$

where λ is then the wavelength of the 'interference radiation'. The time difference is smaller the weaker is the magnetic field because the electron deflection is smaller. Hence, a larger gap produces a weaker magnetic field and a shorter wavelength emission line (figure 4.17(a)). Since $(1/\beta - 1) = (2\gamma^2)^{-1}$ we get

$$\lambda = \lambda_u / 2\gamma^2 \tag{4.18}$$

Harmonics (i) of this fundamental wavelength, λ, also satisfy the interference condition and for $K \geq 1$ have a significant amplitude. A shorter period λ_u moves λ to shorter wavelengths (figure 4.17(b)).

On the axis of the undulator

$$\lambda_i = \frac{\lambda_u}{i2\gamma^2}\left(1 + \frac{K^2}{2}\right) \tag{4.19}$$

Off axis, the wavelength, λ_i, varies as

$$\lambda_i = \frac{\lambda_u}{i2\gamma^2}\left(1 + \frac{K^2}{2} + \gamma^2\theta^2\right) \tag{4.20}$$

Figure 4.17 Calculated undulator emission for 6 GeV (ESRF). (a) For two different gaps, 20 mm and 32 mm, for a fixed period of 48 mm. Opening the gap shifts the fundamental from 2.2 keV to 4.5 keV but seriously decreases the flux in the harmonics. (b) For three different periods for a fixed gap of 20 mm. (c) A comparison of on-axis (through a 1×1 mm² pinhole at 30m) with angle integrated spectrum. Note that, in practical units $K = 0.934 \, B_u(T) \, \lambda_u$ (cm) where $B_u = 1.4 B_r$ exp$(-\pi g/\lambda_u)$ is the peak field, g the gap (B_r is a constant independent of g) and λ_u the spatial period and $E_1 = 0.95 E^2 \, (\text{GeV})/\lambda_u \, (\text{cm})(1 + K^2)$. From Ellaume (1987b, 1989c) with permission of the author and Gordon and Breach Science Publishers SA.

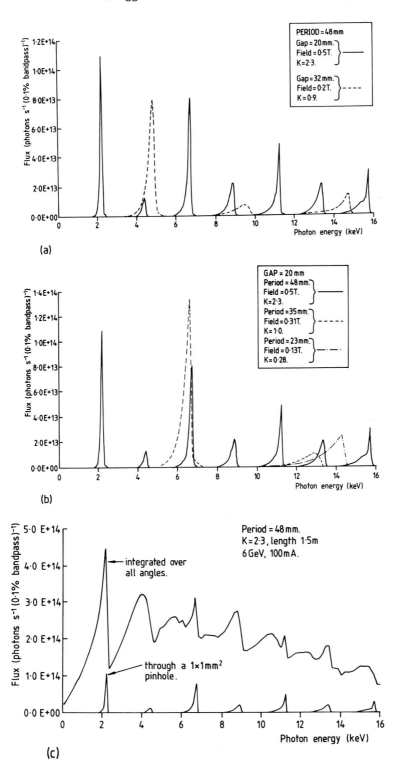

(a)

(b)

(c)

where θ is the angle to the undulator axis.

A comparison of the on-axis (equation (4.19)) with the integrated emission is shown in figure 4.17(c). From equation (4.20) we can see how the λ_i of these 'emission lines' is set by the machine energy since $\gamma=E/mc^2$. The emitted peak has a finite width, $\delta\lambda/\lambda$, with a contribution from the number of magnet periods, N

$$\Delta_0 = \frac{\delta\lambda}{\lambda} \sim \frac{1}{iN} \tag{4.21}$$

There is also a contribution due to the emittance; a large value relaxes the interference condition, broadens the peak bandpass and reduces its height. Also, since the interference condition holds only for the forward direction, a large angular acceptance of an experiment (i.e. >1 mrad in both horizontal and vertical) tends to broaden the peak (figure 4.17(c)). Conversely a pinhole can be used to restrict the angular acceptance of the sample or optics and so sharpen the spectral peak widths and reduce the power loading at the sample (figure 4.17(c)).

Typical overall linewidths, Δ_0, of 0.01–0.1 can be expected. These linewidths can be compared with the linewidths of Cu Kα radiation; the FWHM of Cu K$\alpha_1=4\times10^{-4}$ and of Cu K$\alpha_2=5\times10^{-4}$ (Compton and Allison (1935), table IX-21 p. 745) and the Kα_1–Kα_2 separation $\delta\lambda/\lambda$ is 2.5×10^{-3}.

The obtainable spectral brightness in photons s^{-1} mrad^{-2} (1% bandwidth)$^{-1}$ is given by

$$1.74\times10^{15} N^2 E^2 F_i(K)I_b \tag{4.22}$$

where

$$F_i(K) = \frac{i^2 K^2}{(1+K^2/2)^2}\left[\frac{J_{i-1}}{2}\left(\frac{iK^2/4}{1+K^2/2}\right) - \frac{J_{i+1}}{2}\left(\frac{iK^2/4}{1+K^2/2}\right)\right] \tag{4.23}$$

and where the J_i are Bessel functions. The functions $F_i(K)$ are shown in figure 4.18(a) for several harmonics.

The total flux available at a given wavelength is estimated from the product of the on-axis brightness with the solid angle of the radiation, i.e.

$$N(\lambda) = 1.43\times10^{15} N \frac{F_i(K)(1+K^2/2)I_b}{i} \tag{4.24}$$

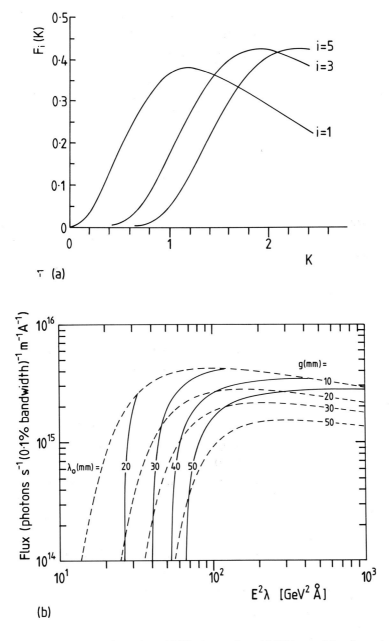

Figure 4.18 (a) The functions $F_i(K)$ (equation (4.23)) used in the calculation of undulator spectra. (b) The flux produced (per
ampere per metre of insertion device) by an undulator as a
function of magnet period, λ_0, and pole gap g; E is the
machine energy (in GeV) and λ is the output wavelength
(in Å), assuming an optimised permanent magnet construction. Figure kindly supplied by R. P. Walker.

A comparison (Walker 1986) can be made of the flux and brightness produced by an undulator at its fundamental on-axis wavelength, λ, to the peak value of a bending magnet source, i.e. the flux ratio is

$$\frac{6.4N}{E} \frac{F_i(K)\ (1+K^2/2)}{i} \qquad (4.25)$$

and the brightness ratio is

$$8.9\ N^2 F_i(K) \qquad (4.26)$$

For a 50-period undulator with $K=1$ for a 1.5 GeV ring the undulator gives a factor of 120 times more flux than 1 mrad from a bending magnet source and a factor of 8200 in brightness. Of course, the wavelengths being compared are quite different. The bending magnet source, with $B=1.5$ T would have $\lambda_c=5.5$ Å while if $\lambda_0=40$ mm the undulator would have $\lambda_1=35$ Å. Figure 4.18(b) shows the flux achievable at the fundamental on-axis wavelength of an undulator, per metre of length, as a function of $E^2\lambda$, with various constant values of period (solid lines) or gap (dashed lines). The solid lines therefore represent the tuning curves for a fixed period device as the gap is varied. At shorter wavelengths the tunability is very limited due to the small K values whereas at longer wavelengths it can be considerable. In practice, tunability at the short wavelengths can be extended by use of higher harmonics. The peak flux is always obtained at the minimum gap that is achievable. At long wavelengths the dependence of the flux on the gap setting is not very great but at short wavelengths the flux achievable is very sensitive to the gap. Hence the minimum gap is used provided the beam lifetime is not adversely affected.

The effect of opening the gap is to reduce the field and hence K so that the fundamental moves to shorter wavelengths (figure 4.17(a), equation (4.19)) but the flux decreases. Conversely to achieve a high flux means closing the gap and to 'prevent' the fundamental moving to long wavelength the machine energy has to be high.

4.10.2.1 Existing undulator radiation sources

The first observations of undulator radiation were made in the visible region of the electromagnetic spectrum in the USSR; on the Pachra 1.3 GeV synchrotron in Moscow (Alferov et al 1979) and the Sirius 1.2 GeV synchrotron in Tomsk (Didenko et al 1979). Permanent magnet undulators have been operated on VEPP-3, on SPEAR (Boyce *et al* 1983) and at Photon Factory (Kohra and Kitamura, pers. comm.). An undulator is working on the SRS for soft X-ray work ($\approx$40 Å (Poole *et al* 1983)). All these devices are built or designed for photon energies below 2 keV.

The extension to high photon energies with an undulator has been achieved with some of the world's existing machines though none of them is dedicated to SR operation. At Stanford a 2 m undulator magnet and associated beam line have been installed on the electron–positron storage ring PEP (figure 4.19(c)). The undulator is designed to produce

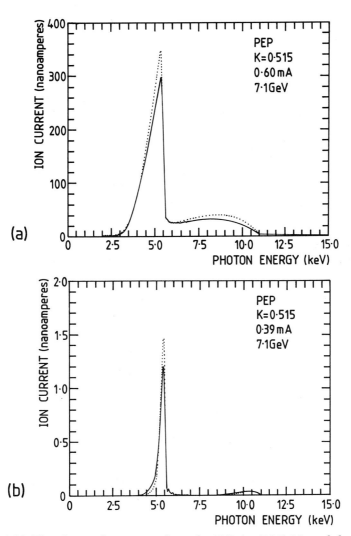

Figure 4.19 The observed spectrum from the PEP (at 7.1 GeV) undulator agrees well with the calculated spectrum (*a*) angle

(c)

Figure 4.19 (*cont.*)

> integrated, and (*b*) on-axis spectrum (i.e. here through a
> 127 μm pinhole, 58 m from the centre of the undulator). The
> solid line represents the experimental data and the dotted
> line represents the theoretical prediction. From Lavender *et
> al* (1989) with permission. (*c*) Aerial photograph of SLAC
> at Stanford showing the PEP ring (circumference 1.2 km),
> with permission.

photons at 12 keV with high intensity when the storage ring is operated
at its nominal energy of 14.5 GeV. Lavender, Brown, Troxel and Coisson
(1989) reported measurements of the spectral and angular properties of
the radiation when PEP was operated in a low emittance mode
(6.4×10^{-9} metre rad) at 7.1 GeV. The spectral properties were in good
agreement with theory; figure 4.19(*a*) shows the angle-integrated spectrum and figure 4.19(*b*) shows the on-axis spectrum (i.e. here through a
127 μm circular pinhole) as measured 58 m from the centre of the undulator with $K = 0.515$. The undulator has a period of 7.72 cm and its on-axis
peak field can be varied from a minimum of 0.10 kG to a maximum of
2.19 kG with K varying as a result between 0.073 and 1.58.

At Cornell a 3.3 cm period Nd–Fe–B hybrid undulator has been successfully operated in the Cornell Electron Storage Ring (CESR). This 2 m

long, 123-pole insertion device is a prototype of an undulator planned for the Advanced Photon Source, APS (figure 4.20). Bilderback *et al* (1989) reported that the undulator produced the expected spectrum at 5.437 GeV with the fundamental X-ray energy emitted varying from 4.3 keV to 7.9 keV corresponding to a change in gap from 1.5 cm to 2.8 cm. Figure 4.21(a) shows the flux through a 100 μm pinhole for the fundamental at 7.9 keV with K at 0.54 (2.8 cm gap) and figure 4.21(b) through a 100 μm pinhole for the third harmonic (12.9 keV) at $K=1.4$ (1.5 cm gap). One experimental feasibility study performed was with the unmonochromatised undulator radiation impinging on a protein (lysozyme) crystal. With 35 mA of electrons stored in a single bunch a diffraction pattern was observed with the X-rays from the single bunch of electrons. The diffraction pattern was recorded on storage phosphors (section 5.4) in an exposure time therefore corresponding to the bunch width (120 ps)!

Figure 4.20 The 3.3 cm period hard X-ray undulator with 61 periods installed in the CESR. The 2.03 m long Nd–Fe–B hybrid undulator was built by Spectra Technology Inc and designed by the Argonne National Laboratory. From Bilderback *et al* (1989) with permission.

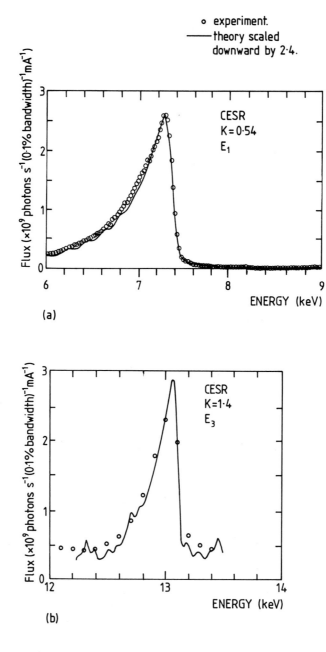

Figure 4.21 The observed on-axis spectra from the undulator shown in figure 4.20 for E_1 and E_3 with (a) $K=0.54$ and (b) $K=1.4$ respectively. The theory curve reproduced the experimental curve if scaled downward by a factor of 2.4. From Bilderback *et al* (1989) with permission.

4.10.3 Comparison of insertion devices with bending magnets as radiation sources

Using the definitions given in section 4.9 for flux, brightness and brilliance it is possible to compare the properties and anticipate the usefulness of these different magnet types as radiation sources. As a basis for comparison we can consider those being designed for the ESRF at Grenoble (Ellaume 1987a,b, 1989c).

The primary purpose of the ESRF 6 GeV storage ring is to provide radiation sources based on insertion devices. The lattice has 32 straight sections, 27 of which will be available for insertion devices. An undulator often requires a low horizontal and vertical divergence of the electron beam to aid the interference condition; it is located therefore in a high β position of the storage ring lattice. For the multipole wiggler a small source size and a larger angular divergence to smooth out the interference are preferred and a low β position of the ring is optimal. Table 4.3 shows the β values and rms dimensions for the generic sources at ESRF, i.e. the two types of bending magnets (0.4 and 0.85 T), a wiggler and an undulator.

Figures 4.22(a)–(c) give the brilliance, brightness and flux for these source types at ESRF (at 6 GeV, 100 mA) for full 5 m representative insertion devices versus bending magnets (from Ellaume (1989c)). The figures show the similar nature of the flux levels emitted but the marked enhancement of the brightness and brilliance of the undulator. The undulator spectra are at specific wavelengths. The continuous curves depicted for the undulators are as a result of varying the gap from the minimum allowed of 20 mm up to infinity (in practice 20 cm or so).

Table 4.3. *ESRF lattice β values and source sizes (from Ellaume (1987a) with permission).*

	Bending magnet source	High β (undulator)	Low β (wiggler)
β_x (m)	2.2	26.6	0.8
β_z (m)	26.8	11.3	3.5
σ_x (mm)	0.16	0.41	0.069
σ_z (mm)	0.129	0.084	0.047
σ_x' (mrad)	0.137	0.015	0.089
σ_z' (mrad)	0.005	0.007	0.013

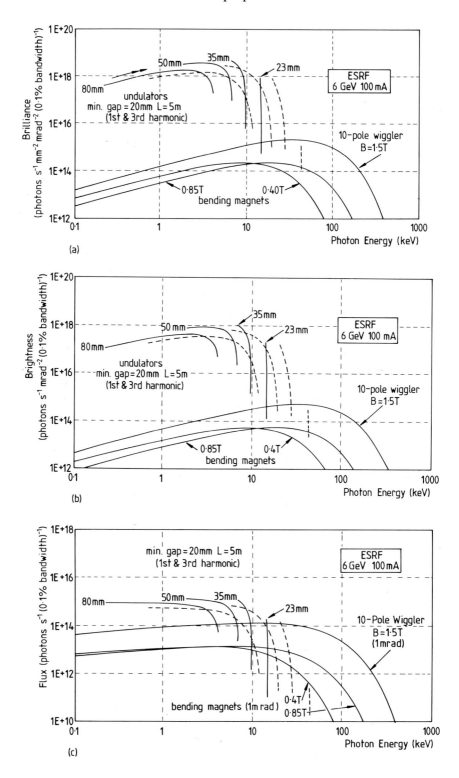

(a)

(b)

(c)

4.11 CRITERIA FOR THE CHOICE AND DESIGN OF SYNCHROTRON X-RAY SOURCES FOR MACROMOLECULAR CRYSTALLOGRAPHY

Parameters of the source that need to be considered include the wavelength range, the emittance (source size × divergence), the stability of the source, the circulating current and the lifetime of the beam. To date the polarisation of the beam has been taken account of in data processing but not specifically designed for a macromolecular crystallography experiment.

Experimental needs dictate the ideal source parameter values. Particularly demanding cases are (a) virus crystal data collection (chapters 6, 10), (b) use of small crystals (chapter 10), (c) Laue diffraction for time resolved studies (chapters 7, 10) and (d) variable wavelength anomalous dispersion studies (chapter 9). These are not strictly partitioned categories of experiment because, for example, virus crystals may well be the subject of time resolved or variable wavelength experiments. However, the categories do rather neatly match what is on offer from the undulator for (a) and (b), multipole wiggler for (c) and bending magnet for (d) on ESRF or APS.

Virus crystal diffraction is characterised by a large number of simultaneously diffracting spots with small angular separation and weak diffraction. The samples are prone to radiation damage: so much so that each sample usually yields one pattern and up to 500 crystals are used per data set. Phase determination (chapter 2) is often achieved by using a low resolution 'spherical' shell based on previously determined structures and phase extension performed by using non-crystallographic symmetry. The high brilliance characteristic of the undulator is very attractive here for such data collection. Ideally, the undulator

Figure 4.22 Calculated spectra for the generic ESRF sources in terms of (a) brilliance, (b) brightness and (c) flux. Several undulator examples are given, i.e. with periods, λ_u of 80, 50, 35 and 23 mm. For each of these the loci of the first and third harmonic emission energies are shown as a continuous function of opening the gap (marked by $\rightarrow\!\!\!\rightarrow$) from the minimum of 20 mm (left-hand edges of the loci) to a large value (right-hand edges of the loci). The tunability range of the undulator is very limited at the shorter period. From Ellaume (1989c) with permission of the author and Gordon and Breach Science Publishers SA.

fundamental should be at 1.0 Å; this will allow flexibility in choice of wavelength because then the third and fifth harmonics will be at wavelengths of 0.3 Å and 0.2 Å respectively and with a significant tunable range on each. The undulator divergence angle is ≈ 0.1 mrad in both the horizontal and vertical. Ultra-short wavelengths may be useful in avoiding radiation damage (chapter 6) – even perhaps to reduce the large number of crystals used from 500 to maybe even one per data set.

Small crystals of proteins and viruses do occur and their diffraction is characterised by weak scattering and radiation damage. Focussing of the divergent beam is required. Even in the case of an undulator, a 0.1 mrad diverging beam will be 6 mm in size, 60 m from the tangent point. In the focussing of such a beam the depth of source (e.g. ≤ 6 m) will produce a significant contribution to the focus width (i.e. 0.1 mrad $\times$ 6 m = 0.6 mm) and demagnification will be required if a microfocus beam is needed (i.e. ≤ 0.1 mm) – see chapter 5. The larger the number of poles in the undulator the larger will be the depth of source contribution to the focus. Certainly very long straight sections are being discussed now, capable of enough poles, e.g. 10^3, that the linewidth of the emission peaks will be of the order, therefore, of 10^{-3}, i.e. in the range where it may be possible to dispense with a monochromator. In some cases, though, it will be necessary to have $\delta\lambda/\lambda$'s in the range of 10^{-4} rather than 10^{-3} and a monochromator will still be required.

Time resolved or perturbation experiments, e.g. on the millisecond timescale, are simplest to design when the sample is held stationary. This requires use of the Laue method (chapters 7, 10). To date it has been demonstrated that with broad bandpass Laue diffraction the data are accurate enough to yield information on small scattering components of a structure, e.g. one water molecule in an enzyme or the hydrogen atoms in a small molecular structure (chapter 7). Wavelength normalisation has been performed here with a single scale factor per spot depending on its wavelength. Since the $\delta\lambda/\lambda$ required to stimulate a spot is ≈ 0.01–0.02 it is essential that the spectrum incident at the sample is both smooth and flat over a $\delta\lambda/\lambda$ of 0.01–0.02 over the full range of λ's that is useful (≈ 0.5–2 Å). It may be that this requirement can be relaxed but it has yet to be demonstrated in terms of molecular structure details. For the present, then, a multipole wiggler offers the requisite bandpass range (0.5 Å $< \lambda <$ 2.0 Å) and smoothness of the spectrum (although the spectral emission has to be examined carefully over the K range, equations (4.18) and (4.19), of a given multipole wiggler). Focussing of the wiggler white light will contain a significant depth of source effect; since the wiggler may be up to 6 m long, then for a 1 mrad divergent beam the horizontal source profile presented will be 6 mm wide and some (e.g. 2:1) demagnification will be needed to push as many polychromatic photons as possible through a 0.2–0.1 mm pinhole collimator. Greater than 2:1 demagnification would produce too large a convergence angle and affect the size of Laue diffraction spots at the detector.

Anomalous dispersion experiments require use of a variable wavelength, and highly precise measurements because of the small size of the effects. Stability of the source and optics is here, therefore, the prime requirement even if this means that the data measurements take many shifts of beam time. A bending magnet on such a machine as the ESRF has considerable virtue for this kind of experiment (chapter 9).

All experiments benefit from a long lifetime of the beam but especially the anomalous dispersion cases because of the length of time required for the measurement.

Dedicated storage rings allow full control over the machine running and installation programme as well as tailoring of the lattice and its emittance for tackling the challenges set by macromolecular crystallography. What SR emittance is required here? A macromolecular crystal has typical values of cross-section size and angular rocking width of 0.3 mm and 0.02° (0.3 mrad) respectively. Hence, a *sample acceptance* parameter can be defined of size × rocking width i.e. typically $(0.3 \times 10^{-3}$ metres$) \times (0.3 \times 10^{-3}$ rad$) = 9 \times 10^{-8}$ metre rad. This sample acceptance should ideally be matched by a similar value for the SR emittance, in the horizontal and vertical.

CHAPTER 5

SR instrumentation

5.1 DEFINITION OF REQUIREMENTS

The instrumentation requirements for macromolecular crystallography at the synchrotron are quite diverse and technically exacting. The diversity arises because of the different classes of experiment, namely:

(a) routine data collection usually at a fixed wavelength;
(b) variable single or multiple wavelength anomalous dispersion measurements;
(c) time resolved crystallography;
(d) small crystals.

There are certainly needs common to each category.

In order to use the synchrotron X-radiation effectively, the 'white beam' of photons diverging from the source must be collected by beam line optical element(s), and brought to a focus with a size approximately equal to the protein crystal size. The sample must be centred in the beam and oriented on a goniostat of some kind. The diffraction pattern has to be measured accurately and efficiently with a detector. In the case of monochromatic experiments the beam line optical scheme will include a monochromator of which there are several common types. There are some special needs for each experimental class.

'Routine' data collection often involves the measurement of relatively weak, high resolution diffraction data. Critical design goals here include a very high intensity at the specimen and a short wavelength (often $\leqslant 0.9$ Å) monochromatised beam to reduce radiation damage, which affects these high angle data first. The detector should ideally pick up the full aperture of the diffraction pattern at one setting and have a high

136

absorption efficiency, low noise, high local and global count rate capacity, good linearity over a wide dynamic range of intensities, be uniform in response and stable with time.

If optimised anomalous dispersion studies are of interest then the spectral resolution, $\delta\lambda/\lambda$, of the incident beam at the sample needs to be narrow enough to take full advantage of any fine structure in an absorption edge; also the setting of the absolute wavelength needs to take account of shifts in the energy of the absorption edge as a function of the absorbing atom valence state (table 9.2 and figure 9.7). Reductions in $\delta\lambda/\lambda$ and of sample size reduce the total counting rate of the diffraction pattern, hence the pattern itself ideally needs to be detected as efficiently as possible, as above, to maximise the information recorded per photon incident onto the sample. The range of λ's needed is defined by the need to access absorption edges usually between 0.6 Å and 1.9 Å (figure 9.3). The range of $\delta\lambda/\lambda$'s needed depends on the XANES of a particular material and lies within the interval $5\times10^{-5}-10^{-3}$; by comparison native protein data can use a $\delta\lambda/\lambda$ as large as 2×10^{-3}.

Time resolved crystallography is at an early stage of development. A reaction or photostimulable effect has to be initiated in a crystal, an interesting point in a time course identified, diffraction patterns measured at several time slices and there has to be careful synchronisation of laser and X-ray flashes. Laue geometry (no monochromator) is being used increasingly for this work. In white beam Laue geometry the spectral bandpass can be set as broad as possible so as to stimulate as large a number of reflections as possible or narrowed to reduce the number for reasons discussed in chapter 7.

Small sample size is a problem afflicting some projects. Diffraction patterns are weak or, put another way, the crystal dies before a significant amount of data can be measured. The sample volumes are in the range $0.3\times0.3\times0.3$ mm^3 to volumes of, say, $0.05\times0.05\times0.05$ mm^3 though this lower limit is somewhat artificial. As in the routine data collection case referred to above a high intensity and short wavelength is beneficial in reducing radiation damage.

In this chapter we will explore a variety of optical components and detectors and their configurations. In addition, various general instrumentation categories will be detailed, e.g. crystal cooling and cryocrystallography, beam monitoring, radiation safety and support facility needs. Finally, a global survey of synchrotron instruments for macromolecular crystallography will be given in a gazetteer.

5.2 MONOCHROMATOR SYSTEMS

5.2.1 Perfect crystals

Over the wavelength range of interest, defined above (section 5.1), perfect silicon or germanium crystals have peak reflectivities approaching 100% (figure 5.1). Also, crystals can be grown to very large lengths (hundreds of millimetres), suitable for different diffraction planes, whereby several mrad of beam can be collected.

The rocking width or Darwin width η^{mono} of a perfect crystal is derivable from dynamical diffraction theory as

$$\eta^{mono} = \frac{e^2}{2mc^2} \frac{\lambda^2}{\varepsilon_0 \pi^2 V_0} \frac{|F(h)|}{\sin 2\theta_B} \tag{5.1}$$

where e^2/mc^2 is the classical radius of the electron, λ is the wavelength, V_0 the unit cell volume, θ_B the Bragg angle, ε_0 is the permittivity of free

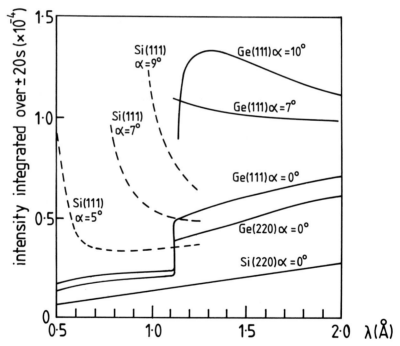

Figure 5.1 Integrated reflecting powers of various perfect crystals as a function of wavelength. The ordinate represents integration of the reflectivity in figure 5.2, in the range of −20 to +20 seconds of arc. From Kohra, Ando, Matsushita and Hashizume (1978) with permission.

space and $|\mathbf{F(h)}|$ the structure factor amplitude. Using Bragg's law $\lambda = 2d \sin \theta_B$, equation (5.1) becomes

$$\eta^{mono} = \frac{d^2}{\varepsilon_0 \pi^2} \frac{e^2}{mc^2} \frac{\tan \theta_B}{V_0} |\mathbf{F(h)}| \tag{5.2}$$

The spectral acceptance of the crystal, $\delta\lambda/\lambda$, is given by

$$\frac{\delta\lambda}{\lambda} = \eta^{mono} \cot \theta_B = \frac{d^2}{\varepsilon_0 \pi^2} \frac{e^2}{mc^2} \frac{|\mathbf{F(h)}|}{V_0} \tag{5.3}$$

From equation (5.3) we see that $\delta\lambda/\lambda$ is approximately constant, though $|\mathbf{F(h)}|$ does vary slowly with λ in the presence of absorption edges. The germanium absorption edge at 1.12 Å considerably reduces the reflectivity below this wavelength.

Table 5.1 and figure 5.2 give the rocking widths of Si(220), Si(111), Ge(220) and Ge(111); in the table, mosaic graphite is included for comparison: although this is common on a conventional X-ray source, graphite is not common at SR sources. The spectral acceptances available from perfect crystals span the range needed to probe XANES or f' at the point of inflection of an absorption edge for optimised anomalous dispersion measurements. Ge(111) has the largest value and is perhaps still a little low for routine protein crystal data collection. Unfortunately, mosaic graphite, which can withstand the white beam, has too broad a rocking width; this, in turn, broadens the incident $\delta\lambda/\lambda$ onto the sample and finally the protein crystal sample rocking width (equation (6.5)) or spot size deteriorates at high Bragg angle.

In practice, perfect silicon and germanium are readily available though germanium of adequate quality is quite expensive. Both

Table 5.1. *The intrinsic resolution of several Bragg reflections from symmetrically cut perfect and mosaic crystals.*

Type	Reflection	$\delta\lambda/\lambda$
Silicon	(111)	1.33×10^{-4}
Silicon	(220)	5.6×10^{-5}
Germanium	(111)	3.2×10^{-4}
Germanium	(220)	1.5×10^{-4}
Pyrolytic graphite	0.5° mosaic spread	3.8×10^{-2}

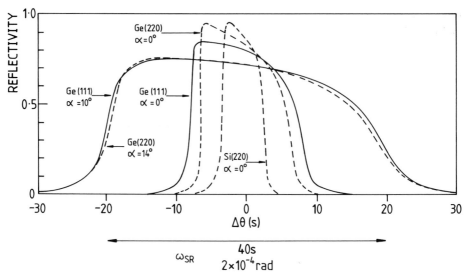

Figure 5.2 Reflectivity curves of various perfect crystal reflections at a
wavelength of 1.55 Å. From Kohra *et al* (1978) with permission. See table 5.1.

materials withstand beam heating very well though the perfect crystal
rocking width is broadened very slightly as a result by a few per cent of
the fundamental (Greaves *et al* 1983b). With the newest generation of
insertion devices thermal limitations need to be considered seriously (see
section 5.2.6).

5.2.2 Possible configurations of perfect crystals; the Du Mond diagram

There are quite a number of possible configurations of perfect single
crystals as monochromators. Detailed reviews have been given by Hart
(1971), Barrington-Leigh and Rosenbaum (1974), Beaumont and Hart
(1974), Bonse, Materlik and Schroder (1976), Hastings (1977) and Kohra
et al (1978). Witz (1969) reviewed the use of monochromators for conventional sources of X-rays. Underwood and Turner (1977) discussed the
use of differently shaped materials for producing specific curvature
optical elements.

The characteristics of crystal monochromators using single or multiple
diffraction are best understood with reference to the diagrammatic
representation of Du Mond (1937). The 'Du Mond diagram' represents
the relationship between the Bragg angle of the monochromator, θ, with
the wavelength reflected. The overall form of the λ–θ plot (figure 5.3) is
defined by Bragg's law. The sine wave has a width which on an angular

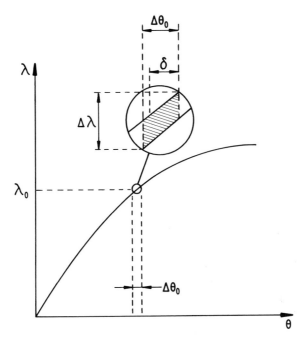

Figure 5.3 The Du Mond diagram for a single crystal relating
wavelength λ and Bragg angle θ of a reflection. The width of
the curve is the crystal rocking width. $\Delta\theta_0$ represents the
range of angles of incidence, e.g. $\Delta\theta_0=0$ at the Guinier position for a curved crystal (see figures 5.6(a)–(d). From
Sauvage (1980) with permission.

scale is the reflection width of the crystal. The fundamental reflected
beam is represented by the shaded area in the λ–θ plot (figure 5.3). The
use of an entrance slit at the monochromator reduces the accepted
divergence angle $\Delta\theta_0$ to δ (see figure 5.3) and consequently also the $\delta\lambda$
reflected. We can also use the Du Mond diagram to represent multiple
crystal arrangements. We will consider only two of these and both are
based on double crystals; the so-called antiparallel $(+,+)$ and parallel
$(+,-)$ arrangements (James 1954). If we define the angle ϕ as the deviation angle of the primary beam after the two successive reflections from
two crystals, then, for the simple case of identical crystals in the $(+,+)$
setting (figure 5.4(a)) we have $\phi=2\theta$, and in the $(+,-)$ setting (figure
5.5(a)) we have $\phi=0°$.

The Du Mond diagram for the double reflected beam in each case is
given in figures 5.4(b) and 5.5(b). The doubly reflected beam is described
by the intersection area of the λ–θ curves for each crystal. In the $(+,-)$
case for identical crystals the two curves are coincident and $\Delta\theta$ (the

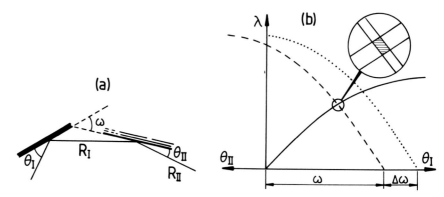

Figure 5.4 (a) The path of the X-ray beam through a (+,+) double
crystal monochromator where, for generality, each crystal is
of a different type so that $\theta_I \neq \theta_{II}$. (b) The Du Mond diagram
for the (+,−) double crystal system; the range of reflected
wavelengths is determined by $\Delta\theta$ the range of incidence
angles in the beam (equation (5.16)). The full curve is for
crystal I, the dashed curve for crystal II and the dotted curve
for crystal II in a parallel setting. From Sauvage (1980) with
permission.

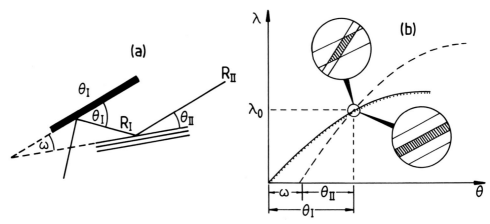

Figure 5.5 (a) The path of the X-ray beam through a (+,−) double
crystal monochromator where, for generality, each crystal is
of a different type so that $\theta_I \neq \theta_{II}$. (b) The Du Mond diagram
for the (+,−) double crystal system; the range of reflected
wavelengths is determined by $\Delta\theta$ the range of incidence
angles in the beam (equation (5.16)). The full curve is for
crystal I, the dashed curve for crystal II and the dotted curve
for crystal II in a parallel setting. From Sauvage (1980) with
permission.

divergence angle of the source) still defines the reflected $\Delta\lambda$. In the (+,+) case the reflecting region is defined by the perfect crystal rocking width alone and is independent of $\Delta\theta$; the (+,+) mode is referred to as the high resolution mode. The (+,−) mode can only operate as a high resolution mode if the source size cross-fire angle is very small (small source size) and narrow monochromator entrance slits are used to limit the divergence accepted from the source; these limit $\Delta\theta$, and hence $\Delta\lambda$, and the use of the entrance slits is at the expense of flux (see figure 5.9).

A monochromator based on a single crystal is favoured for fine focussing. Also, it can operate in a high resolution mode (i.e. limited by perfect crystal rocking width) if it is curved; the curvature of the surface can then compensate exactly for the change in incident angle of the primary beam due to the natural source divergence (the Guinier position, see section 5.2.3).

For SR protein crystallography two main designs for the monochromator system have been utilised to date based on either the bent oblique-cut single crystal or (+,−) parallel double crystal system. In the next sections (sections 5.2.3 and 5.2.4) we describe their properties in detail.

5.2.3 Monochromator based on a single crystal

The primary advantage of this system is that the monochromator may be easily bent in one direction to bring the beam to a line focus. By using a triangular shape, clamping the base and pushing on the apex of the triangle, a perfect cylindrical curvature can be obtained (Underwood and Turner 1977). This, coupled with a Bragg reflection angle of several tens of degrees, allows virtually aberration-free focussing even when demagnifying the source by a ratio of 10:1. This is in contrast with grazing incidence mirror optics (section 5.3) where aberrations are serious away from a 1:1 focussing condition. Whilst the single crystal allows focussing in one direction, point focussing is much more difficult; however, Berreman (1955) has had success with a conventional X-ray source as have Oshima and Tanaka (1981) with lithium fluoride. There is no report of successful point focussing by means of a single crystal with SR.

The single monochromator crystal is usually set to reflect the X-ray beam in a horizontal plane so as to simplify the mechanical design of the rest of the experimental arrangement, which must follow the reflected beam as the wavelength is changed. There is one disadvantage to this due to polarisation losses; for long wavelengths, large 2θ, the horizontal (i.e. parallel) component of the polarised beam, which is the dominant one (figure 4.5), suffers a serious 'attenuation' by a factor of $\cos 2\theta$.

The single crystal can be used in both energy-dispersive or non-energy-dispersive modes simply by changing the curvature. With a flat crystal (radius of curvature $R=\infty$) rays from a point SR source incident at different points along the monochromator length are reflected at progressively different wavelengths (figure 5.6(a)). The angle $\delta\theta$ subtended at the source *point* is given by

$$\delta\theta \;=\; (\text{L}\sin\theta)\,/\,p \tag{5.4}$$

where θ is the monochromator Bragg angle, L is the illuminated length of the monochromator and p is the monochromator to source distance or 'object distance'.

The reflected range of wavelengths, $\delta\lambda$, is

$$\left(\frac{\delta\lambda}{\lambda}\right)_{\text{corr}} \;=\; \delta\theta\cot\theta \;=\; \frac{\text{L}\sin\theta}{p}\cot\theta \tag{5.5}$$

where the subscript 'corr' means λ is correlated with direction (see figure 5.6(a) or (c)).

If $\delta\theta=4\,\text{mrad}$ and $\theta=13.27°$ ($1.5\,\text{Å}$ for Ge(111)) then the $\delta\lambda/\lambda$ reflected, due to the source divergence is 0.017; compare the $K\alpha_1$, $K\alpha_2$ splitting for a Cu $K\alpha$ X-ray tube of 0.0025; the flat monochromator is operating dispersively (in Helliwell *et al* (1982a) this is termed 'underbend').

If we bend the flat crystal there is a curvature which compensates exactly for this change of incidence angle along the crystal (figure 5.6(b)). This is the non-dispersive or 'Guinier setting' or achromatic mode. The diverging beam from the source is now focussed with the image distance p' equal to the object distance p. The reflected wavelength spread is now minimised, the $\delta\theta=0$; equation (5.4) is inapplicable, but there is still the monochromator rocking width contribution (table 5.1) and the source size effect.

If we continue to bend the crystal beyond the Guinier setting ($p'<p$) then the angle of incidence onto the monochromator begins to change along its length and rays converging towards the focus have a correlation of wavelength and direction (figure 5.6(c)). In Helliwell *et al* (1982a)

this is referred to as the 'dispersive setting at overbend'. The range of reflected wavelengths is given by

$$\left(\frac{\delta\lambda}{\lambda}\right)_{corr} = \frac{L}{2}\left|\frac{\sin\theta}{p'} - \frac{\sin\theta}{p}\right|\cot\theta \qquad (5.6)$$

A finite source causes a change in incidence angle at the crystal so that at the focus there is a photon energy gradient across its width (for all curvatures). Placing a slit at the focus to reduce this $\delta\lambda/\lambda$ contribution onto the protein crystal is akin to placing a slit at the tangent point. Use of such a slit is at the expense of flux; with a high brightness source (small source size) such a slit is not needed. At SR sources the object distance p is usually quite large (tens of metres) because of the need for a substantial thickness of radiation shielding around the ring. The non-dispersive setting for such a crystal ($p'=p$) is usable as a non-tunable system but for a tunable instrument it would be impractical to consider a 10 or 20 m optical bench swinging over 50–60° or so. By cutting the surface of the monochromator at an angle, α, to the Bragg planes (figure 5.7 shows so-called compression geometry; a change of sign for α would

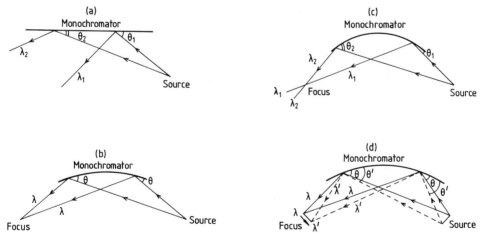

Figure 5.6 The single perfect crystal as monochromator for use with SR: (a) flat crystal; (b) curved crystal for minimum reflected $\delta\lambda/\lambda$ (Guinier setting); (c) curved crystal at overbend; (d) finite source size contribution to $\delta\lambda/\lambda$ in the reflected beam (mono-chromator at Guinier setting). From Helliwell (1984) and reproduced with permission of the Institute of Physics.

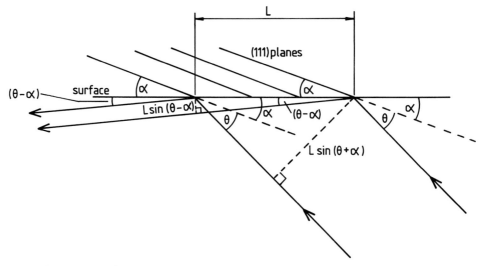

Figure 5.7 Reflection from an asymmetrically cut crystal monochromator. The width of the reflected beam is compressed relative to
that of the incident beam. The asymmetrically cut single
crystal when bent to the 'Guinier' setting (figure 5.6(b))
provides a minimum $\delta\lambda/\lambda$ and a foreshortened focussing distance (equation (5.7)).

make it expanded beam geometry), the non-dispersive setting occurs for
a *foreshortened* image distance, i.e.

$$p'_G = p\frac{\sin(\theta - \alpha)}{\sin(\theta + \alpha)} \tag{5.7}$$

at a radius of curvature R given by

$$\frac{2}{R} = \frac{\sin(\theta - \alpha)}{p'} + \frac{\sin(\theta + \alpha)}{p} \tag{5.8}$$

and equation (5.6) becomes

$$\left(\frac{\delta\lambda}{\lambda}\right)_{corr} = \frac{L}{2}\left|\frac{\sin(\theta - \alpha)}{p'} - \frac{\sin(\theta + \alpha)}{p}\right|\cot\theta \tag{5.9}$$

which for $p'=p'_G$, $(\delta\lambda/\lambda)_{corr}$ is zero.

The angular acceptance of the monochromator η_{acc}^{mono} and the angular width of the reflected beam η^{mono}_{diff} are changed from the symmetric case (i.e. η^{mono} in equation (5.1)) to

$$\eta^{mono}_{diff} = \frac{1}{A}\eta^{mono}_{acc} = \frac{1}{A^{1/2}}\eta^{mono}_{sym} \qquad (5.10)$$

where the asymmetry factor A is given by

$$A = \frac{\sin(\theta - \alpha)}{\sin(\theta + \alpha)} \qquad (5.11)$$

The foreshortened image distance causes a finite source, h, to have a demagnified focussed image given by

$$\frac{hp'}{p} = hA \qquad (5.12)$$

The convergence angle β is given by

$$\beta = \frac{L\sin(\theta - \alpha)}{p'} \qquad (5.13)$$

The demagnification of the source is important if the sample size and source size are widely different. The practical focus width has contributions from aberrations (very small for Bragg focussing optics) and spectral smearing (the latter is quite small for perfect crystals but large for mosaic graphite), i.e.

$$\text{focus width} = (x_1^2 + x_2^2 + x_3^2)^{\frac{1}{2}} \qquad (5.14)$$

where $x_1 = hp'/p$ is the demagnified source term, $x_2 = \eta^{mono}_{diff}p'$ (Lemonnier *et al* 1978) is a smearing term or a term taking account of any mosaic character of a 'perfect crystal' e.g. due to poor etching or material and $x_3 = [L^2\cos(\theta - \alpha)]/8R$ is the optical aberration term. This latter arises because a simple curved surface does not reproduce a point source as a point focus.

Finally, the use of an oblique cut over a symmetric cut produces an overall improvement in *intensity* at the sample of

$$A\frac{1}{A^{1/2}}\frac{\sin(\theta + \alpha)}{\sin\theta} \qquad (5.15)$$

where A is due to the foreshortening/demagnification of the source at the focus, $1/A^{1/2}$ is the reduction factor of the angular acceptance and $[\sin(\theta+\alpha)]/\sin\theta$ is due to the width of the crystal presented to the beam being increased. (This final term is useful if enough mrad of beam are available.)

5.2.3.1 Tuning the monochromator for variable wavelength anomalous dispersion experiments

The Guinier setting can be used to collect data at different specific wavelengths around absorption edges by stepping the monochromator. The angular change needed to cross an edge is small so that the achromatic focussing distance p_G does not change. For each step of the monochromator the reflected beam position is tracked by remote control.

Unfortunately, the spectral resolution is determined by $\eta_{\text{diff}}^{\text{mono}}\cot\theta$ where $\eta_{\text{diff}}^{\text{mono}}>\eta_{\text{sym}}^{\text{mono}}$ (cf. table 5.1); there is also a contribution due to the finite collimator or slit at the focus due to the energy gradient across the focus (figure 5.6(d)). The minimum $\delta\lambda/\lambda$ achievable, using Si(111) with $A^{-1}=10$ is $A^{-1/2}\times10^{-4}$ or 3×10^{-4} assuming a very narrow slit at the focus (say $<100\,\mu$m).

Between absorption edges, e.g. Fe K (1.743 Å) to Mn K (1.896 Å) the change in monochromator Bragg angle is large enough that for a fixed oblique cut, α, the Guinier focussing distance changes significantly. A range of focussing distances and an experimental bench to match are needed. For instance, for $\lambda=1.743$ Å, θ for Si(111) is 16.14° and for an $\alpha=12.5°$ and $p=21$ m, then $p_G=2.78$ m; at $\lambda=1.896$ Å, for identical α, p and Si(111), then $\theta=17.60°$ and $p_G=3.72$ m.

A range of oblique-cut angle crystals are also needed; for instance for the Zn K edge (1.28 Å) the θ for Si(111) is 11.78° and so for an oblique cut of 12.5° the diffracted beam would be reflected into the body of the monochromator. An oblique-cut angle of $\alpha=9°$ would give an asymmetry factor A^{-1} of 7.3 and $p_G=2.87$ m for $p=21$ m at the Zn K edge. For the Au L$_\text{I}$ edge at 0.864 Å, $\theta=7.919°$ for Si(111), so for an oblique-cut angle of 6.75°, then, a $p_G=1.6$ m would be required for an object distance of 20 m. If a p_G of 2.5 m is used, say, then a reduction in the illuminated length of the monochromator to one-third would result in an acceptable $(\delta\lambda/\lambda)_{\text{corr}}$ (equation 5.9)) of ≤0.001. Fortunately, perfect silicon is relatively cheap though it is somewhat inconvenient to change monochromator crystals for different experimental runs. Hence, single crystal monochromators can be tunable with reasonably fine $\delta\lambda/\lambda$ (section 5.2.3.4) but are not rapidly tunable like the double crystal monochromator (section 5.2.4).

5.2.3.2 Energy-dispersive mode for multiple wavelength anomalous dispersion data

In the energy-dispersive setting at overbend there is a correlation of photon energy and direction towards the focus (figure 5.6(c)). By placing a narrow slit at the focus and a protein crystal some way beyond it, each diffraction spot intensity profile produced on rotating the sample is also an energy-dependent profile. By centring the energy profile on an absorption edge the intensity profile will be a combined effect of wavelength-dependent absorption and anomalous dispersion of the metal atom in the protein crystal. The technique has been established experimentally by Arndt *et al* (1982) and Greenhough, Helliwell and Rule (1983) using a known structure involving rhenium (figure 5.8). Theoretically, Greenhough and Helliwell (1982b) and Greenhough *et al* (1983) have derived formulae to predict reflection partiality and spot size and shape in the energy-dispersive mode as well as the $\delta\lambda/\lambda$ achieved at a point within a Bragg, energy-dependent, profile (section 6.2.2).

Though the dispersive mode can be set up with an oblique-cut crystal an advantage of the technique is that a single symmetric-cut crystal ($\alpha=0°$) can be used to span all absorption edges of interest. The total ($\delta\lambda/\lambda$)$_{corr}$ is then controlled by changing the illuminated length of the monochromator (see equation (5.6)) using monochromator pre-slits. The energy-dispersive approach is attractive since, in principle, the full use of anomalous dispersion effects is made (see chapter 9). In applying this method the main difficulty is making an accurate measurement and achieving a fine spatial resolution. For the rhenium small molecule crystal the effects are obvious with film but for a metallo-protein better accuracy is required. Perhaps a CCD (charge coupled device) detector will allow this method to be realised because the CCD offers better detector qualities than film whilst preserving a fine pixel size. This method is similar to the dispersive mode used in *transmission* (i.e. the direct beam) to measure a complete EXAFS spectrum (dispersive EXAFS technique). The energy range needed is, in this case, about 1 keV and requires a considerable overbend of the crystal; even with the bending radii of only a few metres the natural perfect crystal rocking width is maintained since narrow energy width features (XANES) can still be resolved. For example, Matsushita (1980) and Matsushita and Phizackerley (1981) used SPEAR as a white radiation source for feasibility studies with a copper foil. Now dedicated EXAFS spectrometers have been built at SSRL (Phizackerley *et al* 1983), Photon Factory (Matsushita, pers. comm.) and at Daresbury (stations 7.4 and 9.3). In parallel, Flank *et al* (1981) have been developing the technique at LURE. The

Figure 5.8 The polychromatic simultaneous profile technique for complete sampling of absorption edge fine structure. One oscillation photograph is shown; 4.95° scan angle, ten oscillations,
exposure time 2 min, SRS 1.9 GeV 100 mA, energy window
across each diffraction spot 67 eV centred on the dip of f' at
the rhenium L_{III} edge. From Greenhough *et al* (1983).

details given in these papers of the monochromator settings are of
interest.

5.2.3.3 *Bent, oblique-cut, single crystal in a high intensity small sample camera on a low emittance source*

The use of Bragg optics is highly effective in producing aberration-free
demagnification of a source. For a typical protein crystal size of 0.3 mm,

the horizontal source size, at 2.3σ, on the SRS (=10 mm), on DCI (=6 mm), on DORIS (=4 mm), on Photon Factory (=6 mm) and on SPEAR (=4 mm), all need demagnification of 10:1. A new generation of low emittance machines has emerged with horizontal 2.3σ source sizes at the NSLS of 1 mm and the Daresbury high brightness lattice (HBL) of 1.5 mm. The ESRF specification has a 2.3σ source size of 0.2 mm. It is necessary to continue to demagnify the HBL source somewhat to illuminate optimally even the average protein crystal sample (0.3 mm).

From equation (5.14) we can calculate values of x_1, x_2, x_3 as the contributing factors to the focal width at the sample position. In table 5.2 we compare these values for the SRS with the HBL. The ESRF magnets as sources are also included though the source sizes here are probably more optimally double focussed 1:1 by a mirror; for ease of comparison we assume the same focussing optics in each case. Also given in table 5.2 are estimates of the intensity at the sample that could be expected with typical circulating currents and assuming 100% monochromator reflectivity. For completeness in table 5.2 we assume 1:1 vertical focussing by a mirror with 100% reflectivity and accepting the full vertical aperture of radiation.

We can assess the use of Si(111) instead of Ge(111) to reduce the value of x_2 (equation 5.14)); the reduction in the overall focus width does not compensate the loss of flux in using the lower acceptance of silicon over germanium (though for high $\delta\lambda/\lambda$ resolution experiments requiring silicon the intensity would not be reduced as much as the flux). No advantage would be gained in intensity terms by reducing the monochromator length by 2 to reduce the focussing aberration x_3 by 4 since the focal width is not reduced enough to compensate the halving of the flux. Hence, for both x_2 and x_3 it is better to tolerate a slightly broader focus to maximise the flux and the intensity even though the sample seems a lot smaller in cross sectional area than the focus.

5.2.3.4 The low emittance source and bent single crystal combination for fine $\delta\lambda/\lambda$ measurements

We saw in section 5.2.3.1 how we could establish with the bent single crystal its Guinier position and with a fine slit at the focus a minimum incident $\delta\lambda/\lambda$ of $A^{-1/2} \times 10^{-4}$ for Si(111). As we can see from table 5.2 the SRS (pre-HBL) lattice with Si(111) yielded a focus width of 1.29 mm. Hence, if a slit of 0.1 mm width at the focus was used to minimise the contribution to the overall $\delta\lambda/\lambda$ by the focus energy gradient then a loss of flux resulted of 13. With the HBL the focus width is 0.27 mm, and so

Table 5.2. *Details of calculations of intensities on various possible sources: cases 1–9 assume 1:1 vertical focussing mirror + 10:1 demagnifying horizontally focussing germanium monochromator combination: case 9A utilises the natural line-width of the third harmonic without monochromatisation but with point focussing 1:1 mirror. (The symbols are defined in section 5.2.3.) Based on Helliwell and Fourme (1983).*

	1	2	3	4	5	6	7	8	9	9A
	SRS bending magnet	SRS wiggler	SRS HBL bending magnet	SRS HBL wiggler	1982 ESRF bending magnet	1982 ESRF wiggler W_s	1982 ESRF multipole wiggler W_o	1982 ESRF multipole wiggler W_p	1982 ESRF undulator third harmonic	1982 ESRF third harmonic (no monochromator)
2.3σ horizontal source size + depth of source effect (if any) (mm)	12.7	18	1.75	4.0	0.46	0.46	0.16+2.0	1.24+2.0	0.99+0.2	0.99+0.2
2.3σ vertical source size + depth of source effect (if any) (mm)	0.66	0.7	0.31	0.7	0.25	0.25	0.08+0.2	0.15+0.2	0.184+0.2	0.184+0.2
x_1 (mm)	1.27	1.8	0.175	0.4	0.046	0.046	0.216	0.324	0.1	1.19
x_2 (mm)	0.47	0.47	0.47	0.47	0.47	0.47	0.47	0.47	0.47	
x_3 (mm)	0.1	0.1	0.1	0.1	0.1	0.1	0.1	0.1	0.1	
Focal spot size in horizontal $(x_1^2+x_2^2+x_3^2)^{\frac{1}{2}}$	1.36	1.86	0.51	0.62	0.48	0.48	0.53	0.58	0.49	1.19
Focal spot size in vertical	0.66	0.7	0.31	0.7	0.25	0.25	0.28	0.35	0.384	0.384
Focus area (mm²)	0.9	1.3	0.16	0.44	0.12	0.12	0.15	0.20	0.19	0.46

Flux at 1.5 Å (photons s^{-1} · (mrad h)$^{-1}$ · (0.01% $\delta\lambda/\lambda$)$^{-1}$	2.5×10^{11}	3×10^{12}	2.5×10^{11}	3×10^{12}	2.1×10^{12}	3.8×10^{12}	4.2×10^{13}	4.2×10^{13}	4×10^{13}	2×10^{15}
Intensity at 1.5 Å at focus (flux mm^{-2})	2.7×10^{11}	2×10^{12}	1.5×10^{12}	6×10^{12}	1.7×10^{13} (into 1 mrad)	3.2×10^{13}	2.8×10^{14} (into 1 mrad)	2.1×10^{14} (into 1 mrad)	2.1×10^{14} (into 0.1 mrad)	4.3×10^{15} (into 0.1 mrad)
Relative intensity	1	8	6	24	63	119	1037	778	778	15 926
SR power onto optics (W mrad^{-1})	15	60	15	60	68	544	816	816	100 (5% pinhole)	100 (5% pinhole)

Notes:

(1) The intensity of source 1 is about two orders of magnitude more than a conventional Cu Kα anode source.

(2) The horizontal acceptance for sources 1–4 is ~4 mrad, of sources 5–8, ~1 mrad and for source 9(9A) 0.1 mrad.

(3) These calculations are based on the so-called '1982 ESRF': these are reasonable generic sources for W_o, W_p and undulator (2 m length assumed) – but strictly are out of date. These source sizes, fluxes and intensities are still reasonably indicative of ESRF performance. The table is included also for historical interest as this was the basis for assessing the potential of the proposed ESRF in the early 1980s. It also emphasizes the long lead time in a project of this magnitude.

(4) The depth of source effect is very significant for multipole sources (especially wiggler but also undulator) whereby for a divergence angle α and a magnet length L the optics sees a source of projected size $L\alpha$ as well as the intrinsic source size through any period of the magnet. For a 2 m multipole wiggler the horizontal beam divergence angle, α_H, is 1 mrad, and the vertical beam divergence angle, α_v, is 0.1 mrad hence $L\alpha$ is $2\,mm\times0.12\,mm$ in the horizontal × vertical cases. For a 2 m undulator α_H is 0.1 mrad, like α_v so $L\alpha$ is 0.2 mm × 0.2 mm horizontal × vertical.

the loss of flux with the 0.1 mm slit is virtually negligible if one allows for the Gaussian shape of the focus. The HBL allows optimised anomalous dispersion studies with a $\delta\lambda/\lambda$ of 3×10^{-4} with a very high flux at the sample for the single crystal monochromator-based instrument.

5.2.4 Double crystal monochromator in (+,−) (parallel mode) for rapidly tunable work

In the double crystal (+,−) arrangement the emergent monochromatic beam is parallel to and only slightly displaced from the incident white SR beam. It therefore has the advantage over the oblique-cut system of being rapidly tunable. The minimum energy resolution achievable with the (+,−) double crystal system depends critically on the vertical source size and monochromator to source distance (see section 5.2.4.1). As the incident angle is changed by rotation about the axis of the monochromator, the absolute height of the emergent beam moves if the two crystals are joined together as a monolithic device (known as a 'channel cut'). The channel cut is very simple to operate but the exact parallelism of the two crystals allows harmonic reflections to be transmitted. The two crystals can be separately controlled (Hart and Rodrigues 1978; Hastings, Kincaid and Eisenberger 1978; Greaves *et al* 1983b). In this case the parallelism of the crystal planes can be offset slightly by an angle greater than the harmonic rocking width but less than the fundamental rocking width (say 3″ for Ge(333) and 16″ for Ge(111) at 1.5 Å). Harmonic rejection can be achieved in this case. In the separated crystal arrangement one of the crystals can be translated to compensate for the beam translation as the wavelength is changed. This is very advantageous in monochromator–mirror systems where the mirror follows the monochromator.

5.2.4.1 Minimum energy resolution achievable

The energy resolution is determined by the angular size of the source subtended at the monochromator, the rocking width of the crystal and the angular size of the entrance slit of the monochromator subtended at the source (or the natural divergence angle of SR, whichever is the smaller). The geometric factors of source and slit are illustrated in figure 5.9. We have

$$\frac{\delta\lambda}{\lambda} = \left[(\Delta\theta_{source} + \Delta\theta_{slit})^2 \cot^2\theta + \left(\frac{\delta\lambda}{\lambda}\right)^2_{crystal} \right]^{\frac{1}{2}} \tag{5.16}$$

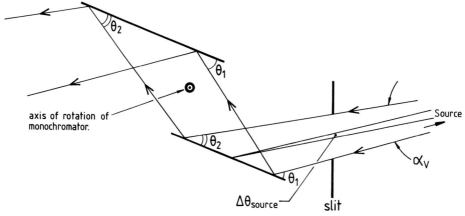

axis of rotation of
monochromator.

$\Delta\theta_{source}$ slit

Figure 5.9 The contribution of source divergence α_v and size $\Delta\theta_{source}$ to the range of energies reflected by a $(+,-)$ double crystal monochromator (see figure 5.5) $(\theta_1\neq\theta_2)$.

where $\Delta\theta_{source}=v/p$ and $\Delta\theta_{slits}=s_v/p$ where v is the vertical size of the storage ring beam and s_v is the vertical separation of the monochromator entrance slits. The double crystal is usually set in the vertically dispersing mode for two reasons. Firstly, vertical source sizes are smaller than horizontal source sizes at SR sources and so $\Delta\theta_{source}$ can be minimised for a given object distance p. Secondly, for long wavelengths (high monochromator Bragg angles) the weak vertical polarisation component is the one attenuated in the vertically dispersing mode (contrast the horizontally dispersing oblique-cut crystal case). If the full length of the first crystal is bathed by the vertically divergent beam, the wavelength reflected varies systematically along the length giving a wavelength direction correlation (see figure 5.6(a)). Since the vertical divergence angle is 0.3 mrad this is only equivalent to 0.001 $\delta\lambda/\lambda$. The use of the slit of size s_v reduces this correlated component. What is the minimum achievable? The SRS (pre-HBL) had a vertical source size of $v=0.66$ mm (machine physics data). At $p=20$ m, the vertical beam width, after diverging from the source is 7–10 mm. If a 1 mm slit is introduced at this position then, taking account of the 0.66 mm source size, for a Si(220) crystal with a rocking width of 2.85×10^{-5} radians at $\lambda=1.743$ Å and $\theta=27.0°$ then:

$$\frac{\delta\lambda}{\lambda} = \left[\left(\frac{0.66+1}{20\,000}\right)^2 \cot^2 27° + (5.6\times10^{-5})^2\right]^{\frac{1}{2}} = 1.77\times10^{-4}$$

We can compare this with the oblique-cut system of $A^{-1/2}\times\delta\lambda/\lambda$ of 3.8×10^{-4} for Si(111) or 1.8×10^{-4} for Si(220) with $A^{-1}=10$. If in the double crystal system we reduce s_v further, at the expense of flux, to 0.2 mm and the HBL with $v=0.31$ (on bending magnet 7), then the estimate above becomes

$$\frac{\delta\lambda}{\lambda} = \left[\left(\frac{0.31+0.2}{20\,000}\right)^2 \cot^2 27° + (5.6\times10^{-5})^2\right]^{\frac{1}{2}} = 7.5\times10^{-5}$$

An improvement over the oblique-cut system is achievable in this case. However, the most important advantage is still the one of being rapidly tunable and without the need for a series of oblique-cut crystals for different absorption edges.

5.2.4.2 Focussing possibilities with a $(+,-)$ separated double crystal system

It is feasible to consider bending the second crystal in the separated crystal arrangement to provide horizontal line focussing (figure 5.10). The introduction of the curvature will cause a slight error in the Bragg angle for the vertically reflected beam from this second crystal but for a small curvature it is negligible. With 2:1 focussing and a horizontal acceptance of 2–3 mrad the resulting convergence angle would be 4–6 mrad. The effect of this cross-fire angle is to broaden the diffraction spots on a film and under typical conditions for a virus crystal 1 mrad is the maximum tolerable. However, for smaller unit cells (100 Å or so) a 6 mrad convergence could be tolerated (with a crystal to film distance ⩽100 mm).

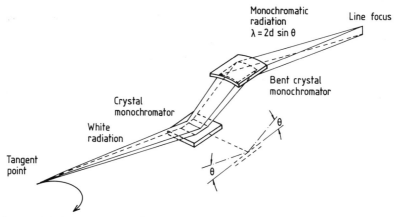

Figure 5.10 Incorporating line focussing into the $(+,-)$ double crystal monochromator. From Bordas (1982) with permission.

A 2:1 demagnified horizontal source produced a horizontal focus on the SRS lattice of 6 mm (HBL ≈1.5 mm). Since an unfocussed 4 mrad beam 25 m from the source would measure 100 mm, this is still a very useful intensity improvement for this rapidly tunable scheme. Between absorption edges the curvature would need to be adjusted. Such a scheme has not been implemented for macromolecular crystallography.

5.2.5 Wide bandpass monochromators

The naturally small rocking widths of perfect crystals of silicon and germanium offer near-optimal performance for: (a) producing finely focussed images of the SR source (x_2 of equation (5.14) is kept small); (b) providing a $\delta\lambda/\lambda$ incident onto a protein crystal sample which is narrow enough to probe the f' minimum in an absorption edge and/or (XANES) features of a bound metal atom; and (c) providing a $\delta\lambda/\lambda$ narrow enough to minimise reflection spot sizes and angular reflecting ranges for very large unit cell virus crystal studies. These monochromators have a high peak reflectivity and are easily bent for focussing. It is important to comment, however, on the possibilities of using some wide bandpass optical arrangements for those circumstances when conditions (b) and (c) above can be relaxed. In some protein crystal studies unit cell size and hence order to order resolution requirements may not be highly demanding and a higher flux onto the sample will be important for studying smaller samples or for time resolved work (see chapter 10). A $\delta\lambda/\lambda$ acceptance range of 0.0025–0.005 is approximately the maximum that can be tolerated in a typical monochromatic case; the exact value depends on the Bragg resolution of data required since a broad $\delta\lambda/\lambda$ affects the high angle reflections the most ($\sim\tan\theta$).

There are three types of wide bandpass arrangements available for consideration: (i) mosaic monochromators; (ii) layered synthetic microstructures; and (iii) transmission/reflection X-ray mirrors or an absorption filter/reflection mirror combination. Type (iii) is discussed in section 5.3.7. The best example of type (i) is, of course, mosaic graphite, which has a reflection efficiency of nearly 30% and the 0.5° rocking width corresponds to a $\delta\lambda/\lambda$ accepted from a white SR source of 0.04. Unfortunately, though the large $\delta\lambda/\lambda$ increases the flux it would also smear out the focus considerably when a focussing distance of a few metres is used (x_2 in equation (5.14)), i.e. for 0.5° over 2 m, then x_2 would be 17.5 mm. With mosaic graphite a flux gain of 100 is available compared with Ge(111) and an intensity gain of 6. However, since Ge(111) can be obtained in longer lengths than pyrolytic graphite, the gain factor, as a focussing element, is actually small and the larger $\delta\lambda/\lambda$ onto the protein crystal sample broadens the high angle protein reflections and thus reduces the mean optical density on a film. For mosaic graphite used on a conventional source the $\delta\lambda/\lambda$ is determined by the Kα_1, Kα_2 emission line separation (2.5×10^{-3} for copper) which is acceptable. Rocking widths lying between perfect Ge(111) and mosaic graphite can be induced by distortion of a perfect monochromator crystal lattice so that the lattice spacing varies as a function of depth. This can be done by mechanically stressing the crystal and by bending (Boeuf *et al* (1978); Kobayashi and Yoshimatsu reported in Kohra *et al* (1978)); a bending radius less than 2 m or so would be needed to achieve this. The rocking width may as a result be broadened by a factor of 15 but the reflectivity is usually degraded by a significant factor (3–4).

A more promising approach to produce a $\delta\lambda/\lambda$ acceptance around 0.005 is based on the 'layered synthetic microstructure' (LSM). This consists of many alternating layers of high and low atomic number materials (e.g.

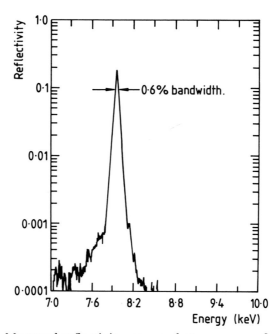

Figure 5.11 Measured reflectivity versus photon energy of a tungsten–carbon LSM: $d=15$ Å, $\theta=51.7$ mrad, $N=260$ layers, $d=67$ Å, $d=8.3$ Å. The fundamental has a peak reflectivity of 19% and a bandpass dE/E of 0.6%. From Bilderback *et al* (1983) with permission.

carbon and tungsten); for N layers sampled by the X-rays the wavelength resolution is $\sim 1/N$ (Nagel, Gilfrich and Barbee 1982). Bilderback (1982) has pointed out that since the reflection is specular then good focussing is possible (i.e. x_2 of equation (5.14) $=0$) if the LSM can be supported on a figured surface. Multilayer structures have also been under development as neutron monochromators (Saxena and Schoenborn 1977) for longer wavelengths. The main area of application for X-rays has also been in the long wavelength region so far. However, Bilderback *et al* (1983) consider the use of tungsten–carbon and molybdenum–carbon LSMs as wide bandpass ($\delta\lambda/\lambda = 0.005$–$0.1$) elements from 5 keV to 30 keV; figure 5.11 reproduces the reflectivity versus energy curve measured on CHESS by Bilderback *et al* (1983) for a 260-layer tungsten–carbon LSM. To summarise, the bandwidth of an LSM is controllable explicitly by the number of layer pairs sampled by the X-rays and by variations in the d-spacing of the microstructure with depth. It is clearly feasible to develop an LSM well matched to some protein crystallographic data collection experiments where high intensity is the important concern as opposed to high spectral or angular resolution.

5.2.6 Thermal problems of monochromators in the most intense SR beams and possible solutions

The high brilliance, insertion-device-based photon beams from the next generation of sources (ESRF and APS) will routinely be in the 1–10 kW range. Hence, the first optical element in the beam will have to perform under a high thermal load.

The effect of thermal load is to distort the first crystal of the monochromator (figure 5.12(a)) whereby the diffracted beam is spread out in both angle and energy. Hence, only some of the X-rays have the correct incident angle and energy to be diffracted by the second crystal, in the case of a double crystal monochromator. In the case of a focussing, single crystal monochromator the quality of the focus will be badly degraded. Both situations require a remedy if advantage is to be taken of the high fluxes available with multipole wiggler or undulator beams. Even on single bump wigglers with relatively modest heat loads there are problems. For example, on the SRS wiggler at Daresbury the heat load is 61 W mrad^{-1} from the central pole at 2 GeV, 300 mA and 5 T; on the station 9.6 protein crystallography station (Helliwell *et al* 1986b) the bent triangular monochromator intercepts up to 5 mrad of beam (150 W after the beryllium window in the front end) and 1 atm of helium gas is

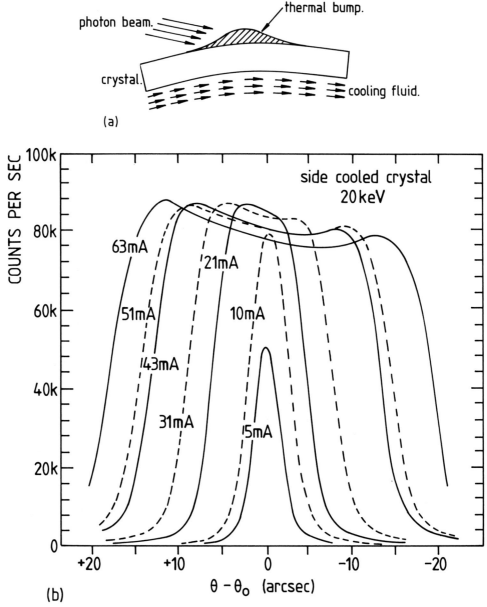

(a)

(b)

Figure 5.12 (a) Schematic drawing of the thermal distortions of a
monochromator crystal subject to high heat loads from
synchrotron beams. (b) Comparison of the rocking curves
for the second crystal in a two crystal monochromator
when the first crystal is cooled only at the side, for different
CHESS electron beam currents and a six-pole electromagnetic wiggler. From Smither *et al* (1989) with
permission.

sufficient to cool the crystal. Helium has a high thermal conductivity and is essentially X-ray transparent. In contrast, for the monochromator *in vacuo* the focus expands significantly reducing the flux through the pinhole collimator.

An example illustrating the magnitude of the problem with higher heat loads is given by Smither *et al* (1989). A two-crystal monochromator on the A2 beam line on CHESS at Cornell intercepts one-third of the beam representing 640 W for 70 mA circulating current and 5.4 GeV from a six-pole electromagnetic wiggler. The photon flux output of this monochromator was linear with current up to 10 mA and then levelled off giving no further improvement as the beam current was increased to 70 mA. Thus a factor of 7 in intensity was lost due to the distortion of the crystal by the heat load (see figure 5.12(b)). Cooling of the first monochromator crystal in a double crystal monochromator seeing such a heat load is obviously critical. Smither *et al* (1989) go on to recommend the use of liquid gallium cooling of silicon crystals in high intensity beams. Liquid gallium has a good thermal conductivity and thermal capacity, a low melting point and a high boiling point. Its very low vapour pressure, even at elevated temperatures, makes it especially attractive in ultra-high vacuum (UHV) conditions. Smither *et al* (1989) analyse the temperature distributions and crystal distortions under a photon beam heat load and compare the performance of liquid gallium versus water. Their analysis considers three kinds of distortions in the crystal monochromator. First, there is the overall bending or bowing of the crystal caused by the thermal gradients in the crystal and, second, there is the thermal bump caused by the expansion of the crystal perpendicular to the surface; these are illustrated in figure 5.12(a). There is an additional distortion due to the change in lattice spacing through thermal expansion caused by the increase in the surface temperature of the crystal.

Another possible solution is to use liquid nitrogen to cool the crystal monochromator. Bilderback (1986) pointed out that the thermal expansion coefficient of silicon is nearly zero at this temperature (figure 5.13(a)) so that distortions could be avoided altogether. Also, the thermal conductivity of silicon can be enhanced by as much as an order of magnitude over its corresponding room temperature value in this low temperature region (figure 5.13(b)). Development of this idea is being pursued by Freund (pers. comm.) for use on ESRF beam lines.

Hart (1990) has analysed the problems of power flow from a line source of heat and the uniform heat flow normal to the flat surface. The results suggest that, using water-cooled silicon at room temperature, special

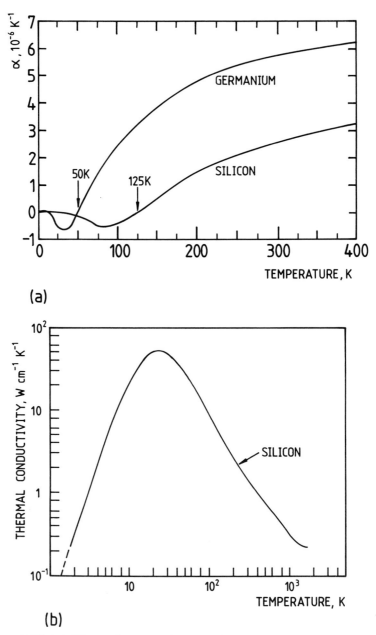

Figure 5.13 (a) Thermal expansion coefficient for silicon and germanium as a function of temperature. The value for silicon goes through a zero point at a temperature accessible to cryogenic techniques. (b) Thermal conductivity as a function of temperature also shows advantageous changes at cryogenic rather than room temperatures. From Bilderback (1986) with permission.

engineering designs could lead to feasible crystal monochromators, for the ESRF for example, operating up to 100 mA current.

5.3 MIRRORS

X-ray mirrors play an essential role at SR sources in focussing applications and/or rejecting high order harmonic reflections from a monochromator.

5.3.1 The phenomenon of total external reflection of X-rays

The refractive index, n, for X-rays is given by $n=1-\delta-i\beta$ where δ, the decrement of the index of refraction, is 10^{-5} and β accounts for any absorption effect. For a perfect, smooth surface and in the absence of absorption ($\beta=0$), Snell's law gives the condition for total external reflection as

$$n = \sin (90 - \theta_c) / \sin 90 \tag{5.17}$$

or

$$1 - \delta = \cos \theta_c \tag{5.18}$$

Since, $\delta \ll 1$ then

$$\theta_c = \left(\frac{e^2}{mc^2} \frac{N}{\pi} \right)^{\frac{1}{2}} \lambda \tag{5.19}$$

If we substitute for δ we have

$$\theta_c \approx (2\delta)^{\frac{1}{2}} \tag{5.20}$$

Here N is the number of free electrons per unit volume of the reflecting material. Table 5.3 gives values of θ_c for different reflecting surfaces. Clearly, the lower the atomic number the smaller N and θ_c for a given λ.

For a range of angles of incidence $\theta(\leq\theta_c)$ X-rays with energies up to a cut-off or critical energy are reflected efficiently whereas more energetic X-rays are not reflected. This is the basis for using a mirror as an efficient harmonic rejector when used with a crystal monochromator.

Table 5.3. *Critical angles and reflectivities for a representative selection of mirror materials for a wavelength of 1.5 Å (adapted from Witz (1969)).*

Material	θ_c (mrad)	R (θ_c)	$\theta_c R$ (θ_c)	$\dfrac{R\ (Cu\ K\beta)}{R\ (Cu\ K\alpha)}$
Crown glass	4.1	0.73	100	0.05
Nickel	6.7	0.65	150	0.02
Gold	9.3	0.45	140	0.13

5.3.2 Vertical aperture

The aperture required to collect the vertically diverging SR beam with natural opening angle of 0.3–0.4 mrad (at 1.5 Å) and 10 m from the source would be 3.0–4.0 mm. For quartz, $\theta_c=3$ mrad and an aperture of 3.00–4.00 mm would need a length L of 1–1.3 m. On the SRS wiggler protein crystallography station (Helliwell *et al* 1986b) the vertical divergence at 1.5 Å is twice that on an SRS bending magnet so an effective mirror length of 2–2.6 m would be needed to collect and focus the beam. These large lengths lead to the use of metal coatings (table 5.3). For example, platinum with a θ_c of 9 mrad at 1.5 Å reduces the lengths needed above to one-third. A single segment metal coated mirror is a suitable choice; fused quartz lengths up to 0.75 m are available and can be coated. Single segment platinum coated mirrors have been used on different instruments by Horowitz and Howell (1972), Hastings *et al* (1978), Helliwell *et al* (1982a), Nave *et al* (1985), Helliwell *et al* (1986b). The alternative solution to subtend the necessary vertical aperture is to use a line of segmented mirrors. Uncoated segmented mirrors have been used by Barrington-Leigh and Rosenbaum (1974), Franks and Breakwell (1974), Hendrix, Koch and Bordas (1979), Rosenbaum and Holmes (1980) and Hashizume *et al* (1982) (the latter three involving 8×20 cm segment systems). Haselgrove, Faruqi, Huxley and Arndt (1977) used a two-segment gold coated mirror on NINA. Bordas (1982) has used multiple segment mirror systems (two gold coated and one uncoated) for three new instruments in HASYLAB on DORIS.

Rosenbaum and Holmes (1980) comment that long single segment mirrors are very expensive. However, if in a system of eight individual segments each is separately motor controlled, the segmented mirror system is the more expensive overall. Moreover, the accurate alignment of each segment one with another is more difficult, especially in the toroid case.

5.3.3 Reflectivity and cut-off properties

If a mirror is to be an effective optical element then the reflectivity must be as close to 100% as possible. Parratt (1954) gives an expression, derived from the Fresnel coefficient for reflection (Compton and Allison 1935), for the reflectivity, R, for an incidence angle, θ, and critical angle, θ_c, for a perfectly smooth homogeneous mirror surface

$$R = \frac{h - (\theta / \theta_c) \sqrt{2} \ (h-1)^{\frac{1}{2}}}{h + (\theta / \theta_c) \sqrt{2} \ (h-1)^{\frac{1}{2}}} \qquad (5.21)$$

where

$$h = (\theta / \theta_c)^2 + \left\{ [(\theta / \theta_c)^2 - 1]^2 + \left(\frac{\beta}{\delta}\right)^2 \right\}^{\frac{1}{2}} \qquad (5.22)$$

Parratt (1954) graphed the values of R for different values of the ratio β/δ as a function of θ/θ_c; see figure 5.14.

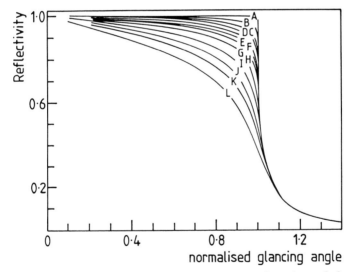

Figure 5.14 Theoretical reflection efficiency as a function of the normalised glancing angle θ/θ_c for a perfectly smooth surface for selected values of β/δ. The shape depends on the absorption of X-rays in the mirror. For no absorption the reflection coefficient is unity for $0 < \theta/\theta_c < 1$ and then drops abruptly for $\theta/\theta_c > 1$. From Parratt (1954). Values of β/δ: A, 0.0; B, 0.005; C, 0.01; D, 0.015; E, 0.02; F, 0.025; G, 0.03; H, 0.04; I, 0.06; J, 0.08, K, 0.10; L, 0.14.

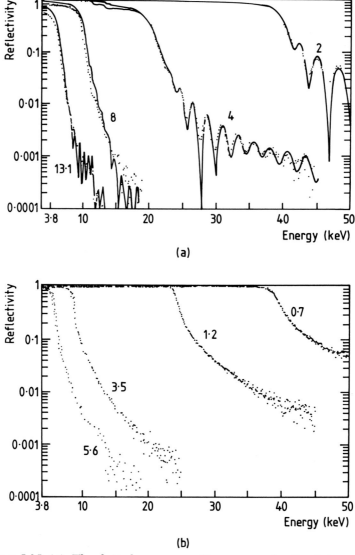

Figure 5.15 (a) The dotted curves are the measured reflectivity at different photon energies of a platinum coated mirror as a function of various grazing angles. The solid curves are the computed reflectivities in each case using a 29 Å rms surface roughness for an air–platinum interface and a 7 Å rms for the platinum–glass interface. The oscillations beyond the cut-off are predictable from the finite thickness of the platinum layer. (b) Measured reflectivity at different photon energies of an uncoated float glass mirror as a function of various angles of incidence. From Bilderback (1981) with permission.

The product of vertical aperture and reflectivity (table 5.3) is important (Witz 1969; Arndt and Sweet 1977). Unfortunately, in practice, an
accurate knowledge of the reflectivity is unlikely to be available for a
practical surface which to some extent is rough and impure (cf. the
assumptions for equations (5.17)–(5.20)) and can deteriorate with time.
Bilderback (1981) reviews theories which take account of thick coatings
as well as surface roughness. The effect of a thick coating is to provide an
oscillatory modulation of the R versus θ/θ_c curve. The effect of surface
roughness is to cause a steeper descent of R as $\theta \rightarrow \theta_c$. In figure 5.15
measured and theoretical reflectivity curves for platinum and float glass
are compared (from Bilderback (1981)); from this we see that the
harmonic rejection performances of platinum and float glass are virtually identical; also the efficient reflection of X-rays is possible up to
38 keV (at the expense of vertical aperture), an important result for
optical design studies for wiggler sources.

5.3.4 Vacuum requirements

The high reflectivity of surfaces such as platinum or float glass mentioned in section 5.3.3 (and the other surfaces discussed later in section
5.3.5) assumes that there is no surface contamination.

Irradiation of the material by an SR beam of significant power (sections 4.1, 4.10.1.3) can cause the deposition of carbonaceous material,
which can gradually worsen reflectivity as time passes, especially at
longer wavelengths. Clearly, the higher the vacuum the better, especially if the mirror is the first optical element in the white SR beam. At the
SRS/HBL, several years of continuous irradiation by the white SR beam of
a platinum coated mirror, on the first protein crystallography instrument
(section 5.6.5.1 and Helliwell et al (1982a), Nave et al (1985)) in UHV
(10^{-8} torr) has not caused a noticeable change of reflectivity or focus
quality. Likewise, the platinum coated mirror on the wiggler protein
crystallography workstation 9.6 (Helliwell et al 1986b) has now had
more than four years of continuous irradiation in UHV without
deterioration.

5.3.5 Choice of materials

Fused quartz has often been used to date either as a reflecting surface
or as a substrate for a metal coating, e.g. Hastings et al (1978) on SPEAR
and Helliwell et al (1982a, 1986b), Brammer et al (1988) on the SRS.

In the future, silicon carbide may be used instead (Franks et al 1983) because of its better thermal rigidity than fused quartz, though it is more expensive. Silicon carbide may prove to be useful with the advent of more powerful synchrotron sources, especially multipole wiggler magnets. Float glass has been used on SPEAR by Webb *et al* (1976, 1977) and on CHESS (Caffrey and Bilderback 1983). Webb *et al* (1977) reported that the float glass deformed considerably on exposure to the SR beam; this is probably due to choice of too thin a section.

The metal coatings used at SR sources for X-ray experiments include platinum and gold (see earlier references in section 5.3.2). Gold is easier to evaporate onto a substrate than platinum. Unfortunately, the degree of adhesion to the substrate is not as good as platinum and so the gold tends to evaporate gradually on exposure to the white SR beam. Arndt and Sweet (1977) discuss the use of nickel coatings, but for conventional sources. On the NSLS rhodium is used with an aluminium substrate; the absorption edges of rhodium are well removed in wavelength from those of interest for anomalous dispersion studies. Platinum ($Z=78$) or gold ($Z=79$) have L absorption edges in the wavelength range of interest (figure 9.3) which is a complication, but for the thickness of the coatings typically used the modulations of the incident intensity at wavelengths around the L absorption edges are correctable with an intensity (I_0) monitor.

5.3.6 Focussing applications

5.3.6.1 Single curvature

Single segment mirrors can be bent by applying equal couples to either end of the block (figure 5.16). For a vertical source size of <0.3 mm unit focussing can be achieved with zero aberrations for all the horizontally accepted mrad. For a grazing incidence angle θ_c, the radius of curvature, R_V, which is needed is given by

$$\frac{2}{R_V} = \frac{\sin \theta_c}{p} + \frac{\sin \theta_c}{p'} \tag{5.23}$$

For unit magnification $p=p'$ then

$$R_V = p / \sin \theta_c \tag{5.24}$$

If we take as an example $p=11$ m and $\theta_c=6$ mrad (platinum, 1 Å cut-off), then $R=1800$ m. For a $\theta_c=3$ mrad (platinum, 0.5 Å cut-off), then R would be 3600 m.

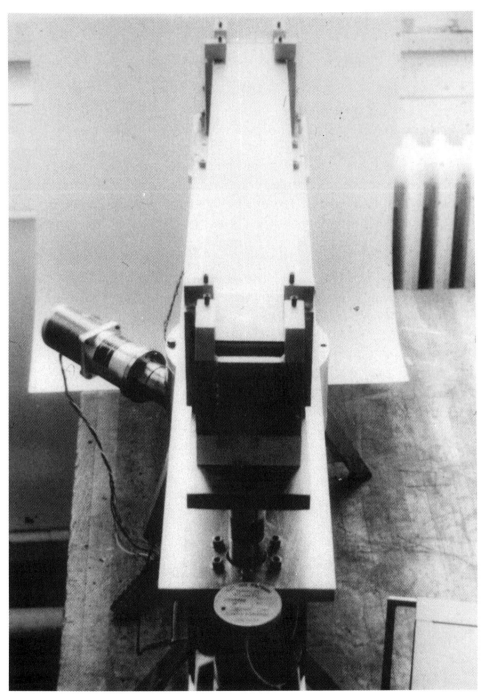

Figure 5.16 An example of a long, single segment mirror and its bender prior to installation in the beam line on the SRS wiggler (station 9.6). From Helliwell *et al* (1986b) and reproduced with the permission of Daresbury Laboratory.

The maximum deflection δ_V at the centre of a mirror of length L is given by the Sagitta relation as

$$\delta_V = L^2 / 8R_V \tag{5.25}$$

For $L=0.6$ m and $R=1800$ m, $\delta_V=25\,\mu$m. The natural deflection of such a segment of, say, 3–4 cm thickness is $5\,\mu$m. Even with such a small deflection as $25\,\mu$m, an accurately curved surface can still be achieved over the full length of the mirror. Accurate focussing in this way becomes more difficult at shorter wavelengths; if we wish to reflect wavelengths greater than, say, 0.5 Å, then, for platinum, $\theta_c=3$ mrad, R is then 3600 m and δ_V is only $19\,\mu$m (for a segment of 0.75 m and $p=11$ m).

In the eight segmented mirror systems of, for example, Hendrix, Koch and Bordas (1979) and Rosenbaum and Holmes (1980) focussing is achieved by slightly adjusting each segment relative to its neighbour and overlapping the reflected beams at the 'focus' position. If the segments are kept flat the source is *not imaged* but the divergent beam is sectioned into eight parts. This design is well suited to a machine with a relatively poor vertical source size. A development of the eight-segment system, made at the Photon Factory, is the firm mounting of each segment on a single mechanical bench; the bending of the bench as a whole maintains the alignment of all segments relative to each other as well as bending them (Hashizume *et al* 1982). This design achieves simplicity of operation and the means to image a small vertical source akin to the bent single long segment.

In the two-segment mirror systems of Barrington-Leigh and Rosenbaum (1974) and Haselgrove *et al* (1977) each segment is adjusted with respect to the other and has individual bending mechanisms.

5.3.6.2 Double curvature

To achieve a point focus the mirror substrate can be ground to a fixed radius of curvature, R_H, in one direction (for the horizontal focussing) and then bent to a curvature R_V (equation (5.24)) for the vertical focussing. The equation for R_H is

$$R_H \doteq R_V \sin^2 \theta_c \tag{5.26}$$

With such simple cylindrical curvatures the overall surface *figure* produced is toroidal and is the simplest way to produce a reasonable point focus. If the system aims to collect a large number of horizontal mrad (e.g. 4–6) the aberrations are significant. These aberrations are best evaluated by computer ray tracing methods – see section 5.3.8.

In addition to the smearing of the focal point size the angular collimation is 'imperfect'. Rays originating from a source point with a horizontal divergence α_H accepted by the toroid will contribute an additional cross-fire angle in the vertical direction given by

$$\alpha_H^2 / 16\theta_c \tag{5.27}$$

(Heald 1982).

As an example, the SRS protein crystallography station 9.5 toroid mirror can be quoted (see section 5.3.8.1 case study, Brammer *et al* (1988)). Here, $\alpha_H \approx 1.4$ mrad at full aperture in practice and θ_c is 3 mrad, hence equation (5.27) yields 0.04 mrad. This is smaller than, but significant, compared with the full vertical acceptance, α_V, of ~ 0.12 mrad. For use of a rapidly tunable double crystal monochromator (downstream from the mirror) an α_V of 0.12 mrad would correspond to a $\delta\lambda/\lambda$ of $\alpha_V \cot\theta$ (e.g. for Si(111), 4.8×10^{-4} or 4.8 eV at 10 keV). For some anomalous dispersion work (e.g. Templeton *et al* (1982a), chapter 9), this is too coarse and a $\delta\lambda/\lambda = 10^{-4}$ is desirable. Slitting down of α_V by a factor of 5 leaves the angular aberration (equation (5.27)) as the dominant term unless some slitting down of the horizontal is performed as well; then satisfactory spectral resolution with good flux can be achieved for most experiments of this type. These calculations can be compared with those in section 5.2.4.1.

To avoid these aberrations a surface would have to be ground to fixed curvature with ellipsoidal or paraboloidal shape (Heald 1982). This is extremely difficult to achieve with any accuracy with the very large values needed for R_V at X-ray reflection grazing incident angles (equation (5.24)). This approach is more feasible for longer wavelengths where θ_c (equation (5.20)) and δ_V (equation (5.25)) are larger and R_V (equation (5.24)) is smaller.

5.3.7 A transmission X-ray mirror

In the total external reflection of X-rays the unreflected beam is absorbed by the substrate. If, however, the substrate is very thin then the transmitted beam can be used, which will have complementary spectral characteristics to those for the reflected beam.

Lairson and Bilderback (1982) discuss their use of 400–10 000 Å thick soap films. By combining this transmission mirror with a reflecting mirror an adjustable bandpass X-ray beam was obtained ($\delta\lambda/\lambda$'s in the range of 12–18% at 13 keV with peak efficiencies of 55–75%). The mean lifetime of the soap films of 1–12 hr is, however, too short to be useful.

An alternative wide bandpass system is to replace the soap film with an attenuator such as aluminium foil which gives a low energy cut-off and is sometimes used in Laue protein crystallography (chapter 7).

5.3.8 Optical ray tracing methods

The principle behind ray tracing is to model, on the computer, the passage of a light ray through a system of optical elements. The light ray is described by a vector and the optical operations of reflection and diffraction are simulated by vector operations. By tracing the paths of many rays, which illuminate the full geometric aperture of each component, it is feasible to:

test different designs on the computer;
evaluate the effect of aberrations;
look for particularly sensitive alignment parameters.

There are a number of ray tracing programs available either based on geometry (e.g. Hubbard and Pantos (1983), Svensson and Nyholm (1985)) or including also finite element analysis routines to simulate thermal heating and distortion of optical elements (Gerritsen, pers. comm.).

The subroutines for mirror surfaces include planes, cylinders (along two different directions), ellipsoids, elliptical cylinders, paraboloids, plane-parabolas, spheres and toroids. Crystal monochromators can be approximated by sets of reflective surfaces but in these account has to be taken of the correspondence between reflected photon energy and incidence angle. Account can be taken of any slits or apertures by determining if a given ray passes through the defining aperture. The target area is the pinhole or collimator entrance hole just upstream of the sample crystal.

The success of the optical focussing scheme can be determined by the fraction of rays successfully brought to the target (pinhole). This fraction takes account of the success of the focussing and the reflectivity of the optical element.

5.3.8.1 Case study: the ray tracing of a toroid reflecting mirror
(Brammer et al 1988)

An extensive set of ray tracing calculations has been made on a toroid mirror for use on the Daresbury wiggler protein crystallography station 9.5. The layout of the station and optics is shown in figure 5.17. The position of the mirror on the beam line is 18.325 m from the tangent point with a focal point at 32.0 m from the tangent point and a grazing angle of incidence of 3.0 mrad (setting the short wavelength cut-off at 0.44 Å for a platinum coating). The radius of curvature for horizontal focussing is 46.9 mm and is machined into the surface of the 75 cm long mirror. The vertical focussing is accomplished by bending the cylinder to the desired radius which is 5221 m.

The ray tracing can be used to illustrate the basic or ideal focussing scheme and then to explore the effects, on the focus, of mirror/source misalignments from the ideal setting. Figure 5.18 shows ray tracing examples including the calculated focal spot produced by a beam from a point synchrotron source when reflected from the toroid mirror accepting 2.5 mrad horizontally by 0.1 mrad vertically. Figure 5.18(a) shows the relationship between the focal spot position to that of the unreflected

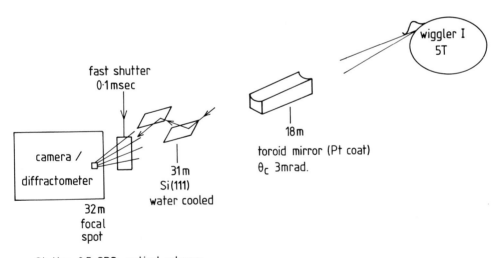

Station 9·5 SRS: optical scheme.

(a)

Figure 5.17 (a) Schematic of the beam line layout on the SRS protein crystallography station 9.5. (b) A view of the toroid mirror under test. From Helliwell (1991) and reproduced with the permission of Daresbury Laboratory.

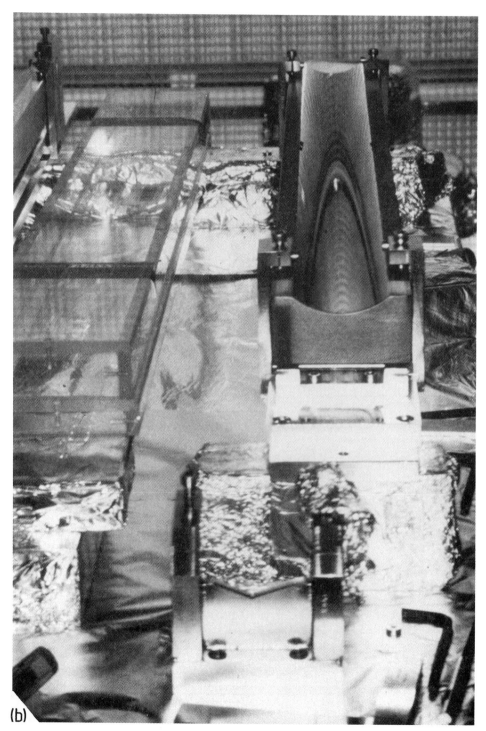

(b)

Figure 5.17 (*cont.*)

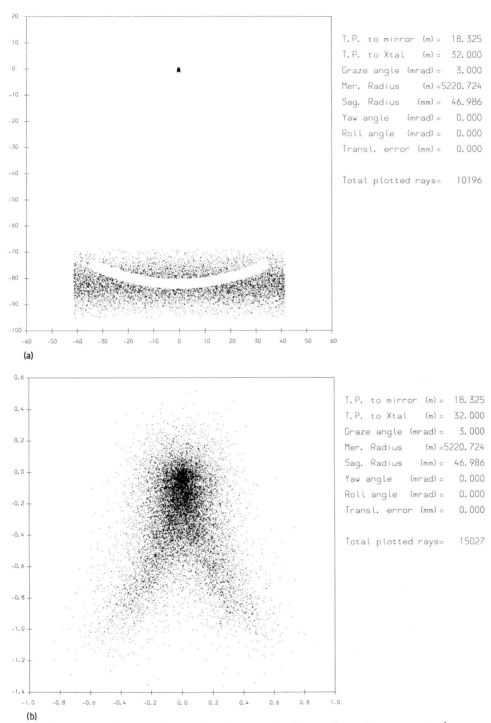

T.P. to mirror (m) = 18.325
T.P. to Xtal   (m) = 32.000
Graze angle (mrad) =   3.000
Mer. Radius   (m) =5220.724
Sag. Radius (mm) = 46.986
Yaw angle   (mrad) =   0.000
Roll angle (mrad) =   0.000
Transl. error (mm) =   0.000

Total plotted rays= 10196

(a)

T.P. to mirror (m) = 18.325
T.P. to Xtal   (m) = 32.000
Graze angle (mrad) =   3.000
Mer. Radius   (m) =5220.724
Sag. Radius (mm) = 46.986
Yaw angle   (mrad) =   0.000
Roll angle (mrad) =   0.000
Transl. error (mm) =   0.000

Total plotted rays= 15027

(b)

Figure 5.18 Ray tracing study of a toroid mirror (from Brammer *et al*
(1988) with permission). (*a*) Cross section of 4 mrad of

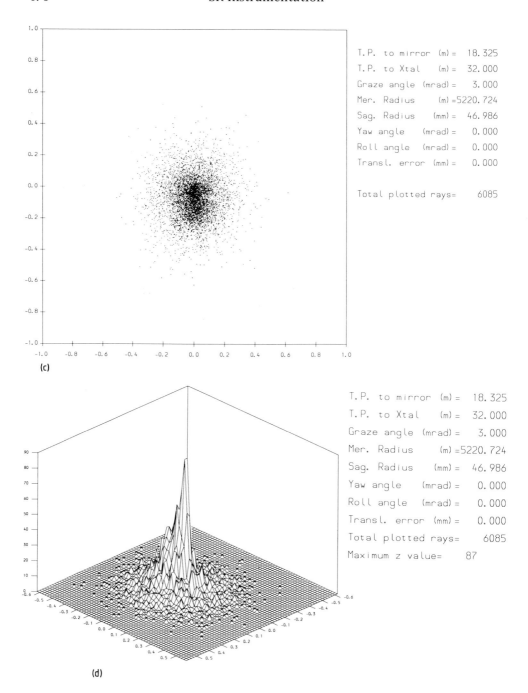

(c)

(d)

T.P. to mirror (m) =	18.325
T.P. to Xtal (m) =	32.000
Graze angle (mrad) =	3.000
Mer. Radius (m) =	5220.724
Sag. Radius (mm) =	46.986
Yaw angle (mrad) =	0.000
Roll angle (mrad) =	0.000
Transl. error (mm) =	0.000
Total plotted rays =	6085

T.P. to mirror (m) =	18.325
T.P. to Xtal (m) =	32.000
Graze angle (mrad) =	3.000
Mer. Radius (m) =	5220.724
Sag. Radius (mm) =	46.986
Yaw angle (mrad) =	0.000
Roll angle (mrad) =	0.000
Transl. error (mm) =	0.000
Total plotted rays =	6085
Maximum z value =	87

Figure 5.18 (*cont.*)
      synchrotron radiation projected to 32 m, the amount col-
      lected by the toroid mirror (the smile) and the position of
      its focus 80 mm above it. All the dimensions given are in
      mm. (*b*) The focal spot in (*a*) is shown at a much larger

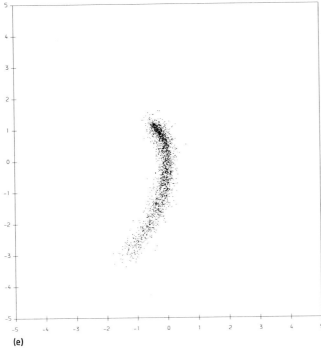

```
T. P.  to mirror  (m) =   18.325
T. P.  to Xtal    (m) =   32.000
Graze angle (mrad) =       3.000
Mer. Radius    (m) =5220.724
Sag. Radius   (mm) =   46.986
Yaw angle   (mrad) =      0.200
Roll angle  (mrad) =      0.000
Transl. error (mm) =      0.000

Total plotted rays=          568
```

(e)

(f)

Figure 5.18 (cont.)
      scale (values given in mm): 50% of the intensity is con-
tained in the centre area of (0.25 mm)^2. (c) If the horizontal beam acceptance is reduced to 1.0 mrad the focal spot
has the shape shown here. Dimensions in mm. (d) This

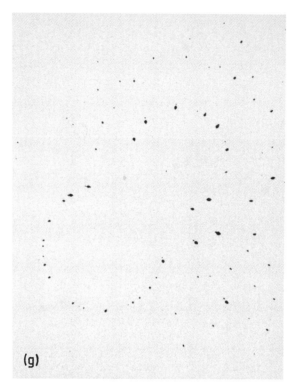

(g)

Figure 5.18 (*cont.*)

    figure depicts the information contained in (*c*) as an inten-
sity surface plot. The *x* and *y* scales are given in mm; the *z*
scale is in arbitrary units. (*e*) The effect on the focal spot
(of the unmasked mirror) when it is rotated only 0.2 mrad
about an axis perpendicular to the mirror reflection surface
(the yaw angle). Dimensions in mm. (*f*) Experimental
scans across the horizontal and vertical foci. Results are
comparable with computer prediction (*c*), when the source
size is included. (*g*) First results: Laue diffraction from copper sulphate (100 μs). Hoorah! Reproduced with the permission of Daresbury Laboratory.

radiation. The acceptance of the mirror (the smile) is evident in the
unreflected beam.

    The quality of the focus can be seen to improve if the horizontal
acceptance is reduced from 2.5 mrad to 1.0 mrad; figure 5.18(*c*) shows the
improved focal spot. A three-dimensional plot of the focus is shown in
figure 5.18(*d*); it is quite easy to see the large concentration of radiation
falling into the focus.

The effect of mirror misalignments can be easily demonstrated. The mirror setting can deviate from ideal in several ways. It can be rotated about an axis perpendicular to the mirror (yaw) or about the optic axis (roll). The roll angle is not as serious as the yaw angle which is extremely critical to the size of the focal spot. Figure 5.18(e) shows the blow-up of the focal spot when the yaw mis-setting error is only 0.2 mrad. Hence, fine motor control is essential with this sort of mirror to permit the adjustment to the yaw angle of very small fractions of 1 mrad. The reason for this sensitivity to yaw is that the vertical radius of curvature is very quickly 'polluted' by the horizontal radius of curvature. Related to the problem of mirror mis-setting is the problem of movement of the source due to fluctuations in electron orbits. A 1 mm horizontal movement of the source is equivalent to 0.05 mrad of yaw for a mirror at 18 m from the tangent point. Such ray tracing simulations illustrate vividly the quality of focussing and the effect of mis-settings of the source or mirror from ideal. The actual performance realised with this mirror is shown in figure 5.18(f) and gives results comparable to figure 5.18(c), when the source size is included.

5.3.9 Thermal considerations

An evaluation of the effect of thermal loading of optical elements is achieved by using finite element analysis computer simulation. Such a computer program calculates what happens over a given number of time steps when a certain heat load is applied to the surface of an optical element such as a mirror. It takes into account the heat conductivity in the mirror bulk and the emissivity of the reflecting and other mirror surfaces. Once the system has reached the calculated equilibrium the thermal expansion is calculated.

5.3.9.1 Case study: thermal simulation of a toroid reflecting mirror
(Brammer et al 1988)

We can consider the fused quartz toroid mirror described above placed in the SRS wiggler beam. For the SRS at 2 GeV and 300 mA and the wiggler at 5 T there is 60 W mrad^{-1} of total energy in the beam. After passage through the beam line beryllium window this drops to $\approx$30 W (horizontal mrad)$^{-1}$. The critical angle of the mirror is set at 3 mrad so as to reflect wavelengths greater than 0.44 Å. For this 75 cm long mirror at 18.325 m from the tangent point at this angle, then, the vertical angular acceptance is 0.12 mrad, whereas the vertical angular aperture

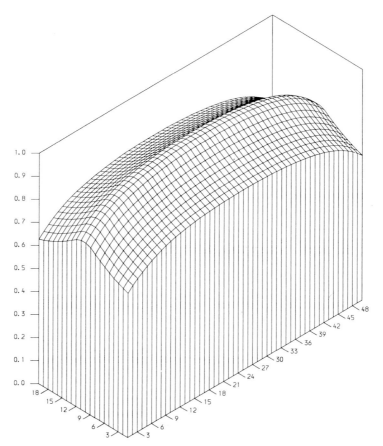

Figure 5.19 Thermal loading study of a toroid mirror (from Brammer *et al* (1988) with permission). The change in mirror thickness over the whole surface is due to thermal loading of the SRS wiggler. The vertical scale is given in micrometres, the *x* and *y* scales are in, arbitrary, 'computing cell' units.

of the beam is 0.36 mrad. The mirror surface therefore receives 10 W mrad^{-1} of which approximately 50% is absorbed, i.e. 5 W (horizontal mrad)$^{-1}$.

A simulation (Brammer *et al* 1988) showed that this mirror had a calculated temperature rise at the very centre of the mirror of 42 °C. Figure 5.19 shows the amount of increase of the mirror thickness as calculated. The maximum increase (between the centre and end of the mirror) is 0.2 μm. This value should be compared with the vertical bend displacement of the centre of the mirror of 13.5 μm for a vertical radius of 5221 m. It is possible to polish such a mirror to a flatness of $\approx$2.5 arcsec, i.e. over a length of 75 cm this is a displacement of 10 μm. This is considerably larger than the thermal effect.

The change in width of the mirror is $\sim 4\,\mu$m and the difference between the top and bottom surfaces is $0.6\,\mu$m. The total calculated length change is $13.5\,\mu$m along the centre of the top surface, $10.5\,\mu$m along the centre of the bottom and 10.4 and $9.7\,\mu$m along the top and bottom of the mirror's outer edges.

All these figures show that the mirror heating, at a 3 mrad grazing angle, should not be a problem. Obviously, on higher power magnets 2–10 kW mrad^{-1} power is emitted, instead of 0.06 kW mrad^{-1}. Thermal effects will then be very significant. Alternative materials to fused quartz will be obligatory. Even finer grazing angles will be needed to reduce the magnitude of the absorbed power; this, of course, creates increasing engineering problems due to the need for more and more precise mechanical alignment.

5.4 DETECTORS

X-ray photons used in crystal diffraction experiments usually have energies in the range 4–25 keV. The X-rays can be detected by a variety of effects which include:

(a) a chemical effect such as the reduction of silver halide to metallic silver in a photographic emulsion (the film method);

(b) scintillation in a crystal of, for example, sodium iodide (used in four-circle diffractometers);

(c) ionisation of a gas leading to the production of electrons and positive ions (the basis of multiwire proportional chamber, MWPC, area detectors);

(d) fluorescence effects whereby the X-rays are converted to visible light which are then detected by a secondary process such as a photocathode (the basis of phosphor coupled television detector systems or image plates);

(e) generation of electron-hole pairs in a semi-conductor (the basis of CCDs).

A useful review of X-ray position sensitive detectors is that of Arndt (1986).

For macromolecular crystallography the detectors that are used can be described as zero-, one- or two-dimensional. The large number of simultaneously diffracting reflections, even for a monochromatic beam and stationary macromolecular crystal, means that a two-dimensional detector is more efficient. Electronic area detectors are now commonly

Table 5.4. *Characteristics of detectors used for macromolecular crystallography at SR sources.*

Type	Pixel size (μm)	Aperture size (mm^2)	Number of pixels	Size of a resolution element i.e. point spread factor at 2% of peak intensity (normal incidence) (μm)	No. of resolution elements	Thickness of sensitive layer
Photographic film	25, 50, 100, 200	125×125	5000^2, 2500^2, 1250^2	grainsize ($\approx$5μm)	(As pixels)	50 μm
IP	100,200	200×200	2000^2, 1000^2	400	500×500	150 μm
MWPC	200–500	100×100, 250×250	500×500, 1000×1000	500	Parallax-dependent ($\sim$100×100)	1 cm
TV(FAST)	100	48×64	512×512	500	$\sim$80×80	20 μm
CCD (direct detector mode)	20	10×10 17×26	512×512 770×1152	40 40	$\sim$250×250 $\sim$385×576	30 μm silicon deep depletion
Single counter (scintillation)			1		1	

Notes:
(1) The ADC sets a limit to the local count rate e.g. for an eight bit ADC operating at video sampling rates for which the leading six bits are used, the maximum ADC reading is 64 recorded in a 20 ms time frame. Hence, for a gain set for weak data (e.g. 1 photon = 1 ADC unit) then a local count rate could be tolerated of 3200 cps. By reducing the gain, however, a much higher local count rate can be tolerated and strong reflection intensities monitored. Ideally, therefore two passes of the experiment have to be performed for each detector position. Each pixel is independent of any other pixel in terms of count rate.
(2) Using somewhat different phosphor thicknesses optimised for each wavelength.
(3) Determined by 16 bit deep memory; up to 10^7 possible if more bits provided.
(4) All characteristics on any given detector not always met simultaneously.

available on conventional X-ray sources (using Cu Kα). At the synchrotron the technical requirements are more formidable. This is due to the high flux, good collimation, small source size and wide wavelength range available. The bunch structure of the storage ring cannot be ignored when counting techniques are used (Arndt 1978). These difficulties have led to the widespread use of photographic film at SR facilities. In comparing different detectors it is necessary to characterise the

Spatial distortion	Count rate (cps)	Absorption efficiency (Wavelength)	Uniformity of response	Dynamic range at one setting of gain (at multiple gain settings)	Energy discrimination
None	∞	~60% (1.5 Å) ~40% (0.9 Å) ~8% (0.33 Å)	Uniform	~256	$\mu \sim \lambda^3$ only
None	∞	~100% (1.5 Å) ~18% (0.34 Å) ~44% (0.33 Å)	<±1%	~10^5	$\mu \sim \lambda^3$ only
Yes	10^6 global 10^4 local	1 atmos of Xe 80% (Cu Kα)	<±1%	~10^5 Note (3)	~20–50%
Yes	Note (1)	>80% (0.9–1.5 Å) Note (2)	Drops to 40% at edge	~100 (>10^3)	
None	Note (1)	40% (at 1.5 Å) 6% (at 0.71 Å)	<±3%	~10^5	10%
No	10^5	~100%		~10^5	40%

properties and behaviour of detectors in various ways. The detectors used for macromolecular crystallography at SR sources and their typical characteristics are given in table 5.4, in terms of various parameters. These include respective values for pixel size, geometric aperture, number of pixels, point spread factor (size of a resolution element), number of resolution elements, count rate, absorption efficiency and dynamic range.

5.4.1 Detector properties

These have to be characterised as follows:

   (i) detection efficiency;
   (ii) dynamic range;
   (iii) count rate (local and global);
   (iv) linearity of response;

(v)   uniformity of response over the active area;

(vi)   spatial distortion;

(vii)   stability of performance with time;

(viii)   energy resolution.

In addition, other important factors are the size and weight and the cost of the system. The flexibility of a system for various uses might also be important.

5.4.1.1 Detection efficiency

This is determined by both the absorption efficiency and the noise, if present, in the detector. The absorption or quantum efficiency of a detector is the fraction of photons absorbed in the active region of the detector. The factor should also be multiplied by a loss factor associated with any losses of photons in an entrance window or inactive layer. For a pure photon counter, in the absence of counting losses, these effects are the only ones of interest because the detector does not introduce any noise.

For detectors other than pure photon counters where the detector introduces some noise the detective quantum efficiency (DQE) is defined as

$$DQE = \left(\frac{S_o}{N_o} \right)^2 \bigg/ \left(\frac{S_i}{N_i} \right)^2 \tag{5.28}$$

where S is the total integrated signal, N is the rms integrated noise and the subscripts o and i refer to output and input respectively.

The DQE is a useful concept because it describes the noise characteristics of the input. Consider, for example, a photon source which obeys Poisson statistics. Suppose that a measurement is required of the output to an accuracy ϱ, i.e.

$$\varrho = N_o / S_o \text{ and } S_i / N_i = M / M^{\frac{1}{2}} \tag{5.29}$$

where M is the number of counts in the input signal of standard deviation, $M^{1/2}$, then

$$\varrho = 1 / (M \cdot DQE)^{\frac{1}{2}} \tag{5.30}$$

For a source delivering a fixed number of photons per unit time, m, then the time, t, needed to make a given measurement of M counts to a desired accuracy ϱ is

$$t = 1/(m \cdot DQE \cdot \varrho^2) \tag{5.31}$$

In the ideal case of a noiseless detector of 100% absorption efficiency the DQE=1. For a noiseless detector but with an entrance window which attenuates the input and/or with a fractional absorption efficiency, then the DQE decreases from 1 to ε where $0<\varepsilon<1$. In this case, the time needed to make a measurement to a desired accuracy, ϱ, for a fixed count rate onto the specimen, increases as ε^{-1}. If, in addition, the detector contributes a noise equivalent to the signal given by η photons, then the DQE is further reduced according to the relation:

$$DQE = \varepsilon M / (M + \varepsilon \eta^2) \tag{5.32}$$

Hence, the exposure time for a given incident count rate and required accuracy is lengthened further.

A specific example will be useful. The DQE is graphed in figure 5.20 for Kodak No-Screen X-ray film (from Gruner, Milch and Reynolds (1978));

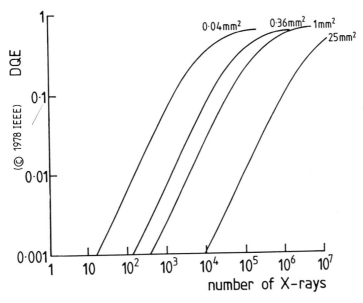

Figure 5.20 The DQE of photographic film (Kodak No-screen) for 8 keV X-rays. The number near each curve indicates the size of the integration area. From Gruner *et al* (1978) with permission, © IEEE.

assuming (*a*) ε to be 0.6 (i.e. for Cu Kα wavelength), (*b*) 10^7 fog grains cm^{-2} with a Poisson statistical distribution and (*c*) each photon absorbed in the emulsion renders one film grain developable. It is seen that at a given dosage, the DQE rises as the integration area falls. The curves quantitatively justify intuition, namely, that weak, diffuse X-ray signals are best captured by confining the signal to a small area, i.e. in the case of diverging X-ray reflections by using a short specimen to film distance. In a similar way it is possible to see immediately a benefit of collimated SR versus an uncollimated conventional X-ray source. For a fixed count in a given diffraction spot the low divergence of the synchrotron beam generates a smaller diffraction spot, e.g. 0.2×0.2 mm^2 ($=0.04$ mm^2) instead of 0.6×0.6 mm^2 ($=0.36$ mm^2) with a conventional X-ray source. The enhancement of the DQE of film used to measure SR patterns, by approximately an order of magnitude, is immediately obvious from figure 5.20. The small diffraction spot sizes achievable with SR also indicate how well ordered protein crystals actually are, i.e. have a small mosaic spread (see figure 2.8(*c*)). Photographic film as a detector is discussed further in section 5.4.2.2.

5.4.1.2 Dynamic range, count rate and linearity of response

These terms refer to the minimum and maximum number of photons per second that can be measured. The dynamic range can be defined as the ratio of the largest to smallest dosage for which the measurement is better than the desired accuracy. Alternatively, it is the range of dose for which the device DQE exceeds a given DQE (corresponding to a given accuracy) in a given measurement. For a discussion see Gruner *et al* (1978). The minimum is determined by the inherent noise of the detector. In the case of a counting device a single X-ray photon can produce a large pulse compared to other effects. For an analogue device it is possible for thermal noise or chemical fog to render single photons unmeasurable. The effect of noise on the weakest measurements can be reduced by digital integration of more than one frame in an electronic area detector. For N summed detector read-out frames the noise grows as $N^{1/2}$ whereas the signal grows as N. Hence, for N summed frames the dynamic range is extended because weaker signals are recorded at a better DQE.

The maximum intensity at which a counter can operate is determined by the dead time set by the electronics. In the case of an analogue device such as the television detector the detector can saturate in terms of the cumulation of charge in a picture element (pixel). The gain of the system can be reduced to allow these strong signals to be measured but to the detriment of weak signals. In the case of film the stronger spots saturate the film which is another way of saying that the response becomes highly non-linear.

It is necessary to distinguish between local and global limits to the maximum recordable count rate. In analogue detectors each pixel behaves as an independent detector. In principle, in these cases, all pixels are capable of recording simultaneously at the maximum intensity. In other devices, such as MWPCs with delay line read-out, the whole detector becomes dead after a photon is absorbed *anywhere* in the detector. The term global counting rate is used to describe this. In addition with gas detectors the ionisation of the gas may saturate due to space charge effects in the case of a strong diffraction spot striking a small volume of gas. This leads to local counting rate problems.

5.4.1.3 Uniformity of response

All position sensitive electronic detectors show long range and pixel-to-pixel variations of response to larger or smaller extents. Calibration of these effects is usually done by providing a smooth and constant illumination across the detector face of X-ray photons with an energy as close as possible to the one of interest for use. The problem with the calibration is trying to achieve a uniform illumination, since a point source would need to be placed infinitely far away from the detector to achieve it, and at a high number of counts captured per pixel. In practice a radioactive Fe^{55} source is placed at a known distance, about 15 cm, from the detector and due allowance for the non-uniform illumination is made. Instabilities in the detector electronics can result in changes of the response efficiency over the detector. This is a fault condition which can be revealed by making repeated measurements of the uniformity of response. The errors in this correction define perhaps the ultimate limit of measurement accuracy with electronic area detectors and therefore their sensitivity in variable λ anomalous dispersion methods (see chapter 9). Uniformity of response is not a problem with a single scintillation counter, film, the CCD with direct detection or with the image plate.

5.4.1.4 Spatial distortion

Electronic area detectors tend to suffer from spatial distortions. The effects can be calibrated for by placing a regular array of pinholes over the detector and illuminating it with X-ray photons from a radioactive source. The detector produces its own array of measured spot positions which deviate from the positions of the holes in the actual array. Hence, analytical or empirical corrections can be made; these need to be at an accuracy level of a fraction (say 0.1–0.2) of a pixel. These distortions also cause variations in response in the sense that some pixels are inherently larger and can therefore 'gather up' more photons than a smaller pixel. This is corrected for in the procedure described in section 5.4.1.3 above. It is essential that the spatial distortion correction is (reasonably) stable with time which is also true of the uniformity of response correction.

5.4.1.5 Energy discrimination

Higher energy photons, once absorbed, deposit more energy than low energy photons. This allows different energy photons to be discriminated and allows a reduction in the background and therefore the variance of the measurement. In table 5.4 the energy discrimination or resolution available for the CCD is 10% and for the MWPC and the scintillation counter $\approx 40\%$. An alternative kind of energy discrimination is available with film or image plates, namely the absorption efficiency varies with wavelength (i.e. the linear absorption coefficient varies as λ^3). Hence, by placing several films one behind the other it is possible for harmonic reflections $(\lambda, \lambda/2)$ from a sample to be 'resolved' by using the measured spot intensity on successive films (see chapter 7). The same applies for the image plate but since each plate is more opaque to X-rays only two plates can be placed in tandem.

5.4.2 Commonly used detectors for crystallography

These are as follows:

(1) scintillation counter;
(2) photographic X-ray film and microdensitometer;
(3) multiwire proportional chamber;
(4) television detector;
(5) CCD;
(6) image plate and scanner.

5.4.2.1 Scintillation detectors

These are used on single counter four-circle diffractometers. The detector is often sodium iodide. The scintillator is used in conjunction with a photomultiplier tube. Thallium activated sodium iodide has an energy resolution of about 40% at 10 keV. Hence, the attractiveness of such a detector lies not only with its counting of individual photons but also its ability, with a pulse height analyser, to discriminate between photons of different energies. Additionally, there is obviously no need for a uniformity of response correction – there is only one resolution element! For macromolecular crystallography the disadvantage is that only one of the many simultaneously diffracting reflections can be measured at any one time. However, new diffractometers with very rapid mechanical settings compensate, to some extent, for this limitation.

5.4.2.2 Photographic X-ray film and microdensitometer

Film is a widely used area detector in X-ray crystallography. It is a very versatile recording medium which allows reasonably accurate measurements to be made. In macromolecular crystallography it has enjoyed a resurgence because of its very high spatial resolution making it the only suitable area detector for many years for largish unit cells. For virus and Laue crystallography, film still occupies a pivotal role because of the high density of spots on the pattern. It gives measurements accurate enough for isomorphous replacement but anomalous dispersion measurements are somewhat limited when recorded with film.

Quantitative measurements are made by means of an automatic scanning microdensitometer. The optical density, D, is obtained in a given small area (raster) of the film by measuring the transmitted, L_T, and incident, L_0, visible light intensities through the film where

$$D = -\ln (L_T / L_0) \tag{5.33}$$

The intensity of a given diffraction spot is given by the integral over the area of a spot. It is essential that the raster area is small enough that D is constant within it. If this is not adhered to, then an error is introduced to the stronger spots as pointed out by Wooster (1964); this arises because the logarithm of the average is not the same as the average of the logarithm and is known as the 'Wooster effect'.

For X-ray photon energies greater than about 2 keV a single photon suffices to render a grain of the emulsion developable. Hence, the standard deviation of an optical density measurement is independent of the grain size of the film (Arndt, Gilmore and Wonacott 1977). For most X-ray films one optical density unit corresponds to about 10^6 grains mm^{-2} (Morimoto and Uyeda 1963). Hence, smaller diffraction spots (say 0.2 mm instead of 1 mm) represent more efficient use of the available number of photons in terms of statistics. This is how the collimation of SR contributes directly to making film intensity measurements more efficient compared with a more divergent, conventional X-ray source. There is a limit to how small spots can be made, ultimately set by the grain size. Hence, intensities of spots as small as $50\,\mu$m or less are increasingly prone to error. Film is of limited use, therefore, for microcrystal intensity measurements compared with, say, the MWPC or television and phosphor detector.

The optical density range of X-ray film is limited between ~0.1 and 2.5. The lower level is set by chemical and cosmic ray fog, which gets worse as the film ages and so determines its useful lifetime of about 6 months (at 4°C). The upper level is set by saturation of the emulsion. Usually, the intensities of spots are digitised in a range of 0–255. To extend the dynamic range several films can be used together in a film pack (or more recently a toastrack arrangement for Laue diffraction). Multiple film arrangements work because the absorption efficiency of film for X-rays, in the wavelength range between 0.6 and 1.8 Å, lies between 20 and 70%. Hence, with multiple films the weaker reflections are exposed on the (front) film while stronger reflections are measured on the lower (rear) films after suitable attenuation. The attenuation or film factor of a double-sided X-ray film is given by

$$I_0 \, / \, I_T \; = \; \exp \mu t \tag{5.34}$$

where I_0 is the incident X-ray intensity, I_T the transmitted intensity, μ is the linear absorption coefficient and t the active film thickness. The absorbed intensity I_A is $I_0 - I_T$ and the absorption efficiency is therefore

$$I_A \, / \, I_0 \; = \; (1 - I_T \, / \, I_0) \tag{5.35}$$

Photographic film contains silver bromide grains and therefore the absorption efficiency jumps at each absorption edge (0.92 Å Br K edge, 0.49 Å Ag K edge). The absorption efficiency of film used for X-ray work is, at 1.5 Å, 0.93 Å and 0.91 Å, 60%, 21% and 44% respectively. The corresponding film factors are 2.5, 1.27 and 1.79. Figure 5.21 shows the variation of these quantities with wavelength. Clifton *et al* (1985) have measured the XANES for photographic film at the bromine and silver edges.

The additional properties of film include the following. It is available in a range of sizes and can be bent to form a cylindrical surface. Microdensitometers usually can handle sizes up to 10"×8". Film has a very high spatial resolution (~12.5–25 μm) and microdensitometer raster

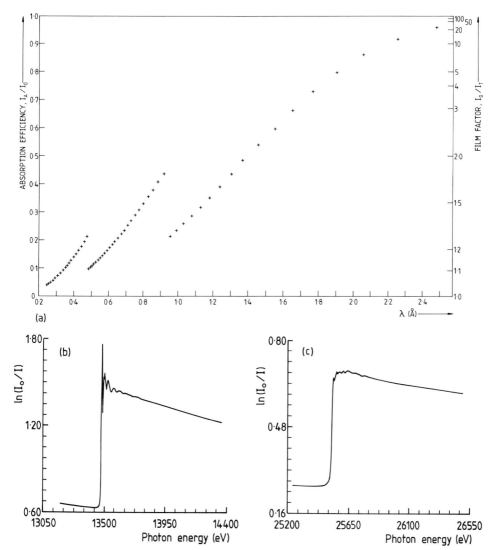

Figure 5.21 Photograph film absorption efficiency (and corresponding film factor) as a function of wavelength. (a) Calculated values (Kodak DEF) over a broad range of wavelengths. Data kindly supplied by A. LeGrand. Figure prepared by S. Weisgerber. (b) Measured at the Br K edge (CEA film). (c) Measured at the Ag K edge (CEA film). (b) and (c) from Clifton et al (1985) with permission of IUCr.

sizes usually available are 25, 50, 100 or 200 μm. A film has a uniform response over its area, when properly developed, and therefore does not need any calibration. Spatial distortion effects are present to a certain extent due to occasional bulging of the film in a cassette. Oblique incidence of a spot onto a film produces an elongation of a spot in the radial direction but is easily calculable. There are no counting rate limitations. By translating the film it can be used as a time resolved or streak detector. A film exposure represents a permanent visible record of a diffraction pattern.

The major disadvantage of film is with respect to the accuracy of the measurement of an individual spot intensity. For the strong reflections the spot is attenuated by perhaps several films. The attenuation factor for each film has to be determined individually owing to slight variations in emulsion thickness and darkroom development. For the weak reflections recorded on the top (front) film the chemical fog may be as large as the weak reflection optical density and hence the DQE for these reflections is deleteriously affected by this fog. Additionally, one can mention that, because of the tediousness of darkroom work, as few films as possible are used. Hence, in the rotation method, for example, as large a rotation angle as possible is used. The rocking width of a reflection is therefore perhaps one-tenth of the rotation angle of the exposure and the X-ray background accumulated over a spot is larger than it needs to be. However, in the case of SR the background tends to be quite low in any case.

Photographic film has been and is used for a variety of macromolecular data collection. It is being replaced by use of other detectors. However, its use for very dense diffraction patterns from viruses and/or Laue patterns is likely to continue.

5.4.2.3 Multiwire proportional chambers (MWPCs)

An MWPC is a chamber containing a large number of anode wires each acting independently as proportional counters. The geometry of the cathode is, however, somewhat different from that of a standard proportional counter. The principles of MWPCs are described by Faruqi (1977).

The detection of photons in a proportional counter relies on the photoelectric effect whereby a photoelectron and an ion are produced on absorption of a photon. An 8 keV photon produces on average 320 ion pairs. The ions drift towards the cathode and the electrons drift towards the anode.

Due to their lower mass the electrons gain energy far more rapidly and then undergo inelastic collisions. At each inelastic collision the number of electrons increases by a factor of 2. If the average number of collisions is n the amplification factor is 2^n; an avalanche initiated by one electron thus consists of 2^n final ion pairs. The output pulse heights are proportional to the incident photon energy.

The MWPC usually consists of three planes of wires as shown in figure 5.22(a). One set of cathode wires is parallel to the anode wires, the wires in the other cathode are orthogonal to it. At the position where an X-ray photon is absorbed and gas ionisation occurs, the resulting electrons drift in the direction of the electric field towards the anode and gas multi-plication takes place. A negative pulse is formed on the anode. Simultaneously, positive pulses are induced on a few of the neighbouring cathode wires in both the cathode planes.

The two-dimensional position of the incident photon can be extracted from the centroid of the induced pulse-height distribution in the two cathode planes. There are two main methods for position determination, the delay line method and the amplifier per wire method.

In the delay line method all the cathode wires are isolated from each other and coupled to a delay line with a well-defined propagation

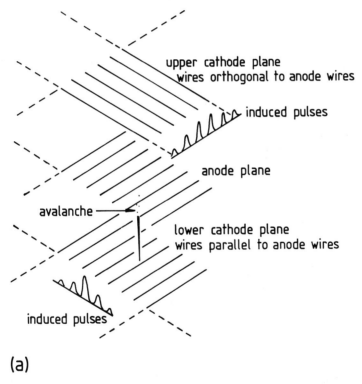

(a)

Figure 5.22 (a) The MWPC usually consists of three planes of wires. The initial avalanche towards the anode plane is accompanied by induced pulses on both sets of orthogonal cathode planes. From Faruqi (1977) with permission. (b) Protein crystal diffraction pattern recorded on an MWPC (flat chamber, 1 atm gas pressure).

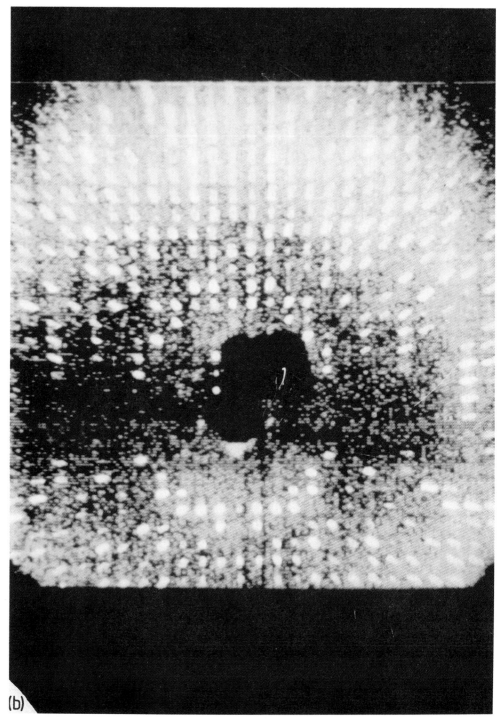

(b)

Figure 5.22 (*cont.*)

velocity. The time delay between the anode pulse (the prompt pulse) and the time when a pulse emerges from the end of the delay line gives a coordinate, say x, of the absorbed photon. Another measurement on the orthogonal delay line gives the second coordinate (i.e. y). The MWPCs of Xuong *et al* (1978), in San Diego, which have been used extensively for macromolecular crystallography (see appendix 2), employ this method.

In the amplifier per wire method the pulses on the orthogonal cathode planes are extracted independently from each wire and the centroid position in x and y is thereby determined. The main advantage of this system is a higher count rate capacity ($\approx 10^6 \mathrm{s}^{-1}$) compared with use of delay lines ($\approx 10^5 \mathrm{s}^{-1}$). This method is used by Kahn *et al* (1982b) at LURE for example.

The overall counting rate in either method is essentially set by the 'dead' time associated with encoding the position of each and every detected photon. It is applicable to the whole of the detector and is referred to as the global counting rate of the detector. There is also a local counting rate whereby a strong reflection can saturate the absorbing gas; this is about 10^4 counts $\mathrm{s}^{-1} \mathrm{mm}^{-2}$ and is determined by space charge build-up in the local region of absorbing gas. This is one of the major limitations of MWPCs at SR sources where, because of the strict collimation of the beam, the reflections have very small sizes. This was an advantage for an analogue device like film (see figure 5.20) but is a disadvantage here!

Another limitation of MWPCs is that photons entering a flat chamber at an angle 2θ to the plane of the detector suffer from a 'parallax' error (figure 5.22(b)). If the sensitive depth of the chamber is g, then the parallax effect is of size $g \tan 2\theta$. For a 10 mm gap and $2\theta = 20°$, then $g \tan 2\theta$ is 3.6 mm. There are three solutions to the parallax problem. Firstly, two or more flat chambers can be used whereby each can tilt towards the sample, on either side of the direct beam. If the tilt angle is α, then the parallax becomes $g \tan(2\theta - \alpha)$. This is the method employed by Xuong *et al* (1978) and is a commercially available system. There are plans to install a second chamber in such a fashion at SSRL, Stanford.

A second possible solution is for the absorbing gas to be pressurised so as to increase the chance for photons to be absorbed within the first few millimetres after entering the chamber. This is the method used in the Xentronics MWPC which is commercially available from Siemens–Nicolet Ltd whereby the gas (xenon) is held at 4 atm behind a curved beryllium window. The point spread factor of the Xentronics detector is ~1 mm at the 2% level of intensity. Since the detector has an aperture of

10 cm there are therefore ~100×100 true resolution elements in this device (see table 5.4 for a comparison between this and other detectors).

The third possible solution is to incorporate a spherical drift space whereby the primary ionising event takes place in a low field region where no avalanching takes place. The electrons drift through a spherically symmetric electric field being brought at normal incidence into the region of the orthogonal cathode and anode wires where gas amplification takes place. This arrangement is known as a 'spherical drift chamber' (Charpak 1982) and is employed at LURE for macromolecular crystallography (Kahn *et al* 1982b, 1986, 1989) – see figure 5.23.

To summarise, the advantages and disadvantages of MWPCs are as follows. The advantages are that true photon counting is realised and there is no detector noise. The noise in the system is set by the very low count rate of the cosmic ray background! The absorption efficiency of 1 cm of xenon at 1 atm is ~80% at 8 keV. MWPCs have been operated over a range of 12–6 keV in photon energy. The chamber is usually relatively uniform in response over the chamber area, compared say,

(a)

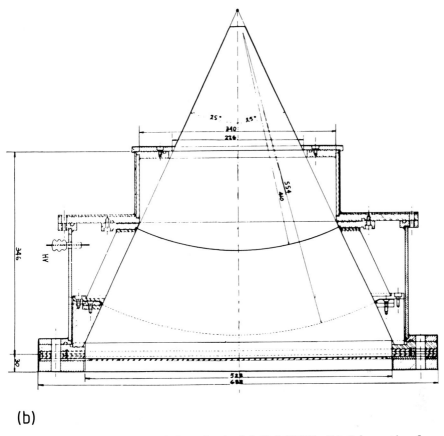

(b)

Figure 5.23 (a) A view of the spherical drift MWPC. (b) Schematic of
the system. From Kahn et al (1986) with permission.

with television area detectors. However, an MWPC response correction
between data frames has been found to be important in experiments at
SSRL, Stanford (Guss et al 1988). The disadvantages of an MWPC are
problems of parallax with flat, 1 atm chambers and fairly limited count
rates. Spot smearing due to parallax gets worse at short wavelengths as
does the absorption efficiency. At SR sources the only detectors successfully operated at wavelengths less than 1 Å are the television area detector at Daresbury (now also Hamburg) and the image plate system at
Hamburg – see next sections. The importance of short wavelength uses in
macromolecular crystallography is dealt with in chapter 6.

MWPCs have been successfully used at LURE and Stanford for multiple
wavelength anomalous dispersion phasing measurements (see chapter
9). An MWPC has also been used to record Laue diffraction patterns at
VEPP-3 (Gaponov et al 1989a,b,c).

5.4.2.4 *Television detector with external phosphor*

Much development has gone into quantitative measurements with area detectors in which the diffraction pattern is formed on an external phosphor which is then fibre optically coupled to a low-light-level television camera (Arndt 1982, 1986; Arndt and Thomas 1982; Kalata 1982; Milch, Gruner and Reynolds 1982; Thomas 1982a,b). Figure 5.24 shows a schematic of the fast-scanning television area detector (from Arndt (1985)) which is the basis of the commercially available Enraf–Nonius system (the FAST). Outlines of how television detectors can be used with SR sources are given by Arndt and Gilmore (1979), Arndt (1984) and Arndt (1990).

The television camera usually embodies an image intensifier coupled via demagnifying optics to a sensitive television camera tube. The light image is converted into a charge image on the target of the camera tube, which is then scanned by an electron beam in the form of a raster containing 625 lines with a field repetition frequency of 25 Hz (in Europe). The output video signal is presented sequentially pixel by

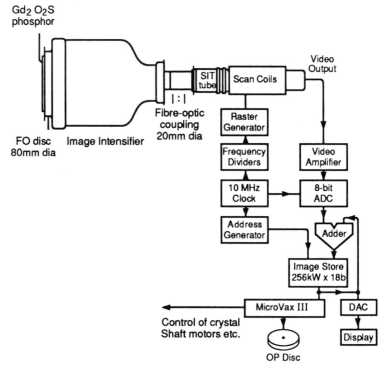

Figure 5.24 Fast-scanning television area detector. From Arndt (1990) with permission of the author and Gordon and Breach Science Publishers SA.

pixel. If this signal is sampled 512 times per line the resultant read-out time per pixel is 100 ns and the sampling frequency is 10 MHz. The period of integration in the camera target is 40 ms. The accumulation of a statistically significant number of X-ray photons usually requires a longer time period. There are three methods of doing this. The fast-scan, slow-scan and counting methods.

Firstly, in the fast-scan method many television frames are digitally added into an image store. Usually 1000 frames are used to make up one image. The range of intensities that can be digitised is set by the number of bits in the ADC (analogue-to-digital converter) working at video rates, e.g. if six bits are used this gives a 'dynamic range' of 64:1. The effect of scanning many frames is to improve the accuracy within this range. For example, if a pixel returns values of, say, 32, 31, 34, 33, etc, up to 1000 measurement values, then on *average* the intensity reading in that pixel is, say, 32.15. To examine stronger or weaker intensities, i.e. outside the basic range of one image of 64:1, the gain of the image intensifier and camera have to be varied.

Secondly, in the slow-scan method the scanning electron beam of the camera is switched off for an exposure period, e.g. as long as 1000 s. During this period an integrated charge is formed on the camera tube target, which is then read out in the usual raster scan fashion. In this method (Milch *et al* 1982) a slow scan is then made using about 20 s per image; the read-out circuits can thus have a narrow bandwith and produce a very good signal-to-noise ratio. The main disadvantage of the system is that it has a low duty cycle (i.e. ratio of exposure time to elapsed time).

Thirdly, digital mode is used when the X-ray intensities are very low. Specially designed circuits detect the charge image produced by a single X-ray photon (Kalata 1982).

The X-ray phosphor is a critical component of these systems. It must have the following properties:

a high X-ray absorption efficiency;
be thin in the interests of fine spatial resolution;
be uniform;
a high fluorescence conversion efficiency;
a short decay time constant (e.g. for Gd_2O_2S this is $\sim$1 ms)

Needless to say some of these requirements are conflicting; e.g. the thinner the phosphor the less efficient it is. A particular advantage of this sort of system is that it relies on commercially available components.

Overall these systems have the advantage of possessing a high absorption efficiency (>50%) even at short wavelengths (e.g. <1 Å) and suffer from negligible parallax. The point spread factor at the 2% level on the FAST system is ≈0.5 mm giving about 100×100 true resolution elements over the detector aperture (table 5.4). The system can accommodate high count rates (and the sharp diffraction spots produced with SR). There are some disadvantages to the system. There is a limited dynamic range of acceptance of intensities (64:1) although this 'window' can be reset by changing the gain of the system; an accurate measurement per pixel is achieved by, for example, scanning many frames. There is a relatively small overall aperture of 64 mm×48 mm; only at short wavelengths can the detector be set well back from the crystal to reduce the background under a spot. There is a large non-uniformity correction (50% at the edges to 100% in the centre, relatively speaking). Although this is correctable it does obviously degrade the absorption efficiency over a large fraction of the detector surface.

The FAST system (of Enraf–Nonius) is installed at the Daresbury wiggler and on DORIS in Hamburg. The Daresbury system has been especially useful for radiation sensitive macromolecular crystals where the combination of the short wavelength beam and the much higher sensitivity of the detector over photographic film is particularly beneficial (Glover, Helliwell and Papiz 1988). It has also made possible microcrystal studies of small molecules where the short wavelengths are essential to measure reflections to high resolution and where the high intensity of the beam compensates for the small crystal volume (Andrews et al 1988). The Hamburg system has been used successfully to record protein crystal Laue diffraction patterns; here the high count rate of the system, the high absorption efficiency over a wide wavelength range (0.5–2.0 Å) and lack of parallax were particularly beneficial (Bartunik and Borchert 1989).

5.4.2.5 CCD

The CCD is a semi-conductor detector. It can be used as an X-ray detector (for reviews see Allinson (1982), Milch et al (1982) and Allinson (1989)) either by direct illumination of the X-rays onto the silicon or by conversion of the X-rays in a phosphor to visible light which is then incident onto the silicon. The previous section dealt with phosphor coupled systems based on television cameras. The CCD can replace the television camera. This is the basis of the system being developed by Strauss et al (1987) and Westbrook (1988); see figure 5.25. An alternative is to use direct illumination onto a so-called deep depletion CCD. CCD area detectors usually come with 512×512 pixels each of ~10 μm×10 μm or up to 27 μm×27 μm, i.e. overall area 5×5 mm^2 or up to ~14×14 mm^2. Larger devices have been made, as specials, such as the Tektronix 2000×2000 pixels device, with pixel size 27 μm. The main limitation therefore is that they are small to the extent that in a routine data collection system the sample and source sizes would need to be scaled down to a point where the pattern size and image content (diffraction spots!) match the area available and the pixel sizes. However, many manufacturers are competing with each other to make CCDs and so prices are falling. Although it is true that the general trend is in the production of even smaller devices (miniature televisions, etc), nevertheless companies are providing a wide range of products. Hence, CCDs for X-ray diffraction applications are under development, especially at the Argonne National Laboratory (Strauss *et al* 1987; Westbrook 1988, also by Gruner pers. comm.) and in Europe (Allinson, Colapietro, Helliwell, Spagna and co-workers). Preliminary experiments with a CCD to image a part of a Laue pattern in 80 ms for time resolved Laue diffraction from protein crystals have been published (Allinson *et al* 1989).

Since the CCD is one of the most promising X-ray diffraction detectors of the future we will now discuss it in more detail. The photons impinging on a CCD generate a set of free electrons and holes in a target pixel (figure 5.26). If a small voltage is maintained on this pixel, these charges separate to yield an electric charge on the pixel that is proportional to the energy of the incident photon. A CCD array consists of two registers;

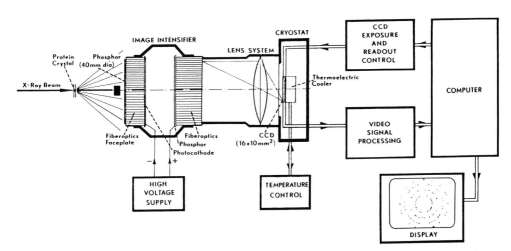

Figure 5.25 The CCD phosphor-based system being developed by Strauss *et al* (1987) and Westbrook (1988) with permission.

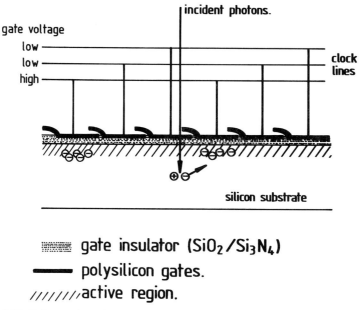

Figure 5.26 Schematic of the operation of the CCD. From Allinson (pers. comm.) with permission.

a horizontal (or serial) register and a vertical (or parallel) register. The parallel register is the two-dimensional array of photosensitive pixels, each of which contains a measurable charge to be read. The serial register is used to unload these charges during read-out. After exposure of the parallel register to light and its consequent development of a charge distribution, the image, represented as an array of charges, is shifted out by repeatedly shifting rows of the parallel register onto the serial register, which can be read out by shifting its charges serially into an ADC. When the serial register is empty, it is refilled by shifting all rows of the parallel register down one pixel. The slowest step in reading the CCD array is the ADC. A 12-bit ADC could process 5 million pixels/s (i.e. a 512×512 array in 50 ms).

Each pixel has an upper limit to the charge it can hold. This capacity is generally about 10^6 electrons and increases proportionally to pixel physical size. With direct detection a 17 keV photon will generate about 200 or so electrons (i.e. 3.65 eV per electron–hole pair and about 5% absorption efficiency for a 30 μm deep depletion device – this efficiency is somewhat less than that of photographic emulsion at this photon energy). Hence, one pixel could accommodate about 5000 17 keV photons before saturation.

At this point one can highlight the divergence of designs into (a) use of X-ray to visible light phosphor conversion schemes (Strauss *et al* (1987) and figure 5.25) and (b) direct detection schemes (Allinson (1989) and figure 5.27). In (a) the absorption efficiency is very good but at the expense of a poorer point spread factor (<0.5 mm) and introducing a phosphor decay time constant. In (b) the absorption efficiency is low but the point spread factor is very small (<0.1 mm) and, of course, compared with film the image is live! Scheme (a) has been used to gather up the whole of a less dense image. Scheme (b) has been used to look at parts of dense images such as a Laue pattern. The CCD is referred to also in section 5.4.3 on detectors for time resolved crystallography.

5.4.2.6 Image plate/storage phosphor and scanner

In this detector scheme there is a phosphor screen which temporarily stores the X-ray image and an image reader or scanner which converts the latent X-ray image stored in the phosphor into a digital signal (via

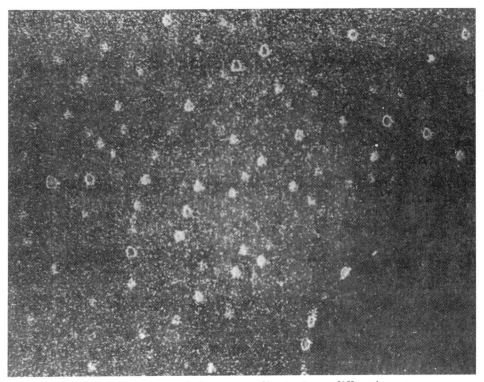

Figure 5.27 Protein crystal (concanavalin-A) Laue diffraction pattern recorded on a CCD detector with direct detection. From Allinson *et al* (1989) with permission.

an ADC) per pixel (figure 5.28). There are two main commercial companies that have conceived and developed this method, primarily for
diagnostic radiography. These are, in alphabetical order, Fuji (who use
the name 'imaging plate') and Kodak (who use the name 'storage phosphor'). A patent dispute between these two companies, now resolved,
held up the free sale of these materials for several years. We will refer to
the sensitive screen as the IP (image plate).

The IP sensitive medium is composed of fine photostimulable phosphor
crystals, BaFBr:Eu^{2+}, typically $150\,\mu m$ thick combined in an organic
binder. This phosphor is able to store a fraction of the absorbed incident
energy in the form of F-centres (also known as colour centres) when
irradiated by X-rays, ultra-violet light, electrons, etc. The stored image
decays with time with a half-life of approximately 8 hr (at room
temperature). On illumination by the visible light (e.g. from a laser) the
trapped electrons at the F-centres are liberated to the conduction band,
return to the Eu^{3+} ions, converting them to excited Eu^{2+} ions which then
relax and emit light. The emitted light is in the range where the
quantum efficiency of a photomultiplier tube is high. After laser illumination there is a residual X-ray image in the IP which has to be properly
erased before the next exposure. This is done by exposing the plate, for
about a minute, to a dose of visible light. Figure 5.28 outlines the
principles of the detector system.

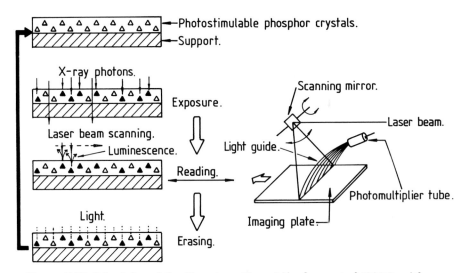

Figure 5.28 Principles of the IP system. From Miyahara *et al* (1986) with
permission.

A typical size of an IP is $200 \times 250 \, mm^2$. The phosphor is deposited on a *flexible* plastic plate. The point spread factor is about 0.4 mm at the 2% level. Hence, the number of active resolution elements is very large ($\sim 500 \times 500$). The absorption efficiency for X-rays is high (figure 5.29(a)) and there is virtually no chemical fog akin to film. Figure 5.29(b) compares the (noise/signal)$_{out}$ versus exposure level for film with that for the IP at 8 keV and 17 keV respectively. At high intensities ($\geq 10^4$ photons $(0.1 \, mm)^{-2}$) this ratio for the two is about the same. At low intensities (e.g. $\leq 10^3$ photons $(0.1 \, mm)^{-2}$) the IP performs much better. Hence, for identical exposure times and diffraction pattern intensities, a much better, i.e. more precise, measurement can be made for the weaker data which is a marked fraction of a typical protein crystal data set. Alternatively IP exposures can be only one-tenth of those with film. This is a somewhat misleading statement however. For example at 8 keV and 10^4 photons $(0.1 \, mm)^{-2}$ the exposure times for film and IP are the same for the same (noise/signal)$_{out}$. At 17 keV, however, there is a gain of ~ 2 for 10^4 photons $(0.1 \, mm)^{-2}$.

The IP is considered to be uniform in response and free from spatial distortion. Descriptions of the IP/SP and its applications can be found in Miyahara *et al* (1986) and Amemiya *et al* (1988a).

The IP is certainly a very important replacement of photographic film. The IP combines a large aperture, fine point spread factor (free of Wooster effect) and a very high dynamic range ($\geq 10^5$). Compared with film it is free from chemical fog, does not need wet/darkroom processing, is reusable and is more absorbent, especially at short wavelengths. Like film it is, however, an 'off-line device' – it needs scanning, a process which takes about 2–15 min (depending on the manufacturer). The light collection efficiency of the scanner is crucial; light can easily be lost by factors of up to 5. This reduces the effectiveness of the medium considerably.

The first protein structure to be solved using IP synchrotron data was that of ω-amino acid:pyruvate aminotransferase using Weissenberg geometry at the Photon Factory; the scanner used was the BA-100 of Fuji (Watanabe *et al* 1989).

An IP reader has been constructed in Hamburg and is now used routinely on the X-11 and X-31 synchrotron instruments of EMBL for macromolecular crystal data collection. More recently, Rigaku have announced a dual IP scanner system for macromolecular crystallography.

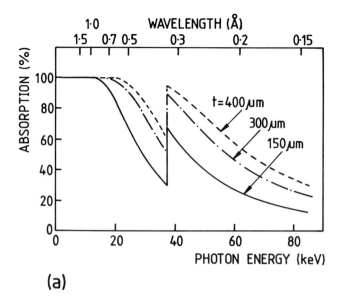

(a)

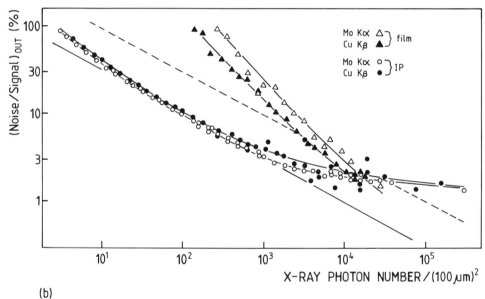

(b)

Figure 5.29 The IP. (*a*) The absorption efficiency as a function of
wavelength and phosphor thickness, *t*. From Amemiya *et al*
(1988a) with permission. (*b*) (Noise/signal)$_{out}$ versus
exposure level for film and IP. The dashed line is for a hypothetical photon counter of 10% efficiency. The solid line is
for an ideal detector. (*c*) Measured line spread functions
along (i) the laser scanning direction, and (ii) the IP scan-

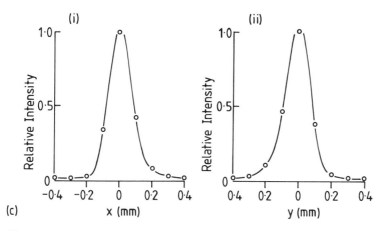

(c)

Figure 5.34 (cont.)
   ning direction (PSL=photostimulated luminescence). (b)
   and (c) From Miyahara et al (1986) with permission.

5.4.3 Possible detectors for time resolved crystallography

One of the most challenging requirements for detectors is in the area of time resolved crystallography. Potentially, in an extreme case, colossal volumes of image data will be produced. For example, for a 1 ms time resolution, a time course of 100 ms and an image of at least 512×512 pixels there will be at least 26 million pixels. Time resolutions of 1 μs are contemplated at APS and ESRF for a broad bandpass Laue pattern (see chapter 7). On CHESS an undulator source was used to produce a narrow bandpass Laue pattern in 120 ps. These are daunting time scales and data volumes.

Realistically, one can presently imagine looking at only a small part of a changing diffraction pattern, e.g. 16×512 pixels with the time dimension dealt with in the electronic hardware (electronic detector case) or mechanical translation. An interesting point in time can then be identified during a time-dependent process such as a reaction in a crystal. The reaction in the crystal (different sample) can then be repeated by using a large aperture detector to record the Laue pattern at key times. To map several key positions on a time course would then require further repeats of the experiment. Clearly, this is a challenging experiment. With slow reactions (>hours) catalysis in the crystal has been studied with the rotation method of data collection and diffusion of reactants into the crystal (Hajdu et al (1987a); see chapter 10). With faster reactions the Laue method is needed so as to reduce the exposure time. Reactions can be initiated with a light flash (Schlichting et al 1989, 1990).

Reactions can be monitored in the crystal by visible spec-
trophotometry, provided a chromophore is involved in the reaction.
Because this will not generally be the case there is a need for a time
resolved detection of the diffraction pattern, or at least, part of it. We
now describe the options currently envisaged or realised.

5.4.3.1 CCDs

In the CCD (described in section 5.4.2.5) charge shifting between
pixels can be done extremely rapidly ($1\,\mu s$) and quantitatively (essentially without loss or noise). Because these shifts can be performed in
parallel in all columns simultaneously, images can be moved with great
speed from one part of a CCD chip to another. This allows the CCD to
store several images before being read by the ADC. If the image formed
covered only, say, one-sixteenth of the surface area, one could shift each
image out of the light onto a darkened part of the CCD while the next
image was being recorded. Thus with intervals of a few microseconds,
one could rapidly acquire 16 time resolved images of a rapidly changing
pattern (based on Clarke (pers. comm.) and Westbrook (1988)). In this
mode the phosphor decay time would be critical (e.g. Gd_2O_2S at $1\,ms$
would be too slow for many studies at the APS or ESRF). Direct detection
would not suffer from this problem.

5.4.3.2 'Cinematic' cassette changer

This is the first of two systems developed for time resolved measurements of X-ray diffraction patterns using an IP detector by Amemiya *et
al* (1989) – the second is described in the next section. The operating
principles of this cassette changer are similar to those of a movie camera.
It consists of an IP exchanger, a mechanical X-ray shutter and a control-
ler. Figure 5.30 shows a schematic drawing of the IP exchanger. IPs of
$126\times126\,mm^2$ are exchanged one by one at the position for X-ray
exposure within a time of $0.2\,s$, during which time the mechanical X-ray
shutter in front of the specimen is closed. The minimum exposure time is
$0.1\,s$; thus the system permits 3.3 exposures per second. A maximum of 40
IPs can be mounted in the bin. The reproducibility of positioning the IPs
for X-ray exposure is within $\pm50\,\mu m$ in the vertical and horizontal. During exposure the IP in the recording position is, of course, stationary. The
IPs are subsequently read in the normal way on the IP scanner.

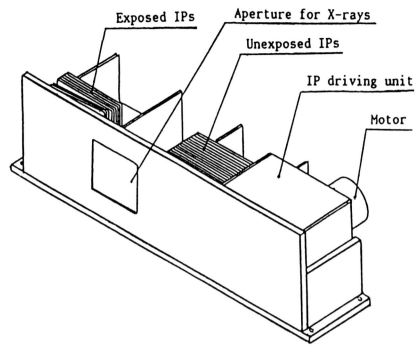

Figure 5.30 Schematic drawing of an IP exchanger used to record time resolved X-ray patterns, based on cinematic principles. Forty two-dimensional diffraction patterns can be recorded at a maximum rate of 3.3 exposures per second. From Amemiya *et al* (1989) with permission.

5.4.3.3 Streak camera

This involves translating the film or IP a certain distance at a pre-set constant speed during an exposure (e.g. see Bilderback, Moffat and Szebenyi (1984)). Diffraction spots are then drawn out into streaks. This has been used by Moffat, Bilderback, Schildkamp and Volz (1986a) to record the radiation damage behaviour of a lysozyme crystal in Laue geometry. Sakabe (pers. comm.) has also used this method but with a shutter synchronised with the exposure; in one test five separate Laue exposures were recorded on one IP (see figure 10.11). These methods allow vast quantities of diffraction data to be collected as a function of time. Their limitation is that two-dimensional Laue patterns are already very dense without displacing the detector during the exposure.

A variation on this scheme is to look at a thin stripe of the diffraction pattern. Amemiya *et al* (1989) describe a rotating drum with a 1080 mm circumference on which a large IP of 200×1000 mm is mounted. There is a receiving slit placed in front of the IP; the vertical aperture can be set

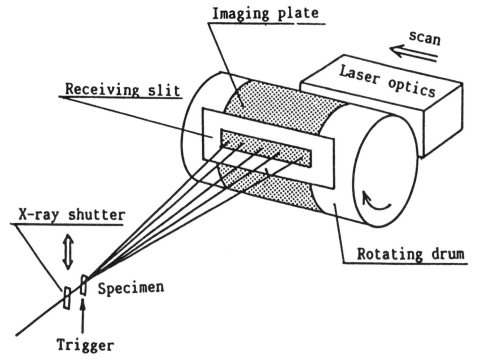

Figure 5.31 Schematic drawing of a rotating drum with IP used to
measure the time course of one-dimensional X-ray patterns
based on a streak camera method. One-dimensional time
resolved X-ray patterns can be recorded with a best time
resolution of $23\,\mu s$ for a total duration of up to 46 ms. From
Amemiya *et al* (1989) with permission.

to 0.5, 1.0, 2.0, 3.0, 4.0 and 5.0 mm. The rotation speed of the drum can be
pre-set from 20 to 0.1 rps. Hence, one-dimensional time resolved X-ray
patterns can be recorded with a *best* time resolution of $23\,\mu s$ for a total
duration of up to 46 ms. This system (figure 5.31) has so far been used
only for muscle diffraction but could be used for Laue diffraction in a
manner akin to that described for the CCD in the previous section.

5.4.4 Data acquisition and computer hardware

All two-dimensional detectors, film, MWPC, television, CCD or IP,
end up producing a two-dimensional image of, say, 512×512 pixels, each
of a certain depth (6, 8, 12 or 16 bits). There are then many images either
resulting from many orientations of the sample as it is rotated in the
monochromatic method or, in the time resolved Laue method, many
time slices (but with fewer orientations needed). Hence, a 'data set' is

made up of 100–1000 frames each of 512^2–1024^2, 8–16 bits of pixel data, i.e. 2.1×10^8–1.7×10^{10} bits of information. The real time needed to record this 'data set' will be a few hours at most (even very short exposures of μs will be interspersed with periods of inactivity whilst changing IPs, for example). These volumes of data and the data rate present formidable problems for the data acquisition computer in terms of disc space, graphics display and monitoring as well as initial data analysis. The computer also controls the careful interlacing of sample rotation (in routine data collection) or reaction initiation (in time resolved work) and the X-ray shutter.

5.4.5 Data analysis software

The final measurement required for each spot ($hk\ell$) is its intensity. Appendix 1 describes in detail the relationship between the coordinates of diffraction spots, in all the commonly used diffraction geometries and the crystal parameters. In outline the stages of data analysis involve:

(a) determination of the crystal parameters and orientation;

(b) prediction of the individual spot coordinates taking account of spatial distortions of the detector;

(c) evaluation of the integrated intensity for each spot, taking account of any non-uniformities of response of the detector;

(d) application of Lorentz and polarisation and wavelength normalisation (Laue case) corrections (see chapters 6 and 7).

For time resolved studies, with the ever increasing power of modern computers one can perhaps ultimately envisage data reduction and Fourier transform calculations being performed on-line and thereby generating electron density and displays of molecular details in (pseudo) realtime.

5.5 GENERAL INSTRUMENTATION

5.5.1 Crystal cooling

Many samples are stable at only low temperatures, say 0–4 °C, and temperature sensitive samples are normally kept cold by a gas stream, sometimes with an enclosure (Hovmoller 1981; Bartunik and Schubert

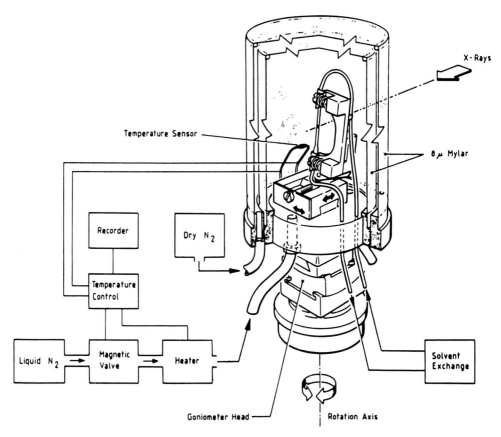

Figure 5.32 Crystal cooling device for temperatures down to 100 K.
From Bartunik and Schubert (1982) with permission.

1982 and figure 5.32). These systems should require as little attention as possible during an SR experiment, but apparatus requirements are no different than for work on a conventional source. Cooling can often considerably increase the amount of data which can be collected per crystal due to the reduction of radiation damage. Temperatures much lower than 0°C are needed for cryoenzymology applications (section 3.1.2) or cryocrystallography (see next section)

5.5.2 Cryocrystallography

There has been an upsurge of interest in using rapid freezing of bio-crystals to liquid nitrogen temperatures so as to eliminate X-radiation damage in the cooled crystals. It has been shown that it is possible to cool biocrystals to cryogenic temperatures without damage. In order to achieve this, care has to be exercised in using appropriate conditions (for separate discussions see Hope (1988) and references cited therein and Dewan and Tilton (1987)). Two approaches are possible based on mounting the crystal on a glass fibre or retaining the glass capillary. In both methods the sample must be transferred into the cold stream as quickly as possible. The agreed procedure, both in terms of rapid cooling of and ice prevention within the protein crystal, is first to deflect the cold gas stream temporarily while the crystal is being brought near its final position on the diffraction apparatus, and then with the crystal in place suddenly restore the flow of the cooling gas. Hence, abrupt cooling results which appear to be critical to preserving the integrity of the crystal. After freezing it is obviously possible to collect data over a fairly prolonged time period; frosting around the sample is prevented by having a warm outer gas stream coaxial with the inner cold gas stream. Frosting, however, can persist even with this precaution when the crystal is mounted in a capillary rather than on a fibre (Dewan and Tilton 1987).

On conventional X-ray sources with their large angular beam divergences, reflection rocking widths are similar at low temperature to room temperature scans. However, tests with a low divergence synchrotron beam clearly reveal an increase in mosaic spread of all samples that survive the freezing process. The lifetime of a successfully frozen crystal does appear to be very long in the X-ray beam. The best results seem to be obtained with the crystal mounted on a fibre followed by freezing rather than with the crystal mounted in a capillary. The initial diffraction resolution limit of a frozen crystal has been observed to be better, the same and worse! It is possible that refinement of techniques may allow resolution limits to be obtained which are consistently the same or better than at room temperature or, alternatively, with more examples it may be possible to correlate sample behaviour with solvent content or sample volume. It is commonly observed that unit cell dimensions decrease on cooling; Hope (1988) quotes a decrease in unit cell volume $\approx 3.5\%$ as a result of cooling being normal for molecular crystals. In addition, Hope (1988) observed a deviation in cell parameters between different examples of frozen crambin crystals of up to $\pm 0.06 \text{ Å}$ in values of $a=40.76$, $b=18.49$, $c=22.3$ ($\beta=90.61°$). This suggests that the method of isomorphous replacement for phasing may be diffraction resolution limited when using frozen crystals (Haas and Rossmann 1970). This is not a problem when using crystallography to study a related series of a previously solved structure, such as engineered mutants of a given protein in the same crystal form.

Spectacular use of cryocrystallography has been made to avoid an impasse reached in data collection at or near room temperature from ribosome crystals (Hope *et al* 1989). At room temperature these samples decay in the X-ray beam (monochromatic λ in the range 0.9–1.5 Å) dramatically making data collection beyond 18 Å essentially impossible. At liquid nitrogen temperatures an initial resolution of 5 Å is preserved essentially indefinitely. Mosaic spreads are unchanged between room temperature and cryotemperatures but are quite large anyway ($\approx 3°$). The intensity of the synchrotron beam is used to make the data collection time manageable.

Another application of cryotechniques is likely to be in those cases of protein microcrystals where radiation damage at room temperature is marked (Helliwell 1989). Again, the cryotechniques can be applied to preserve the sample resolution limit and the synchrotron intensity used to compensate for the weak scattering due to the small volume of the crystal.

It is worth while noting that in the electron microscopy of biological samples there is considerable interest in reducing radiation damage using liquid helium temperatures. Improvement factors between 87× (Knapek and Dubochet 1980; Dietrich *et al* 1980) and 4× (Lamvik, Kopf and Robertson 1983) have been reported. Henderson (1990) has compared the behaviour of frozen crystals in X-ray and electron beams.

5.5.3 Monitoring of incident intensity

Applications of SR are often critically dependent on flux and much care is often given to assessing the feasibility of an experiment on the assumption of a given flux at the sample. An optical instrument, once calibrated and commissioned so that the measured performance is as expected, needs a permanent flux monitor. Phizackerley *et al* (1980) incorporated an ion chamber in the collimator on their MWPC area detector system. A simple but effective design for an Arndt–Wonacott camera was suggested by Bartunik, Clout and Robrahn (1981) and this has been adapted at LURE (Fourme, pers. comm.) and the SRS (Helliwell *et al* 1982a). After beam injection an instrument can be refocussed and realigned. At a dedicated SR facility only small adjustments are required usually. The flux can be monitored until the expected value for the given machine energy and current is reached. The flux incident at the sample is also important for providing a reference scale for reflections measured in a data collection run since this flux will vary over a period of time. In the oscillation film method the time over which one oscillation is made is kept short enough for the flux to be constant; ten or 100 oscillations are made to give a well-exposed diffraction photograph.

The gas ionisation chamber is used with a current amplifier. A potential of about $100-300\,V\,cm^{-1}$ is applied across two parallel plates inside the chamber. The X-rays passing through the chamber are photoelectrically absorbed to produce fast photoelectrons and either Auger electrons or fluorescent photons. The voltage applied across the chamber sweeps electrons and ions apart and these are collected at the plates. The average energy required to produce an electron–ion pair in argon, for example, is 24.4 eV. The number of X-rays stopping in the detector can be calculated from the active volume of the chamber, the gas pressure and the X-ray absorption cross section of the gas that is used.

5.5.4 Sample absorption corrections

Absorption corrections to allow for different crystal and mother liquor absorption paths can be made in a variety of ways (reviewed by Bartels (1977)). The more accurate methods aim to establish a transmission surface for a given sample and incident wavelength. On the oscillation camera, for instance, this is determined by measuring the attenuation of the direct beam, monitored on a film as a drawn-out streak, as a function of the spindle angle ϕ and the 'precession' angle μ. Because of the very high intensity of the incident beam at a synchrotron source highly attenuating filters need to be used for this film method to work. Instead the film can be replaced by an ionisation chamber (Helliwell *et al* 1984b). Of course, without the constraints of trying to access specific absorption edges, an incident wavelength of 0.9 Å (or less) can be chosen to reduce absorption effects almost completely whilst taking advantage of the bromine absorption edge if film is used (see chapter 6).

5.5.5 Anomalous dispersion studies: absorption edge position and magnitude of white line as a function of metal atom oxidation state

In optimised anomalous dispersion studies of a metal atom in a protein crystal, the absolute energy position of the absorption edge needs to be calibrated. If an electronic area detector is used the wavelength setting can be made to maximise a Friedel intensity difference whilst stepping the monochromator. Fourme and Kahn at the Sicily workshop in 1982 showed a very nice plot of a Friedel difference for a single independent reflection versus wavelength which was exactly what one might expect for the terbium L_{III} edge for terbium substituted parvalbumin.

If film is used as detector this control is, of course, not available. Instead an absorption edge scan needs to be made. Since the metal concentration in the protein sample is very low, it is necessary to use several scintillation counters aimed at the crystal sample to pick up the crystal sample fluorescence.

5.5.6 Incident radiation wavelength calibration

Even with native data collection, knowledge of the wavelength is needed in pattern prediction. This is critical if unit cell dimensions are not known accurately, perhaps due to sample to sample variations. The wavelength can be set spectroscopically by wavelength scanning across an absorption edge for a standard sample such as a pure metal foil (Helliwell *et al* 1982a). The point of inflection of the linear absorption coefficient corresponds to the wavelength tabulated for the pure element (see, for instance, *International Tables for X-ray Crystallography*). Alternatively, a standard sample (crystal, powder or wire) can be used and the wavelength calibrated by X-ray diffraction. This method requires an accurate measurement of sample to detector distance.

5.5.7 Independent operation and radiation safety

A very important development has been the move to independently operating workstations at most facilities in contrast to the earlier instruments developed at DESY (Rosenbaum, Holmes and Witz 1971) and NINA (Haselgrove *et al* 1977), which were enclosed in one experimental hall with several other instruments such as EXAFS spectrometers; therefore, the working arrangements of each group interfered directly so that efficiency of use was not high. The use of separate safety hutches for each instrument has additional advantages over and above independent operation. An individual hutch can be refrigerated somewhat and the temperature regulated. Also, it is easier to maintain a clean environment and to provide protection against biohazards. These advantages have been advocated by Staudenmann, Hendrickson and Abramowitz (1989).

5.5.8 Support facilities: sample preparation, sample pre-alignment and film development

Apart from the experiment on the beam line, ancillary facilities and equipment contribute significantly to the success of an experiment and are important in the areas of sample preparation, sample pre-alignment and film development. The requirements for establishing samples safely on the experiment and developing films properly are no different than for conventional source work. At a synchrotron, however, the data collection strategy also includes attempts to make most effective use of beam time. Hence, 'beam time savers' include the following: (1) Use of a Donnay optical analyser for pre-alignment using the polarisation properties of a crystal sample. (2) Pre-alignment of samples on a precession camera and a conventional source can avoid hold-ups in the developing of setting photographs on the beam line. Polaroid used on the SR beam can also reduce this time. (3) Use of goniometer heads which can rotate on their own spindle, reduce the time to centre the sample on an instrument. This is mainly a concern on all motor-driven cameras and diffractometers with a slow drive and without a clutch slip. (4) An alternative to options (1) and (2) is a cheap image intensifier for quick setting of crystal orientation with SR. (The alignment aids (1), (2) and (4) are not needed with an electronic area detector.) (5) For film work, bulk film hanger frames are essential if the exposed films (or even just a fraction of them) are to be developed at a rate matching output.

5.6 GAZETTEER OF SR WORKSTATIONS FOR MACROMOLECULAR CRYSTALLOGRAPHY

There is now a considerable number of SR instruments which have been or are being used for protein crystallography at various SR sources around the world (table 5.5 documents the existing ones). In this section a survey is made of the various instruments. An extensive discussion of mirror-monochromator cameras used for small angle diffraction at DESY, NINA, SPEAR, VEPP-3 and DORIS has been given by Rosenbaum and Holmes (1980).

The intensity at the focus provided by each of the instruments can be compared with the intensity of a laboratory rotating anode (nickel filtered, Elliott GX6) beam behind a 0.3 mm diameter collimator of 9×10^8 photons s^{-1} mm^{-2} (Harmsen, Leberman and Schultz 1976).

The order of the instruments quoted tries to follow a temporal line of development of SR sources rather than of individual instruments; in this way several instruments at one storage ring or synchrotron are presented together, even though some will still be under construction, so as to give a coherent view of what is available at each source.

Table 5.5. *Gazetteer of existing stations (and parameters) for macromolecular crystallography. Based on Helliwell (1984); updated in 1990 under the auspices of the International Union of Crystallography's Commission on Synchrotron*

Source	Station	Magnet	λ_c or E_c	Station or mono horizontal acceptance (mrad)	Mirror configuration	Mono configuration
Stanford (SSRL) SPEAR (3.0–3.5 GeV)	(1)[a] I–5AD	Bending	4.7 KeV at 3 GeV	1.0	None	Double crystal Si (111)
	(2) VII–I	1.8 T eight-pole wiggler	10.8 keV at 3 GeV	1.0	58 cm long platinum coated fused silica bent for vertical focussing	Bent triangle Ge (111) or Si (111)
Paris (LURE) DCI (1.72 GeV)	(1) D41	Bending 1.5 T	3.4 Å	2.0 (mono γ_H) at $\lambda = 1.4$ Å	None	Bent triangle Si (111) 10° cut
	(2) D23	Bending 1.5 T	3.4 Å	1.0	None	Fixed exit double crystal Si or Ge (111) or (220)
Hamburg DORIS (3.5–5.5 GeV)	(1) X–11	Bending	0.45 Å	2.1 mrad at 0.9 Å 2.8 mrad at 1.5 Å	8×20 cm quartz bent	Bent triangle Ge (111) 7° cut
	(2) X–31	Bending	0.45 Å at 5.3 GeV 1.6 Å at 3.5 GeV	3.3 mrad	Ditto but gold plated toroid curved	Channel cut Si (111)
Daresbury SRS-HBL (2 GeV)	(1) 7.2	Bending 1.2 T	3.9 Å	4 mrad	Fused quartz, bent for vertical focussing 1:1 58 cm long, $\theta_c = 3.5$ mrad	Bent triangle Ge (111) 200 mm 6.75–14.75°
	(2) 9.6	5 T wiggler three-pole	0.9 Å	3 mrad	Platinum coated fused quartz 75 cm long, $\theta_c = 3$ mrad	Bent triangle Si (111)
	(3) 9.7	5 T wiggler three-pole	0.9 Å	2 mrad	none	none
	(4) 9.5	5 T wiggler three-pole	0.9 Å	1.2 mrad	Toroid platinum coated point focussing ($\theta_c = 3$ mrad)	Double crystal Si (111)

*adiation (C Syn R) with the help of P. Phizackerley, K. S. Wilson, C. Nave,
. W. Thompson, D. Bilderback, R. Sweet, S. Popov, B. Tolochko, N. Sakabe,
. Kvick and R. Fourme, to whom I am very grateful.*

ocus size $"\times V$ mm^2 (.3 σ)	Quoted λ range (Å or eV)	Usual operational λ or E	$\delta\lambda/\lambda$	Intensity[b] (photons s^{-1} mm^{-2})	Detector/Ref/ Notes
)×3.0 eam size 20 m)	5500–14 500 eV	Near to 7.1 keV or 9 keV or 12.7 keV	2×10^{-4}	$\sim2\times10^{8}$ at 3 GeV, 80 mA (7 keV)	Four circle diffractometer with a San Diego Multiwire Systems Area Detector
×0.2 nin)	6000–13 000 eV	1.54 Å or 1.08 Å	$\sim10^{-3}$	$\sim2\times10^{11}$ at 3 GeV, 80 mA (8 keV)	Rotation + film. IP system planned.
8 mm H)	1–2 Å	1.4 Å	10^{-3}	2×10^{10} at 310 mA	Oscillation camera Kahn *et al* (1982a) Lemonnier *et al* (1978) Kahn *et al* (1986)
mm H)	0.95–2.2	tunable		5×10^{9} at 310 mA at $\lambda = 1.4$ Å	MWPC (350 kHz) 185 500 pixels of 1.1 mm^2. 50% dedicated
2×0.64	0.7–1.5 Å	1.488 Å 0.9 Å		$>10^{11}$ at 5.3 GeV, 45 mA	IP or oscillation camera film
8×2.6	0.6–2.5 Å	1.488 Å 0.9 Å	2.3×10^{-3}	6×10^{9} at 3.7 GeV, 100 mA	IP or oscillation camera film
5×0.3 HBL)	1.0–3.0 Å	1.488 Å	4×10^{-4}	6×10^{11} at 200 mA	Oscillation + film
5×0.3 HBL)	0.5–1.5 Å	0.9 Å	4×10^{-4} Si (111)	1.3×10^{12} at 200 mA	Oscillation + film Enraf–Nonius FAST TV area detector
	0.2–2.6 Å	White beam $(0.2 \leqslant \lambda \leqslant 2.6$ Å$)$		10^{10} in $\Delta\lambda/\lambda$ 1.5×10^{-4} at 1 Å at 200 mA	Film/Laue
3×0.4	0.45–2.6 Å	Tunable or focussed white beam $(0.5 \leqslant \lambda \leqslant 2.6$ Å$)$	1.5×10^{-4}	3.6×10^{11} in $\Delta\lambda/\lambda$ 1.5×10^{-4} at 1 Å at 200 mA	Film; IP planned Laue + anom. disp + routine data coll.

Table 5.5. (*cont.*)

Source	Station	Magnet	λ_c or E_c	Station or mono horizontal acceptance (mrad)	Mirror configuration	Mono configuration
Cornell (CHESS) (5.2–5.6 GeV)	(1) A1	6-pole EM wiggler	29.4 keV	2.25 mrad	Vertical focussing platinum coated	Bent Ge(111)
	(2) F1	24-pole wiggler	23.5 keV	1.75–1.0 mrad	Vertical focussing platinum coated	Bent Si (111)
Japan Photon Factory (PF) (2.5 GeV)	(1) 6A₂	Bending	4 keV	4 mrad	Bent quartz	Bent Si (111) 0–16.5°
	(2) 14A	Vertical wiggler	20.8 keV	1.28 mrad	Platinum coated toroid	Double Si (111) crystal (422) (553)
Brookhaven Biology Dept NSLS	(1) X-12C	Bending magnet	3 Å	2 mrad	Rhodium coated mirror (1:1) for vertical and horizontal focussing	Double Si (111) crystal
Novosibirsk VEPP-3	(1) 5–A	Wiggler (2 T)	3.5 Å	2 mrad	None	Si (111) or Ge (111) 8° cut
	(2) 2–C	Wiggler (2 T)	3.5 Å		None	
Frascati ADONE (1.5 GeV)		Wiggler (6-poles, 1.85 T)	4.5 Å	1 mrad	None	Si (220) 31 m from tangent Ge (220) Si (111)

Notes:

[a] Beam I–5AD; AD stands for Area Detector experimental enclosure. Also, please note that an Enraf–Nonius CAD–4 diffractometer can be mounted in a separate radiation enclosure (station I–5) on the same beam line when station I–5AD is not used; the energy range is 2800–3000 eV with Ge (111), Si (111) or Si (220) monochromator crystals.

[b] For comparison purposes: a Cu Kα (1.54 Å) rotating anode generator operating at 1.6 kW with a 0.2×2 mm² electron focus delivers an intensity (ph/s/mm²) into a 0.3 mm² beam cross section, with a $\delta\lambda/\lambda$ of 2.5×10^{-3}, of 9×10^{8} (Harmsen *et al* 1976). In 1990 commercially available rotating anode generators can operate at up to 5× higher electrical powers in a similar electron focus area.

5.6.1 SPEAR (Stanford)

5.6.1.1 SPEAR-1

The first experimental tests in the applications of SR in protein crystallography were reported by Phillips *et al* (1976, 1977) (see section 10.1) and were made on this instrument, which was constructed by Webb and coworkers (Webb 1976; Webb *et al* 1976, 1977).

The mirror consisted of a 1.2 mm piece of float glass and the monochromator was Si(111) with an oblique cut of 8.5°. The monochromator able 5.5. (*cont.*)

cus size $\times V$ mm^2 (3σ)	Quoted λ range (Å or eV)	Usual operational λ or E	$\delta\lambda/\lambda$	Intensity[b] (photons s^{-1} mm^{-2})	Detector/Ref/ Notes
$\times 0.3$	–	8.05 keV fixed	10^{-3}	5×10^{12} at 5.43 GeV	Oscillation + film + IP/SP
0.3	7–25 keV	7–14 keV	10^{-3} (or 2×10^{-4})	$(7\text{–}10)\times10^{12}$ at 5.43 GeV	Oscillation + film + IP/SP
2	5–25 keV	–	10^{-3}	–	Weissenberg + IP
2	5–85 keV		2×10^{-4}		Four-circle diffractometer
$\times 0.6$	7.5–13 keV	tunable	$10^{-3}\text{–}10^{-4}$	8×10^{10} at 160 mA	Enraf–Nonius FAST TV area detector
	0.8–1.8 Å		10^{-3}	10^{11} at 100 mA	Oscillation + film
	4–25 keV	–	–		Laue, MWPC (Baru *et al* 1978, 1983)
	3–30 keV	8 keV	10^{-4}	10^{9} at 70 mA	Four-circle diffractometer (Huber)

was bent to a logarithmic spiral since this is the theoretical shape needed to image a point source exactly to a point focus; actually, since the aberration is very small for a simple cylindrical curvature (see equation (5.14) and the component x_3) for the small lengths of crystal available (compared with the focussing distances involved), the use of the special bending device seems an unnecessary complication.

A flux of 6×10^8 photons s^{-1} with a focus of 0.5×0.5 mm^2 was recorded for SPEAR at 3.7 GeV, 20 mA (this is an intensity of 2.4×10^9 photons s^{-1} mm^{-2}). Webb *et al* (1977) reported a ten-fold discrepancy between the calculated and the measured values and that this was due to poor monochromator performance. Boeuf *et al* (1978) argued, however, that the loss of flux must have been elsewhere due either to poor mirror reflectivity or instrument misalignment.

This instrument had its own radiation hutch, beam shutter and inter-locks and was therefore independent in its operation from other experiments. This was a major advantage over the early instruments for muscle diffraction at DESY or NINA.

5.6.1.2 SPEAR-2

Phillips, Cerino and Hodgson (1979) established a four-circle diffract-ometer (Enraf–Nonius CAD4) at SPEAR. The diffractometer is mounted 'on its side' so that reflections can be measured in the vertical plane to minimise polarisation losses. The whole system is enclosed in a small hutch mounted on wheels. The system can therefore be 'fed' by an SR beam from any of the available optics on SPEAR. It has been used on the optical system of Hastings *et al* (1978); this consists of a platinum coated fused quartz mirror giving 1:1 double focussing followed by a channel-cut monochromator. A $4.2 \times 2.4 \, \text{mm}^2$ FWHM (full width at half maxi-mum) focus was reported by Phillips *et al* (1979) with a measured inten-sity of 4×10^{10} photons $\text{s}^{-1} \, \text{mm}^{-2}$ at 3.7 GeV and 20 mA SPEAR operating conditions. The minimum $\delta\lambda/\lambda$ was 10^{-3}. The diffractometer has since been used without a focussing mirror, i.e. just the double crystal mono-chromator and a fine slit to give a $\delta\lambda/\lambda=10^{-4}$ (Templeton *et al* 1982a). The diffractometer system has been used principally for very accurate small molecule crystal measurements as a function of λ close to the absorption edges of elements in various crystal structures. From the already known structures (Templeton *et al* 1982b) least squares refine-ment of the dispersion coefficients f' and f'' were made, for example, of caesium (Phillips *et al* 1978), praseodymium (Templeton, Templeton and Phizackerley 1980), gadolinium and samarium (Templeton *et al* 1982a). The polarisation dependence of the f' and f'' for vanadium was also measured by Templeton and Templeton (1980) – further examples are given in chapter 9.

Phillips and Hodgson (1980) used the diffractometer to measure data on a small protein gramicidin-A maximising the f'' of caesium, which were combined with data on a potassium derivative measured at Cu $K\alpha$ (see section 9.7.1). The use of a single channel detector and limited allocation of beam time has restricted the application of the system in protein crystallography. However, a film carousel (Enraf–Nonius) is available for use when the detector arm is swung out of the way; it has, however, been rarely used (Phillips, pers. comm.).

5.6.1.3 SPEAR-3

An MWPC area detector diffractometer (figure 5.33) has been set up on beam line 1–5 for protein crystallography at Stanford (Phizackerley *et al*

(a)

Figure 5.33 (a) The MWPC area detector diffractometer at Stanford.
(b) Schematic of the system – see section 5.6.1.3 for details.
From   Phizackerley,   Cork   and   Merritt   (1986)   with
permission.

1980). The system provides the advantages of an area detector of
simultaneous acquisition of reflections with the advantage of photon
counting for accurate measurements.

A double crystal monochromator providing a $\delta\lambda/\lambda$ of 10^{-4} is followed
by a flat platinum coated mirror for harmonic rejection. A Huber five-
circle diffractometer is mounted to allow the $30\times30\,cm^2$ two-dimen-

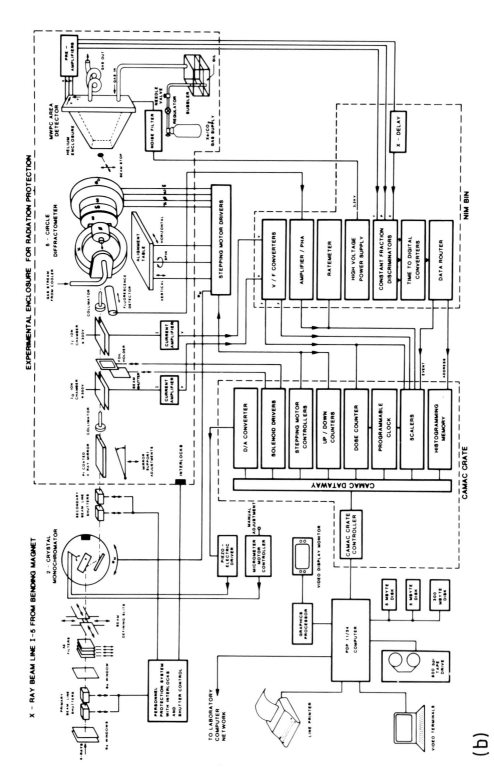

Figure 5.33 (cont.)

(b)

sional MWPC detector to swing in a vertical plane; the 2θ maximum is >90°. A conventional X-ray set is available for commissioning and test purposes in the absence of an SR beam. The instrument has been used primarily for optimised anomalous dispersion for the location of heavy atoms and phasing of protein crystal reflections by multiwavelength methods.

In particular the structure determination of cucumber basic protein has been made using MAD data recorded on this instrument around the Cu K edge (Guss *et al* 1988, section 9.7.5). This protein crystal structure had previously defied solution for many years because of problems associated with preparing heavy atom derivatives.

5.6.1.4 SPEAR-4

An oscillation camera film data collection facility has been established at Stanford. This station (beam line VII–1) uses 1 mrad of beam from an eight-pole multipole wiggler (Winick and Spencer 1980). The optics consists of a bent Ge(111) triangular monochromator followed by a bent metal coated mirror. The small number of available mrad of beam is easily compensated by the number of poles in the wiggler. The station has been used extensively. Examples of structures reported using data collected on this station are given in chapter 10.

5.6.2 DESY (Hamburg)

A mirror-monochromator camera was established on the DESY synchrotron for small angle diffraction experiments by Rosenbaum *et al* (1971) and Barrington-Leigh and Rosenbaum (1974). The instrument was used at a fixed wavelength of 1.5 Å. The system was used to evaluate the synchrotron as an X-ray source for SR protein crystallography by Harmsen *et al* (1976). Their work encouraged the use of DORIS as a source for protein crystallography (see sections 5.6.4 and 10.1).

The X-ray mirror on the camera consisted of two 20 cm segments of fused quartz each of which had a separate bending mechanism. The monochromator was either a bent quartz crystal (101 plane) or Ge(111), each with an oblique cut of 7°; the crystals were rectangular with equal couples applied at each end to obtain the necessary curvature (same principle as the mirror bender).

A focal spot of 200 μm diameter was achieved with an object distance for mirror and monochromator of 30 m and a variable image distance of 1–3 m. With Ge(111) as the crystal, at 1.5 Å, and DESY operating at

7.2 GeV, 10 mA, a flux of 7×10^8 photons s^{-1} was available (Rosenbaum and Holmes 1980). This converts to an intensity of 1.75×10^{10} photons s^{-1} mm^{-2} at the focus. The system is no longer used.

In the context of oscillation camera data processing for the monochromator at 1.5 Å with a demagnification of 10–30 the system would be far from the Guinier setting; the exact $(\delta\lambda/\lambda)_{corr}$ (equation 5.9)) would depend on the length of the crystal monochromator illuminated (see section 6.1).

5.6.3 DCI (Paris)

5.6.3.1 DCI-1

Lemonnier *et al* (1978) established a focussing monochromator camera in 1976 consisting of a bent triangular Ge(111) crystal with 10° oblique cut. The object and image distances are 15.5 and 1.7 m respectively; the Guinier focussing condition, for these parameters, occurs at 1.4 Å wavelength which is the position at which most data have been collected. The small length of the monochromator of 70 mm, which accepts 1.7 mrad of beam, means that any small error in the nominal oblique-cut angle will not cause a significant deviation from the Guinier focussing condition (see section 6.1). The measured horizontal focus was 0.56 mm and the flux through a 0.3 mm diameter collimator for DCI at 1.72 GeV, 120 mA was 1.4×10^9 photons s^{-1} at 1.54 Å (Lemonnier *et al* 1978) corresponding to an intensity of 2×10^{10} photons s^{-1} mm^{-2}. Currents at injection of 300 mA are obtained routinely with long lifetimes and at 1.4 Å wavelength the intensity is now 5×10^{10} photons s^{-1} mm^{-2} (Kahn *et al* 1982a).

The instrument is tunable; the other oblique-cut crystal available is 5° Ge(111) (Lemonnier *et al* 1978). This was the first oscillation camera film data collection service at a synchrotron. One of the earliest protein crystal data sets to be collected was from single crystals of tyrosyl t-RNA synthetase (Monteilhet, Fourme and Blow 1978). Many data sets have now been collected (summaries of which are given in Fourme (1978, 1979), Fourme and Kahn (1981), Kahn *et al* (1982a), Bartunik, Fourme and Phillips (1982) and Wilson *et al* (1983)).

One of the first virus crystal data collection runs took place on DCI-1 (Usha *et al* 1984, see section 10.5.1).

5.6.3.2 DCI-2

An MWPC-based area detector diffractometer is under development by Kahn *et al* (1980). The system is intended for use primarily for optimised anomalous dispersion applications. The MWPC is based on a radial drift chamber designed to overcome the problems of parallax associated with flat chamber MWPC systems (Charpak *et al* 1974). The 4θ Bragg angle range accepted, which is fixed, in both hroizontal and vertical directions is 89° (i.e. a resolution limit of 2 Å at 1.5 Å incident wavelength).

The monochromator is a Ge(220) channel cut. A focussing double crystal monochromator is under development (Goulon and Lemonnier, unpublished work, see section 5.2.4.2). Multiple wavelength data have been collected on terbium parvalbumin at the L_{III} absorption edge of terbium (Kahn *et al* 1985) and the structure solved (see 9.7.4).

5.6.3.3 DCI-3

A superconducting wiggler magnet is being installed at DCI. Facilities for macromolecular crystal data collection include Laue diffraction.

5.6.4 DORIS (Hamburg)

5.6.4.1 DORIS-1 (X-11)

Encouraged by the success on DESY, Harmsen and Rosenbaum (described in Rosenbaum and Holmes (1980) and Rosenbaum (1980)) established a mirror-monochromator camera on DORIS which was used for small angle diffraction and protein crystallography.

X-11 originally employed a mirror consisting of 8×10 cm quartz segments and a 180 mm Ge(111) 7° oblique-cut monochromator 22 m from the source. The mirror segments were flat and so the vertically dispersing beam was sectioned into eight pieces and by adjusting each segment the eight reflected images were overlapped at the sample position. No attempt was made to image the source because each segment was flat (see section 5.3.6.1). A measured flux at 1.5 Å of 10^{10} photons was reported by Rosenbaum and Holmes (1980) for DORIS at 3.7 GeV and 10 mA with a focus of 1 mm^2. (On an identical optical set-up for small angle diffraction, X-13, Hendrix *et al* (1979) reported a flux of 5×10^{10} photons s^{-1} at 1.5 Å for DORIS operating at 4.6 GeV and 20 mA).

Bartunik and Bartels established the X-11 instrument as a centre for routine protein crystallographic data collection from about 1980. Bartunik *et al* (1981) added an intensity monitor and computer control to the Arndt/Wonacott oscillation camera. Bartunik and Schubert (1982) devised a cooling chamber for a wide range of temperatures and incorporating a flow cell (Wyckoff *et al* 1967) – figure 5.32.

The use of the 7° cut Ge(111) with $p=22$ m and p_G lying in a range of 2– 3 m means that the Guinier position (important for the simplicity of predicting reflections, see section 6.2.1) lies at a wavelength of ≈ 1 Å.

Bartunik *et al* (1982) give a summary of the oscillation film data collected on DORIS on X-11 up to that time.

The optical system was upgraded in 1985/6, especially with provision of better mirror vacuum to 10^{-5} Torr. In addition the alignment system was improved and automatic alignment software incorporated. In 1988/9 an on-line IP detector system was introduced for routine use. Figure 5.34(a) shows a schematic of the optics (Wilson 1989).

5.6.4.2 DORIS-2 (X-31)

At the HASYLAB at DORIS a rapidly tunable instrument is available for protein crystallography. The optical design (Bordas and Koch, reported in Bordas (1982)) consists of a channel-cut monochromator, 16 m from the tangent point, followed by eight segmented mirrors. Each mirror is configured so that a horizontal focus is achieved. The beam in the vertical is 'focussed' by superimposing each of the reflected beams at the sample. The intensity is lower than X-11 because of the larger focus. It was one of the first rapidly tunable spectrometers for protein crystallography available in Europe.

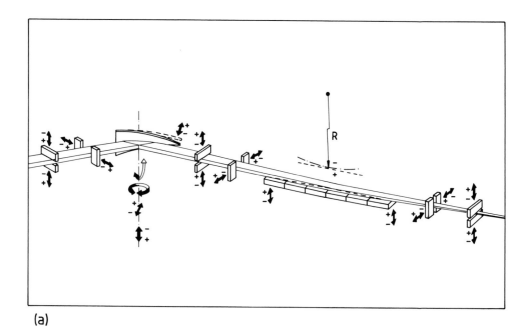

(a)

Figure 5.34 Schematic of the optical layout on (*a*) X-11.

Figure 5.34(*b*) shows a schematic of the optical layout (Wilson 1989). The on-line IP scanner (figure 5.34(*c*) is moved between X-11 and X-31 as required (see figure 10.4).

5.6.5 SRS (Daresbury)

5.6.5.1 SRS-1 (PX 7.2)

This instrument (station 7.2 using the Daresbury nomenclature) provides oscillation camera film data collection facilities which commenced in 1981 (Helliwell *et al* 1982a). The instrument was shared with small angle diffraction applications for several years. There is a bent platinum coated fused quartz single segment mirror providing 1:1 vertical focussing (Nave *et al* 1985). For high intensity applications the monochromator used is a 200 mm long bent triangular Ge(111) crystal with a 10.4° cut. The Guinier focussing position (for p=20.9 m, p'_G=2.52 m) is established for a wavelength of 1.488 Å (calibrated spectroscopically with a pure nickel foil). Until 1985, the measured FWHM focus size at the SRS was 1.1×0.3 mm^2 and the intensity behind a 0.3 mm diameter collimator was 10^{11} photons s^{-1} mm^{-2} at 1.488 Å with the SRS at 2 GeV,

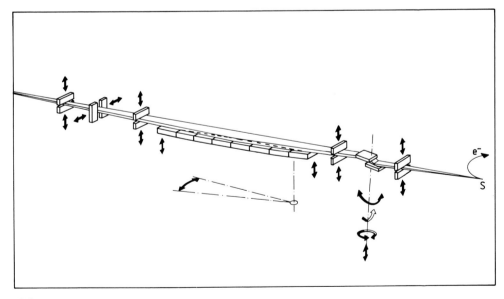

(b)

Figure 5.34 (*cont.*)
    (*b*) X-31 at EMBL, Hamburg.

(c)

Figure 5.34 (*cont.*)
     (*c*) the Hamburg IP scanner: the optical and mechanical
     components. The round white disc is the imaging plate. The
     rectangular object is the reading head which moves radially
     while the IP is rotating. Figures kindly supplied by K. S.
     Wilson and J. Hendrix with permission.

250 mA. The advent of the high brightness lattice at the SRS in 1985 has
enhanced the intensity by a factor of ~5.

    For this type of instrument, particular attention has been given to the
prediction of reflections taking account of the SR geometry to allow
proper partiality calculations to be made in oscillation camera data
processing packages (see section 6.2 and Greenhough and Helliwell
(1982b)).

An energy-dispersive approach to optimised anomalous dispersion crystallography has been tried for the first time on this instrument (Arndt *et al* 1982; Greenhough *et al* 1983).

For experiments over a range of wavelengths a range of oblique-cut Si(111) crystals is available (6.75°, 10.25°, 12.5° and 14.75°). Optimised anomalous dispersion studies have been carried out on Fe–cytochrome C_4 (see section 9.7.2), and Mn–pea lectin (Einspahr *et al* 1985). The minimum $\delta\lambda/\lambda$ achievable with a Si(111) oblique cut, 10:1 demagnification is 3×10^{-4} (see section 5.2.3.1).

A large quantity of protein crystal 2 Å resolution data has been collected over the years; details can be found in the SERC, Daresbury Laboratory Annual Reports (e.g. see section 10.2.2.1).

5.6.5.2 SRS-2 (PX 9.6)

On the wiggler beam line at the SRS another station for protein crystallography has been available since 1984 (station 9.6) (Helliwell *et al* 1986b). The unique advantage of the wiggler source has been in giving a high photon intensity at short wavelengths ($0.5 \leqslant \lambda \leqslant 1.5$ Å). A popular mode of the instrument is the use of a point focussed monochromatic beam of $\lambda = 0.9$ Å (e.g. see section 10.2.1.3)

The large number of mrad available on the wiggler afforded the geometric floor space to operate the instrument either as a side scattering, horizontally dispersing system or as a pseudo 'end of line' station with a vertically dispersing monochromator or with white beam for Laue work. A platinum coated 75 cm fused quartz single segment mirror operates as a vertically focussing 1:1 element. The usual monochromator is a bent triangular Si(111) crystal with an oblique cut of 6.75° providing a wavelength of 0.9 Å with a Guinier position, $p'_G = 2.1$ m ($p = 21$ m); the focussed intensity being about 10^{12} photons s^{-1} mm^{-2}. Between ~0.85 Å and 1.1 Å it is possible to vary the wavelength without change of monochromator, using the monochromator pre-slits to restrict the $(\delta\lambda/\lambda)_{corr}$ contribution. This λ range encompasses the absorption L edges of many of the commonly occurring heavy atom derivatives (e.g. mercury, gold, platinum), see section 9.7.3. For native data collection using a λ of 0.9 Å absorption errors and radiation damage are significantly reduced (see, e.g., Acharya *et al* (1989) section 10.5.1.2) and better data merging statistics are obtained compared with the use of a longer wavelength. Additionally, long crystal to film distances can be used thus reducing the background under a spot; the spot itself hardly increasing in size due to the fine collimation of the beam whereas the X-ray background decreases as the inverse square of the distance (see section 6.6).

The original synchrotron Laue diffraction patterns from protein crystals recorded at Daresbury using a broad bandpass were conducted on this instrument with the monochromator removed (see Helliwell (1984)). Some preliminary multiwavelength experiments with a silicon double crystal monochromator (Si(111) triangle removed) were conducted. The growth of the Laue and MAD experiments has led to two further stations at Daresbury (SRS-3 and SRS-4).

Detection is either via photographic film using the oscillation camera (figures 5.35(b) and (c)) on SRS-2 or via the television area detector diffractometer, FAST, of Enraf–Nonius (figures 5.35(d) and (f)). The latter has been especially beneficial for microcrystal studies (see, e.g., Andrews et al (1988)) and radiation sensitive protein crystals (see, e.g., Glover et al (1988) and section 10.2.1.4).

The advent of the high brightness lattice at the SRS in 1985 enhanced the focussed beam intensity by a factor of ~3.

5.6.5.3 SRS-3 (PX 9.7)

The expansion of the Laue diffraction programme of work at Daresbury and the competition for time on SRS-2 (station 9.6) led to the use of an adjacent, vacant, station on the wiggler (station 9.7) from 1985. This is shared with other energy-dispersive experiments (non-protein crystallography). Hence, approximately one-third of the time is given to protein Laue work. The beam is not focussed owing to the proximity of the hutch to the ring; the sample is at 15 m from the tangent point. The beam is taken from the outermost portion of the fan of radiation from the wiggler.

5.6.5.4 SRS-4 (PX 9.5)

This instrument (station 9.5) is currently under development (figures 5.17 and 5.18) at Daresbury on the wiggler line for macromolecular crystallography (Thompson and Helliwell 1986; Brammer 1987; Brammer et al 1988; Habash et al 1990). It is primarily intended for very rapid Laue and rapidly tunable monochromatic experiments. The sample position is at ~32 m from the tangent point. A point focussing mirror is situated at 18 m and a rapidly tunable monochromator at 30 m (in the hutch).

5.6.6 CHESS (Cornell)

5.6.6.1 CHESS (A1 station)

A point focussed instrument has been constructed by Caffrey and Bilderback (1983). It consisted originally of a bent Si(111) monochroma-

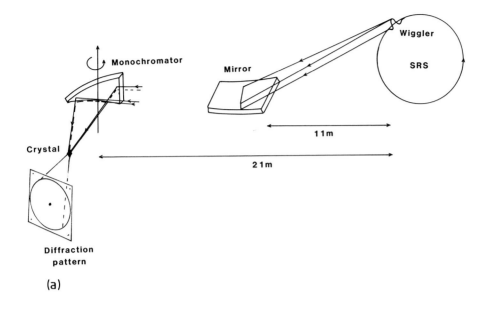

(a)

(b)

Figure 5.35 (*a*) Daresbury station 9.6 beam line optics (schematic).
(*b*) Monochromatic setting.

Figure 5.35 (*cont.*)
   (*c*) Laue mode. (*d*) The television area detector diffractometer of Enraf–Nonius (FAST TV) at the SRS wiggler
   protein crystallography station 9.6.

Figure 5.35 (*cont.*)

   (*e*) Final experimental set-up on the SRS wiggler protein
   crystallography workstation 9.6 at Daresbury; view of the
   inside of the hutch including views of the Enraf–Nonius
   oscillation camera and the FAST TV diffractometer. (*f*)
   Close-up view of the FAST TV diffractometer. Reproduced
   with the permission of Daresbury Laboratory.

tor followed by a float glass mirror. The measured focus width was
1×1.5 mm (horizontal$\times$vertical) and the flux was 5×10^{10} photons s^{-1}
(an intensity of 3.3×10^{10} photons s^{-1} mm^{-2}) for CHESS operating at
5.2 GeV, 18 mA. To enhance the intensity at the focus the optics were

upgraded; a Ge(111) monochromator and a platinum coated mirror were added to enhance the flux and intensity to 10^{12} photons s^{-1} mm^{-2}. The station is not tunable. This instrument has been used extensively for virus crystal data collection by the oscillation film method (see, e.g., Rossman *et al* (1985) and section 10.5.1.1). Film data on very small crystals of bovine growth hormone (Bell, Moffat, Vonderhaar and Golde 1985, see 10.3.2) were also made possible on this instrument and also of the histocompatibility antigen HLA-A2 (Bjorkman *et al* 1987).

5.6.6.2 *CHESS wiggler*

A new diffraction station for macromolecular crystallography is under construction (Schildkamp, Bilderback and Moffat 1989). The beam line will be powered by a 24-pole hybrid permanent magnet wiggler with a critical energy of 25 keV. A focussing Ge(111) monochromator, which handles a specific heat load of 10 W mm^2, will have a range of tunability covering 7–15 keV. The expected brilliance is about 10^{13} photons s^{-1} mm^{-2} mrad^{-2} (0.1% $\delta\lambda/\lambda^{-1}$). The station is to be equipped with an oscillation camera for use with X-ray film of 5″×5″ or 8″×10″ or Kodak storage phosphor plates. Of special note is that the entire diffraction station, its control area, a biological preparation area and a darkroom are to be embedded in a biological safety containment up to USA safety standard 'BL3'. This is to allow diffraction studies of virulent strains of viruses and other biohazards, which could not previously be studied at SR sources, without causing major disruption to the normal laboratory procedure.

5.6.7 Photon Factory

At the Photon Factory data collection facilities for macromolecular crystallography have primarily been centred on either Weissenberg geometry or diffractometry.

5.6.7.1 *Weissenberg camera station*

This camera is situated on a Photon Factory bending magnet (BL-6A2) and accepts 4 mrad of beam. The critical energy of the beam is 4.0 keV. The use of Weissenberg geometry (appendix 1) has the advantage of reducing the number of individual exposures compared to the rotation/oscillation method and therefore avoids producing a significant number of partial reflections. The wide angular range produces a larger background over an individual diffraction spot than would be the case if it was measured only over its rocking width. However, the natural collimation of the SR beam and the large crystal to IP distance used in the camera means that the background is very small in any case. Multilayer-line screens can be used (Sakabe 1983; Sakabe *et al* 1989).

The beam line optics consists of a bent Si(111) monochromator. Various crystals with different oblique-cut angles are available, namely $\alpha = 0°$, $6.0°$, $7.8°$, $9.5°$, $11.4°$, $13.7°$ and $16.5°$ to allow an energy range of 5–25 keV to be covered, by appropriate choice of crystal. There is a bent plane fused quartz mirror (Satow, Mikuni, Kamiya and Ando 1989).

The station and camera (with IPs) – see figure 5.36 – are used for the collection of macromolecular crystal data (see, e.g., the structure determination of ω-amino acid pyruvate transferase (Watanabe *et al* 1989) see plate Al.d.

Figure 5.36 The synchrotron Weissenberg camera station for macromolecular crystallography (BL-6A2) at Photon Factory, Tsukuba. Figure kindly supplied by Professor N. Sakabe and reproduced with permission.

5.6.7.2 Four-circle diffractometer station (Satow and Iitaka 1989)

This station (BL-14A) is situated on the Photon Factory vertical wiggler and accepts 1.28 mrad of beam. The critical energy of the beam is 20.8 keV. Because of the vertical polarisation state of the beam the diffractometer has been mounted in the 'conventional manner', namely with the counter arm swinging in the horizontal plane. The optics consists of a double crystal monochromator (Si(111) or Si(331) or Si(553) to cover energy ranges 5.1–19.1 keV, 12.9–48 keV and 22.7–84.5 keV respectively) and a bent cylindrical mirror for toroidal focussing (platinum coated fused quartz). As well as the diffractometer there is also a rotation camera.

The diffractometer has been used, for example, for MAD measurements at the selenium K edge for the selenobiotin streptavidin (Hendrickson et al 1989) see section 9.7.6.

5.6.8 NSLS (Brookhaven, USA)

The variety of X-ray sources and their parameters at NSLS are detailed in Hsieh et al (1983).

5.6.8.1 Chemistry and Biology Department beam lines

Workstations for protein crystallography have been set up by the Departments of Chemistry and Biology at the Brookhaven National Laboratory. The optical system design for the Chemistry beam line is based on a spherical collimating mirror, followed by a two-crystal (+,−) monochromator and finally by a bent cylindrical mirror (Hastings et al 1983). The instrument combines a liquid helium cryostat Huber diffractometer, for work on valence electron density studies of small molecules, with an oscillation camera for protein crystallography. The diffractometer may be upgraded to include a multichannel electronic area detector.

The Biology Department beam line includes a station for protein crystallography (with Supper oscillation camera and FAST TV diffractometer) and a station for small angle diffraction (with a three-circle goniostat and MWPC electronic area detector). The latter station may be available for optimised anomalous dispersion crystallographic studies. The optical design for each consists of a bent pre-mirror, double crystal monochromator and bent post-mirror; the mirrors have rhodium coatings (Wise and Schoenborn 1982).

5.6.8.2 Participating research teams

Naval Research Laboratory   At the NSLS a large number of 'participating research teams' (PRTs) have established beam lines on the NSLS and a number of instruments use diffractometers as well as oscillation cameras. One of these, for instance, is the Naval Research Laboratory (NRL) and National Bureau of Standards PRT; as part of this effort, NRL is constructing a materials analysis X-ray beam line. The beam optics design consists of a platinum coated copper pre-mirror dynamically bent to approximate a parabolic cylinder, followed by a fixed-exit double crystal monochromator and a platinum coated fused silica cylinder bent to approximate an ellipsoid. The hutch contains a six-circle diffractometer (Kirkland, Nagel and Cowan 1983).

Howard Hughes Medical Institute   The Hughes Foundation is funding a new beam line dedicated to structural analyses of biological macromolecules primarily through crystallography. The facility is under construction at the X4 port of the NSLS and will comprise three beam lines. The first line is devised to apply the MAD method. The second line will be devoted to rapid and routine diffraction measurements using the rotation method. The centre of the available fan of radiation is for the third line, it is planned to serve first as a monitor of the whole X-ray beam vertical stability but will be equipped ultimately for Laue experiments to perform dynamical studies of proteins.

   The detailed plans for this port can be found in Staudenmann *et al* (1989).

5.6.9 ADONE (Frascati)

   This instrument is situated on the wiggler magnet beam line of the ADONE storage ring. The wiggler parameters are $\lambda_c=4.5$ Å (for ADONE at 1.5 GeV, magnetic field 1.85 T), with five full and two half-poles. The electron beam size is $\sigma_x=1.8$ mm, $\sigma_y=0.66$ mm with $\sigma_x'=0.16$ mrad and $\sigma_y'=0.026$ mrad; in the horizontal two sources can be seen ~11 mm apart (Burattini *et al* 1983). There is a Si(220) double crystal monochromator 31 m from the tangent point. The sample is 33.75 m from the tangent point. There is a four-circle Huber diffractometer mounted with the ω-axis horizontal (2θ detector moving in the vertical plane); see figure 5.37. A useful advantage of this beam line is the highly parallel nature of the beam in the vertical direction especially. The angular width con-

Figure 5.37 The four-circle diffractometer at the ADONE ring, Frascati. Figure kindly supplied by M. Colapietro and reproduced with permission.

tributed to an ω-scan of a protein crystal Bragg reflection due to source and monochromator characteristics is $(8''+3.5'')=11.5''$ $(0.003°)$ at 8 keV. This is useful, *inter alia*, for probing protein crystal perfection using the fine angular step on ω of $0.0025°$, see figure 2.8(c). The instrument has been developed by Colapietro and Spagna and coworkers.

5.6.10 Novosibirsk (USSR)

5.6.10.1 VEPP-3 station 5-A

A service facility for oscillation film camera data collection is being established on the VEPP-3 storage ring (Popov *et al* 1989). The beam line optics consists of an $8°$ oblique-cut Si(111) or Ge(111) triangular-shaped crystal. The photon flux at the sample for 100 mA with a vertical mirror and germanium crystal at 8 keV is calculated to be 10^{12} photons s^{-1} mm^{-2} on VEPP-3. Data collection is via an Arndt–Wonacott rotation camera. The station is tunable over a range of wavelengths between 0.8 and 1.8 Å with $\delta\lambda/\lambda=10^{-3}$. For example, oscillation camera data on catalase has been collected to 1.8 Å resolution.

5.6.10.2 VEPP-3 station 2-C

This is a Laue diffraction station (figure 5.38) primarily for studying solid state chemical reactions (e.g. dehydration of calcium sulphate) (Gaponov *et al* 1989a,b,c). The station is fed by a 2 T single bump wiggler with a bandpass of energies between 4 and 25 keV. A two-dimensional MWPC is available (Baru *et al* 1978, 1983) with 128×128 channels of size 2 mm×0.7 mm and a counting rate of 3.0 MHz at 50% losses.

The station has been used to record Laue diffraction patterns of protein crystals on photographic film (Popov, pers. comm.).

5.6.10.3 VEPP-4

It is planned to set up a Laue diffraction station on the VEPP-4 ring to provide a broader bandpass and higher intensity than the VEPP-3 station 2-C referred to above. For the solid state chemical reaction studies a new two-dimensional multiwire X-ray detector is planned with 256×256 channels, with each channel ~1.5 mm in size.

5.6.11 PEP (Stanford)

PEP is a colliding-beam storage ring that has been used for synchrotron research parasitic on high energy physics runs (Bienenstock, Brown, Wiedemann and Winick 1989). PEPs main features as an SR source are a high particle energy (up to 16 GeV), long straight sections (six are 117 m

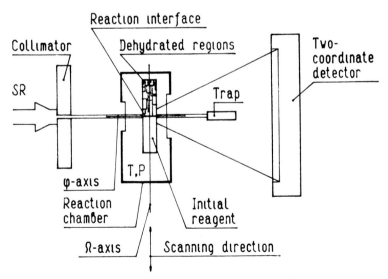

Figure 5.38 The VEPP-3 station 2-C for studying solid state chemical reactions. From Gaponov *et al* (1989b) with permission.

long), large circumference (2200 m) and low bending magnet field (0.32 T at 16 GeV). It has operated in a low emittance mode with an emittance of less than 12 nm rad at 8 GeV.

Undulators are installed in two straight sections at present with emission lines in the energy range 12–42 keV (Lavender *et al* 1989). Experience gained on these lines will be relevant to the ESRF and APS developments.

The potential for use of PEP as an SR source in general and for macromolecular crystallography in particular is discussed in the proceedings of the workshop held in 1987 (Coisson and Winick 1987).

5.6.12 ESRF (Grenoble)

The ESRF machine (details can be found in Laclare (1989)) is due to come on-line in 1993 or so. Three instruments have been proposed by the macromolecular crystallography community. For a discussion see Helliwell (1987) in the ESRF Red Book.

5.6.12.1 ESRF-1

This uses a multipole wiggler and will have operational modes for focussed Laue diffraction work and monochromatic experiments. The small source sizes should allow an equivalently small focal spot from a grazing incidence mirror system. Exposure times in the microsecond range for a macromolecular crystal should be feasible. Depending on the current achieved in single bunch mode it may be possible, at least for smaller unit cell sizes, to record a Laue pattern from one of the single bunches with an intrinsic time resolution therefore of the bunch width. (Feasibility experiments of this kind have been conducted at CHESS but on an undulator (Szebenyi *et al* 1989).)

5.6.12.2 ESRF-2

This utilises an X-ray undulator with the fundamental interference X-ray line being at a wavelength $\lambda \approx 1.0$ Å. The horizontal and vertical source sizes and divergences should be extremely small at about 0.2 mm and 0.1 mrad in both the horizontal and vertical. This high brilliance should be especially effective for very large unit cell data collection.

5.6.12.3 ESRF-3

It has been proposed that an ESRF bending magnet be used for MAD macromolecular crystallography experiments. This source of X-rays has a low emittance compared with the national machines. This instrument will provide a European centre for the development and application of this technique.

5.6.13 APS (Argonne National Laboratory, Chicago)

The APS (Moncton, Crosbie and Shenoy 1989) has published plans for two macromolecular crystallography beam lines (Westbrook 1988).

5.6.13.1 APS-1

This is for multiple energy anomalous dispersion (MEAD) data collection. It is proposed that this uses a multipole wiggler of 20 periods, critical energy 9.8 keV and a calculated flux at this energy of 4.5×10^{14}.

5.6.13.2 APS-2

This is for 'routine' data collection from macromolecular crystals. It is proposed that this uses a bending magnet as source with critical energy of 19.5 keV.

At Argonne development of a CCD-based area detector for macromolecular crystallography is under way (Strauss *et al* 1987; Naday *et al* 1987); see section 5.4.2.5.

Plans are developing rapidly for macromolecular crystallography on APS. These involve different consortia e.g. from Chicago University (Moffat and Schildkamp pers. comm.), and from industry (N. Jones pers. comm.).

CHAPTER 6

Monochromatic data collection

The use of focussed, monochromatised radiation at the synchrotron has so far yielded the most results in terms of biological molecular structure compared with the other methods being developed. This is readily explained because of the ease with which the monochromatic diffraction data measured at the synchrotron have been processed with existing computer programs for data from monochromatic, emission line, laboratory X-ray sources. In contrast, the Laue method, although it is being very actively developed at the synchrotron (chapter 7), had been abandoned in the home laboratory. Hence, the monochromatic method is covered first in this book. In appendix 1 details are given of the various monochromatic diffraction geometries. These geometries are:

(a) monochromatic still exposure;
(b) rotation/oscillation geometry;
(c) Weissenberg geometry;
(d) precession geometry;
(e) diffractometry.

Quantitative X-ray crystal structure analysis usually involved methods (b), (c) and (e) although (d) has certainly been used. Photographic film is being replaced by use of electronic area detectors or, even more recently, the IP.

At the various synchrotrons all these geometries have been exploited for macromolecular crystal data collection as they have also on conventional X-ray sources. Once the polychromatic synchrotron X-ray beam has been rendered monochromatic the single crystal data can be measured and processed as for a conventional X-ray source. These standard procedures will be discussed briefly before moving on to cover the synchrotron specific aspects. These latter include SR instrument smearing effects and polarisation corrections. There is also a variety of wavelength-dependent effects which include geometric factors as well as sample absorption, sample radiation damage and heating.

6.1 FUNDAMENTALS

In chapter 2 the relationship was given between the electron density $\varrho(x,y,z)$, which describes the contents of the unit cell of the crystal, and the set of structure factors $\mathbf{F}(\mathbf{h})$. The amplitude component, $|\mathbf{F}(\mathbf{h})|$, is related to the intensity in the diffracted beam by a formula derived by Darwin (1914). The total energy in a diffracted beam, from a particular reflecting plane $(hk\ell)$ for an ideally mosaic crystal rotating with constant angular velocity ω through the reflecting position is

$$E(hk\ell) = \frac{e^4}{m^2 c^4 \omega} \, I_0 \lambda^3 \, LPA \, \frac{V_x}{V_0^2} \, |\mathbf{F}(\mathbf{h})|^2 \tag{6.1}$$

where I_0 is the intensity of the incident X-ray beam of wavelength, λ, P is the correction for polarisation, L is the Lorentz factor correction for the relative time spent by the RLPs in the reflecting position, A is a correction for absorption of the sample (P, L and A are different for each reflection), V_x is the volume of the crystal sample and V_0 is the volume of the unit cell.

Various factors in equation (6.1) can be merely regarded as constants of proportionality for all the reflections in a given data set, namely

$$\frac{e^4}{m^2 c^4 \omega} \, I_0 \lambda^3 \, \frac{V_x}{V_0^2} \tag{6.2}$$

provided the following conditions are met:

   (a) the angular rotation speed of a crystal is one constant value;
   (b) the incident intensity is constant;
   (c) one fixed wavelength is used for a given data set;
   (d) the crystal is fully bathed in a uniform X-ray beam so that the volume of crystal illuminated is constant for any orientation angle of the sample;
   (e) the unit cell volume is fixed as it will be for the case of one particular type of crystal.

Condition (b) is not always met. At the synchrotron the incident intensity varies with time as the circulating beam current decays or the focal spot drifts slightly off the collimator entrance hole due to source or optical instabilities. A simple method to overcome these effects is to oscillate the sample quickly over a fixed angular range (i.e. as in the rotation photographic method – see appendix 1) so that over the time required to sweep through one oscillation angle period the beam intensity is constant to some acceptable level (e.g. better than 0.5%). Alternatively, the beam intensity is monitored as a function of time and written to a computer file. This latter method is used for electronic area detector or diffractometer data collection at the synchrotron.

The remaining factors in equation (6.1) have to be calculated for each reflection separately, i.e. *LPA*, the Lorentz, polarisation and absorption corrections respectively. The following relationship will now be considered in more detail:

$$E(hk\ell) \propto LPA \, | \, \mathbf{F}(\mathbf{h})|^2 \qquad (6.3)$$

It is most important to note that the Lorentz factor is essentially proportional to $1/\sin\theta$, i.e. to $1/\lambda$ and therefore $E(hk\ell)\propto\lambda^2$ when account of this is taken. The total energy in the diffracted beam is estimated by summing the scattered counts over the angular reflecting range of rocking width of the reflection $hk\ell$ (e.g. see figure 2.8(c)). A correction for the background in the vicinity of a reflection or spot is made by subtracting a number of counts, for a given angle $\delta\theta$, from the total counts at each setting.

An ideally imperfect crystal, for which equation (6.1) holds, is defined as being made up of many small blocks randomly misaligned with respect to each other (figure 2.8(b)). Each block is perfectly crystalline and has an intrinsic rocking width, $\eta_{hk\ell}$, due to the non-infinite number of unit cells contributing to the block. An expression for $\eta_{hk\ell}$ is derivable from the dynamical theory of X-ray diffraction, as (given in SI units):

$$\eta_{hk\ell} = \frac{1}{\pi^2\varepsilon_0} \frac{d^2 e^2}{mc^2} \frac{\tan\theta_{\mathrm{B}}}{V_0} \, | \, \mathbf{F}(\mathbf{h})| \qquad (6.4)$$

where d is the interplanar spacing for the particular Bragg plane in question, θ_{B} is its Bragg angle, V_0 the unit cell volume and $|\mathbf{F}(\mathbf{h})|$ the structure amplitude. The proportional increase in $\eta_{hk\ell}$ due to large $|\mathbf{F}(\mathbf{h})|$ can be understood in the following simple way. If the top plane of atoms in the block reflects 1%, say, of the incident X-ray intensity then it will need approximately 100 layers for the beam to be completely reflected. In such a case the spectral linewidth (cf. diffraction grating theory) would be ~1/100. For a stronger reflection the incident beam would penetrate fewer layers and the angular width would increase. In the case of an organic crystal the scattering strength per layer of atoms is weak so that many layers are penetrated and $\eta_{hk\ell}$ is very small (e.g. calculations using equation (6.4) for a 100 Å unit cell protein crystal indicate a few arc seconds for $\eta_{hk\ell}$ (Helliwell 1988).

The blocks making up a 'mosaic' crystal, in the model of an ideally imperfect crystal, are, however, slightly misaligned in angle with respect to one another. The overall angular misalignment is the so-called mosaic spread, η, intrinsic to the crystal sample and where $\eta > \eta_{hk\ell}$. Typical values of η are 0.1° for a protein crystal but at the synchrotron values as small as 0.01–0.02° for η for particular crystals have been observed (figure 2.8(c)). Less well-ordered samples have η's in the range 0.5–1.0° or even larger.

The rocking width, ϕ_R, measured for a given reflection, ignoring $\eta_{hk\ell}$ above as a very small effect, is determined by η and the X-ray beam divergence angles and spectral bandwidth. Additionally, ϕ_R is a function of the reciprocal coordinates (see appendix 1, figure A1.3) of a given reciprocal lattice point (RLP). Generally, ϕ_R increases gradually with increasing θ (due to the effect of spectral smearing) and dramatically the nearer the RLP is to the rotation axis (becoming infinite on the axis). The actual expression for ϕ_R depends on the instrument (i.e. the X-ray beam parameters) and on the diffraction geometry and will be discussed below.

The measurement of $E(hk\ell)$ requires the full rotation of the crystal sample over the rocking width ϕ_R. Alternatively, in the rotation/oscillation method, a given diffraction photograph is recorded over an angular range several times larger than ϕ_R. Then a subset of the reflections is fully stimulated and a subset is partially stimulated. It is feasible to add intensities of partials on contiguous films. With diffractometry methods the rocking width intensity profile of a reflection is measured over finely subdivided intervals.

6.2 INSTRUMENT SMEARING EFFECTS

6.2.1 Prediction of partiality and angular reflecting range

There are important differences in the diffraction geometry with SR optical systems compared with a conventional laboratory arrangement (Greenhough and Helliwell, 1982a,b). In particular, at an SR source the incident spectral bandwidth can be varied over a wide magnitude, the direction of an incident ray is correlated with the photon energy (in some situations) and a highly asymmetric beam cross-fire usually exists. With a conventional source experimental arrangement consisting of either a nickel filter or a flat graphite crystal the spectral bandwidth is determined by the $K\alpha_1$, $K\alpha_2$ split ($\delta\lambda/\lambda=2.5\times10^{-3}$ for Cu $K\alpha$); each ray direction carries this spread and the beam cross-fire is symmetric. In some applications, focussing mirrors are used with a conventional source to produce adequate spot resolution for virus work (Harrison 1968) and a more asymmetric beam cross-fire may then exist.

Greenhough and Helliwell (1982b) produced a theory for the general setting of an instrument based on the bent, triangular monochromator (section 5.2.3). For routine data collection the monochromator focussing distance is set at the Guinier condition so that $(\delta\lambda/\lambda)_{corr}=0$ (equations 5.7) and (5.9)). To use this sort of instrument at different wavelength settings it is not always possible to realise the Guinier setting. In such cases the monochromator illuminated length is restricted (using slits) so that $(\delta\lambda/\lambda)_{corr}$ is always kept less than ≈0.001 (section 5.2.3.1). For this reason and the interest in the dispersive setting (section 5.2.3.2) the general treatment is important and is summarised here. Table 6.1(a) gives parameters and values at different settings of such an instrument. It may be noted in passing that Schoenborn (1983) has made a derivation, for the neutron diffraction case. This treatment is similar to the SR case because both radiations are polychromatic in nature. For the horizontal dispersing case, we may construct a model for formulating the interaction of an RLP and the Ewald sphere for the prediction of reflections and angular reflecting range (figure 6.1). The angular reflecting range ϕ_R for the protein crystal sample is given, in the case of a horizontal rotation axis (the most common), by

$$|\phi_R|\approx[L^2(\delta d^{*2}+\zeta\gamma_H)^2+\gamma_V^2]^{\frac{1}{2}}+2\varepsilon_s L \qquad (6.5)$$

where L is the Lorentz factor, γ_H and γ_V are the incoming horizontal and vertical cross-fire angles at the sample, δ is $\frac{1}{2}(\delta\lambda/\lambda)_{corr}$ (equation 5.6) or (5.9)), d^* is as given in figures 2.11(b) and (c) and ε_s is given by

$$\varepsilon_s=\frac{d^*\cos\theta_{hk\ell}}{2}\left[\eta+\left(\frac{\delta\lambda}{\lambda}\right)_{conv}\tan\theta_{hk\ell}\right] \qquad (6.6)$$

where η is the mosaic spread and $(\delta\lambda/\lambda)_{conv}$ is given by $\eta_{diff}^{mono}\cot\theta_{mono}$ (equation (5.10)) (Greenhough and Helliwell 1982a,b).

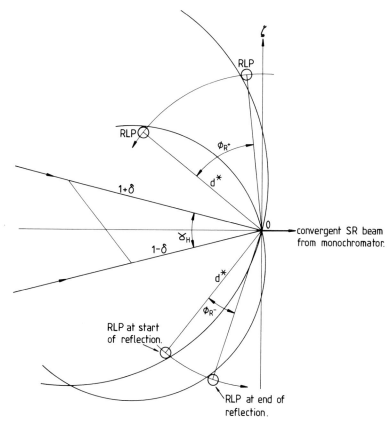

Figure 6.1 Prediction of partiality and angular reflecting range of a
given RLP. The most commonly used instrument so far for
monochromatic data collection is that based on the bent,
triangular monochromator described in section 5.2.3. In its
general setting (figure 5.6(c)) the beam convergence angle
γ_H has associated with it a 'correlated' $\delta\lambda/\lambda$ term (equation
(5.9)) shown here as $\delta=\frac{1}{2}(\delta\lambda/\lambda)_{corr}$. Usually the instrument is
set at the Guinier setting for which $\delta=0$ or γ_H is limited in
value (by slits) to reduce the value of δ to ~0.001 (section
5.2.3.1). The RLP is not actually a 'point' but is finite in size
because of sample mosaic spread. A given $hk\ell$ plane is therefore in the reflecting position, not just instantaneously as
depicted in figure 2.11(b), but over an angle ϕ_R. The figure
here shows two RLPs, in the zero level, both at d* from the
origin of reciprocal space but on opposite sides ($\pm\zeta$) of the
beam with reflecting ranges ϕ_{R+} and ϕ_{R-}. A vertical rotation
axis is shown as the example for clarity (compared to a
horizontal axis). Table 6.1(b) shows values for ϕ_R (at $\pm\zeta$)
for Guinier and non-Guinier settings of an instrument (Table
6.1(a)). Extracted from Greenhough and Helliwell (1982b).

Table 6.1. *Instrument smearing effects: Example instrument settings based on a bent triangular monochromator (section 5.2.3) focussing in the horizontal plane and a horizontal rotation axis of the crystal sample.*

(a) *Beam cross-fire angles and contributions to the spectral spread for a fixed focal distance p' = 2.4 m and variable wavelength. Parameters as in the notes below.*

$\lambda(\text{Å})$	$\theta_{mono}(°)$	$(\delta\lambda/\lambda)_{corr}$	$\eta_{diff}^{mono}\cot\theta_{mono}$	$(x_H/p')\cot\theta_{mono}$	$(L\sin r)/p'$ (mrad)
1.474	13.04	0.0	0.0010	0.0005	3.88
1.608	14.24	0.0033	0.0008	0.0005	5.62
1.75	15.53	0.0061	0.0006	0.0005	7.49
2.08	17.90	0.0100	0.0005	0.0004	10.92
2.25	20.14	0.0128	0.0004	0.0003	14.14
2.5	22.49	0.0150	0.0003	0.0003	17.50

Notes:
(1) $\eta_{diff}^{mono} = \eta^{mono}(\sin i/\sin r)^{\frac{1}{2}} = 45''$, $=0.22$ mrad ($\eta^{mono} = 16''$, for Ge(111) monochromator).
(2) $(\delta\lambda/\lambda)_{corr}$ defined in equation (5.9). $(L\sin r)/p'$ in equation (5.13) (L=length of monochromator illuminated).
(3) Oblique cut angle of monochromator, $\alpha, = 10.37°$; $i=(\theta_{mono}+\alpha)$, $r=(\theta_{mono}-\alpha)$ see equations (5.7)–(5.11).
(4) x_H is the horizontal width of beam accepted by the collimator in front of the sample, here assumed to be 0.3 mm.

(b) *Reflecting ranges (°) for p'=2.4 m and horizontal rotation axis with the monochromator set at the Guinier condition (λ=1.474 Å) and at overbend (λ=2.008 Å).*

	$[L^2(\delta d^{*2}+\zeta\gamma_H)^2+\gamma^2_v]^{\frac{1}{4}}$		$d^* \cos\theta_{hk\ell} [\eta+(\delta\lambda/\lambda)_{conv} \tan\theta_{hk\ell}]L$		ϕ_R (°)	
	λ=1.474 Å	λ=2.008 Å	λ=1.474 Å	λ=2.008 Å	λ=1.474 Å	λ=2.008 Å
2.5 Å resolution						
ζ=0.7		2.78		0.40		3.18
0.5	0.44	0.93	0.27	0.17	0.71	1.10
0.3	0.14	0.56	0.15	0.13	0.29	0.69
0.1	0.04	0.34	0.13	0.12	0.17	0.46
−0.1	0.04	0.17	0.13	0.12	0.17	0.29
−0.3	0.14	0.01	0.15	0.13	0.29	0.14
−0.5	0.44	0.24	0.27	0.17	0.71	0.41
−0.7		1.14		0.40		1.54
4 Å resolution						
ζ=0.4		1.18		0.20		1.38
0.3	0.34	0.69	0.21	0.14	0.55	0.83
0.2	0.15	0.45	0.14	0.12	0.29	0.57
0.1	0.07	0.28	0.12	0.12	0.19	0.40
0	0.00	0.15	0.12	0.11	0.12	0.26
−0.1	0.07	0.02	0.12	0.12	0.19	0.14
−0.2	0.15	0.12	0.14	0.12	0.29	0.24
−0.3	0.34	0.31	0.21	0.14	0.55	0.45
−0.4		0.66		0.20		0.86
8 Å resolution						
ζ=0.2		0.97		0.18		1.15
0.15	0.33	0.57	0.19	0.13	0.52	0.70
0.1	0.15	0.36	0.13	0.12	0.28	0.48
0.05	0.07	0.20	0.11	0.11	0.18	0.31
0	0.00	0.07	0.11	0.11	0.11	0.18
−0.05	0.07	0.06	0.11	0.11	0.18	0.17
−0.1	0.15	0.20	0.13	0.11	0.28	0.32
−0.15	0.33	0.39	0.19	0.12	0.52	0.52
−0.2		0.73		0.18		0.91

Notes:
(1) Sample mosaic spread set at 0.1°. (Left-of-film reflection has +ζ.)
(2) *L* here is the Lorentz factor. For other symbols see text.

In order to decide whether a reflection is partial or not for a given crystal orientation the sample reflection rocking width Δ or a spherical reciprocal lattice volume of radius E can be compared with a unit Ewald sphere; these are given by:

$$E = \frac{1}{2}\left[(\delta d^{*2} + \zeta \gamma_H)^2 + \frac{\gamma_V^2}{L^2} \right]^{\frac{1}{2}} + \varepsilon_s \qquad (6.7)$$

with

$$\Delta = 2E / d^* \cos \theta_{hk\ell} \qquad (6.8)$$

At the Guinier position, $\delta=0$ and the equations for ϕ_R and E reduce to the conventional source type (with asymmetric cross-fire) (Greenhough and Helliwell 1982a). Away from the Guinier position ϕ_R (top of film)$=\phi_R$ (bottom) but ϕ_R (left)$\neq\phi_R$ (right) for comparably placed reflections. At overbend (δ positive) ϕ_R (left)$>\phi_R$ (right) with the reverse true at under-bend. Large left–right differences in reflecting range can thus occur if $(\delta\lambda/\lambda)_{corr}$ is as large as 0.01 (Greenhough, Helliwell and Rule 1983). Table 6.1(b) gives values of ϕ_R for an example diffraction pattern but at two different instrument settings, namely Guinier and dispersive settings.

6.2.2 Special case: energy profiles

In section 5.2.3.2 the means for establishing an energy profile across a crystal sample reflection was discussed (figure 5.8). The theory developed for the prediction of partiality in the SR case (section 6.2.1) can be extended to give an expression for the $\delta\lambda/\lambda$ at any point in the Bragg reflection profile. It can be shown (Greenhough et al 1983) that the wavelength resolution is given by

$$\frac{\delta\lambda}{\lambda} = \frac{\partial\delta}{\delta_{corr}} = \frac{2}{(\partial H)_{corr}}(|(\partial H)_\eta| + |(\partial H)_t| + |(\partial H)_\omega^E| + |(\partial H)_{foc}^E| + |(\partial H)_{\gamma_V}|) \qquad (6.9)$$

where $(\partial H)_{corr}$ is the spot width along the direction of the energy profile due solely to δ_{corr}. The other ∂H components are the energy smearing factors due to sample mosaic spread, η, and thickness, t, monochromator rocking width ω, the energy gradient across the focus and the vertical divergence angle in the beam γ_V. The theory attempts to 'map' a diffraction spot in terms of the range of energies that contribute to each point within the spot from the various possible smearing factors. Values of $\delta\lambda/\lambda$ of 10^{-3} are achievable; not as good as the step scanning method but the technique has the advantage of sampling simultaneously all possible f', f'' values near to an absorption edge. With such a method the 'information content' of a diffraction pattern is enhanced considerably. To record the data a better detector than film is required in terms of dynamic range whilst preserving a fine spatial resolution; the CCD (section 5.4.2.5) may well be appropriate.

6.2.3 Prediction of spot size and shape

The same theory allows almost exact prediction of spot size and shape around a film. Even for $\delta_{corr}=0$ a very definite spot shape can be predicted around a film due to the exact collimation of the SR beam.

By setting $\delta_{corr}=0$ and limiting the cross-fire angles with slits, the final size of a spot is *controllable* (at the expense of flux at sample). With photographic film, with its small grain size, the order to order resolution virtually has no limit so that unit cells well in excess of $1000\,\text{Å}$ (virus case) are tractable with a reasonable flux at the sample (see figures 10.6(b) and 10.18). A practical consequence is that to avoid problems due to the Wooster (1964) effect with photographic film requires the use of small raster sizes (50μm or even 25μm) on automatic densitometers.

6.3 LORENTZ AND POLARISATION FACTORS

On the right-hand side of relation (6.3) are the L, P and A factors. The Lorentz factor, L, takes account of the relative time each reflection spends in the reflecting position. It depends on the precise diffraction geometry used. For example, in the rotation method

$$L = \frac{1}{(\sin^2 2\theta_{hk\ell} - \zeta^2)^{\frac{1}{2}}} \tag{6.10}$$

(i.e. L is roughly proportional to $1/\lambda$).

In the case of an unpolarised incident beam (such as a conventional X-ray source beam without monochromator) the polarisation factor is given by

$$P = \tfrac{1}{2}(1 + \cos^2 2\theta_{hk\ell}) \tag{6.11}$$

The form of this correction can be understood in the following way. For any reflecting plane it is possible to resolve the incident beam into states of polarisation parallel and perpendicular to the plane whereby

$$I_\| = \tfrac{1}{2}I_0 = I_\perp \tag{6.12}$$

After reflection $I_\perp$ is attenuated by $\cos^2 2\theta_{hk\ell}$ so that the overall reflected beam intensity

$$I_R = I_\|^R + I_\perp^R = I_\| + I_\perp \cos^2 2\theta_{hk\ell}$$

$$= \tfrac{1}{2}(1 + \cos^2 2\theta_{hk\ell})I_0 \tag{6.13}$$

The introduction of a monochromator on the conventional X-ray source means that the beam incident onto the sample is already partially plane polarised. At the synchrotron the incident beam itself is partially plane polarised. This more complicated situation is now dealt with.

6.3.1 Polarisation corrections for the synchrotron case

The polarisation state of the synchrotron beam has been discussed in section 4.4. The polarisation state of the beam at the sample is determined by the fraction of the parallel and perpendicular components reflected by the monochromator (there may be several reflections, i.e. two for a double crystal monochromator), and the relative fraction of the parallel and perpendicular components incident on the mono-chromator from the source. This depends on the vertical angular aperture of the source that the sample sees and can be calculated from the source characteristics; the size of the vertical aperture is dependent on whether a focussing mirror is used as this increases the aperture subtended by the sample at the source.

The earliest consideration of the polarisation correction with SR was for the precession camera and formulae were derived by Phillips *et al* (1977).

For the oscillation camera placed on a horizontally dispersing mono-chromator Kahn *et al* (1982a), working from Azaroff (1955), derived the polarisation correction P as

$$P = P_0 - P' \tag{6.14}$$

where

$$P_0 = \frac{(1 + \cos^2 2\theta_{hk\ell})}{2} \qquad (6.15)$$

and

$$P' = \frac{\tau'}{2} \cos 2\psi \sin^2 2\theta_{hk\ell} \qquad (6.16)$$

where

$$\tau' = \frac{\alpha(1 + \tau) - (1 - \tau)}{\alpha(1 + \tau) + (1 - \tau)} \qquad (6.17)$$

$\theta_{hk\ell}$ is the sample reflection Bragg angle for a given $hk\ell$, ψ is the azimuthal angle in the film, α depends on the monochromator crystal ($\alpha = \cos 2\theta_M$ for a perfect crystal, θ_M is the monochromator Bragg angle) and

$$\tau = \frac{I_\| - I_\perp}{I_\| + I_\perp} \qquad (6.18)$$

is the source-dependent feature since $I_\|$ is the flux delivered by the source with a parallel component of polarisation and $I_\perp$ is the vertical component. We assume that a bent perfect crystal remains perfect so that dynamical theory still applies; for the typical bending radii used (>20 m), this assertion is quite reasonable.

For an imperfect crystal, α would be $\cos^2 2\theta_M$. Azaroff *et al* (1974) discussed the possibility of an intermediate power of n in $\cos^n 2\theta_M$ between 1 and 2 for intermediate monochromator crystal states. This would apply if graphite is used, for instance, when tests for the mosaic character of the monochromator crystal would be needed.

6.4 ABSORPTION OF X-RAYS

An X-ray beam passing through a material suffers absorption and its intensity is attenutated. The absorbed X-rays cause thermal heating and radiation damage; we will discuss these in detail in section 6.5. In the context of the current discussion the derivation of $|\mathbf{F}_{hk\ell}|$ from $E(hk\ell)$ requires an absorption correction to be applied. The simplest situation is for the case of a spherical crystal (without a capillary) completely bathed in a uniform X-ray beam. In such a case all the reflections would be equally reduced in intensity. The situation of such a spherical crystal never exists in macromolecular crystallography.

For a beam of monochromatic X-rays passing through an isotropic material the transmitted beam has intensity

$$I = I_0 \exp(-\mu x) \tag{6.19}$$

where I_0 is the incident intensity, μ is the linear absorption coefficient and x is the path length through the material.

For a material consisting of a number of elements, N, the overall mass absorption coefficient μ_m is given by

$$\mu_m = \sum_{i=1}^{N} g_i (\mu_m)_i \tag{6.20}$$

where g_i is the mass fraction of the ith element and $(\mu_m)_i$ the mass absorption coefficient of the ith element, and

$$\mu = \varrho \mu_m \tag{6.21}$$

where ϱ is the density. Usually, ϱ is expressed in $g\,cm^{-3}$, μ_m in $cm^2\,g^{-1}$ and hence μ in cm^{-1} (although mm^{-1} is common). Values of μ_m and ϱ are given in appendix 3.

The mass absorption coefficient, μ_m, varies with wavelength according to the following relationship, in the absence of elemental absorption edges,

$$\mu_m = a\lambda^3 + b\lambda^4 \tag{6.22}$$

where a and b are constants of proportionality. This is the so-called Victoreen relationship and is dominated primarily by the λ^3 term.

There are discontinuities in the wavelength dependency that occur at elemental absorption edges. A remarkable example in protein crystallography is the case of ferritin. This protein stores up to 4000 iron atoms. X-ray crystallography of these samples required use of a $\lambda >$ Fe K edge or a very short λ to avoid pronounced absorption effects and subsequent radiation damage (Smith, Helliwell and Papiz 1985; Smith *et al* 1989b). For all other cases metal atoms are present in proteins at trace element concentrations (e.g. 1 or 2 metal atoms/25 kD). Between any two absorption edge wavelengths the Victoreen relation is followed, although the values of a and b will be different for each wavelength region. A problem related to that of ferritin was the study of Cs-DNA by fibre diffraction using SR from VEPP-3 (Skuratovskii, Kapitonova and Volkova 1978). With Cu Kα radiation, absorption effects of the caesium made Cs-DNA patterns of reasonable quality impossible to obtain. By analysing the problem of a cylindrical fibre and variable wavelength, the calculated optimum wavelength of 1.2 Å was used and gave interpretable patterns. These data enabled the caesium ion positions to be determined (Skuratovskii, Volkova, Kapitonova and Bartenev 1979).

The effect of absorption on the reflection intensities is obviously to reduce them; in the absence of a correction this would affect the estimation of individual atomic temperature factors in refinement. The absorption corrections applied to individual reflections are usually different from one to another. In the case where a single data set is made up of several crystals (because of radiation damage), each of variable shape, then the lack of an absorption correction will leave systematic errors in the data. Even when a single crystal is used for a complete native data set and an identically shaped and mounted crystal is used as the heavy atom derivative, the lack of an absorption correction can seriously affect the measured isomorphous or anomalous differences.

Considerable thought has gone into devising absorption corrections. With the synchrotron, when there is not a requirement to use a particular wavelength for anomalous dispersion (see chapter 9), then as short a wavelength as possible is used to minimise absorption variations – see figure 6.3 – (and reduce radiation damage). This is a point that will be detailed later.

Consider a small volume element, dV, of a crystal (figure 6.2) and let us define the path lengths of the primary (t_p) and secondary or reflected (t_s) beams as these beams enter and leave the crystal, to and from dV, respectively. The attenuation of the secondary beam originating at dV is given by

$$I = I_0 \exp[-\mu(t_p + t_s)] \tag{6.23}$$

For the whole crystal this formula is integrated over all volume elements dV in the crystal.

For small molecules mounted on a glass fibre, the crystal can often be described by indexing the faces of the crystal and estimating their distance from a common point. In such a case a numerical integration can be used. However, this approach is not valid when the crystal is mounted in a capillary surrounded by mother liquor.

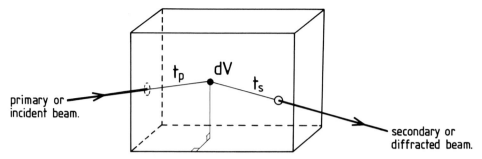

Figure 6.2 The analytical calculation of a crystal sample correction
involves the integration over the whole illuminated volume
of the crystal. dV is a volume element, t_s and t_p are the path
lengths through the crystal sample of the secondary (diffracted) and primary (incident) beams. In macromolecular
crystallography this calculation is never done because the
crystal is bathed in a blob of mother liquor, which may well
decrease or swell during the experiment. A better method is
to use short wavelengths (e.g. 0.9 Å or in future 0.33 Å) – see
figure 6.3.

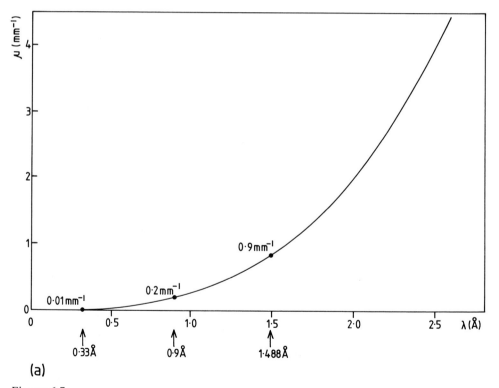

(a)

Figure 6.3

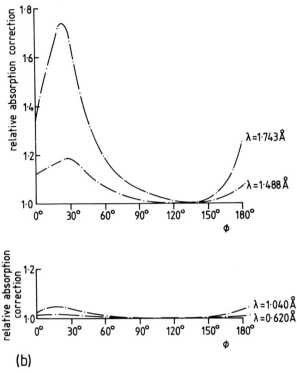

(b)

Figure 6.3 (*cont.*) (*a*) The variation of linear absorption coefficient
($\mu\,mm^{-1}$) with wavelength for proteinaceous material. The
wavelengths marked are those typically used for routine data
collection at SR sources (e.g. 1.488 Å at station 7.2 SRS, 0.9 Å
at station 9.6 SRS) and 0.33 Å as a future standard
wavelength. (*b*) The absorption correction measured for a
protein crystal rotated about an axis, ϕ, for four different
wavelengths 1.743 Å, 1.488 Å, 1.040 Å and 0.620 Å. Extracted
from Helliwell *et al* (1984b).

Alternatively, an experimental measurement can be made of the relative attenuation of a particular diffracted beam constantly set in the
diffraction condition as the crystal is rotated, for example. In this case
relative transmission factors can be calculated as a function of the
angular setting. A different method of getting the absorption reference
curve is to measure the transmission of the primary beam through the
crystal; this requires that the size of the primary beam is smaller than the
smallest dimension of the crystal (Helliwell *et al* (1984b) and figure
6.3(*b*)). Now for the reflection which is to be corrected, the angles of its
primary and secondary beams are known. The relative transmission factors for the primary, T_p, and secondary, T_s, beams can be determined

from the reference curve. The overall transmission factor is then assumed to be

$$T_{p,s} = (T_p + T_s) / 2 \qquad\qquad (6.24)$$

according to North, Phillips and Matthews (1968) or

$$T_{p,s} = (T_p T_s)^{\frac{1}{2}} \qquad\qquad (6.25)$$

according to Kopfmann and Huber (1968). These corrections may be difficult to measure and apply. A major obstacle is when the crystal sample rapidly deteriorates in the X-ray beam and sample size varies. This is common in macromolecular crystallography.

The range of the magnitudes of the correction that have to be applied to a protein crystal for Cu Kα wavelength can be estimated based on a μ of $\approx 1\,\mathrm{mm}^{-1}$ and crystals up to 1 mm thick, from equation (6.19). For comparison at a shorter wavelength, 0.9 Å, μ is reduced to $0.36\,\mathrm{mm}^{-1}$. Clearly, the absolute and relative absorption effects decrease considerably for the shorter wavelength example (see figure 6.3).

6.5 RADIATION DAMAGE AND SAMPLE HEATING

Radiation damage and sample heating arise from the absorption of X-rays in the sample. The sensitivity of the specimen to these effects depends markedly on the temperature of the specimen. Initially comments will be restricted to room (or near room) temperature where protein crystals maintain their liquid-like nature in the solvent channels going through the crystal.

Ionisation processes occur resulting in free radicals, which are potent agents for chemically damaging biological macromolecules. These free radicals have a natural diffusion rate and, hence, radiation damage is dose-rate-dependent, i.e. a larger total dose can be tolerated by a sample if that dose is delivered at a higher rate.

There are data available on dose rate and dose tolerances. Doses of 10^9 rad destroy the structure of organic molecules and 10^6 rad kill living cells (Bordas and Mandelkow 1983). These authors discuss the case of microtubule polymerisation; levels of 30 000–50 000 rad delivered at rates of 2000 rad min^{-1} with medical X-ray sources impair this function whereas on DORIS in solution scattering experiments 1.5×10^6 rad delivered at a rate of 60 000 rad min^{-1} are needed, i.e. an increase of 30 in the rate has increased the total dose that the sample can stand by 30. How does this estimate for a solution compare with a crystal? Blake and Phillips (1962) discuss the effects of 50×10^6 rad of total exposure with a conventional X-ray source on crystals of myoglobin after which only 20% of the protein molecules in the crystal were unaffected by radiation damage. Wilson *et al* (1983) found that with phosphorylase b crystals on a synchrotron source, at least 100× more intense than a conventional source, then 5× more data could be collected per sample. The 'dose rate enhancement' appears to be less for crystals than for solutions; or, more likely, the criteria of sample damage may be different for the different techniques.

Sample heating in SR beams as a consideration was introduced by Stuhrmann (1978); his example is cited in detail in section 6.5.4.1. Helliwell and Fourme (1983) considered radiation damage and sample heating in evaluating the usefulness of the prospective fluxes at the specimen that might be anticipated using the ESRF. Helliwell and Fourme (1983) and Helliwell (1984, pp. 1470–3) discuss the need to go to shorter X-ray wavelengths (e.g. 0.5 Å), to reduce the fraction of absorbed photons, and to use cryotemperatures, with frozen crystals mounted on a copper fibre, to limit the temperature rise experienced by the sample. In this way, frozen microcrystals of protein of size $\approx 10\,\mu$m should be successfully studied on the ESRF. This application of the ESRF was further discussed in Helliwell (1989).

Cryocrystallography of biological macromolecules has been developed by Hope (1988) – section 5.5.2 – as a generally applicable method. A striking success has been its use in ribosome crystal structure studies (section 10.5.2). Hope (1988) has observed that improvements in diffraction pattern lifetimes of at least 1000-fold can be obtained by cooling crystals to liquid nitrogen temperature. Henderson (1990) has compared the behaviour of frozen protein crystals in X-ray versus electron beams; the central conclusion of this paper is as follows. Henderson (1990) 'predicts that, however low the temperature of the specimen, X-ray doses of about 2×10^7 Grays (1 Gray=100 rad) will always destroy the crystalline diffraction from protein crystals'.

Doubts exist about how widely applicable the use of cryotechniques will be for X-ray data collection. Often the mosaicity of the specimen is increased on freezing, which can lead to serious overlap of reflections due to increased spot size and/or angular reflecting range. Whether this can be tolerated or not depends on the size of the unit cell. Hence, finding ways of curtailing radiation damage whilst continuing to operate at room temperature is still important. The use of shorter and shorter wavelengths is tractable with more brilliant (i.e. higher flux more collimated) beams. These methods rely on the intrinsic perfection of a given protein crystal, a property which is indeed remarkable (see figure 2.8(c) and Helliwell (1988)).

A formulation is now given to describe the effect of the interaction of X-rays with protein crystals.

6.5.1 Absorbed heat

The heat absorbed by the sample per second is given by

$$\partial H / \partial t = I_0 d^2 [1 - \exp(-\mu d)] E \times 1.6 \times 10^{-19} \qquad (6.26)$$

where I_0 is the incident intensity in photons s^{-1} mm^{-2} (horizontal mrad)$^{-1}$, d is the size of an isodimensional protein crystal sample in mm, $1-\exp(-\mu d)$ is the absorbed fraction of X-rays, E is the photon energy in eV and 1.6×10^{-19} converts the absorbed eV s^{-1} into W. In the small sample case, we can rewrite equation (6.26) as

$$\partial H / \partial t = I_0 d^2 \mu d E \times 1.6 \times 10^{-19} \qquad (6.27)$$

since $1-\exp(-\mu d)=\mu d$. Note that since $\mu \propto \lambda^3$ and $E \propto 1/\lambda$, then

$$\partial H / \partial t \propto I_0 d^3 \lambda^2 \qquad (6.28)$$

We see that use of a small λ can reduce considerably $\partial H/\partial t$.

6.5.2 Temperature rise in thermal isolation (adiabatic case)

For a given mass of sample, m, with specific heat capacity, s, the rate of increase in temperature, T, with time, t, is

$$\frac{\partial T}{\partial t} = \frac{1}{ms} \frac{\partial H}{\partial t} \qquad (6.29)$$

Putting $m=\varrho V=\varrho d^3$ and substituting for $\partial H/\partial t$ (equation (6.27)) gives

$$\frac{\partial T}{\partial t} = \frac{1}{\varrho d^3 s} I_0 \mu d^2 dE \times 1.6 \times 10^{-19} \qquad (6.30)$$

and

$$\partial T \, / \, \partial t \propto I_0 \lambda^2 \qquad\qquad (6.31)$$

It is interesting that $\partial T/\partial t$ is independent of d, the sample size; an approximation which breaks down for large samples (i.e. $1-\exp(-\mu d)\neq\mu d$).

6.5.3 Sample radiation lifetime

We can write for the radiation lifetime of a sample, τ,

$$\tau = N_s \, / \, I_0 d^2 \mu d \varepsilon \qquad\qquad (6.32)$$

where N_s is the number of molecules or unit cells in a sample $(d/\text{unit cell})^3$, $I_0 d^2 \mu d$ is the number of absorbed photons s^{-1}, i.e. $\mu d \sim 1-\exp(-\mu d)$ and ε is a radiation damage factor. In such a simple model we can argue that each *absorbed* photon will damage a certain *number* of protein molecules (ε is therefore a dimensionless quantity). ε is a function of dose, exposure time, dose rate, incident λ, temperature, sample and protein variability and radiation protectants. For instance, Blake and Phillips (1962) assumed that ε was 70 for 8 keV photons incident onto a myoglobin crystal at room temperature with the exposure times utilised on a conventional sealed tube source (intensity at the specimen $\approx 10^8 \, s^{-1} \, mm^{-2}$.

With SR, damage is reduced, e.g. with phosphorylase on DCI by 5 (Wilson *et al* 1983). In addition, in some cases with cooling it is reduced by a further factor of 2–5. Hence, with 8 keV photons and typical exposure times of a few minutes ε may be ~ 2.8–7. Finally, from equations (6.32) and (6.22)

$$\tau \propto 1 \, / \, I_0 \lambda^3 \varepsilon \qquad\qquad (6.33)$$

τ is improved with a lower λ provided ε does not increase with decreasing λ; τ is independent of sample size d.

6.5.3.1 SRS bending magnet, $\lambda=1.5\,\text{Å}$ (8 keV photons) at 10^{11} photons $s^{-1}mm^{-2}$ (2 GeV, 200 mA)

Sample, $d = 0.1\,\text{mm}$; $\mu = 0.9\,\text{mm}^{-1}$ Unit cell $= 100\,\text{Å}$

$$\partial H / \partial t = 10^{11} \times (0.1)^3 \times 0.9 \times 8266 \times 1.6 \times 10^{-19}$$
$$= 1.19 \times 10^{-7} \text{ W}$$

$$\frac{\partial T}{\partial t} = \frac{1}{1 \times (0.01)^3 \times 4.2} \times 1.19 \times 10^{-7}$$
$$= 0.028 \text{ K s}^{-1}$$

$$\text{Lifetime, } \tau = \frac{(0.1/100 \times 10^{-7})^3}{10^{11} \times (0.1)^3 \times 0.9 \times \varepsilon}$$
$$= \frac{11\,000}{\varepsilon} \text{ s}$$

For $\varepsilon=5$ the lifetime would be $\approx 2200\,\text{s}$.

The exposure time per degree of crystal sample rotation here would be typically $\approx 100\,\text{s}$ per degree, i.e. each sample will allow $22°$ of total rotation to be recorded before it dies.

6.5.3.2 SRS wiggler, $\lambda=0.9\,\text{Å}$ at 10^{11} photons $s^{-1}mm^{-2}$ $\mu=0.2\,\text{mm}^{-1}$, otherwise sample parameters as in section 6.5.3.1.

$$\partial H / \partial t = 10^{11} \times (0.1)^3 \times 0.2 \times 13\,778 \times 1.6 \times 10^{-19}$$
$$= 4.4 \times 10^{-8} \text{ W}$$

$$\frac{\partial T}{\partial t} = \frac{1}{1 \times (0.01)^3 \times 4.2} \times 4.4 \times 10^{-8}$$
$$= 0.01 \text{ K s}^{-1}$$

$$\text{Lifetime, } \tau = \frac{(0.1/100 \times 10^{-7})^3}{10^{11} \times (0.1)^3 \times 0.2 \times \varepsilon}$$
$$= \frac{50\,000}{\varepsilon} \text{ s}$$

For $\varepsilon=5$, the lifetime would be $10\,000\,\text{s}$.

The exposure time per degree of crystal sample rotation here would be $100 \times (1.5/0.9)^2 = 278\,s$ per degree, i.e. $36°$ of rotation of the sample before it dies.

6.5.3.3 A source of SR photons with $\lambda = 1.5\,\text{Å}$ at 10^{14} photons $s^{-1}\,mm^{-2}$

Sample, $d = 0.1\,mm$; $\mu = 0.9\,mm^{-1}$ Unit cell $= 100\,\text{Å}$

$$\partial H / \partial t = 10^{14} \times (0.1)^3 \times 0.9 \times 8266 \times 1.6 \times 10^{-19}$$
$$= 1.19 \times 10^{-4}\,W$$

$$\frac{\partial T}{\partial t} = \frac{1}{1 \times (0.01)^3 \times 4.2} \times 1.19 \times 10^{-4}$$
$$= 28\ K\ s^{-1}$$

$$\text{Lifetime,}\ \ \tau = \frac{(0.1 / 100 \times 10^{-7})^3}{10^{14} \times (0.1)^3 \times 0.9 \times \varepsilon}$$
$$= \frac{11}{\varepsilon}\,s$$

For $\varepsilon = 5$ the lifetime would be $\sim 2.2\,s$.

The exposure time per degree of crystal sample rotation here would be about $0.1\,s$ per degree, i.e. per sample $22°$ of rotation before it dies. Obviously at the higher intensity one might hope that ε may be less than 5 in the same way that $\varepsilon = 5$ for 10^{11} was a lot less than $\varepsilon = 14$, say, at 10^9 photons $s^{-1}\,mm^{-2}$.

6.5.3.4 ESRF multipole wiggler, $\lambda = 0.5\,\text{Å}$ at $\approx 10^{14}$ photons $s^{-1}\,mm^{-2}$ (6 GeV, 100 mA)

Sample, $d = 0.1\,mm$; $\mu \approx 0.033\,mm^{-1}$ Unit cell, $100\,\text{Å}$

$$\partial H / \partial t = 10^{14} \times (0.1)^3 \times 0.033 \times 24\,800 \times 1.6 \times 10^{-19}$$
$$= 1.31 \times 10^{-5}\,W$$

$$\frac{\partial T}{\partial t} = \frac{1}{1 \times (0.01)^3 \times 4.2} \times 1.31 \times 10^{-5}$$
$$= 3.1\ K\ s^{-1}$$

$$\text{Lifetime, } \tau = \frac{(0.1/100 \times 10^{-7})^3}{10^{14} \times (0.1)^3 \times 0.033 \times \varepsilon}$$

$$= \frac{303}{\varepsilon} \text{ s}$$

For $\varepsilon=5$ the lifetime would be ~60 s.

The exposure time per degree of crystal sample rotation here would be ≈ 0.9 s per degree (i.e. $0.1 \times (1.5/0.5)^2$) from the Darwin formula, i.e. each sample will allow 67° of rotation before it dies.

6.5.3.5 ESRF undulator harmonic, $\lambda=0.33\,\text{Å}$ at 10^{14} photons s^{-1} mm^{-2} (6 GeV, 100 mA)

Sample, $\mu = 0.01\,\text{mm}^{-1}$

$$\frac{\partial H}{\partial t} = 10^{14} \times (0.1)^3 \times 0.01 \times 37\,576 \times 1.6 \times 10^{-19}$$

$$= 6.0 \times 10^{-6} \text{ W}$$

$$\frac{\partial T}{\partial t} = \frac{1}{1 \times (0.01)^3 \times 4.2} \times 6.0 \times 10^{-6}$$

$$= 1.4 \text{ K s}^{-1}$$

$$\tau = \frac{(0.1/100 \times 10^{-7})^3}{10^{14} \times (0.1)^3 \times 0.01 \times \varepsilon}$$

$$= \frac{1000}{\varepsilon} \text{ s}$$

For $\varepsilon = 5$ the lifetime would be 200 s

Exposure time, $0.1 \times (1.5/0.33)^2 = 2.0$ s per degree of rotation. Amount of data per crystal=100° of rotation.

Table 6.2 summarises these results for a range of intensities and at several different wavelengths emphasising how the model calculations indicate the increase in the amount of data recordable per crystal at shorter wavelengths.

6.5.4 Other models

6.5.4.1 The Stuhrmann model

Stuhrmann (1978) performed a model calculation for the temperature

Table 6.2. *Model calculations suggest that short wavelengths increase the amount of data that can be recorded per crystal at room temperature. These calculations actually underestimate the benefit if an ε=5 is used for all intensities.*

Example source/ beam line	Wavelength of the beam (Å)	Intensity at the sample (photons s^{-1} mm^{-2})	Absorbed heat $\delta H/\delta t$ (W)	Adiabatic temperature rise $\delta T/\delta t$ (K s^{-1})	Radiation lifetime $\varepsilon\tau$ (s)	Typical exposure time per degree (s)	Number of degrees of data per sample (if ε=5)
SRS Station 7.2	1.5	10^{11}	1.19×10^{-7}	0.028	11 000	100	22
SRS Station 9.6	0.9	10^{11}	4.4×10^{-8}	0.01	5000	278	36
ESRF multipole wiggler	1.5	10^{14}	1.19×10^{-4}	28	11	0.1	22
	0.5	10^{14}	1.31×10^{-5}	3.1	303	0.9 (Note (4))	67
ESRF undulator	0.33	10^{14}	6.0×10^{-6}	1.4	1000	2.0	100

Notes:
(1) Typical protein crystal sample of size $(0.1)^3$ mm^3 assumed and 100 Å unit cell.
(2) ε may well be less at the higher intensity levels rather than the constant value of 5 assumed here in this table.
(3) An identical absorption efficiency of the detector is assumed in these calculations for 0.5 Å and 1.5 Å λ's. Hence, to exploit these benefits of changing λ truly needs an efficient detector in each wavelength range. For example, at 0.9 Å film is 40% efficient and the IP 80% whereas at 0.33 Å wavelength film is only 8% efficient in absorbing photons whereas the IP is 44% efficient.
(4) Due allowance has been made for the difference in scattering efficiency of the sample at these wavelengths.

rise of a 1 mm^3 myoglobin sample in a beam of 10 keV photons of flux 10^{11} photons s^{-1} mm^{-2} *assuming 100% absorption of this beam.* In fact, $1-\exp(-\mu d)$ would be 0.32 so that this approximation is a fair one for such a sample size at an energy of 10 keV. The calculated temperature rise, in thermal isolation was 0.038 K s^{-1}. Also, he estimated that for 10^{16} myoglobin molecules in that sample it would take 3 hr to cause absorption processes in about 10% of the molecules for 10^{11} absorbed photons s^{-1}. (This corresponds to an assumed damage factor $\varepsilon=10$.)

6.5.4.2 The Bordas model

Bordas (1982) suggests that from experience of biological samples in small angle experiments on DORIS there is a limitation to the intensity one can use for a typical specimen due to radiation damage effects, i.e.

$$I_{max} = \frac{1}{\varepsilon} \times (10^{12} \text{ or } 10^{13}) \text{ photons s}^{-1} \text{ mm}^{-2}$$

For example, on the SRS bending magnet, at the focus, I_{max} is $\sim 2.7 \times 10^{11}$ photons s^{-1} mm^{-2} which gives a τ of 20 s. The corresponding τ is 600 s on our model (from table 6.1), i.e. $\varepsilon=30$ is being assumed by Bordas.

6.5.5 Relevance of these calculations

The use of increased intensities is necessary for a variety of experiments including use of shorter wavelengths for which the scattering is weaker. The sample volume can also be reduced for an increased intensity with a fixed exposure time. However, a situation is reached whereby the sample lifetime is less than the exposure time. In contrast, for a fixed sample volume both the exposure time and sample lifetime reduce *pro rata*. This is obviously the basis under which kinetic experiments can be contemplated.

6.6 THE USE OF SHORT WAVELENGTHS IN DATA COLLECTION

The positive benefits of using a short wavelength are several fold. The reduction in sample absorption (figure 6.3) has three benefits. Firstly, the high resolution reflections especially are attenuated less than they would otherwise be, i.e. the overall temperature factor of the sample is less; this is important for macromolecular model refinement. Secondly, sample to sample variations in absorption are reduced; this improves data consistency (reduced merging R's) and improves the accuracy of the estimation of isomorphous and anomalous differences. Thirdly, as we have seen in detail in section 6.5, the sample lifetime is enhanced, although this effect depends to some extent on an individual sample's sensitivity to radiation. Additionally, the geometry of the diffraction, coupled with the collimation of the beam, allows longer crystal to detector/film distances to be used for a given resolution limit (d spacing). The signal to noise of the measurement is thereby enhanced (figure 6.4).

The penalty of using short wavelength radiation is the reduction in the scattering efficiency of the sample. The wavelength-dependent factors in the Darwin formula are $\lambda^3 L$, where L is the Lorentz factor which varies approximately as $1/\lambda$. Hence, use of 0.9 Å instead of 1.54 Å reduces the scattering efficiency by a factor of ≈ 2.9.

The absorption efficiency of the detector/film also reduces with λ according to a factor $\exp(-\mu t)$. Photographic film, for example, reduces in absorption efficiency between 1.54 Å and 0.9 Å by a factor of ~1.35 (see figure 5.21). In table 5.4 the absorption efficiencies of various detectors as a function of wavelength are compared. Note that for film the Br

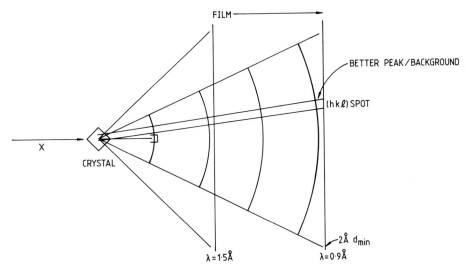

Figure 6.4 The use of short wavelengths also benefits the signal to noise of a diffraction spot intensity measurement. The detector can be moved further from the crystal for a given resolution limit. As a result the background per unit area on the film is reduced according to the inverse square law, excluding the diffuse scattering. The diffraction spot hardly increases in size because of the beam collimation. The reduced scattering at short wavelengths (equation (6.1)) is compensated for by the strength of the SR beam intensity.

Table 6.3. *Projects benefiting from the use of short wavelength synchrotron radiation.*

Sample	SR source	Detector	λ (Å)	Unit cell parameters a / α	b / β	c (Å) / γ (°)	Space group	Resolution limit	Reason for using SR	Reference
Beef despenta-peptide insulin	SRS	FAST	0.9	52.7	26.2 93.4	51.7	C2	1.3	HR	Holden in Glover et al (1988)
Chloramphenicol acetyl transferase	SRS	Film	0.90	74.5 92.5			R32	1.75	HR	Leslie, Moody and Shaw (1988), Leslie (1990)
FMDV	SRS	Film	0.9	345			I23	2.9	RRD	Acharya et al (1989)
Glycogen phosphorylase b	SRS	Film	0.9	128.5		116.3	$P4_32_12$	1.9	RRD, RAE	Oikonomakos et al (1987), Sprang el al (1988)
Spinach ribulose-bisphosphate carboxylase (rubisco)	SRS	Film	0.87	157.2		201.3	$C222_1$	2.4	RRD	Andersson et al (1989)
Rh. rubrum rubisco	SRS	Film	0.87	65.5	70.6 92.1	104.1	$P2_1$	1.7	HR	Andersson et al (1989)

P21 (Val–12) ras oncogene protein	SSRL	Film	1.08	83.2		105.1	$P6_522$	2.2	HR	Tong et al (1989)
B. cereus phospholipase C	SRS	Film	0.88	89.93		73.99	$P4_32_12$	1.5	HR	Hough et al (1989)
R-state glycogen phosphorylase	SRS	Film	0.88	119.0	190.0 109.35	88.2	$P2_1$	2.8	Tetramer in asymmetric unit	Barford and Johnson (1989)
A. niger α-amylase	SRS	Film	0.9	81.1	98.3	138.0	$C222_1$	2.1	HR, RAE	Boel et al (1990)
Partially oxygenated T-state haemoglobin	SRS	Film	1.0	95.8	97.8	65.5	$P2_12_12_1$	1.5	HR	Waller and Liddington (1990)

Notes:
(1) SR source acronyms are as given in Table 4.1.
(2) HR = high resolution study for model refinement; RRD = reduced radiation damage;
     RAE = reduced absorption error use of a short wavelength. FAST is a tradename of the Enraf–Nonius television area detector.

K edge at 0.92 Å and the silver bromide in the emulsion do allow specific enhancement of the absorption if a $\lambda \leq 0.92$ Å is chosen.

Experience with the television detector system (Enraf–Nonius FAST) on the focussed wiggler beam line at Daresbury using a λ of 0.9 Å suggests that the intensity of the diffraction patterns is often somewhat too strong for the detector. Use of even shorter wavelengths than 0.9 Å will reduce the strength of the pattern whilst giving further reduction in absorption errors and enhanced sample lifetime in the beam, etc, outlined above.

At the insertion device machines such as the ESRF and the APS an undulator with a fundamental or third harmonic emission in the 1.0 Å region is realistic and will provide exceptional low divergence beams in this beneficial short λ regime. An even shorter wavelength than 1.0 Å would be attractive for the user but could be achieved only in higher harmonics of the undulator emission. Some possibilities are put forward in the next and final section of this chapter. Table 6.3 provides a compilation of solved structures based on short wavelength data collection in macromolecular crystallography.

6.7 POSSIBLE USES OF VERY SHORT AND ULTRA-SHORT WAVELENGTHS IN MACROMOLECULAR CRYSTALLOGRAPHY WITH ULTRA-RADIATION SENSITIVE SAMPLES

The previous sections (sections 6.5 and 6.6) outlined the basis of using shorter wavelengths in reducing absorption errors and prolonging crystal sample lifetime. The calculations made in section 6.5 indicate that there is no optimum wavelength as such but that it is always better to work at shorter and shorter wavelengths. In doing so more data may be measured per sample. The penalty of shorter wavelengths is that the exposure time increases, for a fixed incident intensity and identical detector. This increase in exposure time can be compensated for by using a higher incident beam intensity. Hence, Helliwell and Fourme (1983), suggested the use of a multipole wiggler on the ESRF at a monochromatised wavelength of 0.5 Å (see Helliwell (1984), p. 1472). Such a wavelength one could define as very short in contrast to short (~0.9 Å) or ultra-short (~0.3 Å). Hence, the use of a short wavelength of 0.9 Å, at SRS, Daresbury or EMBL, Hamburg on the single bump wiggler and bending magnet respectively, represents about the shortest wavelength that it is possible to use, consistent with a reasonable exposure time, at these sources.

A promising application of multipole wigglers (at 0.5 Å) or undulators (i.e. a harmonic at ~0.3 Å) is to solve the problem of data collection from very radiation sensitive crystals. For example, in virus data collection several hundred crystals are needed for structure determination. Of course, once the basic structure is known, a greatly reduced amount of data and far fewer crystals are needed in drug binding (difference Fourier) studies.

Test experiments need to be made with these kinds of samples at these wavelengths, i.e. 0.5 Å on a multipole wiggler and 0.3 Å on an undulator (harmonic). The intensity of these beams will compensate for the increase in exposure time resulting from the λ^2 effect of the Darwin formula (section 6.1). At 0.5 Å the absorption efficiency of an IP is still reasonable (50%) and at 0.33 Å (37.5 keV) it is ~44%. The Ba K edge at 0.331 Å usefully enhances the stopping power of an IP from 19% just above the edge to 44% just below it. By comparison photographic film (Kodak DEF) would only absorb 8% of the photons at 0.33 Å.

The crystal to IP distances for 0.5 Å and 0.3 Å are, of course, increased. This is advantageous because of the inverse square law effect in reducing the background under the diffraction spot. However, a crystal to film distance of, say, 0.5 m with a 1 mrad divergence beam would lead to a sizeable increase in the diffraction spot size (i.e. by 0.5 mm). On an undulator, however, beam divergences are intrinsically ~0.1 mrad and so the spot size over a 0.5 m distance would only increase by 0.05 mm due to this effect (table 6.4). Mosaic spreads of specimens also need to be narrow but ~0.1 mrad is a quite reasonable expectation for these samples (Helliwell (1988) and Colapietro, Helliwell, Spagna and Thompson, unpublished, see figure 2.8(c)).

The beautiful combination of properties of SR is so often what is important in making an experiment work. In this context the undulator harmonic would provide a very high intensity at ultra-short wavelength (0.3 Å) with a very small divergence beam.

Ten years ago it would have seemed inconceivable that the structure of viruses would be solved using data collected at short wavelengths like 0.9 Å. After all in the home laboratory Mo Kα (0.71 Å) is reserved solely for unit cells up to ≈20 Å and Cu Kα (1.54 Å) for macromolecules. Yet 0.9 Å data collection on today's bending magnets and wigglers is commonplace. It is not unreasonable to consider data collection from radiation sensitive samples like virus crystals, in future, using an undulator harmonic at 0.33 Å with an IP placed 0.5–1.0 m from the crystal.

Table 6.4. *Promising possibilities with very short wavelength multipole wiggler and ultra-short wavelength undulator harmonic radiation for ultra-radiation sensitive samples.*

	λ (Å)	d_{min} (Å)	θ (degrees)	IP radius (cm)	D (cm)	αD (mm)
Very short wavelength (multipole wiggler)	0.5 0.5	3.0 2.0	4.780 7.181	10 10	59.4 39.1	0.59 0.39
Ultra-short wavelengths (undulator harmonic)	0.33 0.33	3.0 2.0	3.153 4.732	10 10	90.5 59.99	0.091 0.06

Notes:

(1) αD is the product of the typical divergence angle (0.1 mrad undulator, 1 mrad multipole wiggler) and the crystal sample to plate distance.

(2) A protein crystal mosaicity in the 0.1 mrad (0.006°) range is required if the long crystal to plate distances are to be realised. As documented by Helliwell (1988) and Colapietro, Helliwell, Thompson and Spagna (unpublished work) protein crystals at room temperature do have rocking widths of 0.01–0.02° (0.17–0.34 mrad) FWHM.

CHAPTER 7

The synchrotron Laue method

The original X-ray diffraction experiment was based on an idea of von Laue and conducted by Friedrich and Knipping (Friedrich *et al* 1912). It earned von Laue the Nobel Prize for Physics in 1914. The basis of the idea was that if X-rays were electromagnetic waves then their wavelengths might be of the same order as the interatomic separation in crystals and diffraction would be observed. The original diffraction photograph was from a crystal of copper sulphate.

The essential feature of the Laue method, as it became called, is that the incident X-ray beam is polychromatic and the crystal sample is held stationary. All the X-rays emitted by the emission tube and passing through the tube exit window are allowed to impinge onto the sample; no special filtering or monochromatisation is employed. The Bremsstrahlung continuum and the characteristic emission lines constitutes the incident spectrum of X-rays. This beam hits the stationary crystal and the spots making up the diffraction pattern arise from the different wavelengths. A given reflecting plane in the crystal extracts from the beam the particular wavelength which allows constructive interference or reflection to occur. In contrast to the angular rocking width of a reflection in the monochromatic rotating crystal method, in the Laue method each reflection is stimulated by a small range of wavelengths whose mean wavelength lies somewhere in the broad range of incident wavelengths. The Laue method is described in the book by Amorós, Buerger and Amorós (1975).

The first analyses of simple crystal structures were made using Laue photographs when W. L. Bragg deduced the structure of sodium chloride and other alkali halide crystals (for a historical review see Bragg (1975)). Subsequently, Bragg primarily used the monochromatic X-ray beam to illuminate crystals in the spectrometer designed and built by his father W. H. Bragg. Particular exponents of the use of the Laue method were Wyckoff (1924) and Pauling. However, the Laue method at that time was subject to various limitations and its use lapsed as a method for determining molecular structure as it became superseded by the monochromatic rotating crystal method. In any case, the weakness, by four orders of magnitude, of the Bremsstrahlung radiation relative to the characteristic lines naturally meant that the Laue method was of limited use.

The universal spectral curve of SR, however, ideally lends itself to Laue geometry since a broad band of wavelengths of high intensity is emitted by the synchrotron electron beam. As a result, a very large number of reflections can be recorded in a remarkably short exposure time in a Laue experiment; this has led to a revival of the Laue method as a means for quantitative structure analysis, especially for kinetic studies and the analysis of structural perturbations. Time resolved investigations include the study of enzyme or zeolite catalysis in crystals. The perturbations which are of interest include the effects of electric fields, pressure and temperature on various materials.

In the monochromatic rotating crystal method it has been feasible to record complete data sets from a protein crystal on a timescale of half an hour or less using an SR source (see chapter 10). Ultimately, the total data collection time in the monochromatic method is set by the mechanical overheads of the angular rotation speed of the crystal and by the necessity to replace film cassettes or transfer detector images to computer mass store and to 'refresh' the detector.

However, in the Laue method the data collection time can be set only by the exposure time. Hence, with an unfocussed SR beam, exposure times are of the order of a few seconds for a near-complete data set, in favourable cases. With the use of an undulator source, and focussing mirror optics, exposure times as short as 10^{-10} s for a single Laue pattern have been achieved using the CHESS source at Cornell (Szebenyi *et al* 1989). Hence, very fine time resolution diffraction experiments are feasible with SR even on relatively weakly diffracting systems like protein crystals. It should be mentioned that Rabinovitch and Lourie (1987) have used the conventional source Bremsstrahlung from a flash X-ray generator to study a dynamic perturbation induced by an electric field in a strongly diffracting crystal.

This chapter concentrates on the development of the Laue method, and the way in which it is employed, at SR sources.

7.1 HISTORICAL PERSPECTIVE

The Laue method was formerly considered unsuitable for intensity measurement and structure determination for several major reasons, which were:

(a) The multiplicity or overlapping orders problem (Bragg 1975, p. 137; Wyckoff 1924, p. 142; Amorós *et al* 1975, p. 13; Steinberger, Bordas and Kalman 1977, p. 1260).

(b) The wavelength normalisation problem (Bragg 1949).

(c) In the case of a protein crystal, the radiation sensitivity of the sample in the SR beam (Blundell and Johnson 1976; Greenhough and Helliwell 1983; J. C. Phillips, pers. comm.).

For a discussion of these objections see Helliwell (1989).

Additionally, in Laue geometry only cell ratios can be determined from the angular coordinates of spots. The absolute cell parameters therefore have to be provided from a monochromatic study. However, if the intensities of spots are also considered, then there appears to be scope to solve the problem by using knowledge of λ_{min} (section 7.2.3) and the bromine K edge in the photographic film as wavelength markers.

7.1.1 The multiplicity problem

According to Bragg (1975), p. 137:

> X-ray analysis started with the Laue photograph. It is too hard to attach a quantitative significance to the intensity of the spots, which are due to the superposition of diffracted beams of several orders selected from a range of white radiation.

According to Wyckoff (1924), p. 142:

> In general, reflections to be used in establishing the structure of a crystal should involve wavelengths between the low wavelength limit, λ_{min} and $2\lambda_{min}$. Otherwise, if $n\lambda > 2\lambda_{min}$ an observed reflection may be partly of one and partly of another order. Only when it is known in some other way that planes of particular types do not give reflections in the first one or more orders can higher values of $n\lambda$ be safely employed in intensity comparisons.

These objections are fundamental in the sense that an energy-sensitive detector with the additional properties of high spatial resolution, high count rate and high absorption efficiency is not available.

7.1.2 The wavelength normalisation problem

According to Bragg (1949):

> The deduction of the crystal structure from the appearance of the
> Laue photograph is a complicated process, because the intensities of
> the spots do not depend upon the structure alone. They depend also
> upon the strength of the components in the continuous range of the
> original beam to which they are respectively due, and each spot may
> be composed of several orders superimposed. They are also
> influenced by the different blackening effect of radiation of different
> wavelengths and complications arise here owing to the absorption
> of the X-rays by the silver and bromine in the photographic plate. In
> spite of these difficulties, the Laue photograph can be made a sound
> method of analysis, and has, for instance, been used with striking
> success by Wyckoff. Advantageous features are the ease and
> certainty with which indices can be assigned to the spots, and the
> wealth of information represented by a single photograph. Never-
> theless, the methods which employ monochromatic radiations are
> more direct and powerful.

Each reciprocal lattice point can give rise to a reflection with some
wavelength within the experimental bandwidth. A variety of
wavelength-dependent factors affect the measured reflection intensity.
These factors include the SRS spectral profile, the effect of optical elements, sample scattering efficiencies, absorption of components in the
beam and detector response. In addition, the presence of anomalous
scatterers in a crystal introduces a wavelength variation in the structure
amplitudes and, incidentally, in the phases (see chapter 9).

7.1.3 Radiation damage

The expectation was that SR would lead to increased radiation
damage (Blundell and Johnson 1976). However, it has not been found to
be a problem in fact. On the contrary, with monochromatic SR, radiation damage can often be reduced (as discussed in chapter 6). In the
case of polychromatic radiation the beam intensity is strong enough that
the heating of the sample has to be considered as well as radiation
damage. Based on early experiences at the NINA synchrotron it was
concluded that a protein crystal sample would not withstand the white
beam (Greenhough and Helliwell 1983). However, the use of a
somewhat restricted bandwidth by Moffat, Szebenyi and Bilderback
(1984) gave hope that protein crystals might survive in the full white

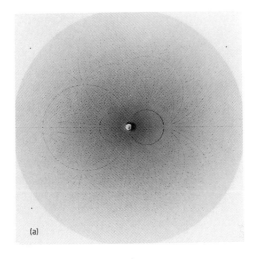

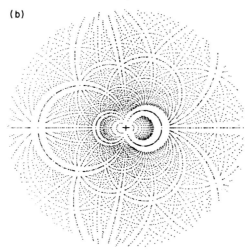

Figure 7.1 (*a*) Pea lectin synchrotron Laue diffraction pattern recorded
on photographic film on station 9.6 of the SRS and then in (*b*)
the corresponding predicted Laue pattern is shown followed
by those parts of the pattern which are (*c*) energy overlap
spots, (*d*) spatial overlap spots, (*e*) spots which are both
energy and spatial overlaps and, finally, those that are (*f*)
neither energy nor spatial overlaps. From Helliwell (1985)
with the permission of Elsevier. See Table 7.1, note 3 for
experimental conditions (crystal to film 95 mm)

beam. As a result of extensive experience with many protein crystals
using monochromatic synchrotron data collection facilities at
Daresbury, Helliwell, in a collaboration with Einspahr and Suddath,
found a protein crystal which was very stable in the beam. This protein

(c)   (d)

(e)   (f)

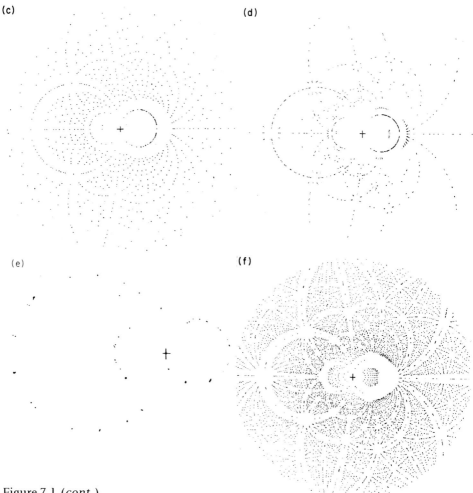

Figure 7.1 (*cont.*)

crystal was pea lectin. Pea lectin crystals were the ideal test case, therefore, to be placed in the full bandwidth beam and several Laue exposures
of a single crystal of pea lectin were successfully recorded at Daresbury
(Helliwell (1984, 1985) and figure 7.1).

It seems that a substantial part of the radiation damage in proteins and
viruses (and some other crystals, too) is a function of time from the first
exposure to the beam. The synchrotron Laue method allows complete
data recording in a shorter time from first exposure than any other
method. Of course, protein crystals differ greatly in their susceptibility to
radiation damage. For example, some survive a single exposure while, at
the other extreme, cases such as pea lectin can survive many exposures.
Indeed, it now seems that many protein crystals (>50%) are able to yield
high quality synchrotron Laue diffraction photographs.

7.2 DIFFRACTION GEOMETRY

For a stationary crystal and white radiation with $\lambda_{max} \geq \lambda \geq \lambda_{min}$, the RLPs whose reflections can be recorded lie between the Ewald spheres of radii $1/\lambda_{max}$ and $1/\lambda_{min}$. These spheres touch at the origin of the reciprocal lattice (figure 7.2(a)), and the wavelength at which any individual RLP diffracts is determined by the reciprocal radius of the Ewald sphere passing through it.

Table 7.1 gives the computed distribution of recorded RLPs in wavelength intervals for pea lectin as the example for various crystal to film distances. The 34 mm distance corresponds to the position where a film of radius 59.3 mm must be put to intercept an RLP of d spacing 2.6 Å stimulated by a wavelength of 2.6 Å (i.e. $\sin\theta_{acc} = 2.6/2 \times 2.6$). Long crystal to film distances tend to be used in practice to reduce the effect of spatial overlaps. At such distances the angular acceptance of the film is (substantially) reduced. Hence, the angular change between setting of

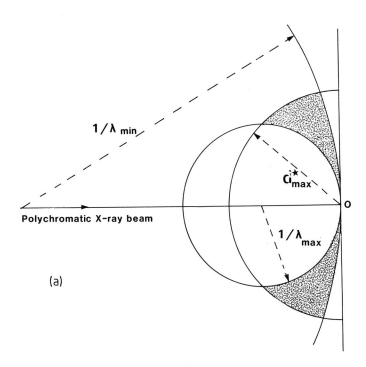

(a)

Figure 7.2 (a) Laue diffraction geometry showing the accessible region of reciprocal space between the Ewald spheres associated with λ_{min} and λ_{max} and the sample resolution limit d^*_{max}; O is the origin of reciprocal space.

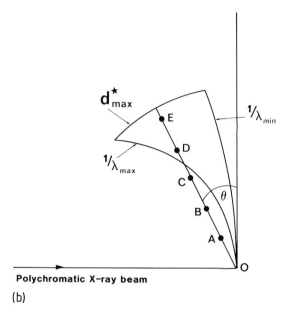

Polychromatic X-ray beam

(b)

Figure 7.2 (b) A ray with n orders inside the d^*_{max} sphere can have a
recorded multiplicity $m<n$ where $(n-m)$ RLPs are outside the
accessible region. The diagram shows the case of $n=5$ and
$m=2$. Only the upper section of the volume of resolution of
the accessible region is shown. From Cruickshank, Helliwell
and Moffat (1987) with the permission of the IUCr.

the sample has to be reduced in order to survey reciprocal space (the
total number of exposed photographs increased). Table 7.2 shows how
several settings of the sample are needed to capture the unique set of
RLPs; the example chosen has Laue symmetry mmm as an intermediate
case between $\bar{1}$ symmetry and m$\bar{3}$m symmetry. Just a few Laue photographs can contain a substantial fraction of the unique RLPs in the case
of high symmetry crystals. Better detector arrangements, such as the
toastrack (figure 7.6(b), section 7.2.2), will reduce considerably the
impact of the spatial overlap problem whilst allowing a larger angular
aperture to be measured at each sample setting. As a result fewer settings
will be necessary in all cases. The choice of a general orientation of the
crystal also leads to a greater percentage of the unique RLPs being captured, at the expense of fewer symmetry related reflections. The latter
are often needed for wavelength normalisation however.

   The spot coordinate on the film is related to the RLP coordinate as
shown in figure 7.3. There is also a sample resolution limit d^*_{max} ($=1/$
d_{min}), so that no reflections are recorded from RLPs outside a sphere
centred at the origin with radius d^*_{max} (shown in figure 7.2). The acces-

Table 7.1. *Distribution of recorded RLPs in each wavelength interval as a function of* θ_{acc}. *From Cruickshank, Helliwell and Moffat (1987).*

			θ_{acc} (°)	30	20.1	16.0	12.7	8.2
Maximum λ in each bin (Å)	Crystal to film distance (mm) μ			34 1.0	70 0.69	95 0.55	125 0.44	200 0.29
0.593				1018	1018	1018	1018	1018
0.737				970	970	970	970	970
0.880				977	977	977	977	729←
1.023				996	996	996	996	386
1.167				973	973	973	970←	213
1.310				1001	1001	1001	731	137
1.453				1011	1011	1008←	480	87
1.597				964	964	747	309	51
1.740				1008	1008	546	223	42
1.883				975	928←	388	152	25
2.027				982	685	284	112	28
2.170				991	548	233	107	20
2.313				963	370	145	56	9
2.457				993	330	137	44	9
2.600				984	245	95	42	10
Total				14806	12024	9518	7187	3734
Number *lost* due to θ cut				0	2782	5288	7619	11072
Percentage *lost* due to θ cut (%)				0	19	36	52	75

Notes:
(1) The symbol ← shows the first bin in each histogram affected by the θ cut.
(2) The specific case treated is that for pea lectin (Helliwell 1985).
(3) Experimental conditions for computer prediction:
$\lambda_{min} = 0.45$ Å, $\lambda_{max} = 2.6$ Å; $\phi = 30°$, $\delta\phi_x = 0.1239°$, $\delta\phi_y = 0.1282°$,
$\delta\phi_z = -0.6254°$; $d_{min} = 2.6$ Å; reciprocal cell: $a^* = 0.0200322$ Å^{-1},
$b^* = 0.0166559$ Å^{-1}, $c^* = 0.007447$ Å^{-1}. Film radius = 59.3 mm.
(4) Each wavelength bin is 0.144 Å wide.
(5) $\mu = \sin\theta_{acc}/\sin\theta_{max}$ where θ_{acc} is the acceptance angle of the film and θ_{max} the maximum Bragg angle.

sible region of reciprocal space, which is cylindrically symmetrical about the incident X-ray beam, may be further limited by experimental restrictions on the scattering angles. We will discuss in this section the occurrence of energy overlapped reflections in a spot, the problems with spots that are spatially close or overlapped (figure 7.1(c) and (d)), nodal spots, blank regions of photographs and, finally, the use of the gnomonic projection.

Table 7.2. *The use of different settings of a sample to capture the symmetry unique set of RLPs. From Elder (1987).*

Spindle	Add. fract symm. uniq. captured	Fract. symm. uniques not recorded	Fractional overlap with previous spindles				
			1	2	3	4	5
		1.000					
45.0	0.582	0.418					
0.0	0.257	0.162	0.307				
67.5	0.059	0.103	0.372	0.355			
90.0	0.029	0.074	0.288	0.437	0.341		
22.5	0.014	0.060	0.372	0.365	0.312	0.341	0.000

Notes:
(1) Pea lectin. Crystal to film distance 34 mm; Laue symmetry mmm.
(2) Crystal perfectly set with mirror plane perpendicular to the rotation axis.
(3) This table ignores the effect of spatial (and energy) overlaps. Hence in practice although the symmetry uniques are captured on film, those that can be processed exclude the overlaps. In practice therefore, longer crystal to film distances are used and a larger number of spindle settings are needed than indicated here (e.g. by a factor of 2). The use of the toastrack, for example, will largely avoid the need for this, however.

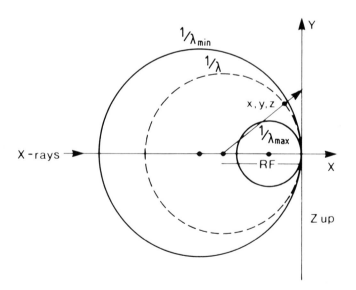

Figure 7.3 The coordinate geometry of Laue diffraction. The RLP at (x,y,z) lies on the Ewald sphere of radius $1/\lambda$ and causes a spot on the Laue diffraction pattern. From Helliwell *et al* (1989b) with the permission of IUCr.

7.2.1 Energy overlaps

A Laue diffraction spot can be composed of several RLPs and wavelengths such that for an RLP corresponding to a spacing d and a wavelength λ the RLPs with $d/2$, $d/3$, etc., will also be stimulated by the respective wavelengths $\lambda/2$, $\lambda/3$, etc. That is, these orders of a Bragg reflection are exactly superimposed. This was the problem referred to by Bragg and Wyckoff as mentioned above. Such a spot has a potential multiplicity determined by the resolution limit. However, the recorded multiplicity may be less. For example, figure 7.2 (b) shows a ray with five orders inside the d^*_{max} sphere, but of which only two are within the accessible region. The ray and corresponding Laue spot are therefore of actual multiplicity 2. However, contrary to the statements of Bragg and Wyckoff, the pattern is dominated by single wavelength, single RLP spots even if there is an infinite bandpass ($\lambda_{max}=\infty$, $\lambda_{min}=0$).

If we consider a given reflecting plane of spacing d there is an associated set of Miller indices (h,k,ℓ). Now $2h,2k,2\ell$ may be beyond the resolution limit and h,k,ℓ may also have no common integer divisor. Hence, the Laue spot that results will contain h,k,ℓ only.

It can be shown (Cruickshank et al 1987) that the probability that a randomly chosen RLP has no common integer divisor is

$$Q = \left(1-\frac{1}{2^3}\right)\left(1-\frac{1}{3^3}\right)\left(1-\frac{1}{5^3}\right)\left(1-\frac{1}{7^3}\right)\left(1-\frac{1}{11^3}\right)\cdots \qquad (7.1)$$

$$= 0.83191\ldots$$

To explain this expression one needs to understand the properties of integers. The probability that an integer, h, is divisible by another integer, p, to produce an integer is $1/p$. The probability that three integers h, k and ℓ are each divisible by the same integer, p, is $1/p^3$ and the probability that (h,k,ℓ) do not have a common integer divisor, p, is $1-1/p^3$. To exclude the possibility of any value of p being a divisor the product series giving $Q=0.832$ involves the infinite set of terms, based upon prime numbers only. It will be noted that the proof assumes that h, k, ℓ are random integers lying between $-\infty$ and $+\infty$. However, to take a case pertinent to small molecule crystallography, if h, k, ℓ were restricted to maximum values of 10, then the terms from $1-1/11^3$ onwards would be omitted. But the product is quickly convergent, and even when only the lowest primes 2 and 3 are considered, the first two terms yield a product 0.843. In macromolecular crystallography the maximum integers are appreciably greater than 10 and any approximation involved in truncating the series for Q is even smaller. Thus for all crystallographic purposes it is a very good approximation to use Q. This probability is independent of the position of an RLP in reciprocal space.

In the case of $\lambda_{max}=\infty$ and $\lambda_{min}=0$ all RLPs lying between $d^*_{max}/2$ and the origin of reciprocal space will be recorded as part of a multiple Laue spot. For those RLPs lying between d^*_{max} and $d^*_{max}/2$ there is a probability Q that they will be recorded as singles; this region is 7/8 of the resolution sphere.

Hence, a proportion of all RLPs$=(7/8)Q=72.8\%$ lie on single rays and a proportion $7/8=87.5\%$ of all rays (Laue reflections) are single rays. Similarly, $2(1/2^3-1/3^3)Q=14.6\%$ of all RLPs lie on double rays, whereas $(1/2^3-1/3^3)=8.8\%$ of all rays are double rays. In the case of a more restricted bandwidth, the experimental situation, the proportion of singles increases, see figure 7.4.

A probability map can be constructed for the general case (finite λ_{max} and λ_{min}) giving details of the likelihood of RLPs being recorded as single, double, triple, etc; see figures 7.5(a), (b), (c). A striking feature of the

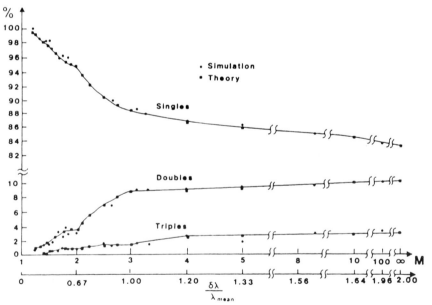

Figure 7.4 The variation with M ($=\lambda_{max}/\lambda_{min}$) of the proportions $p(1)$, $p(2)$ and $p(3)$ of RLPs lying on single, double and triple rays for the case of $\lambda_{max}<2/d^*_{max}$. From Cruickshank *et al* (1987) with the permission of IUCr.

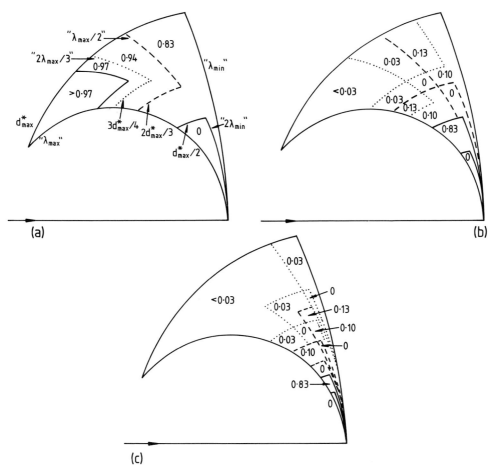

Figure 7.5 The probability of an RLP being (a) single, (b) double and
(c) triple in different regions of reciprocal space. The
wavelength labels indicate which Ewald sphere is appropriate. The actual radius of each Ewald sphere is the reciprocal
of the label shown. The labels for the resolution spheres, e.g.
d^*_{max} are true distances on the diagram. From Cruickshank *et
al* (1987) with the permission of the IUCr.

probability map for singles, for example, is the region of zero probability
between $d^*_{max}/2$, and the origin, on the one hand and $2\lambda_{min}$ and λ_{max} on
the other.

Moreover, since the number of RLPs sampled is given by

$$\frac{\pi d^{*4}_{max}(\lambda_{max}-\lambda_{min})}{4V^*} \text{ for } \lambda_{max}<2/d^*_{max} \qquad (7.2)$$

where V^* is the reciprocal unit cell, the variation of d^* to the fourth power shows that the volume of low resolution data sampled is actually quite small.

The number of RLPs within the resolution sphere is

$$\frac{4}{3} \frac{\pi d^{*3}_{max}}{V^*} \tag{7.3}$$

The Laue method therefore appears very effective at sampling data between d^*_{max} and $d^*_{max}/2$. These RLPs are largely recorded as singles but, of course, their intensities need wavelength normalisation if they are to be used for structure analysis.

The effective sampling of the region between $d^*_{max}/2$ and the origin of reciprocal space needs a method to achieve the separation of the component reflections in a spot. There are several possible methods available to achieve this. The process is often referred to as wavelength or energy deconvolution. These methods include:

(a) The use of multiple films to record a Laue pattern. The variation of the film absorption factor with wavelength means that the different components of (and wavelengths in) a Laue spot are attenuated at different rates (Zurek, Papiz, Machin and Helliwell 1985; Helliwell *et al* 1989b).

(b) In the case of high symmetry it is possible that an RLP recorded in one place as a component of a 'double' spot may also occur elsewhere as a 'single' spot. In this case, the other component of a double may therefore be derived. Alternatively, in the case of multiples an RLP may occur in several multiples, stimulated by different wavelengths and hence recorded with different intensities. From these, it may be possible to extract the intensities uniquely associated with each RLP. Obviously, if there are a maximum of n RLPs in a spot there would need to be at least n unique observations.

The first method has been used and reasonable merging R factors obtained for doubles although not as good as for singles (Zurek *et al* 1985; Helliwell *et al* 1989a).

In any event, quantitation of Laue photographs is a tractable problem because the pattern is dominated by single wavelength spots (table 7.3).

Table 7.3. *The multiplicity distributions of recorded Laue rays as a function of θ_{acc}. From Cruickshank* et al *(1987).*

		Number recorded				
	θ_{acc} (°)	30	20.1	16.0	12.7	8.2
	Crystal to film					
Multiplicity	distance (mm)	34	70	95	125	200
of Laue ray	μ	1.0	0.69	0.55	0.44	0.29
1		12 630	9 863←	7 524	5 431	2 742
2		704	695	649←	559	282
3		121	122	109	101	64←
4		43	43	41	35	25
5		21	21	18	16	12
6		3	3	2	2	1
7		7	7	6	6	5
8		3	3	3	3	1
9		1	1	1	1	1
10		1	1	1	1	0
18		1	1	1	1	1
Total number of RLPs		14 806	12 024	9 518	7 187	3 734
Total number of Laue rays		13 535	10 760	8 355	6 156	3 134

Notes:
(1) The symbol ← indicates when a sizable cut on the singles, doubles, triples has occurred.
(2) Experimental conditions and symbols are as for Table 7.1.

7.2.2 Spatial overlaps

The Laue diffraction spots on a film are of a certain size dependent on factors such as crystal mosaic spread, beam divergence and collimator size (Andrews, Hails, Harding and Cruickshank 1987; Helliwell *et al* 1989a). If a spot encroaches on its nearest neighbour then these Laue spots may be classed as spatially overlapping. The fraction that are spatially overlapping spots can be large (e.g. >50%) in the protein crystal case and even higher fractions can be found in the case of virus Laue patterns (e.g. compare figure 1 and figure 3 of Bilderback *et al* (1988)). Computational approaches to the problem of evaluating their intensities are described as spatial deconvolution. Obvious experimental approaches involve the use of large detector sizes and long crystal to detector distances; this is feasible when the sample is well ordered and because of the tight collimation of the X-ray beam. This, of course, may well not be the case in a kinetic experiment when the sample may disorder slightly at certain stages, as has, in fact, been observed (Hajdu *et al* 1987b).

It can be shown (Cruickshank, Helliwell and Moffat 1991) that:

(*a*) The largest average spatial density of spots occurs at

$$\theta_c = \sin^{-1} \frac{\lambda_{min}}{2d_{min}} \qquad (7.4)$$

(*b*) The largest local density of spots occurs along the arcs approaching nodal spots. (Nodal spots are of low h, k, ℓ values or multiples thereof and are associated with principal zones of the lattice.)

(*c*) The nearest neighbours to a nodal are single (see also Jeffery (1958)).

(*d*) Most overlaps occur between singles. This explains why the population of spatial overlaps is essentially distinct from the population of energy overlaps (figure 7.2(*c*), (*d*) and (*e*)).

The spatial overlap problem is not as fundamental an obstacle as the energy overlap problem once appeared to be. In small molecule Laue

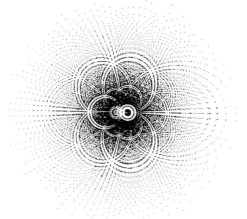

(a)

Figure 7.6 Spatial overlaps dominate a protein crystal Laue pattern when a standard film is set to accept all stimulated spots. The example shown here in (*a*) is a computer prediction equivalent to figure 7.1(*b*) for the pea lectin case but the film is now set at 34 mm from the crystal. From Cruickshank *et al* (1991) with the permission of IUCr.

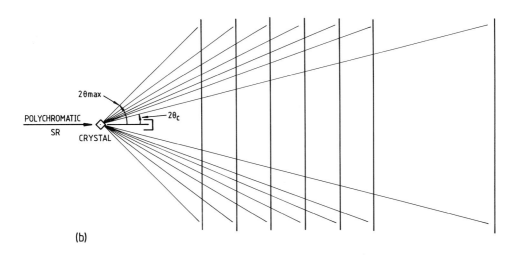

(b)

Figure 7.6 (*cont.*) In (*b*) the 'toastrack' arrangement of films is shown schematically. The closest film to the crystal accepts the whole pattern whereby the outer parts (high θ, long λ) contain spots that are not too densely packed. The inner (low θ, short λ) parts of the front film would not be quantified because of too high a density of spots. Instead these would penetrate to the rear films and expand to fill the available aperture but with a greatly reduced density of spots. From Helliwell (1991) with the permission of Elsevier.

patterns there are virtually no spatial overlaps. However, in macromolecular crystallography it is a matter for serious concern (Cruickshank *et al* (1991) and figure 7.6(*a*)).

Table 7.4 shows how, with a flat film set at various distances from a pea lectin crystal (as an example), the numbers of Laue spots, RLPs and overlaps vary. This table shows that there is an optimum distance (64 mm for this unit cell and wavelength bandpass). This optimum arises because as the film is moved further back to reduce the number of spots lost to spatial overlaps eventually this is counterproductive because more spots are lost as they pass beyond the edges of the film. Table 7.5 explores how this optimum distance varies with the spot-to-spot resolution feasible in the data processing computer program. Table 7.6 likewise examines these dependencies for a hypothetical crystal with doubled, pea lectin, cell dimensions, which has eight times more stimulated RLPs.

An experimental method to solve a significant part of the spatial overlap problem has been proposed. By arranging successive films in a special cassette with a pre-set gap between them (e.g. 10 mm) the front film

Table 7.4. *The variation with crystal to plate distance of the numbers of Laue spots, RLPs and overlaps. From Cruickshank et al (1991).*

Crystal to plate distance (mm)	No. of Laue spots intercepted by the plate	No. of stimulated RLPs intercepted by the plate	No. of RLPs in energy overlaps	No. of RLPs in spatial overlaps	No. of RLPs in energy and spatial overlaps	No. of RLPs free from energy or spatial overlaps	% of RLPs intercepted and free from overlaps
		A	B	C	D	$E = A-B-C+D$	
120	6826	7894	1848	420	0	5626	35.9
95	8802	10010	2082	1214	0	6714	42.8
80	10304	11564	2184	2152	52	7280	46.5
64	12046	13352	2260	3758	236	7570	48.3
50	13502	14814	2268	6144	640	7042	44.9
34	14358	15670	2268	10408	1524	4518	28.8

Notes:
(1) Total number of stimulated RLPs 15670 (see note (4)).
(2) Spatial overlap criterion used was spot centre-to-centre of 0.35 mm or less.
(3) Perfectly set crystal of pea lectin, incident beam along **c**, PL 0° orientation; $\lambda_{min} = 0.45$ Å, $\lambda_{max} = 2.6$ Å, $d^*_{max} = 1/d_{min} = 1/2.6$ Å^{-1}; plate radius 59.3 mm.
(4) The 34 mm distance was set by the need for a 59.3 mm radius plate to accept the maximum stimulated Bragg angle $\theta_m = \sin^{-1}(\lambda_{max}$ $d^*_{max}/2) = 30.0°$. The 95 mm distance was that chosen in the original experiments of Helliwell (1984, 1985).

Table 7.5. *Variation of optimum crystal to plate distance as a function of spot to spot resolution distance. From Cruickshank et al (1991).*

Spot to spot resolution distance (mm)	Optimum crystal to plate distance (mm)	No. of RLPs free from energy or spatial overlaps	% of RLPs intercepted and free from overlaps
0.2	50	10788	68.8
0.3	60	8532	54.4
0.35	64	7570	48.3
0.4	70	6656	42.5
0.5	80	5174	33.0

Notes:
(1) Total number of stimulated RLPs 15 670.
(2) Pea lectin 0° orientation.
(3) λ_{min}, λ_{max}, d^*_{max} and plate radius as in Table 7.4.

Table 7.6. *Variation of optimum crystal to plate distance as a function of spot to spot resolution distance for doubled cell dimensions. From Cruickshank et al (1991).*

Spot to spot resolution distance (mm)	Optimum crystal to plate distance (mm)	No. of RLPs free from energy or spatial overlaps	% of RLPs intercepted and free from overlaps
0.2	100	33164	26.5
0.3	170	15878	12.7
0.4	220	8831	7.0
0.5	290	5262	4.2

Notes:
(1) Total number of stimulated RLPs 125356.
(2) Pea lectin 0° orientation.
(3) λ_{min}, λ_{max}, d^*_{max} and plate radius as in table 7.4.

allows quantitation of the long λ, high θ spots and on the back film(s) the short λ spots can penetrate and expand to fill the whole aperture. Such an arrangement is referred to as a film toastrack (see figure 7.6(b)).

7.2.3 Nodal reflections and blank areas

A visually striking feature of Laue photographs is the existence of nodal spots (figure 7.7). Around the nodal is a blank area (Jeffery 1958).

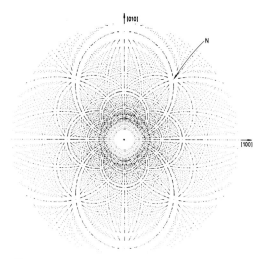

Figure 7.7 A predicted Laue pattern of a pea lectin crystal set with a zone axis parallel to the incident, polychromatic X-ray beam. There is a pronounced blank region in the centre of the film. The spot marked N is one example of a nodal spot. See section 7.2.3. The crystal to film distance is 64 mm. Other experimental conditions are as described in table 7.1. From Cruickshank *et al* (1991) with the permission of IUCr.

The nearest neighbouring spot to a nodal spot is a Laue spot whose Miller indices (h', k', ℓ') are related to those of the highest integer indices in the nodal spot (e.g. $nh, nk, n\ell$) such that, for example, $h' = (n+1)h$. For this reason (h', k', ℓ') cannot have a common integer divisor and (h', k', ℓ') must be stimulated by a single wavelength. Such spots are useful for establishing the resolution limit of the pattern for a given exposure time and sample, with due recognition, of course, that for particular spots the structure amplitude may be near zero as a result of the molecular transform. A survey of several nearest neighbour spots to nodals in a Laue pattern can therefore be used to establish the resolution limit of the data, if the pattern has been indexed and the cell dimensions are known.

The nodal spots lie at the intersection of curves of spots. Each curve corresponds to the intersection with the film of a cone, the surface of which contains the incident beam direction. With a flat film the intersections are conic sections.

If the crystal is oriented so that a well-populated principal zone is normal to the incident beam, then the Laue photograph will show a pronounced blank region in the centre of the film. The radius of the blank region is determined by the minimum wavelength in the beam and the magnitude of the reciprocal lattice spacing parallel to the X-ray beam (see figure 7.7) (Jeffery 1958). For example, consider the case where the X-ray beam is perpendicular to the $hk0$ zone, then

$$\lambda_{min} = |\mathbf{c}|(1 - \cos 2\theta)$$

where

$$2\theta = \tan^{-1}(R/D) \tag{7.5}$$

and R is the radius of the blank region (see figure 7.7) and D is the crystal to film distance. If λ_{min} is known, then an approximate value of $|\mathbf{c}|$ for example, can be estimated and vice versa.

7.2.4 Gnomonic and stereographic transformations

A useful transformation of the flat film Laue pattern is the gnomonic projection. This converts the pattern of spots lying on curved arcs to points lying on straight lines. This is advantageous in determining the soft limits λ_{min}, λ_{max} and d_{min} (Cruickshank, Carr and Harding, to be

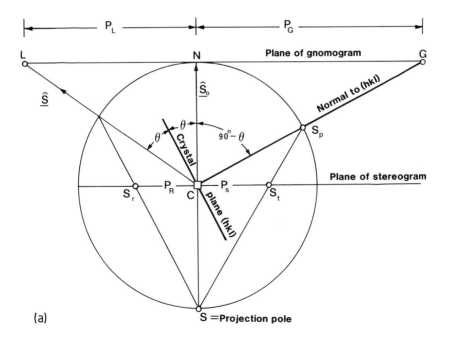

(a)

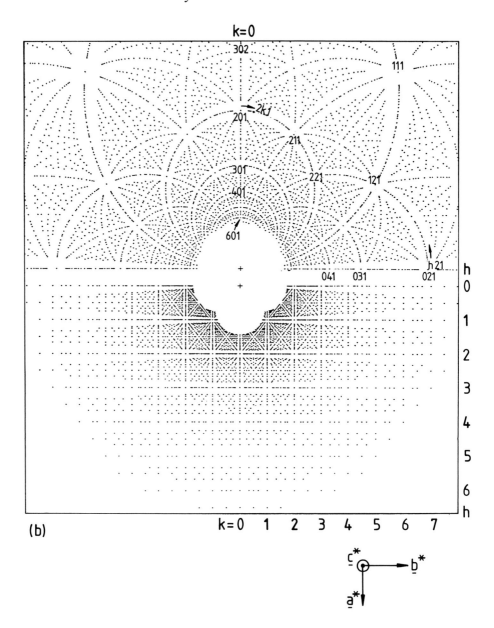

(b)

Figure 7.8 (a) Geometrical principles of the spherical, stereographic and gnomonic projections. From Evans and Lonsdale (1959) with permission of the IUCr. The gnomonic projection is the most useful. (b) A gnomonic projection predicted pattern (bottom) and its equivalent predicted Laue pattern (top) for an orthorhombic crystal (of pea lectin), as an example, aligned with the incident beam along **c** (**c***).

published) and the unit cell and orientation for an unknown crystal (Carr, Cruickshank and Harding, to be published). The stereographic projection can also be used. Figure 7.8 shows the graphical relationships involved for the case of a Laue pattern recorded on a flat film, between the incident beam direction SN, which is perpendicular to a film plane and the Laue spot L and its spherical, stereographic and gnomonic points S_p, S_t and G and the stereographic projection S_r of the reflected beams. The transformation equations are

$$P_L = D \tan 2\theta \qquad (7.6)$$

$$P_G = D \cot \theta \qquad (7.7)$$

$$P_S = D \frac{\cos \theta}{(1 + \sin \theta)} \qquad (7.8)$$

$$P_R = D \tan \theta \qquad (7.9)$$

7.3 REFLECTION BANDWIDTH AND SPOT SIZE

The reflection or spot bandwidth is the narrow range of wavelengths extracted from the overall, much broader bandpass required to stimulate fully the RLP. It is directly akin to the rocking width in the monochromatic rotating crystal method.

The overall spot bandwidth is determined by the mosaic spread and horizontal beam divergence (since $\gamma_H > \gamma_V$).

$$\left(\frac{\delta\lambda}{\lambda} \right)_{hk\ell} = (\eta + \gamma_H) \cot \theta_{hk\ell} \qquad (7.10)$$

where η is the sample mosaic spread, assumed to be isotropic. γ_H is the horizontal cross-fire angle, which in the absence of focussing is $(x_H + \sigma_H)/p$ where x_H is the horizontal sample size, σ_H is the horizontal source size and p is the sample to the tangent point distance. Similar terms apply for x_V in the vertical direction although, generally, at SR sources $\sigma_H > \sigma_V$. When a focussing mirror element is used γ_H and/or γ_V are convergence angles determined by the focussing distances and the mirror aperture.

The size and shape of the diffraction spots vary across the film. The radial spot length is given by convolution as

$$(L_R^2 + L_c^2 \sec^2 2\theta)^{\frac{1}{2}} \qquad (7.11)$$

and tangentially as

$$(L_T^2 + L_c^2)^{\frac{1}{2}} \tag{7.12}$$

where L_c is the size of the X-ray beam (assumed circular) at the sample, and

$$L_R = D\sin(2\eta + \gamma_R)\sec^2 2\theta \tag{7.13}$$

$$L_T = D(2\eta + \gamma_T)\sin\theta\sec 2\theta \tag{7.14}$$

and

$$\gamma_R = \gamma_V \cos\psi + \gamma_H \sin\psi \tag{7.15}$$

$$\gamma_T = \gamma_V \sin\psi + \gamma_H \cos\psi \tag{7.16}$$

where ψ is the angle between the vertical direction and the radius vector to the spot.

For a crystal that is not too mosaic the spot size is dominated by L_c. For a mosaic or radiation damaged crystal the main effect is a radial streaking arising from η, the sample mosaic spread (figure 7.9). For a discussion see Andrews *et al* (1987).

7.4 ANALYSIS OF LAUE DATA AND WAVELENGTH NORMALISATION

The wavelength normalisation problem was introduced in section 7.1.2. The measured intensity of a single Laue spot or of an energy deconvoluted spot needs correction or normalisation for a variety of wavelength-dependent factors if data are to be obtained comparable to monochromatic data. The wavelength at which a reflection is stimulated affects the measured spot intensity through a variety of processes:

(*a*) The intensity spectrum which is incident at the sample. The calculated spectrum (see chapter 4) is modified at the long wavelength end by absorption due to beam line windows or any airpaths in the Laue camera. Also, the short wavelengths are cut off if a reflecting mirror is used (and this may be the case for focussing or to aid more precise definition of spot multiplicity).

(*b*) The source polarisation state varies with wavelength (see Papiz and Helliwell (1984) for the SRS example).

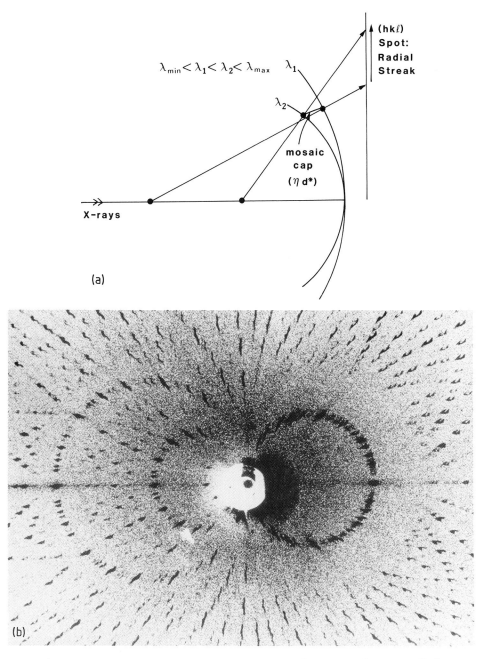

Figure 7.9 The effect of a large mosaic spread is to produce a radial streaking of the spots in a Laue pattern. (*a*) geometric construction; the wavelength labels indicate which Ewald sphere is appropriate. The actual radius of each Ewald sphere is the reciprocal of the label shown (e.g. $1/\lambda_1$): (*b*) Laue pattern from a sample with pronounced mosaic spread after prolonged irradiation.

(c) Sample absorption also truncates the long wavelengths compared with the short wavelengths.

(d) Film response increases at long wavelengths. The bromine and silver K absorption edges at $0.92\,\text{Å}$ and $0.49\,\text{Å}$, respectively, enhance the film absorption efficiency in a wavelength region where it is otherwise decreasing (Clifton *et al* 1985).

(e) Anomalous scatterers in the structure affect the magnitude (and phase) of the structure factors. The wavelength dependence of this is dealt with in detail in chapter 9.

(f) A factor λ^4 in the formula for the integrated intensity (Zachariasen 1945; Kalman 1979).

In addition, there is the Lorentz factor $(\sin^2\theta)^{-1}$ (Zachariasen 1945; Buras and Gerward 1975; Kalman 1979).

The recorded Laue intensity (strictly, an integrated power) is given for the reflection $hk\ell$ (abbreviated to the symbol **h**) by (Campbell *et al* (1986) and references cited therein)

$$I_{\rm L}(\mathbf{h}) = \left(\frac{e^2}{mc^2}\right)^2 \frac{\mathrm{d}I}{\mathrm{d}\lambda} \lambda^4 \frac{1}{2\sin^2\theta} \frac{V}{V_0^2} PAD\,|\mathbf{F}(\mathbf{h})|^2 \qquad (7.17)$$

Here, $\mathrm{d}I/\mathrm{d}\lambda$ denotes the spectral intensity distribution of the incident X-ray beam; V is the volume of sample illuminated; V_0 is the sample unit cell volume; θ is the Bragg angle for the reflection **h**; P is the polarisation factor; A is an absorption correction for the sample in its capillary and D is a detector sensitivity and obliquity factor. Quantities such as P, A and D vary with any or all of λ, θ and $\mathbf{x}$, the position of the diffracted beam on the detector; the spectral intensity distribution is, in general, not precisely known in advance; and the detector may suffer from spatial distortion and non-uniformity. Thus equation (7.17) may be written as

$$I_{\rm L}(\mathbf{h}) = KG\,(\lambda,\,\theta,\,\mathbf{x})\,|\mathbf{F}(\mathbf{h})|^2 \qquad (7.18)$$

where K is a constant. Assuming that all quantities which depend on more than one variable are factorable, then $G(\lambda,\theta,\mathbf{x})=f(\lambda)g(\theta)j(\mathbf{x})$, and

$$|\mathbf{F}(\mathbf{h})|^2 = [Kg(\theta)j(\mathbf{x})]^{-1}\,[f(\lambda)]^{-1}\,I_{\rm L}(\mathbf{h}) \qquad (7.19)$$

It is reasonable to assume that K, $g(\theta)$ and $j(\mathbf{x})$ are known, to a very good approximation. Thus, the quantifying of Laue diffraction patterns depends almost entirely on determination of $f(\lambda)$, known as the wavelength normalisation curve or λ-curve.

The methods which have been used to derive a wavelength normalisation, $f(\lambda)$, curve are:

(i)   The use of a silicon standard crystal and a film in a pilot study on aluminium phosphate (Wood, Thompson and Mathewman 1983). In this method the silicon crystal is rotated over a continuous angle and a given reflection sweeps out a continuous streak. By digitising the film streak the intensity variation of the reflection gives the λ-curve.

(ii) The use of monochromatic data as a reference set for the same sample (Campbell et al 1986).

(iii) The use of symmetry equivalent reflection intensities measured at separate wavelengths (Campbell et al 1986) (see figure 7.10). Similar information is also available when the same reflections are measured on several film packs with different crystal orientations.

Methods (ii) and (iii) have generally been the ones used to process protein and small molecule crystal Laue data. The protein crystal processing statistics are almost as good as monochromatic oscillation camera data (Machin and Harding 1985; Helliwell et al 1986b, 1989b; Machin 1987; Temple and Moffat 1987). Method (iii) breaks down in the presence of strong anomalous scattering; in such a case either method (ii) should be used or a λ-curve derived by method (iii) from a non-anomalously scattering crystal structure; alternatively, in trying to derive a data set that is completely equivalent to a monochromatic one with many measurements of a reflection at different wavelengths it should be possible, in principle, to derive the wavelength-independent and -dependent parts of the structure factor separately using the method of Karle (1980) (see chapter 9).

In the case of monitoring a sample of known structure via the Laue pattern it is possible to use the *fractional* differences whereby the wavelength-dependent corrections cancel. This *difference ratio method* can be used with time resolved or perturbation measurements. A simplification, in principle, of the time resolved Laue experiment over its static counterpart now appears: only fractional changes in intensities are needed (Bilderback et al 1984). In the static Laue experiment, the relationship between the measured Laue intensity I_L of the reflection $\mathbf{h}$ and the structure amplitude $|\mathbf{F}(\mathbf{h},0)|$ is given by

$$I_{\mathrm{L}} = c\ (\lambda,\mathbf{r})\ |\mathbf{F}(\mathbf{h},0)|^{2} \tag{7.20}$$

where λ is the wavelength that stimulates the reflection $\mathbf{h}$, $\mathbf{r}$ is the direction of the diffracted beam and $c(\lambda,\mathbf{r})$ is a function of the numerous experimental variables mentioned earlier. Now consider the time resolved Laue experiment and write (Moffat $et\ al$ 1986b):

$$F_{\mathrm{L}}(\mathbf{h},\lambda,t) = I_{\mathrm{L}}^{\frac{1}{2}} = c^{\frac{1}{2}}\ (\lambda,\mathbf{r})\ |\mathbf{F}(\mathbf{h},t)| \tag{7.21}$$

Hence,

$$\Delta F_{\mathrm{L}}(\mathbf{h},\lambda,t) = F_{\mathrm{L}}(\mathbf{h},\lambda,t) - F_{\mathrm{L}}(\mathbf{h},\lambda,0)$$
$$= c^{\frac{1}{2}}(\lambda,\mathbf{r})\ [|\mathbf{F}(\mathbf{h},t)| - |\mathbf{F}(\mathbf{h},0)|] \tag{7.22}$$

and

$$\Delta F_{\mathrm{L}}(\mathbf{h},\lambda,t)/F_{\mathrm{L}}(\mathbf{h},\lambda,0) = [|\mathbf{F}(\mathbf{h},t)| - |\mathbf{F}(\mathbf{h},0)|]/|\mathbf{F}(\mathbf{h},0)| \tag{7.23}$$

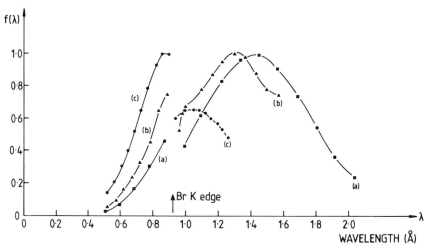

Figure 7.10 λ-curves determined under various beam line conditions and from protein crystal Laue film data: (a) SRS station 9.6 prior to mirror installation (Helliwell 1984; Helliwell $et\ al$ 1989b); pea lectin. (b) SRS station 9.6 after mirror installation; mercury α-amylase crystal derivative. (c) SRS station 9.7, no mirror, with 50μm aluminium filter; cubic con-canavalin-A. Each curve on an arbitrary relative level. From Helliwell $et\ al$ (1989a). Reproduced with the permission of the American Institute of Physics.

which may be written as

$$[|\mathbf{F}(\mathbf{h},t)| - |\mathbf{F}(\mathbf{h},0)|] = |\mathbf{F}(\mathbf{h},0)| \, \Delta F_{\mathrm{L}}(\mathbf{h},\lambda,t) / F_{\mathrm{L}}(\mathbf{h},\lambda,0) \qquad (7.24)$$

Here, the proportionality constant $c(\lambda,\mathbf{r})$ is time-independent and drops out, as first noted by Bilderback *et al* (1984) and used by Hajdu *et al* (1987b). $|\mathbf{F}(\mathbf{h},0)|$ may be quantitated in the Laue experiment itself if $c(\lambda,\mathbf{r})$ has been determined, or it may have been obtained in a prior monochromatic experiment. A time-dependent difference Fourier will utilise $[|\mathbf{F}(\mathbf{h},t)| - |\mathbf{F}(\mathbf{h},0)|]$ as coefficients to yield $\Delta\varrho(t)$, the change in electron density as a function of time.

A basic weakness of the difference ratio method is that each coefficient involves a division by $F_{\mathrm{L}}(\mathbf{h},\lambda,0)$, which will very often be weak or even zero, leading to a large error in the coefficient. In the earlier methods, where the λ-curve is explicitly determined, only the strong intensities in the pattern are used; hence a reflection is scaled by a factor determined from these more precise measurements alone.

There are also practical problems with the difference ratio method if $c(\lambda,\mathbf{r})$ is not strictly constant. It is essential to maintain the crystal in a fixed orientation during the sequence of photographs. In practice, this can be very difficult to do. Hence, for all these reasons the method based on explicitly deriving a λ-curve from symmetry related reflections (Campbell *et al* 1986) is nearly always used rather than the difference ratio method.

7.5 EXPERIMENTAL PARAMETERS AND INSTRUMENTATION

From the above considerations we can see that the relevant experimental parameters are:

$$\left. \begin{array}{c} \lambda_{\max} \\ \lambda_{\min} \end{array} \right\} \frac{\lambda_{\max}}{\lambda_{\min}}, \; \lambda_{\max} - \lambda_{\min}$$

$$\left. \begin{array}{l} \delta\lambda / \lambda \text{ for a given spot} \\ \text{spectral smoothness} \end{array} \right\}$$

detector angular acceptance;
detector point spread factor (may be a function of λ);
detective quantum efficiency (λ).

7.5.1 The source

In the optimisation of the method one can say that the broadest possible bandpass should be used since this leads to the largest number of
stimulated RLPs. Note that a 0.1 Å bandwidth at short wavelength contains the same number of RLPs as the same bandwidth at long
wavelength. However, the multiplicity distribution is determined by
$\lambda_{max}/\lambda_{min}$ and so if the bandwidth is increased this should be to the long
wavelength end. Consideration of the pea lectin data processing results
(see table 7.7) suggests that a high flux over the bandwidth
$\approx 0.6 < \lambda < 1.6$ Å is optimal.

To achieve the short wavelength may require use of a wiggler in the
case of a relatively low energy machine. The use of a multipole device
also leads to further reductions in exposure time. However, even a
multipole wiggler has some undulator character, i.e. has a long
fundamental wavelength of emission and so the smoothness of the emitted spectrum needs to be checked in a given case.

Ideally, a source is required which delivers the highest flux in a
wavelength range at least between 0.6 Å and 1.6 Å and with a smooth
spectrum.

7.5.2 Optics

A focussing mirror is obviously attractive if the beam size, in the
absence of focussing, is much bigger than the collimator size (Brammer
et al 1988). For example, for a 2×0.2 mrad beam divergence the beam
size at 15 m from the source has an area of 30×3 mm^2 compared with a
collimator pinhole area of $\sim 0.2 \times 0.2$ mm^2. This is a size mismatch of
2250! The effectiveness of focussing depends on the source size and therefore the focal spot size. On a storage ring like Daresbury the source size is
$\sim 3 \times 0.3$ mm^2 whereas the ESRF is $< 0.2 \times 0.2$ mm^2 (FWHM values). The
optical element has, of course, to withstand the heat load in the beam.
Alternatively, if the sample could be placed much closer to the source,
the divergent beam would be smaller, but it is still likely to be considerably larger than a focal spot from a mirror; at the SRS the closest distance
is ~ 8 m and at the ESRF 30 m. Another advantage of a mirror is that any
small changes in beam source position will have a smaller effect on the
stability of the λ-curve when a mirror, with its large collection aperture,
is used.

The use of a reflecting mirror also allows a sharper definition of λ_{min}.
This is important in the certainty of the classification of a given Laue

Table 7.7. *Merging R-factors (on intensity) in per cent binned as a function of wavelength in the upper diagonal matrix. The lower diagonal matrix shows the number of single Laue reflections contributing to each R-factor. Overall merging R, 9.0%. Pea lectin synchrotron Laue data (Helliwell 1985; Helliwell et al 1989).*

⟨λ⟩ Bin	0.5 Å 1	0.59 2	0.68 3	0.77 4	0.87 5	0.98 6	1.10 7	1.22 8	1.35 9	1.45 10	1.57 11	1.68 12	1.80 13	1.92 14	2.04 15
1	10.1 (465)	10.7	13.5	11.7	14.8	14.1	11.9	10.2	11.3	9.9	12.4	9.7	9.9	13.6	23.3
2	469	5.9 (774)	7.6	9.0	9.4	7.2	7.4	7.8	9.1	8.4	8.8	7.9	8.3	9.0	14.4
3	349	745	6.2 (945)	7.5	7.6	5.8	7.5	6.8	8.3	7.2	7.8	8.7	7.0	9.1	10.0
4	257	503	898	5.4 (985)	7.5	6.8	7.3	6.9	7.3	6.7	8.1	10.1	9.0	11.7	12.1
5	256	421	515	911	6.6 (990)	8.7	8.8	9.6	9.3	9.3	9.9	11.5	10.5	13.0	13.7
6	286	352	371	471	910	5.4 (1273)	6.1	7.0	8.0	8.1	7.6	10.2	7.3	10.5	15.5
7	304	382	340	450	439	1012	5.0 (1425)	6.4	8.2	8.1	8.9	8.9	7.8	9.2	14.7
8	290	375	450	468	467	552	1376	5.5 (1594)	6.9	6.9	7.0	8.1	8.6	10.0	18.4
9	285	367	441	472	534	609	712	1532	5.8 (1868)	6.6	6.7	8.3	7.8	13.8	9.3
10	318	321	401	447	424	468	532	783	1638	5.7 (1572)	6.8	6.9	8.0	8.1	11.3
11	236	249	279	259	320	331	307	451	545	1122	5.9 (1138)	6.7	7.5	8.0	17.8
12	225	197	215	153	182	219	223	305	338	430	794	5.8 (831)	7.2	9.2	11.4
13	110	167	136	146	154	164	175	222	206	227	327	560	6.9 (547)	8.0	12.5
14	112	128	86	103	99	80	123	112	134	152	133	252	381	7.4 (351)	10.1
15	63	70	85	47	40	69	70	75	53	76	92	103	132	248	8.9 (244)

spot as single or double, etc. Figure 7.10 compares λ-curves for various source and optics options at the SRS. Figure 7.13(b) provides the 'acid test' for data quality with a mirror and shows difference Fourier and Patterson maps for Laue and monochromatic data which compare favourably. Moreover, figure 7.14 uses phases derived from mercury derivative Laue data recorded with mirror; the three sections indicate that the anomalous differences are very significant measured in Laue geometry with mirror. Exposure times without focussing are typically a second or so for a protein crystal broad bandpass Laue pattern. With point focussing millisecond exposure times are possible for sources like SRS and microsecond (to picosecond) should be possible with multipole sources.

7.5.3 Detectors

The characteristics of a Laue pattern are:

exceedingly high global count rate over the whole area of the detector (e.g. $>10^9$ photons s^{-1} has been realised and up to 10^{15} should be realised at ESRF or APS);
very high local count rate in a given spot (e.g. a maximum of 10^6–10^{12} photons s^{-1} per spot);
a high density of spots;
a wide range of wavelengths.

This places extreme demands on the detector!!

Photographic film has been used primarily to date. With film, high global and local count rates are not a problem. The spatial resolution is excellent and is not degraded at short wavelength, in contrast to an MWPC, where spot smearing becomes serious at short wavelengths. The film absorption is reasonable over a wide wavelength range; the bromine and silver absorption edges at 0.92 Å and 0.49 Å, respectively, signifi- cantly enhance this in the short wavelength range. The variation of film absorption with wavelength allows different components in a multiple spot to be estimated; for the case of deconvoluted doubles versus singles data compare figure 7.13(a) (i) with 7.13(a) (ii). The film microdensit- ometers that are commonly available allow the scanning of films up to $10''\times8''$ (250×200 mm^2). However, to date, $5''\times5''$ films (125×125 mm^2) have largely been used. This is often because of space constraints on the beam line.

The Fuji image plate has been used at Photon Factory (Amemiya *et al* 1987) and the Kodak storage phosphor on CHESS (Bilderback *et al* 1988).

An advantage of these over film is that the absorption efficiency is higher (Amemiya *et al* 1987), particularly in the short wavelength region. Also, the intrinsic noise is lower (compared with film chemical fog), the DQE is also better and the dynamic range is larger. However, spatial resolution is substantially worse than film (Amemiya *et al* 1987). Film and IPs are recording media which are off-line.

In a time resolved study a reaction has to be initiated in a crystal and then an interesting instant of the ensuing time course has to be identified (section 10.5). Amemiya (pers. comm.) has designed and built a device for rapid recording of one-dimensional muscle diffraction data on an IP, complete with a reader unit. A one-dimensional slit is placed in front of the IP which is rotatable. A time resolution of $23\,\mu s$ is achievable over a total time course of $100\,ms$ (or slower). The same arrangement could, of course, use films, but with decreased sensitivity. Such a device would allow important instants in a time course to be identified to allow recording in two dimensions in a separate experiment.

In the case of Laue data collection from a microcrystal the signal to noise of the pattern has to be optimised. In monochromatic experiments (Andrews and Helliwell, unpublished) on a $30\,\mu m$ crystal of lysozyme a $200\,\mu m$ collimator was used and final collimation provided by motorised Huber slits; the diffraction pattern was monitored in real time on the FAST television diffractometer. The setting of the collimation in a Laue camera for microcrystal work also requires an electronic area detector. However, such a detector needs a very fine point spread factor. A number of phosphor coupled detectors, employing pre-storage gain and CCD read-out have been suggested (e.g. Strauss *et al* (1987)). Though such systems can provide a wide aperture through demagnifying optics, they suffer from degradation of the point spread factor, and, more importantly, poor overall quantum efficiency (Allinson and Greaves 1988). Allinson *et al* (1989) used a CCD imaging device in direct detection mode to monitor in real time a small portion of the diffraction pattern (figure 5.27). The diffraction spot size of $\sim 200\,\mu m$ was measured on the detector as $180\,\mu m$, i.e. $\sim 9 \times 9$ pixels. The detective quantum efficiency of the CCD for Mo $K\alpha$ (0.71 Å) was measured as $\sim 6\%$ by comparison with a scintillation counter on a Hilger and Watts four-circle diffractometer. Radiation damage to the detector during these experiments was not apparent. Extensive tests on radiation damage in CCDs demonstrates that such a direct detection mode is feasible and the cumulative damage, after a number of experiments, can be annealled (Magorrian and Allinson 1988). It is envisaged, therefore, that a CCD can allow careful setting of the collimation in microcrystal work. It has also been proposed (West-brook, pers. comm., Clarke, pers. comm.) that the CCD can be used in a time frame mode as a linear detector.

The main principle behind detection schemes for Laue diffraction pat-terns is the use of a hybrid arrangement to provide the necessary range of options, namely a large aperture plus on-line facility, for film and CCD, respectively.

7.6 RESULTS ESTABLISHING THE CREDENTIALS OF THE METHOD

SR Laue data were recorded from a crystal of aluminium phosphate and analysed by Wood *et al* (1983). The *R*-factor on *I* in the Laue data refinement was 19% compared with a conventional monochromatic sin-gle crystal study (Thong and Schwarzenbach 1979) with an *R*-factor of 2.2% on *F*. For the Laue data 20% of the measurements were rejected on the basis of bad intensity agreements. Of the eight refined positional parameters six were within 2σ and only one outside 3σ. The temperature factors were somewhat less well determined. This was considered to be due to the significant extinction and multiple diffraction effects for a hard material such as aluminium phosphate.

Laue data from a protein crystal were first recorded by Moffat *et al* (1984, 1986a) who used a somewhat restricted wavelength range to avoid the multiplicity problem. It was established that rapid exposure times were feasible.

Broad bandpass Laue data were recorded from a crystal of the protein pea lectin using the SRS wiggler (Hails *et al* 1984; Helliwell 1984, 1985). These data were used in a pilot study to assess protein crystal Laue data quality. This protein crystal is relatively radiation insensitive. Overall merging statistics for the Laue data after wavelength normalisation (Campbell *et al* 1986) were almost as good as monochromatic data. The merging *R*-factors on intensity for Laue and monochromatic pea lectin data were respectively 8.2% and 6.3%. The mean fractional difference on *F* of the Laue data scaled to monochromatic data was 12.9% to 2.6 Å resolution. These results are summarised in Helliwell *et al* (1989b). Table 7.7 shows the merging *R*s for the Laue data as a function of wavelength. The best quality data occur in the λ range $0.59\,\text{Å} < \lambda < 1.57\,\text{Å}$. At the short wavelength end the detector becomes increasingly transparent and at the long wavelength end sample absorption increases. Both effects weaken the diffraction spot. Sample absorption also introdu-ces systematic errors at long wavelengths.

A difference Fourier map was calculated for the enzyme glycogen phosphorylase b using synchrotron Laue data (Hajdu *et al* (1987b, 1988), see figure 7.11). The map showed electron density at the binding site for bound maltoheptose in phosphorylase b crystals. The phosphorylase b difference Fourier maps calculated for the Laue case used fractional intensity changes (difference ratio method) as discussed earlier in section 7.4.

All the studies described below use the wavelength normalisation method of Campbell *et al* (1986). The structure of an organo-metallic compound was solved with Laue data whereby a Patterson synthesis yielded iron and rhodium positions (Harding *et al* (1988) and figure 7.12). The unit cell parameters came from monochromatic photographs. It was then possible to locate the remaining atoms in the subsequent difference Fourier syntheses. The final *R*-factor was quoted as 0.14 (from

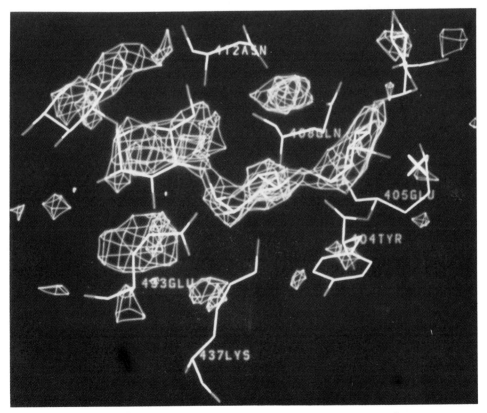

Figure 7.11 Electron density map from Laue data using the difference ratio method. From Hajdu *et al* (1987b) with permission, from *Nature* **329**, 178–81 Fig 3a. Copyright © 1987 Macmillan Magazines Ltd.

SHELX – Sheldrick (1976)). A second organo-metallic (rhodium complex) was solved by Patterson and difference Fourier syntheses by Clucas, Harding and Maginn (1988) with an R of 0.16 on F for reflections with $F \geqslant 6\sigma(F)$.

A heavy atom derivative data set was analysed for the protein glucose isomerase by Farber *et al* (1988). The difference Fourier map showed that the heavy atom positions for the Laue and monochromatic data agreed. However, the isomorphous difference Patterson calculated from the Laue data was uninterpretable.

A mercury heavy atom derivative data set was analysed for the protein α-amylase by Helliwell *et al* (1989a) (see figure 7.13). The Laue singlet spots stimulated by wavelengths less than 1.0 Å were used to take advantage of the enhanced anomalous signal due to the Hg L edges. The Laue data yielded 'Laue' SIROAS phases (single isomorphous replacement with optimised anomalous scattering) capable of yielding the heavy

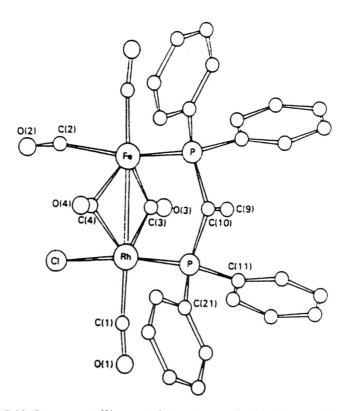

Figure 7.12 Organometallic crystal structure solved using wavelength normalised Laue data. From Harding *et al* (1988) with permission.

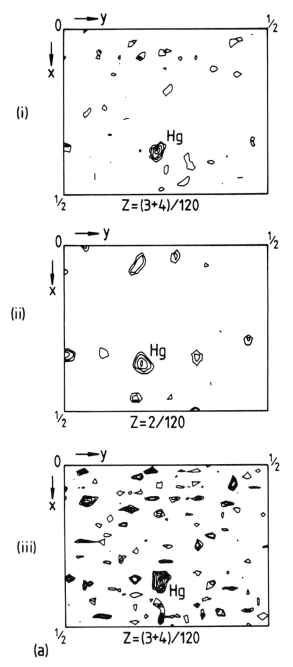

(a)

Figure 7.13 (a) Difference Fourier map calculated with coefficients $(F_{LaueHg} - F_{Nvmono})$ $\exp i\alpha_{(Pb+Pt)mono}$ using wavelength normalised data for (i) singles (3 Å differences), (ii) the low resolution component of energy deconvoluted doubles (6 Å differences); in (iii) monochromatic Hg data (3 Å differences).

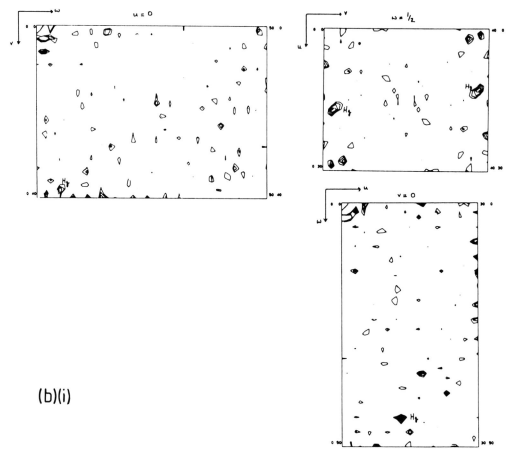

(b)(i)

Figure 7.13 (*cont.*) The Laue maps (i) and (ii) compare very favourably with (iii), the monochromatic map. From Helliwell
et al (1989a) with the permission of the American Institute
of Physics. (*b*) (i) Difference Patterson maps based on

atom positions of independent lead and platinum derivatives (figure
7.14). The isomorphous difference Patterson calculated from the Laue
data was as good as from the monochromatic data (figure 7.13(*b*)).

A small molecular structure has been solved with Laue data alone
using direct methods phasing (MULTAN). This was a chlorine complex
of 14 non-hydrogen (Gomez de Anderez *et al* 1989) atoms (figure 7.15).
The same crystal was used throughout and the SHELX *R*-factors (on *F*)
for Mo Kα, Cu Kα and Laue were respectively 6.23%, 6.45% and 8.19%.
The anomalous differences for this $P2_12_12_1$ non-centrosymmetric crystal
gave a consistent indication of the hand of the molecule for each data
set. The unit cell parameters for the Laue analysis came from the mono-

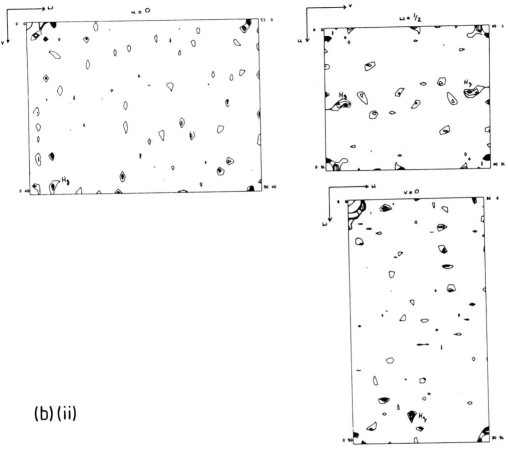

(b) (ii)

Figure 7.13 (*cont.*) ($F_{\text{LaueHg}} - F_{\text{LaueNv}}$) compared with (ii) the equivalent monochromatic Harker sections for the mercury amylase case. The Laue maps are as interpretable as the monochromatic ones.

chromatic measurements. The missing reflections due to harmonic overlap did not hinder the direct methods phasing process.

An organic complex of 29 non-hydrogen and 20 hydrogen atoms has been studied by M. Helliwell *et al* (1989) (figure 7.16). Again, MULTAN solution using the Laue data was possible. The SHELX *R*-factors (on *F*) for the Mo Kα and Laue data sets were respectively 4.60% and 5.29%. The rms deviation of the two structures is 0.010 Å. In each case the 20 *hydrogen atoms* appeared in difference Fourier syntheses after anisotropic temperature factor refinement of the non-hydrogen atoms. This work illustrated that the Laue data processed according to the procedures described by Helliwell *et al* (1989b) are good enough for weak

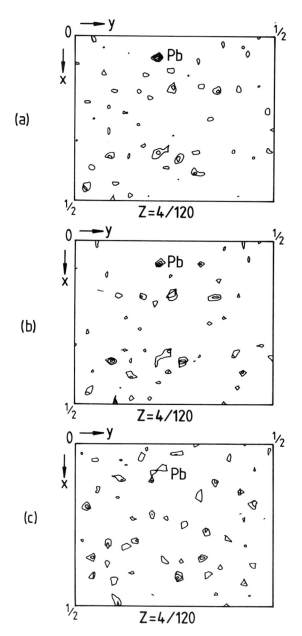

Figure 7.14 Difference Fourier map calculated with coefficients $(F_{Pbmono} - F_{Nvmono})$ exp$i\alpha_{HgSIROASLaue}$: (a) correct hand; (b) no hand; (c) incorrect hand. The mercury wavelength normalised Laue (3 Å) data were selected only if the stimulating wavelength was $\leqslant 1$ Å so as to yield an optimised anomalous dispersion signal from the Hg L absorption edges. From Helliwell *et al* (1989a) with the permission of the American Institute of Physics.

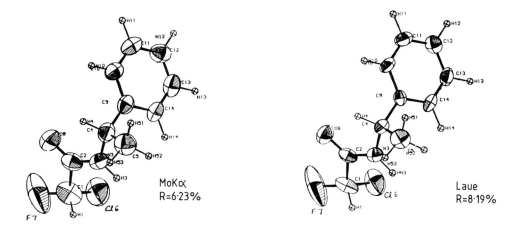

Figure 7.15 Views of a chlorine small molecule crystal complex solved
with wavelength normalised Laue data and MULTAN used
with defaults and comparison with the monochromatic
result from the same crystal. From Gomez de Anderez *et al*
(1989) with the permission of IUCr.

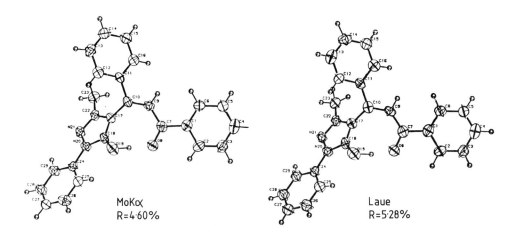

Figure 7.16 Views of an organic small molecule crystal solved with
wavelength normalised Laue data and MULTAN and comparison with monochromatic result. The 20 hydrogen atoms
were located from difference Fourier syntheses after
anisotropic temperature factor refinement of the 29 non-
hydrogen atoms. This work illustrates the sensitivity of the
Laue data to weak scattering centres such as hydrogen and
shows it is as good as monochromatic data. From M. Helliwell *et al* (1989) with the permission of IUCr.

scatterers like hydrogens to be revealed and this compares favourably with monochromatic methods.

A study on a low pH form of carbonic anhydrase (Lindahl *et al* 1990) has shown that a sensitivity to the scattering of 1 water molecule in 29 000 Da of enzyme is possible with the synchrotron Laue method (figure 7.17).

Calcium binding to tomato bushy stunt virus has been studied using the synchrotron Laue method (Hajdu *et al* 1989). The non-crystallographic symmetry of the virus crystal aided the clarity of the final map in showing up the calcium sites. A novel spatial deconvolution algorithm was used to obtain intensity estimates of the individual diffraction spots (Shrive, Clifton, Hajdu and Greenhough 1990).

The above studies are clearly establishing the credentials of the Laue method to yield accurate intensity and structural data. In chapter 10 the

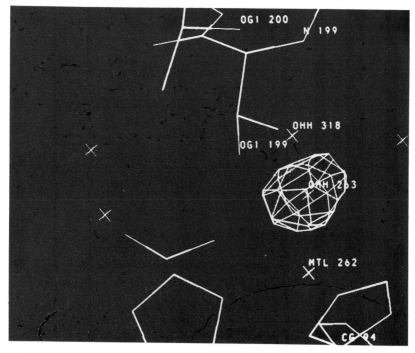

(a)

Figure 7.17 The sensitivity of the Laue method proven in the macromolecular case: (*a*) a single water molecule bound to the zinc ion in carbonic anhydrase; (*b*) an overall view of the carbonic anhydrase fold. From Lindahl *et al* (1990) with permission.

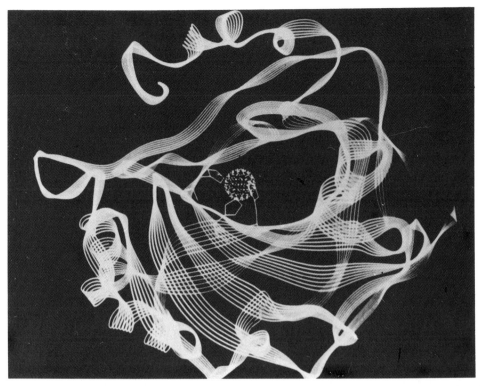

(b)

Figure 7.17 (*cont.*)

first time resolved synchrotron Laue experiment, performed by Schlichting *et al* (1990) to study GTP hydrolysis in the ras p21 protein, is described in some detail.

CHAPTER 8

Diffuse X-ray scattering from macromolecular crystals

Not all the diffracted photons from a crystal end up in the Bragg reflections from specified ($hk\ell$) planes. Indeed, for quite a large number of macromolecular crystals the non-Bragg diffraction or diffuse scattering is strong in intensity. The diffuse scattering is due to a breakdown in the periodicity of the crystal and carries information on the mobility and flexibility of the molecules in the crystal. There are text-books describing diffuse scattering from small molecule crystals such as Amorós and Amorós (1968) and Wooster (1962).

At the synchrotron the long exposure times used in the measurement of the high resolution Bragg data on film or IP also automatically give the diffuse scattering and reveal a diversity of diffuse background patterns from different crystals. These observations have stimulated considerable interest in trying to understand and interpret these features. Of great interest to the molecular biologist is the relationship between macromolecular structure and function. Recent years have shown that besides the static/time-averaged structural information, appreciation of the molecular flexibility and dynamics is essential. Usually this information has been derived from the crystallographic atomic thermal parameters and also from molecular dynamics simulations (see, e.g., McCammon (1984)) which yield individual atomic trajectories. A characteristic feature of macromolecular crystals *compared to small molecule crystals*, however, is that their diffraction patterns extend to quite limited resolution even employing SR. This lack of resolution is especially apparent in medium to large proteins where diffraction data may extend to only 2 Å or worse, thus limiting any analysis of the protein conformational flexibility from refined atomic thermal parameters. It is precisely these crystals where flexibility is likely to be important in the protein function.

An example of a monochromatic diffuse scattering pattern, figure 8.1, observed in studies of protein crystals was that of 6-phosphogluconate dehydrogenase (Helliwell 1977a; Helliwell *et al* 1986a); it is thought that this could be due to a hinge-bending motion of the domains in this molecule; computer graphics simulations have been used to assess the feasibility of such correlated motions. The case for hinge-bending flexibility in proteins is reviewed by Dobson (1990) following evidence for hinge bending in T4 phage lysozyme based on the analysis of several different crystal structures of this molecule by Farber and Matthews (1990). A second most intriguing (Laue) photograph showing pronounced diffuse scattering is shown in figure 8.1(*b*).

Interpretations of the continuous diffuse scattering in terms of waves of displacement along molecular filaments in tropomyosin (Boylan and Phillips 1986), correlated rigid body displacements in lysozyme (Doucet and Benoit 1987) and 'liquid-like' atomic displacements and range of coupling in insulin (Caspar, Clarage, Salunke and Clarage, 1988) have been put forward. The importance of interpreting the diffuse scattering as applied to proteins has been pointed out by Artymiuk (1987, 1988). A general, quantitative, approach to the refinement of correlated inter- and intramolecular displacements from diffuse scattering intensities of molecular crystals is being developed (Moss and Harris, to be published). An estimate of the thermal vibration mean square amplitudes for typical protein crystals has been made using estimates of the longitudinal speed of sound from laser generated ultra-sound measurements and of the transverse speed from Young's modulus estimates (C. Edwards *et al* 1990). This estimate is consistent with that made via a computational analysis of inter- and intramolecular vibrations of bovine pancreatic trypsin inhibitor by Diamond (1990).

The diffuse scattering background, which nearly always occurs in a diffraction pattern, may arise from several sources including:

(*a*) thermal diffuse scattering;
(*b*) static disorder scattering;
(*c*) solvent disorder;
(*d*) Compton scattering;
(*e*) fluorescence;
(*f*) scattering from mounting tubes;
(*g*) air scattering;
(*h*) intrinsic film fog or detector noise.

The static or dynamic displacement of atoms in crystals causes a break-

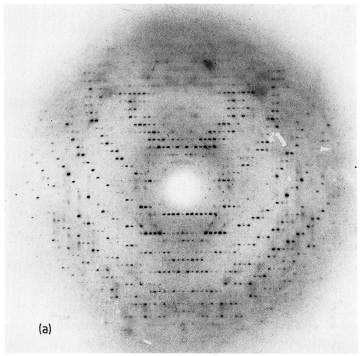

(a)

Figure 8.1 (*a*) Monochromatic, oscillation diffraction pattern recorded from a crystal of sheep liver 6-phosphogluconate dehydrogenase. This system was the one that stimulated the author's interest (Helliwell 1977a) in trying to explain the non-Bragg scattering (and perhaps, even, to use the information – such was the scepticism in the 1970s!).

(*b*) Polychromatic, stationary crystal (Laue) pattern of sodium chlorate showing unusual features around diffraction spots. No fully satisfactory explanation for these features has yet been found, but they are probably related to defects in the crystal lattice. Similar, but much weaker, features have been seen in the Laue patterns of some other sodium chlorate crystals. The crystals are unusually sensitive to radiation and after two or three exposures the spots become large and diffuse. If the crystal had high mosaic spread the diffraction spots would be radially elongated ellipses; note that here the elongation is not always radial, and the density is not exactly elliptical. (Crystal $140 \times 140 \times 120 \mu$m, exposure time 1 s, SRS at 2 GeV, 137 mA, 0.18 mm Al attenuator in beam, crystal–film distance 47.8 mm.) This figure and caption for part (*b*) kindly provided by R. J. Rule and M. M. Harding.

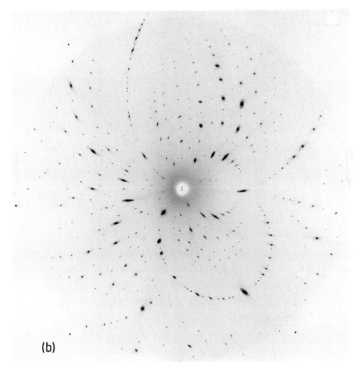

(b)

Figure 8.1 (*cont.*)

down of translational symmetry, leading to a reduction in the Bragg intensities at high resolution and the appearance of diffuse scattering at and between the reciprocal lattice positions. In macromolecular crystals the diffuse scattering is often quite strong, rich in detail and apparently distinctive to the crystal. It represents a potentially valuable source of information regarding atomic displacements.

Static disorder arises when unit cells exist with different arrangements of the time-averaged positions. Static orientational disorder occurs in molecular crystals where molecules or flexible domains or side groups take up different orientations breaking the translational symmetry. Dynamic disorder arises from thermal vibrations and is present in all crystals. Two types of lattice vibrations may be distinguished, acoustic modes due to the propagation of ultrasonic waves in the crystal and optic modes of vibration such as are observed in infra-red and Raman spectra. Ultrasonic vibrations give rise to thermal diffuse scattering which peaks primarily at the reciprocal lattice positions and is observed characteristically as a feature around the Bragg peak (figure 8.2). Optic mode vibrations along with other disorder modes give rise to diffuse scattering which is distributed continuously but non-uniformly throughout

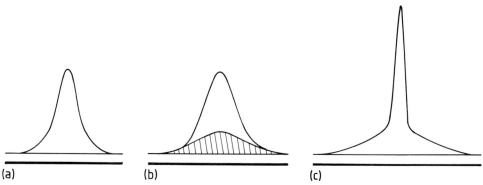

Figure 8.2 Schematic diagram illustrating reflection profiles for (*a*) an ideal Einstein solid and (*b*) a real solid, showing the contribution to the reflection of the acoustic diffuse scattering component. (*c*) This illustrates the use of very fine collimation to distinguish the Bragg peak from the broader diffuse scattering shoulder, minimising the diffuse scattering contribution to the integrated intensity. From Glover *et al* (1988) with permission.

reciprocal space and is sometimes observed as lines or streaks in a diffraction pattern. Reduction of temperature can be used to diminish the thermal diffuse scattering component. Cooling to liquid nitrogen temperature of protein crystals by rapid freezing is now enjoying increased success (section 5.5.2) and its use in the study of diffuse scattering will be of considerable interest in the future.

In this chapter a variety of X-ray diffraction patterns is given to illustrate, for a reasonably broad range of macromolecular crystals, the diversity of patterns that do occur. There then follow sections on the diffuse diffraction ring (section 8.3) and the nature of the acoustic scattering at/near to the Bragg peaks (section 8.4). Analysis of diffuse scattering effects is at a preliminary stage and is not dealt with in detail. We start, in the next section, with a description of neutron diffraction techniques which have been used considerably in the case of smaller unit cell crystals in the past and might be thought to be immediately applicable to macromolecular crystals now.

8.1 NEUTRON STUDIES OF DIFFUSE SCATTERING

Diffuse scattering has been extensively studied using neutrons. There is one distinct advantage that neutrons possess over X-rays when used to study crystal vibrations. The energy of these vibrations is quantised and the quanta of excitation are known as phonons. In the inelastic processes of thermal diffuse scattering of X-rays or neutrons, phonons are either created or destroyed. With X-rays, the change in energy of the diffracted beam has been too small to be measurable, until the advent of synchrotron X-radiation. However, with thermal neutrons the relative change is large and vibration frequencies are easily measured as a function of the wave-vector of the vibrational mode (Willis and Pryor 1975). The energy of thermal neutrons corresponds to a frequency of about 6 THz which is of the same order as that of low frequency phonons. By contrast a quantum of X-ray energy is some six orders of magnitude larger. It is from such studies with thermal neutrons that phonon dispersion curves can be constructed. By comparison, X-rays have a frequency of 3×10^6 THz, and a $30 \, \text{cm}^{-1}$ lattice vibration has a frequency of 1.0 THz. In addition, neutron scattering enables static disorder scattering to be studied quantitatively by removing the inelastic thermal diffuse scattering (Boysen, Frey and Jagodzinski 1984). Neutron diffraction is thus an important technique complementary to X-ray diffuse scattering. However, there are difficulties in applying neutron diffraction to the study of macromolecular diffuse scattering. For macromolecules, the large incoherent background scattering from hydrogen, the difficulty of exchanging the bound hydrogens (Wlodawer 1980) and the large size of crystals required for neutron experiments argue against the use of the technique.

8.2 EXAMPLES OF DIFFUSE DIFFRACTION PATTERNS IN MONOCHROMATIC GEOMETRY

There is a wide range of diffuse scattering patterns. This can be illustrated by the following examples (from Glover, Harris, Helliwell and Moss (1991)). These are drawn from a series of oscillation and stationary crystal diffraction patterns from a variety of protein crystal samples, all of which exploit the high intensity and small beam divergence of synchrotron X-ray beam lines. The size and diffracting power of the crystals vary widely as do the observed diffuse scattering features. Bovine ribonuclease-A (RNAse), 6-phosphogluconate dehydrogenase (6-PGDH), glutamate dehydrogenase (GDH) (the latter courtesy of D. Rice, Sheffield University), rabbit serum transferrin, avian γ_{II} crystallin have been examined at the Daresbury SRS. The diffraction patterns of t-RNA (met) were recorded by P. Sigler, Chicago, at the Stanford SSRL. The diffraction pattern from avian pancreatic polypeptide (aPP) was recorded at DESY, Hamburg. The diffraction patterns are shown in figures 8.3(a)–(f) and 8.4.

Case 1. aPP, space group C2, $a=34.2$ Å, $b=32.9$ Å, $c=28.5$ Å, $\beta=105.3°$ (Wood, Pitts, Blundell and Jenkins 1977). aPP is a small, 36-residue polypeptide hormone. The crystals have a low solvent content and diffract to extremely high resolution (<0.98 Å using conventional sources). The monochromatic, oscillation diffraction pattern (figure 8.3(a)) shows very little background diffuse scattering and little indication of acoustic scattering. It is a somewhat uncharacteristic macromolecular crystal with most scattered radiation appearing as elastically scattered Bragg reflections, and could be termed a 'hard' crystal.

Case 2. RNAse, space group P2₁, $a=30.45$ Å, $b=38.37$ Å, $c=53.22$ Å, $\beta=106°$ (Carlisle *et al* 1974). RNAse is an enzyme of 10 000 molecular weight and the crystals diffract to 1.2 Å using SR. A monochromatic, still diffraction pattern (figure 8.3(b)) shows distinct acoustic diffuse scattering haloes around many of the Bragg peaks but with little background diffuse scattering observed except for indistinct patches at resolutions above 3.2 Å. The solvent scattering ring is markedly anisotropic and shows considerable inhomogeneity.

Cases 3 and 4. 6-PGDH and GDH: 6-PGDH space group C222₁, $a=72.72$ Å, $b=148.15$ Å, $c=102.91$ Å (Adams *et al* 1977) and GDH space group C2, $a=147.1$ Å, $b=151.3$ Å, $c=94.6$ Å, $\beta=132.75°$ (Rice, Hornby and Engel (1985) and P. Baker, pers. comm.) are both in the category of medium to large proteins of 50 000 molecular weight and show features of both acoustic and continuous diffuse scattering (figures 8.1 and 8.3(c) and (d)). Some evidence of preferential orientation in the continuous diffuse scattering is seen. The diffuse 'solvent' scattering ring is evident. This ring is noticeably anisotropic, particularly in the GDH diffraction pattern (figure 8.3(d)) where the 'solvent' ring is approximately a pentagon extending from 4.5 Å to 2.7 Å resolution.

Case 5. Transferrin, space group P4₃2₁2, $a=b=127.4$ Å, $c=145.4$ Å (Al-Hilal *et al* 1976), molecular weight 80 000. Crystals of transferrin diffract to only about 3 Å but show considerable diffuse scattering, seen as a series of streaks running at approximately 45° to the Bragg layer lines (figure 8.3(e)). These features are strongest between 5.4 Å and 3.3 Å resolution. The diffuse scattering ring builds up at high resolution but

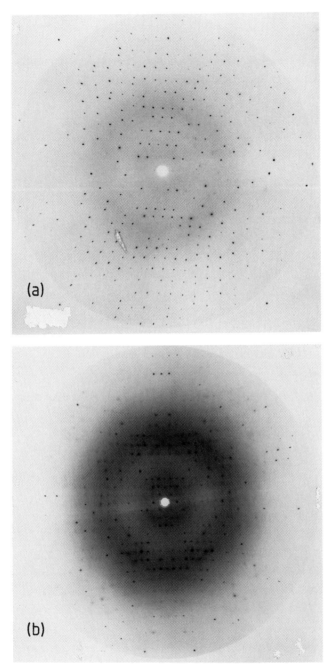

Figure 8.3 Single crystal synchrotron X-ray diffraction patterns recorded from a variety of macromolecular crystals illustrating the variety of diffuse scattering features. (*a*) aPP, resolution limit 1.8 Å, wavelength 1.49 Å, 4.9° oscillation photograph. (*b*) RNAse, resolution limit 1.9 Å, wavelength 1.488 Å, still exposure.

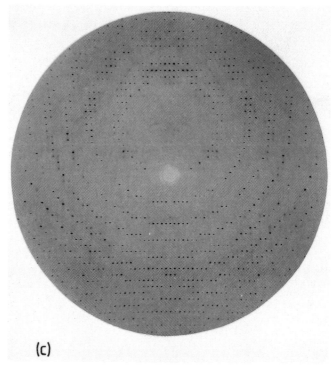

(c)

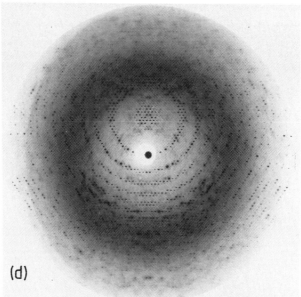

(d)

Figure 8.3 (*cont.*) (*c*) 6-PGDH with bound 2'AMP, 1.4° oscillation photograph, incident wavelength 1.488 Å, resolution limit 2.8 Å. (*d*) GDH, 1.5° oscillation photograph, resolution limit 1.9 Å.

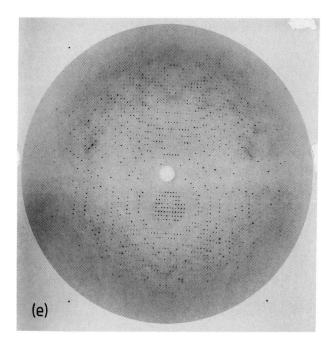

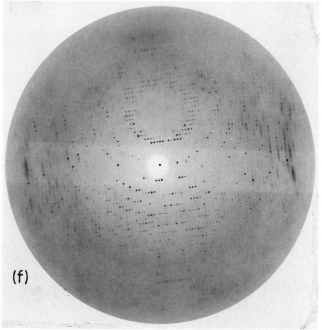

Figure 8.3 (*cont.*) (*e*) Transferrin, 1.0° oscillation photograph, resolution limit 3.3 Å, incident wavelength 1.74 Å. (*f*) t-RNA (met) 1.5° oscillation pattern, wavelength 1.54 Å. From Glover *et al* (1990); figures (*d*) and (*f*) courtesy of D. Rice and P. Sigler respectively with permission.

the inner (low resolution) edges form a distinct diamond shape with the same orientations as the diffuse scattering streaks. Chevrons of diffuse scattering, not associated with the layer lines, are seen along the meridian as the limit of the Bragg diffraction is reached and the diffuse scattering begins to dominate.

Case 6. t-RNA(met) (courtesy P. Sigler, Chicago), space group P6$_4$22, $a=b=113.7$ Å, $c=136.25$ Å. The diffraction pattern obtained from the crystals of t-RNA shows pronounced diffuse scattering. Figure 8.3(f) shows a 1.5° monochromatic oscillation photograph measured with a wavelength of 1.54 Å, the **c***-axis is inclined at approximately 10° to the horizontal. Equatorially oriented strong diffuse streaks can be seen which are approximately 8 mm long and follow the Bragg layer lines. Further streaking occurs perpendicular to these features, again following the layer lines. The strongest diffuse scattering features occur at around 3.3 Å resolution. The diffuse ring peaks at approximately 3.45 Å resolution and forms a diamond shape. Chevrons of diffuse scattering occur at the extremes of the Bragg diffraction, but these features do not follow the layer lines and hence are solely a function of the molecular transform and not the lattice.

Case 7. γ_{II} crystallin, space group P4$_1$2$_1$2, $a=b=58.1$ Å, $c=98.4$ Å (Carlisle, Lindley, Moss and Slingsby 1977), molecular weight 20 000. Crystals of the native protein, an eye lens protein, diffract to 1.4 Å on a synchrotron source. The monochromatic, still diffraction photograph shown in figure 8.4(a) was recorded to 2.25 Å resolution, and has distinct acoustic diffuse scattering but with little evidence of oriented diffuse scattering apart from the diffuse ring. Routine crosslinking of the crystals using 2.5% gluteraldehyde makes them physically more robust and is used before addition of heavy atoms. Unfortunately, the Bragg resolution limit is degraded and the diffuse scattering significantly increases. Figure 8.4(b) and (c) shows the reduction in the Bragg resolution of the monochromatic, still diffraction pattern and the concomitant rise in detail in the diffuse scattering as a result of crosslinking and heavy atom treatment.

8.2.1 Summary of the observations on these examples

In the cases cited above, the diffuse scattering features vary quite widely from crystal to crystal, but usually with a common feature of

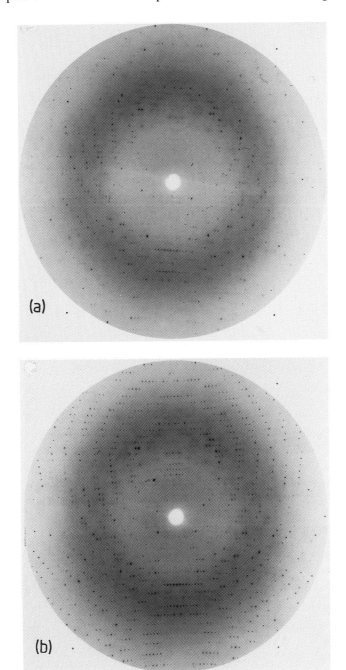

Figure 8.4 Still diffraction patterns recorded from native and modified
crystals of γ_{II} crystallin. (a) Native crystal, (b) cross-linked
crystal (2.5% gluteraldehyde). Resolution limit approx-
imately 2.25 Å with noticeable diffuse scattering.

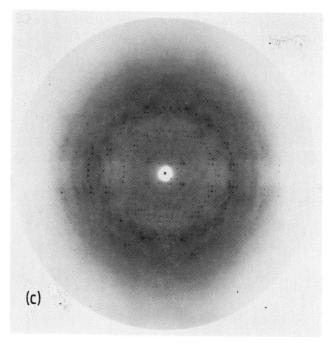

Figure 8.4 (*cont.*) (*c*) Cross-linked crystal soaked in 10 nM thiomersal. Resolution limit 2.8 Å with a drastic increase in diffuse scattering. All patterns were recorded at the SRS Daresbury, incident wavelength 1.488 Å with identical crystal to film distances and exposure times. From Glover *et al* (1990) with permission.

strong diffuse scattering close to the 'solvent' ring. Continuous diffuse scattering features vary from almost absent in the 'hard' aPP case to the dominating effects seen from crystals of t-RNA. Chemical modification in γ_{II} crystallin drastically alters the appearance of the diffraction pattern, degrading the Bragg component and enhancing the diffuse component. The variations in the patterns described above suggest that patterns from other samples will also be distinctive, thus offering specific information on a molecular crystal.

8.3 CONTRIBUTIONS TO THE 'DIFFUSE DIFFRACTION RING'

8.3.1 Experimental results

A strong feature of some of the diffraction patterns is the presence of an anisotropic ring of inhomogeneous diffuse intensity in the regions between approximately 4.5 Å and 2.8 Å resolution. These features are clearly not due to the solvent, either within the crystal or surrounding it, because of the marked anisotropy of the diffraction features. The scattering from a glass capillary (commonly used for mounting crystals), as well as water and glass capillary and water:ethanol and glass capillary, is shown in figure 8.5. The scattering from the glass capillary peaks at approximately $1/d=0.10$ Å^{-1} and 0.24 Å^{-1} whilst the other scattering plots (being dominated by the liquids) show isotropic rings peaking at $1/d$ values of 0.33 Å^{-1} and 0.31 Å^{-1} for water and water:ethanol, respectively. It should be noted that the glass scattering is less than 20% of the intensity of that recorded from the solutions.

The overall envelope of the experimentally observed diffuse scattering is shown in figure 8.6 for both RNAse and γ_{II} crystallin, recorded parallel

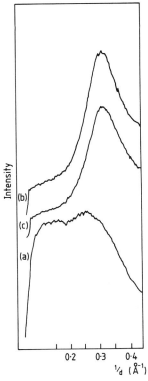

Figure 8.5 The densitometered profiles from the monochromatic diffraction patterns recorded from (a) a glass mounting tube, (b) a mounting tube filled with double distilled water, (c) a mounting tube filled with double distilled water:ethanol mixture (60:40 v/v). From Glover *et al* (1991) with permission.

and perpendicular to the crystal mounting axis. The scattering is clearly seen to be anisotropic and inhomogeneous. It peaks at resolutions outside the regions due to glass or solvent scattering. The observed scattering from macromolecular crystal samples in the 'solvent ring' is therefore not due to solvent scattering but is a result of inelastic scattering events, i.e. the interaction of X-ray photons with dynamic processes occurring in the crystal.

Compton scattering will also occur under the experimental conditions used, but will produce a slowly increasing isotropic background as resolution increases. This background is isotropic and monotonic for the biological molecules and only starts to dominate the Thompson scattering at higher resolutions ($>0.5\,\text{Å}^{-1}$).

8.3.2 Theory

The simplest model of thermal diffuse scattering is one where the crystal is considered to be an 'Einstein solid'. In this approximation all the atoms are vibrating independently as isotropic harmonic oscillators, each atom having the same mean square displacement U. For this model

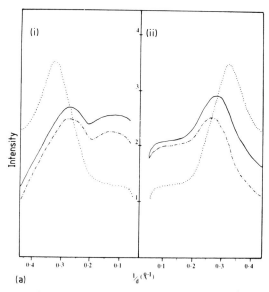

Figure 8.6 (a) Densitometered profiles from the recorded, monochromatic diffraction patterns of (i) RNAse and (ii) γ_{II} crystallin showing the overall envelope of the observed diffuse scattering intensity (I) recorded parallel (full line) and perpendicular (chain line) to the crystal mounting axis. The scattering from water (dotted line) is plotted for comparison.

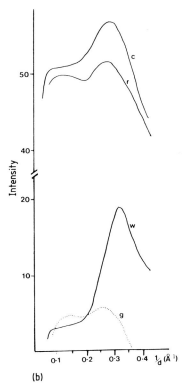

(b)

Figure 8.6 (*cont.*) (*b*) The recorded scattering intensity (*I*) for water 'w', RNAse 'r' and γ_{II} crystallin 'c' scaled to equal scattering volume and incident X-ray intensity. Glass mounting tube scattering 'g' is shown for reference. From Glover *et al* (1990) with permission.

the thermal diffuse scattering is given (Amorós and Amorós 1968) by

$$\sum_j f_j^2 [1 - \exp(-Q^2 U)] \tag{8.1}$$

and is plotted in figure 8.7 ($Q=(4\pi \sin \theta)/\lambda$). It can be seen that the predicted diffuse scattering peaks are between $0.25 \, \text{Å}^{-1}$ and $0.2 \, \text{Å}^{-1}$, depending on U, followed by a gradual fall off as Q increases. It can be seen from figure 8.6(*a*) that the observed diffuse scattering from both RNAse and γ_{II} crystallin do peak but at about $0.29 \, \text{Å}^{-1}$ resolution and decrease more rapidly than is predicted for an Einstein crystal. In particular, the diffuse scattering does not appear to increase relative to the Bragg scattering at resolutions higher than $0.4 \, \text{Å}^{-1}$. This latter observation is in common with most of the samples investigated here, with the exception of the

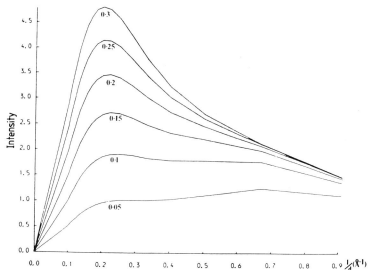

Figure 8.7 The overall form for the intensity of diffuse scattering (I) from an ideal solid, calculated for different values of the mean square displacement U (see equation (8.1)). From Glover *et al* (1990) with permission.

t-RNA crystal where the diffuse scattering clearly dominates at high resolution. Thus, with t-RNA as a possible exception, a protein crystal is a poor approximation to an Einstein crystal.

8.4 ACOUSTIC SCATTERING

The plots (figure 8.8), based on a monochromatic diffraction pattern, show the contoured optical densities at several reciprocal lattice positions and line sections through the peak (integrated with a $50\,\mu m$ half-width). The plots clearly show broad diffuse scattering shoulders to the Bragg peaks, the form of which varies with resolution. At low resolution the shoulders spread least and peak most strongly; the converse is evident at higher resolution. The explanation of these features is at an early stage. Glover *et al* (1991) are able to attribute the shape of these features to single phonon interactions, at least for the low resolution ($6\,\text{Å}$) acoustic diffuse scattering features.

8.5 DISCUSSION AND CONCLUDING REMARKS

The acoustic diffuse scattering is centred at the Bragg positions and can constitute a significant source of errors in integrated intensity measurements. The characteristic fine collimation of SR can be exploited in the collection of data to minimise the acoustic scattering contribu-

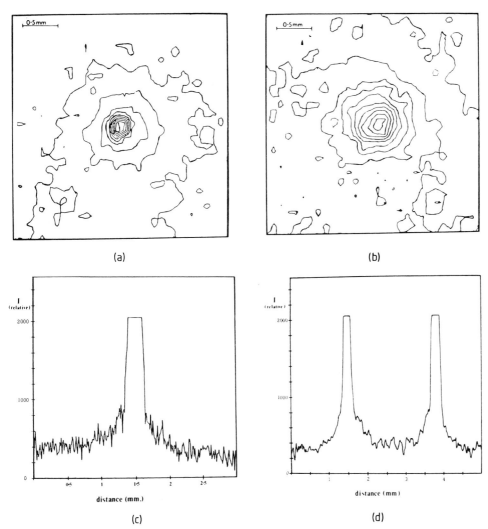

Figure 8.8 The form and distribution of acoustic scattering from a crystal of RNAse taken from a monochromatic diffraction pattern. 3×3 mm^2 boxes contoured according to scattered intensity in the region of (a) a low (5.5 Å) and (b) high (2.8 Å) resolution reflection, both densitometered using a $10 \times 10 \mu$m^2 raster and (c) and (d) line sections through two truncated Bragg peaks superimposed upon a broader shoulder of acoustic diffuse scattering. From Glover et al (1990) with permission.

tions. This improves the accuracy of the intensity data for isomorphous and anomalous difference measurements as well as atomic model refinements.

From even the limited number of patterns presented here, it is clear that the diffuse scattering observed from macromolecular crystals is very rich in detail and in many cases constitutes a significant contribution to the total diffraction intensity. Also, the distribution of diffuse scattering is clearly not uniform in reciprocal space and differs from one macromolecule to another. In particular, oriented streaks are often seen, and the mis-named solvent ring is a region of strong diffuse scattering which is often markedly anisotropic and may contribute more diffracted energy than due to the solvent itself.

These various observations suggest that the continuous diffuse scattering is a potentially useful source of unique dynamic information regarding correlated atomic displacements in a crystal.

The use of the high intensity of SR to measure the weak diffuse scattering and the collimation to 'exclude' the Bragg peaks opens up the possibility of routinely being able to gather the necessary experimental data for detailed analysis of molecular mobility and flexibility. Work here is at a relatively early stage.

Recently, Diamond (1990) has published a theoretical analysis of vibrations in crystals of BPTI (bovine pancreatic trypsin inhibitor). In this study (see figures 8.9(a)−(c)) he is able to separate contributions of intra- and intermolecular mobility (lattice vibrations and disorder) to

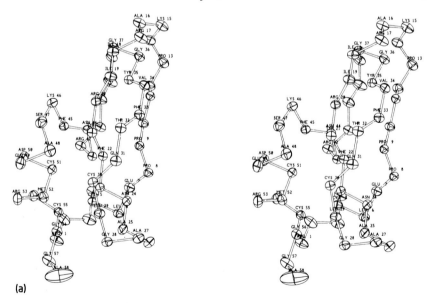

(a)

Figure 8.9 Analysis of the mobility of atoms in a protein crystal. Stereo drawings of BPTI for (a) overall mobility (this includes contributions depicted in (b) and (c) and static disorder component)

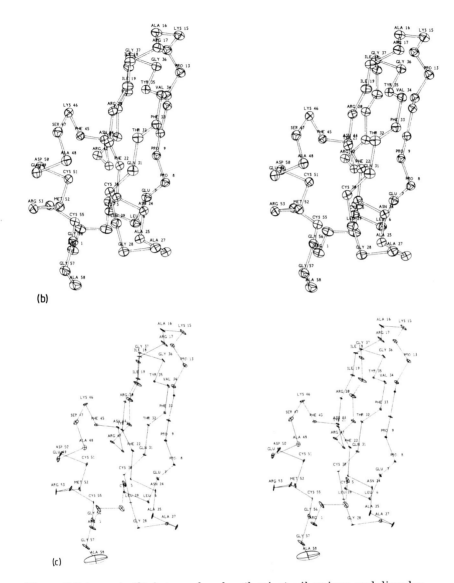

Figure 8.9 (*cont.*) (*b*) intermolecular (lattice) vibrations and disorder, and (*c*) intramolecular vibrations. From Diamond (1990) with permission.

the overall mobility of atoms in this small protein molecule. The theoretical estimate for the intermolecular contribution is consistent with that derived experimentally from laser generated ultra-sound and Young's modulus measurements for ribonuclease (Edwards *et al* 1990).

CHAPTER 9

Variable wavelength anomalous dispersion methods and applications

Anomalous dispersion effects have been used for a considerable time, to some extent, in crystal structure analyses. In combination with a tunable synchrotron X-ray source there is considerable scope for a dramatically increased use and benefit of these small but significant effects.

The technique used most often, at least so far, to determine new protein structures is multiple isomorphous replacement (Green *et al* (1954); chapter 2). Anomalous dispersion (i.e. the f'' component) of each heavy atom has been used in a secondary role to give extra information to improve the phase estimate. More importantly, the anomalous differences based on f'' have been used to determine the handedness of the molecule unambiguously. Anomalous difference Patterson maps (Rossman 1961) have been used to confirm the positions of the heavy atoms, generally determined by isomorphous difference Pattersons, at least in the case of a limited number of heavy atom sites. Whereas the f'' component shows up in the difference between $hk\ell$ and $\overline{hk\ell}$ reflections (for example) at one wavelength, the real component of the anomalous dispersion can only be used if data are collected at more than one wavelength. This has been done using conventional X-ray sources by Hoppe and Jakubowski (1975) in a feasibility study on erythrocruorin. This is an iron containing protein and these workers used Ni Kα (1.66 Å) and Co Kα (1.79 Å) radiations to vary f' and enhance the size of f'' of the iron atom around its K edge (1.74 Å).

The ideas involving use of anomalous scattering and different wavelengths have a fairly long history. Okaya and Pepinsky (1956) suggested the use of a second incident wavelength to resolve the phase ambiguity inherent in use of f'' alone (see equation (2.20)). Mitchell (1957) examined the two-wavelength method of Okaya and Pepinsky (1956) and took the interesting step of separating the normal scattering of all the atoms in the unit cell from the anomalous scattering of one (or a few) atoms. Herzenberg and Lau (1967), writing from Manchester, suggested the use of several wavelengths (i.e. >2) for phase determination and the possible application to proteins using the sulphur scattering. Karle (1967) first presented his algebraic analysis of the structure factor equations using several wavelengths and where the positions of the anomalous scatterers are not known. Karle (1980) expanded this approach and it has formed the basis of the so-called MAD (multi-wavelength anomalous dispersion) method with SR; see section 9.4.

The scope for utilising anomalous dispersion is considerable in conjunction with the synchrotron. SR has a high intensity over the X-ray wavelength range of interest and can access the absorption edges of (a) commonly occurring metals in metallo-proteins, (b) metals which are used in heavy atom derivatives, (c) selenium in specially grown seleno-proteins (see section 9.8) and (d) bromine in brominated nucleotides.

There is an immediate means of using SR in the multiple isomorphous replacement (MIR) method whereby for a particular protein the heavy atoms in question have their f'' values optimised. For example, it is possible to collect data from separate platinum, gold and mercury derivatives at closely spaced wavelengths to maximise the f'' component at the respective L edges (Helliwell 1977a). Use of the L_{III} edge gives $\approx 10e^-$ and the L_I edge $\approx 12e^-$ in each case (ignoring white line effects) in contrast to Cu Kα values of $\approx 7e^-$ for f'' in each. It is straightforward on an SR instrument to access these slightly different wavelengths and collect each heavy atom derivative data set with optimal f''. One can coin the acronym SIROAS (Baker et al 1990) or MIROAS for this approach, single (or multiple) isomorphous replacement with optimised anomalous scattering. This approach has been applied in various cases and clearly makes the electron density map and the hand determination clearer. For the case of these high Z number derivatives the additional advantage of using a short wavelength in accessing the L edges is important because of the reduction of sample absorption errors. The native data should also be collected at similar, short wavelength as well. The very slight differences in the sample absorption surface that remain can be taken care of by local scaling of the derivative to native data sets.

Another application of anomalous dispersion and SR is the case of a protein containing a cofactor of two metal atoms of similar atomic number. In such a case the native electron density map would not clearly indicate which metal atom is which. However, the anomalous dispersion f'' values could be made markedly different by suitable choice of wavelength. One such study was that of pea lectin (Einspahr *et al* 1985), which contains manganese and calcium. Another study was that of copper zinc superoxide dismutase (Kitigawa *et al* 1987).

These examples involve the use of f'', and not f', at chosen wavelengths from the SR spectrum. However, both f'' and f' are used in the so-called MAD phasing method referred to above. In order to stimulate the maximum f'' and f' values requires use of a minimum of three wavelengths. Two of these are very close together at the absorption edge to maximise f'' and minimise f' (remember it is negative). The third wavelength is well removed from the absorption edge and so provides a reference data set from which $\Delta f'$ differences can be derived from differences between the data sets. In principle, these three wavelengths can be reduced to two but the maximum change in f' will not be utilised. Clearly, it is feasible to use tube emission lines of different elements on a conventional source to achieve a MAD phasing analysis. As mentioned earlier, Hoppe and Jakubowski (1975) did that experiment for erythrocruorin using its iron atom. The preponderance of iron metalloproteins does mean that there is scope to try and turn Hoppe and Jakubowski's approach into a viable method. Indeed, Xuong (pers. comm.) is doing this with a specially constructed rotating anode (to deliver easily either Ni $K\alpha$ or Co $K\alpha$ radiation) and taking advantage of the data collection accuracy and efficiency of multiwire area detectors that have been realised on conventional X-ray sources. The conventional source set up has the advantage of stability of source intensity and wavelength but without the flexibility of using any desired wavelength.

In this chapter the use of SR and anomalous dispersion will be discussed in some detail, after various aspects of f' and f'' have been covered. The chapter will also give several case studies to illustrate the scope of the various SR methods and approaches. A treatment of the geometric effects of diffraction at different wavelengths and the variation of sample absorption with wavelength was given in chapter 6. Appendix 3 gives tabulations of the wavelengths of the K and L absorption edges for all elements and values of f', f'' over a wide range of λ's for selenium, bromine, platinum and mercury as examples.

9.1 THE DISPERSION COEFFICIENTS f'' AND f'

9.1.1 Unbound atoms

Figure 9.1 shows an idealised plot of f'' and f' each as a function of X-ray wavelength for a K edge (singlet) and L edge (triplet) (from Hoppe

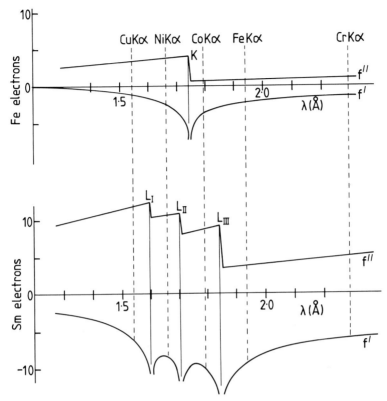

Figure 9.1 Idealised curves for the real (f') and imaginary (f'') terms of the anomalous scattering for K edge (iron) and L edge (samarium) cases. From Hoppe and Jakubowski (1975) with permission.

and Jakubowski (1975)); f'' increases markedly from below to above the edge in energy, while f' varies fairly symmetrically about the edge dipping to its largest negative value at the midpoint of the rise of the absorption edge. The wavelength of this midpoint is the one tabulated in appendix 3 as the wavelength of the pure element absorption edge. If we plot f'' versus f' as a function of photon energy at an absorption edge (figure 9.2), then the variation follows a loop; the maximum of f'' does not occur at the same energy as the minimum of f'. We will refer to this figure again later in terms of the phasing power of the anomalous effects.

Unlike the normal atomic scattering factor, f_0, the values of f' and f'' do not decrease with scattering angle because the radius of the inner absorbing electron shell is much less than the wavelength of the radiation used (table 9.1). This actually aids the signal to noise of the anomalous effects at high resolution (equation (2.22)).

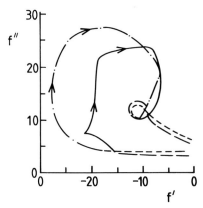

Figure 9.2 Plot of measured values of f' and f'' for the case of metal atoms praseodymium (chain line) and samarium (full line) in the isomorphous crystal salts of sodium (metal) ethylene diamine tetra-acetate octahydrate. The direction of increasing SR photon energy (decreasing SR photon λ) is indicated by the arrows. The total wavelength change was 0.0088 Å for praseodymium and 0.0115 Å for samarium with a $\delta\lambda/\lambda$ at each point of 1×10^{-3}. Based on tabulated data from Templeton et al (1980a) with permission. The total phasing power is proportional to the area of either loop; the effect of a worsened $\delta\lambda/\lambda$ would be to decrease this area. Any instrument slippage leading to a λ change even of 1×10^{-3}, would cause a serious excursion around the loop and dramatic changes in f' and f''; obviously some parts of the loop are more susceptible to a given λ change than others.

Many elements, covering nearly the whole of the periodic table, show the absorption edge effects in the X-ray region (figure 9.3). The proliferation of the edges indicates the widespread opportunities of using optimised anomalous dispersion techniques.

The values of f' and f'' in appendix 3 are based on the calculations of Cromer and Liberman (1970) using relativistic wave functions. These calculations assume that the atoms are free (i.e. unbound). This approach is based on work by James (1948), Parratt and Hampstead (1954), Dauben and Templeton (1955), Templeton (1955) and Cromer (1965). The real component f' can be measured directly by X-ray interferometry.

Siddons (1979) and Hart and Siddons (1981) made measurements of f' on zirconium, niobium and molybdenum and compared them with the Cromer and Liberman (1970) calculated values. These were in dis-

Table 9.1. *The theoretical variation of f' and f" with (sinθ)/λ for a K shell (cobalt) and an L shell (samarium) for Cu K wave-length. The K shell radius is less than the L shell so that fall off with (sinθ)/λ is less significant. In turn, the fall off of f_0 is pro rata much greater because of the large size of the atom compared with the core shell radii. Values of f' and f" for (sinθ)/λ = 0 are from Dauben and Templeton (1955); the corrections for other values of sinθ/λ are derived from Berghuis et al (1955) and Veenendaal et al (1959) (extracted from International Tables of Crystallography III 213–16) with permission. Units of f', f" and f_0 are in e^-.*

| | | (sinθ)/λ (d_{res} Å) | | |
| | | 0 (∞) | 0.4 (1.25) | 0.6 (0.83) |
Element				
Cobalt	f' (f")	–2.2 (3.9)	–2.2 (3.9)	–2.2 (3.8)
	f_0	27.0	14.7	10.3
Samarium	f' (f")	–6.6 (13.3)	–6.6 (12.8)	–6.7 (12.4)
	f_0	62.0	38.6	30.2

crepancy on the long wavelength sides. Cromer and Liberman (1981), as a result, published an amended calculation for those wavelengths where $1.0 < λ/λ_{edge} < 1.06$. These calculations were in closer agreement with the measurements by X-ray interferometry as well as those by diffractometry of known structures (e.g. of caesium by Templeton, Templeton, Phillips and Hodgson (1980b)). The earlier estimates of Cromer and Liberman (1970), tabulated in *International Tables of Crystallography*, are relatively error free at general wavelengths (i.e. by less than $0.05e^-$). Cromer and Liberman (1981) also point out errata in previously published values for germanium, molybdenum, silver and praseodymium. Other measurements of f' and f" using pure materials have been made by Fukamachi and Hosoya (1975) on the gallium K edge by energy-dispersive diffractometry. X-ray interferometry has been used to derive values for the nickel K edge (Bonse and Materlik 1972, 1976), the copper K edge (Freund 1975), arsenic (Fukamachi, Hosoya, Kawamura and Okunki 1977), selenium (Bonse *et al* 1980) and the L edges of platinum (Hart 1980). The last is shown in figure 9.4. A large number of X-ray interferometry measurements on many elements were undertaken on the SRS station 9.7 (Hart and Siddons, pers. comm.) as well as at other SR sources.

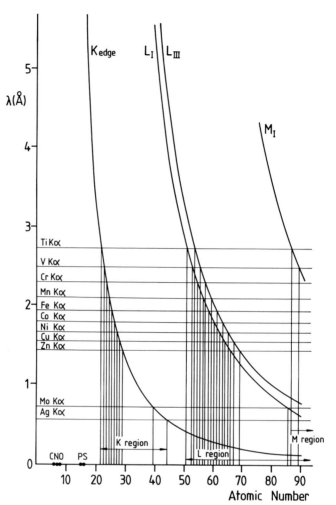

Figure 9.3 Absorption edges (K, L and M) plotted as a function of wavelength showing the potential widespread application of optimised anomalous dispersion since for most elements a K, L or M edge is in the range of wavelengths available from an SR source. The limited range of discrete wavelengths available from conventional line emission sources is also shown. From Hoppe and Jakubowski (1975) with permission.

9.1.2 f'' and f' for bound atoms

The environmental state of an X-ray absorbing atom in a protein, as in most 'materials', is considerably removed from that of a free atom. As a result, considerable 'structure' exists in linear absorption coefficients very close to the absorption edge (see Appendix 4, figure A4.1).

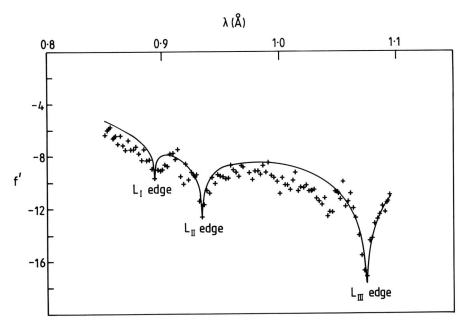

Figure 9.4 L edge f' dispersion coefficient obtained by X-ray interferometry measurements for pure platinum. Note that $|f'|$ is largest for L_{III}. From Hart (1980) with permission.

The presence of this XANES was reported by Coster (1924) who called the features 'white lines' because of their appearance in spectra on photographic plates (cf. figure 5.8 for the polychromatic Bragg spots and the presence of the rhenium white line, from Greenhough *et al* (1983)). The structure near the edge is attributed to exciton formation (Brown, Peierls and Stern 1977) or many-body effects, such as edge singularities arising from multi-electron–hole pair formation (Mahan 1967; Nozieres and De Dominicis 1969), plasmon satellite structures and multiple electron excitations (Fano 1961). Holland, Pendry, Pettifer and Bordas (1978) used the near edge structure to obtain information on the shape of the atomic potential and Greaves, Durham, Diakun and Quinn (1981) to understand metal structure.

For a number of heavy atom derivatives the local environment of the absorbing atom, which dominates the XANES, is preserved even after binding to the protein, e.g. $K_2Pt(CN)_4$ has the platinum firmly surrounded by the four cyanide ligands in a square planar arrangement. The L_{III} spectrum of $K_2Pt(CN)_4$ is shown in figure 9.5, and should be a reasonably close approximation to $Pt(CN)_4^{2-}$ bound to the protein; the $K_2Pt(CN)_4$ can be used as a 'model compound', with its high platinum

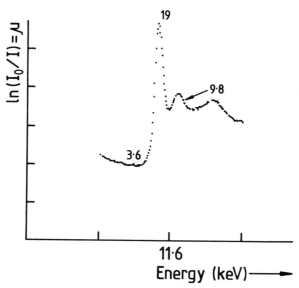

Figure 9.5 XANES scan at the L_{III} edge of platinum with $K_2Pt(CN)_4$ as sample measured on DORIS with a Si(220) double crystal, harmonic suppressing monochromator with 0.7 eV band-width. A large jump in the absorbed intensity (white line) is seen very close to the absorption edge. Estimates of f'' (in e⁻) are given at specific points. From Helliwell, Phillips and Siddons (unpublished). See also figure 9.8.

concentration, to simulate the situation at low platinum concentration in the protein. The wave function calculations of Cromer and Liberman (1970) give values of f'' which reflect the basic 'step function' shape of f'' at the edge. The value of f'' at the XANES of the real material can be estimated by interpolation. A value for f'' of 19e⁻ (by measurement and interpolation) exists at the XANES of $K_2Pt(CN)_4$, whilst the calculated value is 9e⁻ close to the L_{III} edge. By comparison the EXAFS can change f'' by ±1 or 2e⁻. In terms of maximising f'' the XANES is clearly important. Whether any significant XANES is present needs to be investigated for each case. For example, mercury compounds, which are often used as heavy atom derivatives only have small XANES features (Phizackerly, pers. comm.). In the lanthanides the width of the XANES feature is often very narrow (figure 9.6) in $\delta\lambda/\lambda$ and so the demands on the instrument are greater. Lye *et al* (1980) have discussed L edge white line spectra and their use in biological crystallography.

The accurate setting of the absolute λ, as well as $\delta\lambda$, is, of course, very important and must take account of absorption edge shifts. Figure 9.7

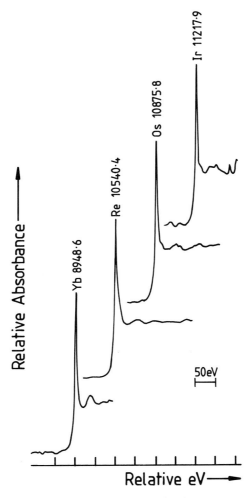

Figure 9.6 The L_{III} absorption edges for ytterbium and several heavier transition metals. The complexes are $Yb(CdPm)_3$, $ReC\ell_3$, $OsC\ell_6(Bu_4N)_2$ and $NaIrC\ell_6$. Note the narrow energy width of the white line. From Lye *et al* (1980) with permission.

shows the variation in edge position as a function of manganese metal valence state. For the permanganate case (valence VII) the edge shift is 20 eV. Table 9.2 indicates the size of the variation both between and also within valence states as a function of the ligand. The effect of a change in ligand gives a smaller edge shift (of the order of 1–3 eV) than valence state. In the case of lanthanide compounds where the XANES features are very large, the complexation of the metal with the protein will have a unique coordination so that no model compound would be available to mimic the protein–metal complex ligand structure. For these and other

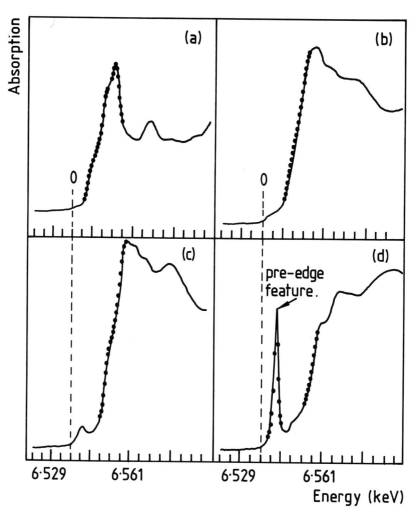

Figure 9.7 XANES spectra of the reference oxides: (*a*) MnO, (*b*) Mn$_2$O$_3$, (*c*) MnO$_2$ and (*d*) KMnO$_4$ representing valence states II, III, IV and VII of manganese respectively and their associated edge shifts. From Brown, McMonagle and Greaves (1984) with permission. See also table 9.2.

examples, therefore, measurement of the absorption edge is carried out using a fluorescence detection system for the protein crystal itself *in situ* on the camera or diffractometer (section 5.5.5).

In certain cases the absorption edge effects may exhibit dichroism whereby there is a change as the crystal orientation is changed. Not only the XANES effects can change but also the edge position itself. Templeton and Templeton (1980, 1982) show these effects clearly in their

Table 9.2. *The energy shifts of the K absorption edges of manganese, iron and copper in selected chemical compounds (from Salem, Chang, Lee and Severson (1978) with permission).*

Manganese compounds		Iron compounds		Copper compounds	
Compound	Energy shift (eV)	Compound	Energy shift (eV)	Compound	Energy shift (eV)
Mn	0.00 reference	Fe	0.00 reference	Cu	0.00 reference
MnO	4.36 ± 0.52	FeB	0.00 ± 0.24	Cu_2O	-0.67 ± 0.6[b]
	5.4[a]	FeO	4.47 ± 0.24	CuI	2.20 ± 0.6[b]
MnF_2	7.57 ± 0.52		9.2[a]	CuO	3.87 ± 0.6[b]
$MnCl_2$	5.84 ± 0.52	FeS	6.5[a]	$CuSO_4$	4.88 ± 0.6[b]
Mn_2O_3	7.81 ± 0.52	FeF_2	6.15 ± 0.24	$CuCO_3$	4.78 ± 0.6[b]
MnO_2	10.73 ± 0.53	$FeCl_2$	7.57 ± 0.24		
	15.2[a]	$FeCl_3$	9.97 ± 0.24		
	9.0[a]	Fe_2O_3	7.53 ± 0.24		
			13.0[a]		
$MnSO_4$	9.7[a]	$FeSO_4$	12.1[a]		
$KMnO_4$	20.0[a]	$Fe_2(SO_4)_3$	9.0[a]		
	18.5[a]		13.9[a]		

[a]From Kawata and Maeda (1973) and references cited therein
[b]From Verma and Agarwal (1968)

investigation of absorption edges in a vanadyl ion and the linear uranyl ion (UO_2^{2-}). Templeton and Templeton (1985a,b) have also studied the dichroism in the pyramidal bromate ion (BrO_3^{-2}) and the square planar tetrachloro platinate II ion ($PtC\ell_4^{2-}$), see figure 9.8, respectively. Templeton and Templeton (1989) have extended these studies to include lithium iodate and so examine the effect of the K level width. These factors and their effects on subsequent phase calculation will not be seen unless high resolution monochromator systems are employed. The effects derive from an asymmetric environment of a bound atom.

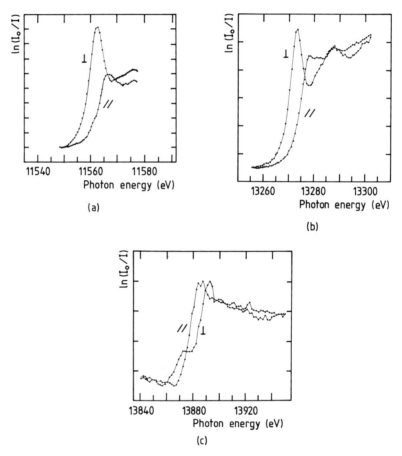

Figure 9.8 Absorption coefficient (on an arbitrary scale) versus photon energy for polarised X-rays with electric vector parallel and perpendicular to the four-fold axis of $PtC\ell_4^{2-}$ near (a) the Pt L_{III} absorption edge, (b) Pt L_{II} edge, (c) Pt L_I edge. From Templeton and Templeton (1985b) with permission.

9.1.3 The magnitude of X-ray wavelength optimised f'' and f'' in protein crystallography

The magnitude of effects at the K edge is considerably smaller than at the L_{III} edge. This reduces the effectiveness with which K edge effects can be applied. Of course, many naturally occurring metallo-proteins contain elements with K edges such as molybdenum, zirconium, copper, nickel, iron, manganese and calcium. Whether the effects are big enough depends on how accurately the data are measured and the concentration of the metal in the protein (equation (2.22)). The metal can be considered a trace element at the concentrations which usually occur. Also, a partial occupancy of the site reduces the Δ_{ANO} signal. The addition at a second site does not help the signal-to-noise problem since the measured signal still needs to be good enough to resolve the effects attributable to the separate distinct sites. In the case of a low concentration the signal to noise can only be improved by minimising the noise (see section 9.9) and by stimulating as large an f'' and f' as possible. Fortunately, the natural metal atom in a metallo-protein is usually tightly bound so that the anomalous effects are not reduced too much by thermal vibration.

The largest values observed for f' and f'' to date are those reported for praseodymium (Templeton *et al* 1980b); $f'=-27e^-$, $2f''=55e^-$ and gadolinium (Templeton *et al* 1982a); $f'=-31.9e^-$, $2f''=62.4e^-$. These values were determined crystallographically using known structures. Hence, the use of anomalous dispersion effects and the accuracy with which diffraction patterns are measured should be seen in the context of isomorphous replacement where $50–90e^-$ can be added at the heavy atom site, and, in addition, the electrons of the atoms attached to the heavy atom also contribute to the scattering. However, the multiwavelength anomalous data will always be exactly isomorphous, an extremely important advantage. Moreover, in the heavy atom derivative case, the metal will be on the protein surface and so have a large temperature factor.

9.1.4 The use of f'' and f' with conventional X-ray sources

The presence of f'' leads to a breakdown in Friedel's law (see equation (2.17)) (Coster, Knol and Prins 1930). Bijvoet (1949) showed that the breakdown of Friedel's law could be used to determine not only the phases of X-ray reflections but also the absolute configuration of molecules. This work was extended by Ramachandran and Raman (1956).

In protein crystallography the effects were used to determine heavy atom positions (Rossmann 1961; Harding 1962; Dodson and Vijayan 1971; Dodson, Evans and French 1975; Mukherjee, Helliwell and Main 1989). For phase determination North (1965) and Matthews (1966) extended the Blow and Crick (1959) formulation (section 2.4.4) for multiple isomorphous replacement to include anomalous scattering. Friedel differences alone have been used to solve the structure of crambin (Hendrickson and Teeter 1981), and a trimeric haemerythrin (Smith, Hendrickson and Addison 1983) without resort to isomorphous data. In the crambin case sulphur was the principal anomalous scatterer though its absorption edge at 5Å was far removed from the Cu Kα radiation used, i.e. it was non-optimal! In both these cases the phase ambiguity of acentric reflections was resolved by taking the phase (of the two possible) which was closest to the heavy atom phase (sulphur or iron). Earlier, Herzenberg and Lau (1967) discussed the use of sulphur anomalous dispersion in protein structure determination because sulphur is a naturally occurring element in proteins.

Ramaseshan (1964) discussed the possibilities of optimising the anomalous dispersion with different X-ray tubes. Hoppe and Jakubowski (1975) used Ni and Co Kα radiation to collect two data sets about the iron edge in erythrocruorin; in this pioneering study, phases were determined with a figure of merit of 64% (mean phase error=50°).

9.1.5 Neutron anomalous dispersion studies

The advent of 'white' neutron sources, before SR facilities, stimulated the use of multiple wavelength techniques and several structures were solved in this way (table 9.3 from Ramaseshan and Narayan (1981)). The technique is limited to only a few nuclei (unlike the ubiquity of X-ray absorption edges) such as ^{113}Cd, ^{149}Sm, ^{151}Eu and ^{157}Gd. The low flux of neutron sources has restricted their use in protein crystallography.

Table 9.3. *A list of almost all the structures that have been solved using neutron anomalous scattering methods.*

Cd $(NO_3)_2 \cdot 4D_2O$	MacDonald and Sikka (1969)
Sm $(BrO_3)_3 \cdot 9H_2O$	Sikka (1969a,b)
NaSm (EDTA)$\cdot 8H_2O$	Koetzle and Hamilton (1975)
Cd (tartrate)$\cdot 5H_2O$	Sikka and Rajagopal (1975)
Cd-histidine$\cdot 2H_2O$	Bartunik and Fuess (1975)
Agua (L-glutamate) Cd (II) H_2O	Flook, Freeman and Scudder (1977)

Schoenborn (1975) successfully phased the structure of cadmium-myoglobin.

9.1.6 Nuclear anomalous dispersion studies

Nuclear energy levels would yield much larger anomalous scattering effects (by 1–2 orders of magnitude). The use of radiation to probe these effects has been suggested (Moon 1961; Raghavan 1961) and pursued experimentally (Black 1965; Mossbauer 1975; Parak *et al* 1976; Bade *et al* 1982) with γ sources. The use of SR over tiny bandwidths (10^{-9} eV) is also being investigated at several SR sources but flux levels have so far been necessarily very low because of the bandwidth.

9.2 THE OPTIMAL WAVELENGTHS FOR PHASE DETERMINATION

There is a best combination of wavelengths which will most effectively determine the phase of each reflection. In figure 9.9 we represent

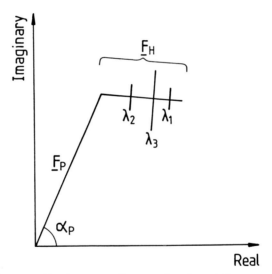

Figure 9.9 The contributions to a Bragg reflection at three wavelengths through an absorption edge; f' and f'' vary rapidly with wavelength but F_P is constant. The Friedel related reflection, mirrored through the real axis is also shown. The wavelengths shown follow the sequence $\lambda_1 > \lambda_2 > \lambda_3$ where λ_2 corresponds to the minimum of f', λ_3 to the white line position (f'' maximum) and λ_1 to a position in the pre-edge region. From Helliwell (1984) with the permission of the Institute of Physics.

the contributions to a Bragg reflection at three wavelengths through an absorption edge. The maximum of f'' is not coincident with the minimum in f' as stressed earlier. The quantities we can measure are F^+_{PH} or F^-_{PH} at any number of wavelengths which we illustrate with the three chosen here.

If we use just two wavelengths and measure

$$F^+_{PH\lambda_3}, \; F^-_{PH\lambda_3} \text{ and } F^+_{PH\lambda_2}$$

(λ_3 corresponds to the white line and λ_2 to the minimum in f') and draw the corresponding phasing circles, we get a well-resolved unique phase (figure 9.10). The centres of these phasing circles are well separated and non-collinear; this is necessary and sufficient for phasing (as has also been pointed out by Templeton *et al* (1980b), and by Ramaseshan and Narayan (1981)).

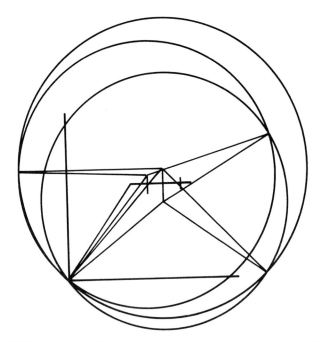

Figure 9.10 The centres for the phasing circles in this Harker diagram are based on the Friedel equivalent pair at the white line (or just on the short λ side of the edge) and one equivalent for the minimum of f' (i.e. two wavelengths in total). The centres are well separated and non-collinear and the unique phase is well resolved. From Helliwell (1984) with the permission of the Institute of Physics.

Between λ_2 and λ_3 we exploit some differences in the f' values. With three wavelengths, however, the change in f' can be optimised. Hence, λ_1 is chosen well removed from the absorption edge ($f' \rightarrow 0$), λ_2 at the point of inflection of the edge (f' most negative) and λ_3 just on the short wavelength side of the edge (f'' most positive).

If, however, we had measured F_{PH}^{+} at three wavelengths on the low energy side of the edge (f' does not reach its minimum and f'' is constant and ≈ 0), then though the centres are well separated they are collinear and a phase ambiguity results. At first sight the variation of f' at three λ's seems to simulate isomorphous replacement with native and two derivatives; figure 9.11 shows that it does not.

The effect of a larger and larger $\delta\lambda/\lambda$ would gradually reduce the size of f'' and f' so that the centres of the phase circles would move closer and the unique intersection of the phase circles could not be defined leading to a phasing error. Likewise, in the plot of f'' versus f' (figure 9.2) the area of the loop decreases as $\delta\lambda/\lambda$ increases (Ramaseshan, pers. comm.). We

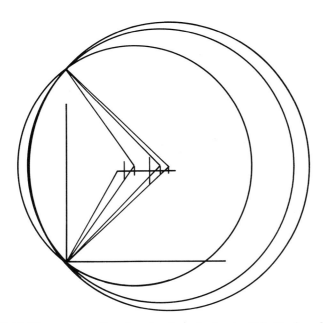

Figure 9.11 The centres for the phasing circles in this Harker diagram are based on measurements of the same member of a Friedel pair but at three separate wavelengths on the low energy side of the edge; f' does not reach its minimum and f''=constant. The centres are well separated but collinear and the determination of the phase remains ambiguous. From Helliwell (1984) with the permission of the Institute of Physics.

can think of this area as being proportional to the 'phasing power' in a given case. By measuring data at a few wavelengths we use a part area of the loop.

Phillips (1978) and Narayan and Ramaseshan (1981) considered schemes to decide which are the optimum wavelengths to use; for the case of three measurements the best ones to choose maximise the area of the triangle which these three points define on the f', f'' plot. The three measurements may be made at two wavelengths (including one Friedel pair) or at three wavelengths.

There are, however, energy-dispersive possibilities which utilise the whole area of the loop to maximise potential phasing power (see section 6.2.2) (Arndt *et al* 1982; Greenhough *et al* 1983). This profile approach is of interest because the discrete wavelength method is susceptible to sudden wavelength changes during collection of diffraction data sets which may result from mechanical or thermal instability of optical systems used at SR sources or movement of the source point. For example, Harada, Yasui, Murakawa and Kasai (1986) choose their wavelengths in a different way from that outlined earlier to take account of such instabilities; λ_1, as usual, was chosen far from the absorption edge in order to obtain the maximum difference of f' from λ_2 or λ_3 but λ_2 and λ_3 were chosen slightly different from λ_{edge}, since the actual bandwidth of the beam may be broader than might be provided by the optics owing to fluctuations of the electron orbit; hence, for the Fe K edge λ_1 was 1.077 Å; λ_2 was 1.730 Å; λ_3 was 1.757 Å.

9.3 TRIGONOMETRIC RELATIONSHIPS OF THE STRUCTURE AMPLITUDE AS A FUNCTION OF WAVELENGTH

There are a number of distinct cases to be considered. In this section these are:

(*a*) SIROAS or MIROAS;
(*b*) F_{PH} as a function of λ (classical treatment).

In the next section (9.4) an alternative to (*b*) is described (due to Karle (1980)).

9.3.1 SIROAS or MIROAS

Single (or multiple) isomorphous replacement with optimised anomalous scattering (Helliwell 1977a; Helliwell *et al* 1984a) can be analysed trigonometrically in an entirely analogous way to Kartha (1975) and as embodied in equations (2.19) and (2.20). The important difference is the experimental one that instead of using a fixed wavelength (Cu Kα) the f'' for each derivative has been optimised using the appropriate SR wavelength. An example of SIROAS is given in section 9.7.3.

9.3.2 F_{PH} as a function of λ (classical treatment)

Equations similar to those given earlier for SIRAS (equations (2.19), (2.20)) (Kartha 1975) can be derived relating measurements made of F^+_{PH} and F^-_{PH} as a function of λ.

In the case of a metallo-protein α_{PH} is the quantity of interest (rather than α_P) and, of course, α_{PH} is itself an explicit function of λ and must be related to a reference wavelength, λ (perhaps where f' and f'' are $=0e^-$). It can be shown from figure 9.9 that

$$\alpha_{PH\lambda_3} = \frac{\pi}{2} + \alpha_H \pm \cos^{-1}\left(\frac{(F^+_{PH\lambda_3})^2 - (F^-_{PH\lambda_3})^2}{4F_{PH\lambda_3}f''_{\lambda_3}}\right) \tag{9.1}$$

where

$$F_{PH\lambda_3} \approx \tfrac{1}{2}(F^+_{PH\lambda_3} + F^-_{PH\lambda_3}) \tag{9.2}$$

and

$$\alpha_{PH\lambda_2} = \alpha_H \pm \cos^{-1}\left(\frac{(F^2_{PH\lambda_1} - F^2_{PH\lambda_2} - f'^2_{\lambda_{2,1}})}{2|f'_{\lambda_{2,1}}|F_{PH\lambda_2}}\right) \tag{9.3}$$

where $|f'_{\lambda_{2,1}}|$ is the difference in f' between λ_2 and λ_1. Equation (9.1) gives two possible values for α_{PH} based on f'' and equation (9.3) two values based on f'. The information in f' and f'' is 'complementary' (or better put, 'orthogonal') in an analogous manner to 'isomorphous' and 'anomalous' data in 'conventional source crystallography' and the equations can be compared directly with those given earlier (equation (9.1) with equation (2.20) and equation (9.3) with equation (2.19)). The change in α_{PH} between λ_2 and λ_1 is given by

$$\alpha_{PH\lambda_2} - \alpha_{PH\lambda_1} = \pm \cos^{-1}\left(\frac{f'^2_{\lambda_{2,1}} - F^2_{PH\lambda_2} - F^2_{PH\lambda_1}}{-2F_{PH\lambda_2}F_{PH\lambda_1}}\right) \tag{9.4}$$

From equation (9.4) we see that in the special case of

$$\alpha_{PH} = \alpha_H \text{ then } f'_{\lambda_{2,1}} = \pm(F_{PH\lambda_1} - F_{PH\lambda_2}) \text{ and } \alpha_{PH\lambda_2} = \alpha_{PH\lambda_1}$$

In this treatment we have considered for simplicity the case of one anomalous scattering type only and where α_{PH} is of interest. In equation (9.1) the λ used is that to maximise f'' and in equation (9.3) the variable λ is set to maximise $|f'|$. Of course, other wavelengths can be used and the equations still apply. If the data sets are measured at widely different λs, especially to exploit f' changes, then the absorption corrections that need to be applied will be widely different for each data set.

The optimised values of

$$F^{+}_{PH\lambda_3} - F^{-}_{PH\lambda_3} \text{ and } F_{PH\lambda_2} - F_{PH\lambda_1}$$

can be used as coefficients in Patterson syntheses for the initial location of the metal atom site in the crystal unit cell and for its subsequent refinement. This information is, of course, needed to calculate α_H for use in equations (9.1) and (9.3). The situation is analogous to the isomorphous difference Patterson (equation (2.9)) to allow location of the derivative site to provide α_H for equations (2.19) and (2.20). This type of 'classical' approach to phasing has been used with multiple wavelength data by Harada *et al* (1986) with cytochrome c' (Fe K edge) and Korszun (1987) with azurin (Cu K edge), for example.

Karle (1967, 1980) has given an algebraic analysis whereby the non-anomalous parts of the atomic scattering factors are separated from the anomalous parts so allowing the quantities explicitly dependent on λ to be distinguished. This is an important theoretical development and is dealt with in the next section in more detail.

9.4 SEPARATION OF ANOMALOUSLY AND NON-ANOMALOUSLY SCATTERING CONTRIBUTIONS: MAD AND KARLE PHASING ANALYSIS

Mitchell (1957) introduced the idea of separating out the normal scattering of a structure from any anomalous scattering from that structure. Karle (1980) presented a detailed theory using the idea of separating the intensities of scattering for the anomalous and non-anomalous parts. Examination of the systems of equations that resulted showed that by appropriate choice of variables, many of the equations resulted in a usefully linear form.

Other theoretical approaches (e.g. section 9.3) have treated separately the anomalously scattering atom. This was done largely because the mathematics paralleled the chemical procedure of heavy atom derivatisation.

In contrast, Karle's detailed theory leads to expressions for the structure amplitudes and phase differences defined in terms of the non-anomalous atomic scattering factors for *all* the atoms in the structure and whereby the *wavelength-dependent parts are treated separately*. The notation in common use is that of Hendrickson (1985), which is essentially the one used below. Compared with equation (2.6) $(hk\ell)$ is now written as $(\mathbf{h})$ and $(hx_j+ky_j+\ell z_j)$ as $(\mathbf{h}\cdot\mathbf{r}_j)$ for brevity to allow the extra nomenclature now needed to be clearer.

For a structure of N atoms containing n anomalously scattering atoms of the same type the structure factor is a function of the wavelength used given by

$$^{\lambda}\mathbf{F}(\mathbf{h}) = \sum_{j=1}^{N} {}^{0}\!f_j \exp(2\pi i\mathbf{h}\cdot\mathbf{r}_j) + \sum_{k=1}^{n}({}^{\lambda}\!f_k' + i\,{}^{\lambda}\!f_k'') \exp(2\pi i\mathbf{h}\cdot\mathbf{r}_k) \qquad (9.5)$$

The meaning of the nomenclature is that the superscript 'λ' refers to wavelength-dependent terms, the superscript '0' refers to wavelength-independent terms. The right-hand side of equation (9.5) involves a summation over N terms, which is wavelength-independent and a summation over n terms, which is wavelength-dependent.

Writing the λ-independent part as ${}^{0}\mathbf{F}_T(\mathbf{h})$ then, for the case considered here of one type of anomalous scatterer A, we can factor out of the right-hand summation the values $({}^{\lambda}\!f_A' + i{}^{\lambda}\!f_A'')$, so

$$^{\lambda}\mathbf{F}(\mathbf{h}) = {}^{0}\mathbf{F}_T(\mathbf{h}) + ({}^{\lambda}\!f_A' + i\,{}^{\lambda}\!f_A'')\sum_{k=1}^{n}\exp(2\pi i\mathbf{h}\cdot\mathbf{r}_k) \qquad (9.6)$$

The subscript 'T' refers to the total structure including the normal scattering part of the n anomalously scattering atoms. This latter part is

$$^{0}\mathbf{F}_A(\mathbf{h}) = \sum_{k=1}^{n} {}^{0}\!f_k \exp(2\pi i\mathbf{h}\cdot\mathbf{r}_k) \qquad (9.7)$$

Again, for the case considered here of one kind only of anomalous scatterer the ${}^{0}\!f_k$ can be factored out as a single value, ${}^{0}\!f_A$, so

$$^{0}\mathbf{F}_A(\mathbf{h}) = {}^{0}\!f_A\sum_{k=1}^{n}\exp(2\pi i\mathbf{h}\cdot\mathbf{r}_k) \qquad (9.8)$$

Combining equations (9.6) and (9.8)

$$^\lambda F(h) = {}^0F_T(h) + \left(\frac{^\lambda f'_A + i \, ^\lambda f''_A}{^0 f_A} \right) {}^0F_A(h) \tag{9.9}$$

This key equation is now used to obtain an expression for the intensity of a reflection, **h**, at a given wavelength, λ.

$$| {}^\lambda F(h)|^2 = {}^\lambda F(h) \cdot {}^\lambda F(h)^* \tag{9.10}$$

Using (9.9) and (9.10)

$$| {}^\lambda F(h)|^2 = | {}^0F_T(h)|^2 + \left(\frac{^\lambda f'^2_A + {}^\lambda f''^2_A}{^0 f^2_A} \right) |F_A(h)|^2 + \frac{^\lambda f'_A}{^0 f_A} \, [{}^0F_T(h) \, {}^0F_A(h)^*$$

$$+ \, {}^0F_A(h) \, {}^0F_T(h)^*] + i \frac{^\lambda f''_A}{^0 f_A} \, [{}^0F_A(h) \, {}^0F_T(h)^* - \, {}^0F_T(h) \, {}^0F_A(h)^*] \tag{9.11}$$

Now expand the structure factors into an amplitude and phase, i.e.

$$\left. \begin{aligned} {}^0F_T(h) &= | {}^0F_T(h)| \, \exp(i\Phi_T) \\ {}^0F_A(h) &= | {}^0F_A(h)| \, \exp(i\Phi_A) \end{aligned} \right\} \tag{9.12}$$

and also introduce the following terms, as shorthand,

$$\left. \begin{aligned} {}^\lambda a &= \left(\frac{^\lambda f'^2_A + {}^\lambda f''^2_A}{^0 f^2_A} \right) \\ {}^\lambda b &= \frac{2 \, ^\lambda f'_A}{^0 f_A} \\ {}^\lambda c &= \frac{2 \, ^\lambda f''_A}{^0 f_A} \end{aligned} \right\} \tag{9.13}$$

Equation (9.11) now becomes

$$| {}^\lambda F(h)|^2 = | {}^0F_T(h)|^2 + \, {}^\lambda a | {}^0F_A(h)|^2 + \frac{^\lambda b}{2} \, [| {}^0F_T(h)| | {}^0F_A(h)| \, \exp(i\Phi_T) \, \exp(-i\Phi_A)$$

$$+ | {}^0F_A(h)| | {}^0F_T(h)| \, \exp(i\Phi_A) \, \exp(-i\Phi_T)]$$

$$+ \frac{i \, ^\lambda c}{2} \, [| {}^0F_A(h)| | {}^0F_T(h)| \, \exp(i\Phi_A) \, \exp(-i\Phi_T) \tag{9.14}$$

$$- | {}^0F_T(h)| | {}^0F_A(h)| \, \exp(i\Phi_T) \, \exp(-i\Phi_A)]$$

using

$$\cos \theta = \tfrac{1}{2}[\exp(i\theta) + \exp(-i\theta)]$$

and

$$\sin \theta = \tfrac{1}{2i}[\exp(i\theta) - \exp(-i\theta)]$$

and introducing $\Delta \Phi = (\Phi_T - \Phi_A)$ then

$$|{}^\lambda \mathbf{F}(\mathbf{h})|^2 = |{}^0 \mathbf{F}_T(\mathbf{h})|^2 + {}^\lambda a\, |{}^0 \mathbf{F}_A(\mathbf{h})|^2$$
$$+ {}^\lambda b\, |{}^0 \mathbf{F}_T(\mathbf{h})|\,|{}^0 \mathbf{F}_A(\mathbf{h})|\cos \Delta \Phi + {}^\lambda c\, |{}^0 \mathbf{F}_T(\mathbf{h})|\,|{}^0 \mathbf{F}_A(\mathbf{h})|\sin \Delta \Phi \qquad (9.15)$$

The equivalent expression to equation (9.15), but for the Friedel mate $(\bar{\mathbf{h}})$ intensity measurement necessitates only a change of sign in the ${}^\lambda c$ term, i.e.

$$|{}^\lambda \mathbf{F}(\bar{\mathbf{h}})|^2 = |{}^0 \mathbf{F}_T(\mathbf{h})|^2 + {}^\lambda a\, |{}^0 \mathbf{F}_A(\mathbf{h})|^2$$
$$+ {}^\lambda b\, |{}^0 \mathbf{F}_T(\mathbf{h})|\,|{}^0 \mathbf{F}_A(\mathbf{h})|\cos \Delta \Phi - {}^\lambda c\, |{}^0 \mathbf{F}_T(\mathbf{h})|\,|{}^0 \mathbf{F}_A(\mathbf{h})|\sin \Delta \Phi \qquad (9.16)$$

(One check of these expressions, for the reader, is to take the 'standard' anomalous difference, at a particular wavelength, by subtracting equation (9.16) from (9.15). The resulting expression may then be compared directly with equation (2.20).)

The following procedure then applies for MAD phasing. Simultaneous equations may be set up by observing $|{}^\lambda \mathbf{F_h}|^2$ and $|{}^\lambda \mathbf{F_{\bar h}}|^2$ at several λ's. By choosing as variables

$$\left. \begin{array}{l} |{}^0 \mathbf{F}_T(\mathbf{h})|^2 \\[4pt] |{}^0 \mathbf{F}_A(\mathbf{h})|^2 \\[4pt] |{}^0 \mathbf{F}_T(\mathbf{h})|\,|{}^0 \mathbf{F}_A(\mathbf{h})|\cos(\Phi_T - \Phi_A) \\[4pt] |{}^0 \mathbf{F}_T(\mathbf{h})|\,|{}^0 \mathbf{F}_A(\mathbf{h})|\sin(\Phi_T - \Phi_A) \end{array} \right\} (9.17)$$

the system of equations is linear in these variables.

In the centric case, $\sin(\Phi_T - \Phi_A) = 0$ and $\cos(\Phi_T - \Phi_A)$ is either $+1$ or -1. In this case the values to be determined are $|{}^0 \mathbf{F}_T(\mathbf{h})|^2$, $|{}^0 \mathbf{F}_A(\mathbf{h})|^2$ and the choice of sign of $\cos(\Phi_T - \Phi_A)$, which are obtained from the observed values of $|{}^\lambda \mathbf{F_h}|^2$ at several wavelengths.

The positions of the anomalous scatterers can be established using the $|{}^0 \mathbf{F}_A(\mathbf{h})|$ values entered as coefficients in a Patterson synthesis or direct methods program. Alternatively, the values of $|{}^\lambda \mathbf{F}(\mathbf{h})| - |{}^\lambda \mathbf{F}(\bar{\mathbf{h}})|$ can be used in similar fashion; this may or may not be more accurate than using the $|{}^0\mathbf{F}_A(\mathbf{h})|$ values. The values of Φ_A can be calculated from the positions of the anomalous scatterers. Hence, Φ_T can be obtained, $|{}^0\mathbf{F}_T(\mathbf{h})|$ is available already and an electron density map calculated. Note that there still remains the problem of the handedness of the anomalous scatterer partial structure. The partial structure and its enantiomorph are both consistent with the $|{}^0\mathbf{F}_A(\mathbf{h})|$'s. Taking the enantiomorph changes globally the sign of the Φ_A phases and thus gives completely different results for the Φ_T set. Hence, both electron density maps are calculated and the decision as to which is the correct one is made by visual inspection (Fanchon and Hendrickson 1990). A summary and general exposition of the MAD method is given in Karle (1989). An example of the MAD method as applied to the case of streptavidin is detailed in case study 9.7.6; other examples of MAD phasing analysis are listed in table 9.6.

9.5 THE USE OF BIJVOET RATIOS RATHER THAN DIFFERENCES

An alternative formalism to using differences has been suggested for phase determination when using anomalous dispersion data which are based on the ratio of Friedel equivalents (Unangst, Muller, Muller and Keinert 1967; Bartunik 1978). Cascarano, Giacovacco, Peerdeman and Kroon (1982) have utilised such a scheme for determining the positions of anomalously scattering atoms in a crystal structure. The use of a ratio removes scaling errors due to absorption. Ramaseshan and Narayan (1981) suggest, however, that with the multiple wavelength method, the same crystal should be used for all the experiments so that scaling may not be a problem and, by using ratios only, an important piece of information will be ignored, i.e. the absolute values of the radius vectors in the complex plane.

9.6 THE EFFECT OF THERMAL VIBRATION AND MOLECULAR DISORDER

The anomalous dispersion effect is associated with the ejection of photoelectrons from inner shell electrons in an atom. The normal scattering describes the interaction of all the electrons in the atom with the X-ray beam. The radial distribution of the electrons in an atom can be calculated using quantum mechanics, originally by Hartree's self-consistent field method (Hartree 1933). In figure 9.12 this distribution is given for rubidium, which has a K edge at 0.8155 Å; the mean radius for

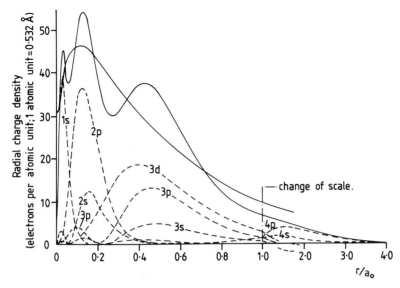

Figure 9.12 The radial charge density distribution of Rb$^+$ (a_0 is a constant$=0.532$ Å). Based on Hartree (1933).

the K shell is only 0.025 Å whereas the mean atomic radius is 0.2 Å. As mentioned in section 9.1.1, a consequence of this is that f falls off quite rapidly with $(\sin\theta)/\lambda$ whereas f' and f'' do not (table 9.1). The Fourier transform of a delta function is finite for all spatial frequencies. In this section consideration is given to the scattering effects in real rather than reciprocal space.

In crystal structure analysis the quantity determined is the electron density in e$^-$/Å^3 (equation 2.7)). Consider a rubidium atom, for example, fixed rigidly at equivalent positions for all unit cells in a crystal. For the anomalous effect due to a K shell of $f'=-1$e$^-$ and $f''=3$e$^-$ (for a λ of 0.7 Å and Rb K absorption edge at 0.8155 Å) and mean inner shell radius of 0.025 Å, the anomalous electron density is 1.92×10^5 e$^-$/Å^3 (for f''). The normal scattering effect due to 37e$^-$ with a mean atomic radius of 0.2 Å is 4.6×10^3 e$^-$/Å^3. Clearly, the average electron density of the anomalous effect is much greater than that due to normal scattering in principle. In practice, thermal vibration and molecular disorder smear out the atomic coordinates and reduce the electron density; the effect is especially marked for the anomalous electron density.

The effect of thermal vibration is to attenuate the crystal reflection intensity by a factor

$$\exp\left(\frac{-2B\sin^2\theta}{\lambda^2}\right) \tag{9.18}$$

where $2B$, the temperature factor, is related to the mean square amplitude of atomic vibrations by the relation

$$2B = 16\pi^2 \overline{u^2} \tag{9.19}$$

The temperature factor $2B$ can be evaluated according to the specific heat theory of Debye–Waller so that

$$2B = \frac{12h^2 T}{m_A k_B \theta^2} \, \phi \left(\frac{\theta}{T} + \frac{1}{4} \frac{\theta}{T} \right) \tag{9.20}$$

and m_A is the atomic mass, k_B is Boltzmann's constant, h is Planck's constant, T is the absolute temperature, θ is the Debye chararacteristic temperature for the crystal and ϕ is a function of θ/T and the term $\frac{1}{4}\theta/T$ takes account of the 'zero point energy of vibration of the lattice' which is a necessary consequence of the quantum theory of the harmonic oscillator.

The effect of the temperature on the intensities of reflection from a complex crystal cannot be taken into account as simply as equations (9.18) and (9.19) suggest. The mean square vibration of each kind of atom must be considered so that we can define a temperature factor for atom j as

$$B_j = 8\pi^2 \overline{u_j^2} \tag{9.21}$$

and equation (2.6) becomes

$$\mathbf{F}(\mathbf{h}) = \sum_j f_j \, \exp\left(-B_j \frac{\sin^2 \theta}{\lambda^2} \right) \exp(2\pi i \mathbf{h} \cdot \mathbf{r}_j) \tag{9.22}$$

At absolute zero the B_j are close to zero (equation (9.20)) and so are the thermal u_j^2 (equation 9.21)). However, for a protein crystal each unit cell is not exactly identical to the next, i.e. there is a statistical population of atomic coordinates causing an effective random disorder for all temperatures. Since the disorder is random equations (9.18) and (9.22) still apply and there are corresponding u_j^2 (equation (9.21)) due to the disorder.

For many proteins the $\langle B_j \rangle$ due to disorder is $10\,\text{Å}^2$ corresponding to a $(\overline{u_j^2})^{1/2}$ of $0.36\,\text{Å}$. If we convolute this 'disorder smearing' with the K shell radius of $0.025\,\text{Å}$ as

$$(0.025^2 + 0.36^2)^{\frac{1}{2}} = 0.361\,\text{Å} \tag{9.23}$$

then the f'' electron density for a rubidium atom subject to such a disorder is $63.6e^- \text{Å}^{-3}$.

The overall effect of smearing of the atomic radius is somewhat less marked *pro rata*, i.e.

$$(0.2^2 + 0.36^2)^{\frac{1}{2}} = 0.41\text{Å} \tag{9.24}$$

with an electron density of $537e^- \text{Å}^{-3}$.

Unfortunately, no protein crystal exists with such a very small molecular disorder for the anomalously scattering atom and which is coolable to absolute zero; if it did then the effective electron density for an anomalous f' or f'' of only a few electrons would have an electron density considerably higher than the normal scattering electron density (by 50–100 times). In a practical case for metallo-proteins with a natural metal cofactor buried in the protein core, the disorder B-factor will be considerably less than the overall disorder factor for the whole protein where the surface residues will have particularly large temperature factors. On the other hand, a heavy atom derivative bound to the surface of the protein will have a large disorder factor.

It is of interest to consider cooling crystals to liquid helium temperatures to maximise the anomalous electron density. Whether the intrinsic benefit of the small inner shell radius can be utilised depends on the residual molecular disorder and is more appropriate for consideration of solvent-free, smaller molecule crystals than of proteins. To take advantage of it in the diffraction pattern, i.e. in reciprocal space, would require reflections measured to very high resolution or small interplanar spacings.

For protein work isomorphous replacement is used primarily for 'low resolution phasing' (say to 3Å); the limit being set by lack of isomorphism and fall of f with $(\sin\theta)/\lambda$. Optimised anomalous dispersion is appropriate for low and high resolution phasing since there is no lack of isomorphism and f' and f'' do not fall off so rapidly with $(\sin\theta)/\lambda$.

9.7 RESULTS ILLUSTRATING THE SCOPE OF THE METHOD: VARIABLE SINGLE WAVELENGTH AND MULTI-WAVELENGTH METHODS

We describe several examples of optimised anomalous dispersion applied to location of metal sites and phasing.

9.7.1 Case study: optimised anomalous dispersion; small unit cell; gramicidin-A (Phillips and Hodgson 1980)

Gramicidin-A is a small (MW 3000) transmembrane channel protein for which attempts were made to solve the structure by direct methods, model building and isomorphous replacement without success (see above reference for details). A derivative containing caesium was prepared but was not isomorphous to the native crystals and so no phase information was obtained. Because the caesium derivative was thought to be of biological significance it was planned to solve the structure using multiple wavelength techniques at SSRL.

The caesium derivative was isomorphous to a potassium derivative. Data were collected on the potassium derivative on a conventional source at Cu $K\alpha$ wavelength. Data were also collected at 2.47 Å at the peak of the absorption of the caesium L_{III} edge where f'' is a maximum. Data extended to a resolution of 3.8 Å. The isomorphous and anomalous differences gave phases, which allowed an electron density map to be calculated; the sense of progression of an α-helix was clearly recognisable from the map.

The 3.8 Å resolution limit was imposed due to the restricted amount of beam time available and the fact that a single counter diffractometer was used. The concentration of caesium in the molecule (1 atom in 3000 D) is very high compared with most other metallo-proteins and so the wavelength induced Friedel intensity change was large.

A related protein gramicidin-S (complexed with urea) was solved by direct methods (Hull *et al* 1978).

9.7.2 Case study: optimised anomalous dispersion large unit cell; cytochrome C_4 (Sawyer, Harding, Gould, Papiz and Helliwell as reported in Helliwell (1984))

Cytochrome C_4, from *Pseudomonas aeroginosa*, is a di-haem protein of 19 000 molecular weight crystallising in space group $P6_122$ with $a=b=62$ Å and $c=174$ Å. A complete data set was collected on a single crystal to 2.25 Å resolution on film at a wavelength of 1.739 Å where f'' was expected to be $7e^-$. At Cu $K\alpha$ this value is $4e^-$. The wavelength was selected using haemin as the model compound with a bent single crystal monochromator (Ge(111), 10.4° cut) focussed at 4.2 m, the achromatic or Guinier setting at this wavelength. The $\delta\lambda/\lambda$ used of 10^{-3} was considered a good match to the width of the 'white line' of haemin of 40 eV; a narrower XANES feature can be tackled with a similar Si(111) monochromator instead of Ge(111). The space group and c-axis mount of the samples allowed Friedel intensity pairs to be measured on the same film; Friedel differences on any of the 15 packs were readily visible by eye. Data quality was very good; R_{merge} $(I)=5.0\%$ varying from 3% at low angle to 8% at 2.3 Å resolution, 90% of the reflections had $I>3\sigma(I)$. The small size of spots at high angle due to the collimation of SR, at the achromatic setting, is well illustrated here. The mean difference between partial and fully recorded reflections was only 0.5% (on I). A Patterson synthesis with Δ_{ANO} values as coefficients gave peaks which could be *readily* interpreted in terms of the iron coordinates known already from a 6 Å conventional source study (figure 9.13(a)). Refinement of the two iron sites assumed that the largest Δ_{ANO} values could be set equal to $2f''$; the least squares line gradient between Δ_{ANO} and $2f''_{calc}$ was 0.62 with an overall R-factor of 37%. Phases were determined using the calculated phase of the iron atoms to resolve the phase ambiguity for this single wavelength data set. The phasing statistics (figure of merit, etc) are given in figure 9.13(b) as a function of $(4\sin^2\theta)/\lambda^2$. These phases readily gave a uranyl binding site in a difference Fourier (figure 9.14) with uranyl isomorphous differences based on an additional data set measured at a $\lambda=1.760$ Å at the SRS. However, these anomalous derived phases did not yield an interpretable protein electron density map. This may be attributed to the weakness of the procedure whereby the iron atom phase was used to resolve the phase ambiguity arising from the single wavelength measurement. It was not possible with this monochromator to ideally access the dip in f' at the point of inflection of the absorption edge. The quality of the film data was as good as might be obtained with film but could clearly be improved on by a digital (photon counting method), although the long c-axis would cause difficulties for a single counter diffractometer or electronic area detector.

An interpretable electron density map has since been obtained using a combination of optimised anomalous and isomorphous phase information modified by solvent flattening techniques (Rule, Harding and Sawyer, unpublished).

9.7.3 Case study: glutamate dehydrogenase (GDH) (Baker *et al* 1990)

These authors describe the use of SIROAS applied to protein crystallography in general, and to glutamate dehydrogenase in particular involving a mercury derivative. The acronym, SIROAS, is an adaptation of the standard Cu Kα acronym SIRAS, single isomorphous replacement

(a)

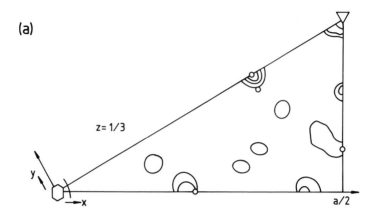

(b)

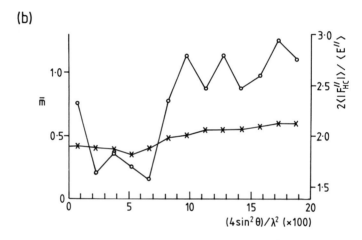

Figure 9.13 (a) A Harker section in the Patterson synthesis calculated using optimised Δ^2_{ANO} terms as coefficients for Fe:cytochrome C_4. The agreement between the observed peaks and the predicted positions is self-evident. All the expected Harker peaks occur in the map and are significantly above background noise.

(b) Phasing statistics using the anomalous differences Δ_{ANO} taken at a single λ at the Fe K edge and the calculated heavy atom phase, α_H, used to resolve the phase ambiguity. Note how the figure of merit *improves* (with $(4\sin^2\theta)/\lambda^2$ as the anomalous difference becomes more significant. The overall figure of merit, $\langle m \rangle$ to 2.3 Å was 0.504; without the use of α_H, $\langle m \rangle$ was 0.37; x, $\bar{m}$; $\circ$, $2\langle|F''_{HC}|\rangle/\langle E''\rangle$. From Sawyer, Harding, Gould, Papiz and Helliwell (unpublished results) and reproduced with permission.

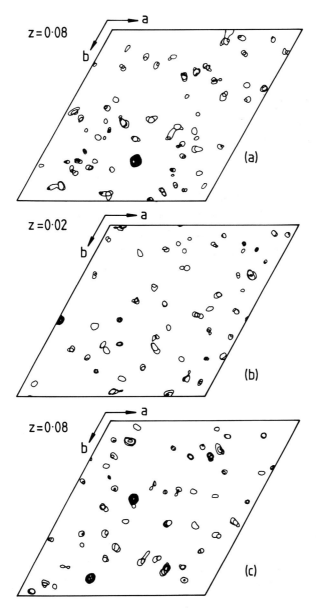

Figure 9.14 (a) Difference Fourier to 3.0 Å resolution using the 'α_H resolved' phases described in figure 9.13 with uranyl isomorphous differences. The signal to noise of the uranyl peak to background is 3:1. (b) and (c) Sections of the difference Fourier to 3.0 Å resolution using the native anomalous differences as coefficients and phases based on the uranium atom alone. The iron sites are clearly revealed, again indicating the quality of the measured anomalous differences. With permission (as figure 9.13).

with anomalous scattering. The wavelength of 0.86 Å was selected on the Daresbury SRS wiggler protein crystallography instrument (9.6) to optimise the mercury f'' at the L_{II} edge at a value of $10.8e^-$. The value of f'' for mercury at 1.54 Å is $7.7e^-$. A wavelength of 0.83 Å would have yielded a value of f'' of $11.8e^-$, however, this wavelength was inaccessible for the 6.7° oblique cut Si(111) monochromator in use on station 9.6. At the L_{II} edge the increase in f'' was therefore 10.8/7.7 more than at 1.54 Å. In addition to this benefit the use of a λ of 0.86 Å reduced the crystal habit absorption effect. For GDH a typical crystal had a rectangular plate morphology of dimensions $1.0 \times 0.7 \times 0.2 \, \text{mm}^3$. The two extreme cases for the absorption effect were with (a) the plate perpendicular and (b) the plate parallel to the X-rays. Hence, at a wavelength close to 1.5 Å, the emergent intensity, I, was $0.8I_0$ for case (a) and $0.4I_0$ for case (b), a change in absorption by a ratio of 2 as the crystal is rotated around its mounting axis (assuming $\mu = 0.87 \, \text{mm}^{-1}$). Whereas, at a wavelength of 0.86 Å, $I = 0.96I_0$ for case (a) and $I = 0.83I_0$ for case (b) a change in absorption by a ratio of 1.15 (assuming $\mu = 0.18 \, \text{mm}^{-1}$). Thus at the shorter wavelength the absorption effect of a native crystal is reduced considerably. This is also true of the capillary holding the crystal and of the mother liquor surrounding the crystal.

In this study the native data were in fact collected at Cu Kα on a diffractometer and the data corrected for absorption using the empirical method of North et al (1968). One isomorphous derivative was prepared using ethylmercury phosphate (EMP) and data measured on film at a λ of 0.86 Å. A similar derivative (i.e. mercury containing) was prepared using 2,3,bis(bromomercurimethyl) tetrahydrofuran (DBMMF) and data measured on film at a wavelength of 1.488 Å. The mean fractional isomorphous difference for the DBMMF derivative was greater than for the EMP derivative (0.247 versus 0.174) as expected on account of the additional scattering arising from the bromine atoms in the DBMMF. Both derivatives showed a similar degree of isomorphism.

A common method used to estimate the quality of the anomalous scattering signal is to compare the empirical anomalous ratio (k_{emp}) as defined by Matthews (1966) as

$$k_{emp} = 2 \left(\langle (F_{PH} - F_P)^2 \rangle / \langle (F_{PH}^+ - F_{PH}^-)^2 \rangle \right)^{\frac{1}{2}} \tag{9.25}$$

with the theoretical ratio, k_{theor} given by

$$k_{theor} = (f_0 + f')/f'' \tag{9.26}$$

Figure 9.15(a) portrays k_{emp} and k_{theor} for both the EMP and DBMMF derivatives. It can be seen that the k_{emp} values of the EMP (short λ) data agree very closely with k_{theor}, whereas for the DBMMF longer λ data the k_{emp} ratio only agrees well at very low resolution and rapidly becomes worse as the resolution increases. Furthermore, the ratio of $\langle 2F''_{H_{calc}}/E'' \rangle$ figure 9.15(b) is better for the EMP also, where $F''_{H_{calc}}$ is the calculated anomalous scattering amplitude for a given reflection and E'' is the size of the discrepancy vector between the observed and calculated value of F''_H (known as the lack of closure error).

This work was performed to solve the structure in Sheffield. From the methods point of view, the comparisons could be improved by using EMP for the short versus long λ data collection runs. Also, the native data themselves could be measured at the short λ (instead of Cu Kα), although the data were corrected for absorption at Cu Kα. A better detector than film at the synchrotron would improve the derivative data.

9.7.4 Case study: the use of MAD with L edge: Tb-parvalbumin (Kahn et al 1985)

The anomalous dispersion effects at L edges are larger than at K edges. Kahn et al used the terbium L$_{\text{III}}$ ($\lambda=1.650$ Å) absorption edge to solve, by the MAD method, the structure of the protein parvalbumin from *Opsanus tau*. Tb binding to the protein in solution was checked by PIXE (proton induced X-ray emission) and fluorescence. Tb-parvalbumin crystals are of space group P2$_1$2$_1$2$_1$ with $a=56.1$ Å, $b=59.4$ Å and $c=27.8$ Å. The use of an area detector is described as essential, for the use of the MAD technique, by Kahn et al (1985). The wavelengths used were $\lambda_1=1.649$ Å, $\lambda_2=1.6469$ Å and $\lambda_3=1.6501$ Å with values of f', f'' determined as $(-14.1, 19.9)\text{e}^-$, $(-12.4, 12.0)\text{e}^-$ and $(-28.0, 10.2)\text{e}^-$, respectively. (Hence, no reference wavelength far from the edge was measured in this study.) For the acentric data the phasing method was based on multiple isomorphous replacement methods (chapter 2 and section 9.3). For the 'constrained-phase' reflections a procedure was developed to treat wavelength-dependent and -independent parts of the $hk\ell$ structure factor. MAD electron density maps based on 2439 phased reflections to 2.3 Å (figure of merit 0.73) were calculated for the two enantiomorphs. One of the maps showed an unambiguous chain tracing on which the atomic coordinates of another parvalbumin could be superimposed (after molecular replacement solution).

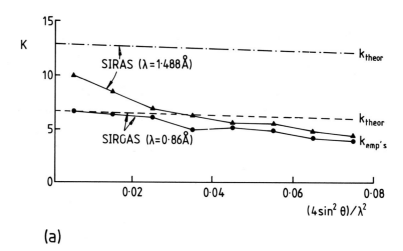

(a)

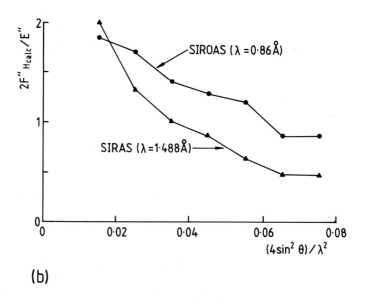

(b)

Figure 9.15 SIROAS on mercury derivative of GDH. (a) A comparison of the theoretical and empirical values of k, the ratio of the size of the isomorphous and anomalous scattering signals as a function of resolution. The curves for SIROAS (mercury at $\lambda=0.86$ Å) are shown for k_{emp} (●) and k_{theor} (- - - -) and for SIRAS (mercury at $\lambda=1.488$ Å) are represented as k_{emp} (▲) and k_{theor} (·—·—). (b) A graph of $2F''_{Hcalc}/E''$ against $(4\sin^2\theta)/\lambda^2$ for SIROAS (mercury at $\lambda=0.86$ Å) derivative (●) and SIRAS (mercury at $\lambda=1.488$ Å) (▲), E'' is the rms lack of closure in the anomalous case. From Baker et al (1990) with permission.

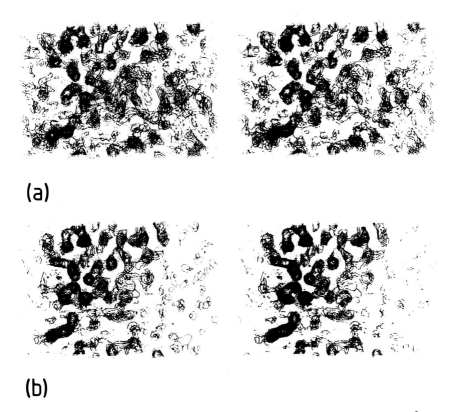

(a)

(b)

Figure 9.16 CBP stereo sections from electron density maps at 3.0 Å
resolution (a) before and (b) after solvent flattening. Seven
successive sections separated by intervals of 1.1 Å along z
are shown. The contour intervals are 1σ, beginning at the 1σ
level (σ being the estimated standard deviation of the
electron density). To produce map (b), MAD phase likelihood distributions were modified by a solvent flattening
algorithm. The data to 3.0 Å resolution were used to generate a molecular envelope corresponding to 35% solvent (the
theoretical solvent content being 47%). The map was calculated after three cycles of phase recombination. From
Guss *et al* (1988) and references cited therein, with permission. Copyright © 1988 by the American Association for the
Advancement of Science, AAAS.

9.7.5 Case study: cucumber basic protein (CBP) (Guss *et al* 1988)

The SSRL area detector (Phizackerly *et al* 1980) has been used to solve
the structure of cucumber basic protein using the MAD method. CBP is a
protein of molecular weight 10 100 Da and contains one copper atom.

Guss *et al* (1988) measured data at four wavelengths about the copper K edge, i.e. 1.2359 Å, 1.3771 Å, 1.3790 Å and 1.5416 Å. Of particular note in the data reduction was the need for an empirical method for coincidence losses to be used, as the analytical dead time correction for the area detector was unreliable at the upper end of the range of counting rates. As a result, merging *R*-factors within the data set at each wavelength were reduced by 20–60%, for the empirical versus the analytical coincidence method of data correction.

The method of Karle (1980) was used for phasing (section 9.4) as implemented by Hendrickson (1985). Prior to fitting a molecular model, the noise in the electron density map was reduced by solvent flattening (figure 9.16). The interpreted map followed by refinement of the model yielded an *R*-factor of 22% at 1.8 Å resolution.

9.7.6 Case study: selenobiotinyl streptavidin (Hendrickson *et al* 1989)

These authors solved the crystal structure of a proteolysed fragment of streptavidin complexed with selenobiotin using the MAD technique based on the Karle (1980) analysis (section 9.4). Streptavidin is a protein of poorly understood function, probably involving antibiotic properties. This crystal contains two seleniums per 252 residues. The study supports the suggestion that biologically produced selenomethyionyl proteins could make the MAD method broadly applicable (Hendrickson (1985) and section 9.8).

The protein complex crystallised in space group $I222$ (or $I2_12_12_1$) with cell dimensions $a=95.20$ Å, $b=105.63$ Å and $c=47.41$ Å (measured with Cu Kα). The complex is a tetramer. Each monomer consists of 126 residues with one selenobiotin bound per monomer. A dimer is in the asymmetric unit of the crystal.

Preliminary experiments were conducted at SSRL (Stanford) using an MWPC area detector (Phizackerly *et al* 1980) and then continued at the Photon Factory in Tsukuba using a single counter four-circle diffractometer. The selenium absorption edge profile was established by measuring the fluorescence from the crystal used for data collection in each case. The point of inflection (f' dip) was calibrated at 11921.1 eV at SSRL and frames of data were collected alternating among four wavelengths: 1.1000 Å, 0.9792 Å, 0.9789 Å and 0.9000 Å over a period of five days. At the Photon Factory the absorption edge dip was calibrated (by a different method from that at SSRL) at 0.9809 Å and data were collected at this wavelength and at 0.9795 Å and at 0.9000 Å. The anisotropy of the anomalous scattering was measured by suitable reorientation of the

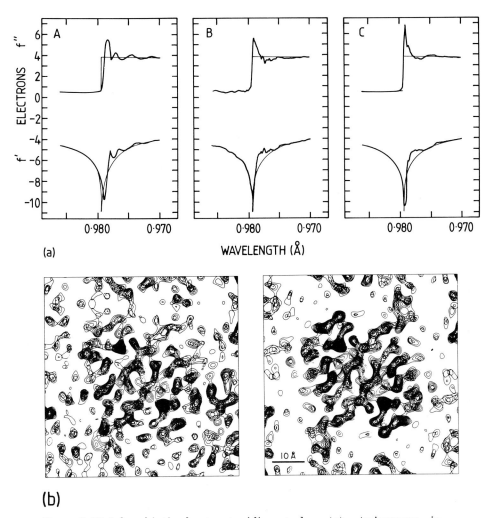

(a)

(b)

Figure 9.17 Selenobiotinyl streptavidin study. (*a*) Anisotropy in anomalous scattering factors for selenium. A–C spectra (thick lines) measured with the electric vector **E** parallel with crystal axes **a**, **b** and **c** respectively. Theoretical values for elemental selenium are shown as thin lines. Values of f'' (upper traces) and f' (lower traces) in electrons are plotted as a function of decreasing wavelength in Å. (*b*) Sections of electron density distributions; left-side MAD phases ($m=0.78$) and right-side map produced after molecular averaging. Resolution 3.3 Å. From Hendrickson *et al* (1989) with permission.

crystal coupled with energy scans across the absorption edge with the high resolution SSRL monochromator (figure 9.17(*a*)). The Photon Factory monochromator energy resolution was limited ($\approx 10\,\text{eV}$). The values

Table 9.4. *Values of f', f" and appropriate λ for the selenobiotin streptavidin experiment. From Hendrickson* et al *(1989) with permission.*

	λ (Å)	f' (e⁻)	f'' (e⁻)
(a) SSRL	0.9000	−1.622	3.285
	0.9789	−7.623	5.083
	0.9792	−9.203	4.382
	1.1000	−1.958	0.617
(b) Photon Factory	0.9795	−6.203	3.663
	0.9809	−8.198	2.058
	0.9000ᵃ	−1.609	3.284

ᵃThese values not given in the paper of Hendrickson *et al* (1989) so values from Sasaki (1989) (see appendix 3.2) have been substituted.

used for the f', f'' at each wavelength needed for use in phase determination were the averages of the anisotropic values. Table 9.4 summarises the f', f'', λ values used at SSRL and the Photon Factory.

The selenium positions were determined from Patterson syntheses based on the SSRL data. The phase evaluation was based on the algebraic analysis of MAD data introduced by Karle (1980), discussed in section 9.4.

An electron density map was calculated for the Photon Factory data and was interpreted. A final model was established and refined to an R-factor of 17.7% using further data to 2 Å. The average phase discrepancy between the MAD phase set and the model phases was 56.9°. This shows that the MAD method is capable of high quality estimation for a structure of this size (2 seleniums in 252 amino acid residues per asymmetric unit). An advantage of the general use of selenium with many proteins (see section 9.8) would be that an instrument could be purpose built for the particular wavelengths illustrated in table 9.4 to stimulate the selenium anomalous dispersion factors. The MAD electron density map is shown in figure 9.17(b) along with the map produced after molecular averaging (i.e. the 'final' map).

9.8 UTILISATION OF MODIFIED PROTEINS AND GENETIC ENGINEERING

Sample preparation technologies are improving considerably. The production of large quantities of very pure proteins by genetic methods is important in improving the quality and size of protein crystals. In addi-

Table 9.5. *SIROAS between a selenomethionine protein and the natural protein.*

λ	f''_{Se} (e⁻)	f'_{Se} (e⁻)	f'_{S} (e⁻)	$(Z_{Se} + f'_{Se}) - (Z_S + f'_S)$ (e⁻)
0.965	3.724	−3.567	0.184	14.25
[0.9794	3.844	−9.79	0.188	8.022]

Notes:
(1) Atomic number of selenium is 34 and of sulphur is 16.
(2) f' and f'' values from Sasaki (1989); the selenium tables are reproduced in appendix 3.2.

tion, the solution of the crystallographic phase problem is also being facilitated in the protein case by the production of seleno-proteins and cysteine mutants; we discuss these options here.

It is possible to express proteins where the sulphur atom in methionine is replaced by a selenium atom. Unlike sulphur the K edge for selenium at 0.9797 Å falls nicely in the range of X-ray wavelengths for which X-ray data collection is routine (unlike sulphur whose K edge is at 5.0185 Å). Methionine has a frequency of 1 in 58 residues on average. This is a sufficient concentration that a significant anomalous dispersion signal should be measurable from a protein in which the methionines are actually selenomethionine. It has been suggested, therefore (Hendrickson 1985), that the crystal structures of all such selenomethionine produced proteins could be tackled by MAD techniques and this would represent, therefore, a general approach for solving the phase problem. Alternatively, since the seleno derivatives are isomorphous with the sulphur counterpart, seleno-proteins could be studied by the SIROAS technique whereby the isomorphous difference is Se–S (i.e. $\approx 14e^-$), and the selenium anomalous difference would be an f'' of $3.844e^-$. The $\Delta f'$ that could be stimulated on selenium alone would be $\approx 7e^-$ maximum (table 9.5 gives an example). Note that SIROAS on Se–S would use a wavelength slightly remote from the Se K edge, on the short wavelength side, to avoid the effect of the large (negative) f'_{Se} exactly at the edge. The disadvantage of the Se–S SIROAS approach would be the need to prepare both crystals. However, the use of a $14e^-$ $((Z_{Se}+f'_{Se})-(Z_S+f'_S))$ difference (table 9.5) instead of $7e^-$ ($\Delta f'$) would extend these methods to cases where the selenium is at a weak concentration.

One potential problem of introducing a relatively large number of anomalously scattering atoms into a protein is that of determining their positions. This, of course, is critical because of the need to calculate the metal atom structure factor for determining the overall phase. A Patterson map could be difficult to interpret. Recently, it has been shown, however, that direct methods can be used in conjunction with anomalous differences to determine the metal atom positions (Mukherjee *et al* 1989).

An alternative scheme to selenomethionine, based on genetic engineering, has been tried for producing mutants of a protein whereby a specific amino acid residue is altered to another, such as cysteine, to which mercury heavy atom compounds bind easily. This has been successfully applied to a previously unknown protein structure, the membrane-pore-forming fragment of colicin-A (Parker *et al* 1989). In this case the amino acid sequence was used to make a prediction of the structure, which was expected to be largely α-helical. Residue number 16 was targeted for mutation from serine to cysteine. At this position it was unlikely that, based on the prediction, the mutation would interfere with helix formation. In addition, since hydrophilic residues are usually exposed on the surface of the protein there would be a reasonable chance that a heavy atom could bind without disrupting the crystal. These expectations were indeed fulfilled and the crystal structure of this fragment of colicin-A was determined successfully using isomorphous replacement (mercury chloride) and conventional source data.

This is a useful illustration that different technologies associated with crystal structure analysis are developing all the time. There is a close interplay between these developments which has led and will lead further to an enhanced rate of crystal structure determination.

9.9 AN ASSESSMENT OF POTENTIAL SOURCES OF ERROR IN VARIABLE WAVELENGTH ANOMALOUS DISPERSION METHODS

The case studies attempt to illustrate the successes and problems of these methods. The MAD experiment itself is technically very demanding of the source and instrument. The stability of the wavelength and the wavelength bandpass is crucial if the optimal values of f' and f'' are to be stimulated and if these sets of values are to hold constant for each reflection measured. Such stability of λ and $\delta\lambda/\lambda$ can be achieved provided the source position is stable and the monochromator is stable under the heat load of the synchrotron beam. The latter is commonly achieved by water cooling of the first monochromator crystal. Although

these possible time-dependent errors can be controlled there are other
time-dependent processes of which account must be taken. These are
the SR beam decay and, of course, radiation damage to the sample is
time-dependent. Ideally, it should be arranged that a given reflection or
batch of a small number of reflections be measured close together in time
at the various wavelengths needed. This requires precise control and
reproducable settings of the monochromator for the wavelengths of
interest. Moreover, it is ideal if the beam position at the sample does not
move as the wavelength is changed; this can be achieved by a fixed-exit
slit double crystal monochromator whereby the second crystal moves as
the angle is changed to compensate for such beam movements. The
detector response should be stable with time over long periods. The
response of wire chambers can vary with time because these detectors
are sensitive to vibrations and barometric pressure changes. Television
systems are sensitive to variations in magnetic field, electrical interference and temperature (the latter affects the thermal noise). Area
detectors suffer, to a greater or lesser extent, from variations in response
over their area. The scintillation counter in a conventional four-circle
diffractometer avoids problems with area detector sensitivity variations
and is stable with time in its response. The same appears to be true with
the IP.

Sample absorption can be a serious source of systematic error which
can affect reflections in a set of equivalents differently or diminish the
intensity of the equivalents in an overall way. The former problem can
be avoided by measuring $hk\ell$ and $\overline{hk\ell}$ only, since the absorption correction is identical for each, provided the sample is always fully bathed in a
uniformly intense beam. If the absorption edge is at short wavelength
(e.g. $\leqslant 1$ Å) then, in the case of protein crystals, absorption variations
are usually negligible for a typical-sized crystal. At longer wavelengths
an absorption correction will become essential.

The counting time per reflection needs to be relatively long in order to
achieve good statistics (e.g. at the 1% level 10^4 counts are needed in the
peak) and so allow accurate differences between reflection measurements to be calculated. The stronger reflections may need correction for
counting/dead time losses in the detector. Longer counting times would
be useful for the medium and weaker reflections. Variable counting
times per reflection are most easily achieved with a single counter four-
circle diffractometer.

Scaling between data sets measured at different wavelengths can be
critical in order to remove any systematic trends in the reflection intensi-

Table 9.6. *Anomalous dispersion studies using synchrotron radiation.*

Sample	SR source	Detector	λ (Å)	Unit cell parameters a / α	b / β	c (Å) / γ (°)	Space group	Resolution limit	Reason for using SR	Reference
Rubredoxin	SSRL	Film	1.74–1.80	64.5		32.7	$R3$	1.7	MAD (Fe)	Phillips *et al* (1976, 1977)
6-Phosphogluconate dehydrogenase	NINA	Film	1.0	72.7	148.2	102.9	$C222_1$	3.0	SIROAS (Pt)	Helliwell (1977)
Gramicidin A-CsSCN	SSRL	Counter	2.47	32.07	52.29	31.20	$P222_1$ or $P2_12_12$ or $P2_12_12_1$	3.8	SIROAS (Cs)	Phillips and Hodgson (1980)
Cytochrome c_4	SRS	Film	1.739 1.760	62.4		174.2	$P6_122$	2.25	OAD (Fe)	Helliwell (1984), Mukherjee *et al* (1989)
Pea lectin	SRS	Film	1.86	50.7	61.2	136.6	$P2_12_12_1$	2.4	OAD (Mn)	Einspahr *et al* (1985)
Parvalbumin	LURE	MWPC	1.6490 1.6469 1.6501	56.1	59.4	27.8	$P2_12_12_1$	2.3	MAD (Tb)	Kahn *et al* (1985)
Cytochrome c'	Photon Factory	Counter	1.077 1.730 1.757	51.6		155.4	$P6_1$	6.0	MAD (Fe)	Harada *et al* (1986)
Azurin	CHESS	Film	1.33 1.50	52.3		99.3	$P4_322$	3.0	MAD (Cu)	Korszun (1987)
Lamprey haemoglobin	SSRL	MWPC	1.6500 1.7380 1.7402 1.8000	44.57	96.62	31.34	$P2_12_12_1$	3.0	MAD (Fe)	Hendrickson, Smith, Phizackerley and Merritt (1988)
Ferredoxin	SSRL	MWPC	1.5000 1.7390 1.7419 1.8000 1.9000	34.56		75.27	$P4_32_12$	5.0	MAD (Fe)	Murthy *et al* (1988)

Protein	SR source	Technique	Wavelength (Å)	a	b	c	Space group	Resolution (Å)	Method	Reference
Cucumber basic blue protein	SSRL	MWPC	1.2359 1.3771 1.3790 1.5416	30.88	46.41	65.57	$P2_12_12_1$	3.0	MAD (Cu)	Guss et al (1988)
Streptavidin	SSRL	MWPC	0.9000 0.9789 0.9792 1.1000	95.20	105.63	47.41	I222	3.1	MAD (Se)	Hendrickson et al (1989)
	Photon Factory	Counter	0.9000 0.9795 0.9809	95.27	105.41	47.56	I222	3.1	MAD (Se)	Hendrickson et al (1989)
ω-Amino acid: pyruvate amino-transferase	Photon Factory	IP	1.004	124.7	137.9	61.5	I222	2.0	MIROAS (Pt,Hg)	Watanabe et al (1989)
Glutamate dehydrogenase	SRS	Film	0.86	147.1	151.3 132.75	94.6	C2	2.5	SIROAS (Hg) RAE	Baker et al (1990)
Octopus dofleini hemocyanin functional unit	Photon Factory	IP	1.2000 1.3779 1.3796	92.6	167.4	59.2	$P2_12_12_1$	2.9	MAD (Cu)	Cuff et al (1990)
Interleukin 1α	Photon Factory	Counter	0.9000 0.9784 0.9795 1.1000	31.97	54.52 108.63	45.63	$P2_1$	3.3	MAD (Se)	Graves et al (1990)

Notes:
(1) SR source acronyms are as given in table 4.1.
(2) Technique acronyms are as given in the text or glossary of symbols. In each technique/acronym example the element symbol is given in brackets.

ties between data sets. A reference set can be calculated by averaging *all* the measurements for one reflection. This improves the counts used to derive each reflection intensity estimate. This approach assumes that, for the purposes of the scaling of batches of reflections, the anomalous dispersion variations are small perturbations which cancel each other in terms of an overall scale factor.

The expected distribution of observed anomalous differences allows unusually large differences to be weeded out and ignored (or reset to some theoretical maximum). The theoretical maximum anomalous difference, for example, is

$$2\sum_j f_j'' \tag{9.27}$$

where the summation is over all the anomalous scatterers.

9.10 CONCLUDING REMARK

In the past, beam time has been limited for the rather accurate measurements required, as has been pointed out by Moffat (1989b). This application of SR is gaining considerable momentum as more beam time and instruments become available to tackle these important molecules. A catalogue of the studies, performed to date, is given in table 9.6.

CHAPTER 10

More applications

10.1 EARLY HISTORY AND GENERAL INTRODUCTION

The first discussions in the literature concerning the applications of SR in protein crystallography were given by Harrison (1973), Wyckoff (1973) and Holmes (1974). The first experimental tests were made on SPEAR by Webb *et al* (1976, 1977) and reported by Phillips *et al* (1976, 1977); precession photographs of protein crystals were obtained with a 60-fold reduction in exposure times over a home laboratory X-ray source (in this case a conventional fine focus Cu Kα tube running at 1200 W) and test data were collected about the iron K edge for rubredoxin and the copper K edge for azurin. The azurin crystal suffered much less from radiation damage in the intense beam than during a longer equivalent exposure on a conventional source. This was the first indication that radiation damage to a protein crystal was less with a more intense X-ray source (figure 10.1). The anomalous dispersion effects using the Fe K edge enabled phases to be determined for rubredoxin with a mean figure of merit of 0.5 (mean phase error of 60°). The anomalous dispersion effects using the Cu K edge were used to confirm the copper sites in azurin utilising phases determined from conventional source data (Adman, Stenkemp, Sieker and Jensen 1978).

A comparison of the intensity of a rotating anode source with the DESY synchrotron (section 5.6.2) by Harmsen *et al* (1976) was favourable enough to encourage data collection apparatus to be established on DORIS by Harmsen and Rosenbaum (described in Rosenbaum and Holmes (1980) and Rosenbaum (1980)). This workstation (X-11) accommodated small angle diffraction as well as protein crystallography.

Stimulated by the early SPEAR results, some preliminary experiments were performed at Daresbury in 1976 shortly before the NINA synchro-

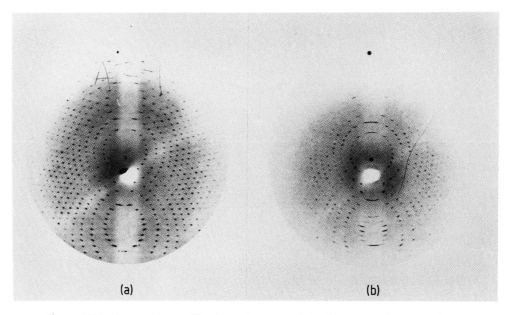

Figure 10.1 Cone-axis oscillation photographs of an azurin crystal
recorded (*a*) at Stanford on SSRL ($\lambda=1.740\,\text{Å}$), and (*b*) with
a Cu Kα sealed tube X-ray source. The SR-based data
extended to higher resolution. Extracted from Phillips *et al*
(1976) with permission.

tron was closed (Helliwell (1977a); local contact J. Bordas) with a
$K_2Pt(CN)_4$ derivative of single crystals of the enzyme 6-phosphoglu-
conate dehydrogenase to optimise its anomalous scattering at the
platinum L absorption edge, with a double crystal monochromator
without focussing; exposure times were too long for the allotted beam
time, only very weak photographs were obtained, and results were
inconclusive, except that focussing was obviously going to be essential if
exposure times with film were to be reasonable. Attempts were also
made to record a Laue pattern from a protein crystal but without success
which was assumed to be because of excessive radiation damage to the
sample with the white beam (but see chapter 7).

On DCI, Lemonnier *et al* (1978) established a focussing monochromator camera in 1976 (section 5.6.3.1) on which oscillation camera data
sets were collected (e.g. from single crystals of tyrosyl t-RNA synthetase
(Monteilhet *et al* 1978) or resolution and radiation damage tests were
conducted (e.g. 6-phosphogluconate dehydrogenase (Helliwell 1979)).

These experiments confirmed reduced exposure times and reductions in the relative amount of radiation damage over that on a conventional source (Fourme 1978, 1979; Kahn *et al* 1982a). In the Soviet Union on the VEPP-3 ring some preliminary data were collected by Mokulskaya (1981) using an electronic area detector in an attempt to optimise the anomalous dispersion of a platinum derivative of pea lectin crystals.

In any discussion of early results involving protein crystallography at a synchrotron source, full acknowledgement should be given to the important developments made by small angle diffractionists in studies with SR of muscle well before protein crystallographers. In particular, Rosenbaum *et al* (1971) published the first work on the use of SR for X-ray diffraction in their studies on muscle.

All these early experiences and results created an explosion of interest from many research groups in the potential of SR within protein crystallography. There is considerable demand for beam time to study many systems on present SR sources and those planned for the next generation, all insertion device machines (see, e.g., table 1.1).

From equation (6.1) we saw how, in principle, small crystal volume and large unit cell would reduce the strength of the diffraction intensities but be compensated for by an increase in I_0, the high source intensity provided by an electron storage ring.

Also, considerable benefits have been derived from the collimation of SR and use of short wavelengths, especially in virus and ribosome crystallography. There is also the unexpected reduced radiation damage effect referred to in the early work on azurin on SPEAR at SSRL. This chapter details all these applications of SR and a variety of case studies is provided within macromolecular crystallography as well as a survey of the literature. The previous chapters dealt with variable wavelength anomalous dispersion methods for macromolecular structure solution. This chapter will deal with all other application areas, namely: high resolution data for refinement, weak diffracters, small crystals, time resolved crystallography and the study of large unit cell crystals (viruses and ribosomes). These studies use either monochromatic or Laue geometry. The Laue geometry is advantageous when speed is of the essence, i.e. in time resolved work (e.g. of enzymes or photolabile proteins) or surveying large numbers of crystals (e.g. viral drug complexes). Monochromatic geometry is advantageous compared to Laue geometry when the sample is somewhat mosaic and/or when the greatest accuracy and resolution is required.

10.2 REDUCTION OF RADIATION DAMAGE: HIGH RESOLUTION, WEAK DIFFRACTION AND CRYSTAL ASSESSMENT

That sample radiation damage should be reduced at an intense SR source seems perhaps somewhat paradoxical. However, phenomenologically there are important time-dependent crystal processes involved in the radiation damage initiated by the exposure to the X-ray beam. These may be chemical, structural or lattice determining effects. Exposure times are reduced using SR to such an extent that these slower damaging processes can be ameliorated; this applies when increasing monochromatic or white beam intensities but eventually sample heating becomes limiting.

The reduction of radiation damage has several benefits. Firstly, the initial resolution or maximum Bragg angle, that can be observed, can seem better with SR. It is not that the data are absent in conventional source work, but that with SR statistically significant data can be collected before radiation decay occurs. The extension of data resolution may also be due, in part, to the well-collimated geometry of the SR beam. Because of this the high angle reflections collected as spots on a film can be smaller than for a conventional source, thus making the average optical density recorded stronger and so more statistically significant (as discussed in chapter 5, see figure 5.20). Secondly, more data per sample crystal can often be collected with SR to the extent that a single sample can yield a complete data set which is a significant advantage.

Considerable further reduction in radiation damage (more data per crystal) has come from the use of short wavelengths as discussed in chapter 6 (table 6.3). (For specific examples see section 10.2.1.3 and also the FMDV virus studies in section 10.5.1.2.)

10.2.1 High resolution

Table 10.1 provides a catalogue of model refinements where SR was critical to yield the high resolution data. This is one of the largest categories of work at SR centres and so several case studies (sections 10.2.1.1–10.2.1.4) are given.

10.2.1.1 Case study: Phosphorylase b (Wilson et al 1983)

Although large crystals can be grown routinely, phosphorylase b presented problems for data collection because of its large subunit molec-

Table 10.1. *High resolution SR data for protein structure model refinements (entries given in chronological order).*

Sample	SR source	Detector	λ (Å)	Unit cell parameters a / α	b / β	c (Å) / γ (°)	Space group	Crystal resolution limit (Å) SR	Cu Kα	SR Refinement R-factor (%)	Reference
Bacterial ribonuclease	LURE SRS	Film Film	1.4 1.488	59.0 —	— —	81.6 —	P3$_2$	2.2	—	—	Mauguen et al (1982) Sevcik, Dodson and Zelinka (1987)
Trypsinogen	DORIS	Film	1.0688	55.4 —	— —	107.8 —	P3$_1$21	1.7	—	19.3	Walter et al (1982)
Glycogen Phosphorylase b, T-state	LURE DORIS SRS SRS	Film Film Film Film	1.4 1.488 0.86	128.5 —	— —	115.9 —	P4$_3$2$_1$2	1.9	2.0	18.7	Wilson et al (1983) Oikonomakos et al (1987) Sprang et al (1988)
Kallikrein	DORIS	Film	1.532	90.2 —	— —	159.4 —	P4$_1$2$_1$2	2.05	2.5		Bode et al (1983)
Human deoxyhaemoglobin	LURE	Film	1.4	63.15 —	83.59 99.34	53.80 —	P2$_1$	1.74	2.5 (2.0)		Fermi et al (1984)
Human T-state oxyhaemoglobin	LURE	Film	1.4	95.8 —	97.8 —	65.5 —	P2$_1$2$_1$2	2.1	3.5	21.0	Brzozowski et al (1984)
Glycolate oxidase	SRS	Film	1.488	148.1 —	— —	135.1 —	I422	2.2	—	21.0	Lindqvist and Brändén (1984, 1985, 1989)
Glyceraldehyde 3-phosphate dehydrogenase	SRS	Film	1.488	82.44 —	124.10 108.98	82.54 —	P2$_1$	1.8	—	17.7	Skarzynski, Moody and Wonacott (1987)
C-phyocyanin	DORIS	Film	1.472	154.6 —	— —	40.5 —	P6$_3$	2.1	—	21.7	Schirmer, Bode and Huber (1987)
Human T-state met haemoglobin	LURE SRS	Film Film	1.4 1.488	95.8 —	97.8 —	65.5 —	P2$_1$2$_1$2	2.1	—	21.0	Liddington et al (1988)
Beef DPI	SRS	FAST	0.9	52.7 —	26.2 93.4	51.7 —	C2	1.3	1.5	—	Holden in Glover et al (1988)
Chloramphenicol acetyltransferase	SRS	Film	0.9	74.5 92.5	— —	— —	R32	1.75	—	15.9	Leslie et al (1988), Leslie (1990)
Serine endopeptidase proteinase K	DORIS	Film	—	68.17 —	— —	108.26 —	P4$_3$2$_1$2	1.5	—	16.7	Betzel, Pal and Saenger (1988a)

Table 10.1 (cont.)

Sample	SR source	Detector	λ (Å)	Unit cell parameters a / α	b / β	c (Å) / γ (°)	Space group	Crystal resolution limit (Å) SR	Cu Kα	SR Refinement R-factor (%)	Reference
Pseudoazurin	DORIS	Film	—	50.0 / —	— / —	98.5 / —	$P6_5$	1.55	—	18.0	Petratos, Dauter and Wilson (1988)
B. cereus phospholipase C	SRS	Film	0.88	89.93 / —	— / —	73.99 / —	$P4_32_12$	1.5	2.8	15.7	Hough et al (1989)
Concanavalin-A (Cd)	SRS	Film	1.488	89.9 / —	87.2 / —	63.1 / —	$I222$	2.0 (1.2)	1.8	24.0 @ 2 Å	Derewenda Z et al (1989) Kalb et al (1988)
β trypsin/inhibitor native complex I complex II	DORIS	Film	1.009 1.009 1.488	63.7 / —	63.5 / —	68.9 / —	$P2_12_12_1$	1.5 1.5 1.85	— — —	16.7 16.9 15.7	Bartunik, Summers and Bartsch (1989)
A. vinelandii lipoamide dehydrogenase	DORIS	Film	1.488	64.1 / —	83.8 / —	192.0 / —	$P2_12_12_1$	2.2	—		Schierbeek et al (1989)
p21 (Val-12) ras oncogene protein	SSRL	Film	1.08	83.2 / —	— / —	105.1 / —	$P6_522$	2.2	—	23.6	Tong et al (1989)
Monoclinic insulin/phenol complex	SRS	Film	1.488	61.36 / —	61.71 / 110.8	47.95 / —	$P2_1$	1.8	2.5	22.0	Derewenda U et al (1989)
Rh. viridis photosynthetic reaction centre	DORIS	Film		223.5 / —	— / —	113.6 / —	$P4_32_12$	2.3	2.9	19.3	Deisenhofer and Michel (1989)
Quinoprotein methylamine dehydrogenase	DORIS	Film		129.8 / —	— / —	104.3 / —	$P3_121$	2.25	—		Vellieux et al (1989)
p-hydroxybenzoate hydroxylase	DORIS		1.482	71.5 / —	145.8 / —	88.2 / —	$C222_1$	1.9	—	15.6	Schreuder et al (1989)

Name	Source	Method					Space group				Reference
Interleukin 1β	DORIS	Film	1.464	54.9	—	76.8	$P4_3$	2.0	2.4	17.2	Priestle, Schär and Grütter (1989)
Rh. rubrum rubisco	SRS	Film	0.87	65.5	70.6 / 92.1	104.1 / —	$P2_1$	1.7	2.0	18.0	Schneider, Lindquist and Lundquist (1990)
Partially oxygenated T-state haemoglobin	SRS	Film	1.0	95.8	97.8	65.5	$P2_12_12$	1.5	—	19.6	Waller and Liddington (1990)
Green alga plastocyanin	DORIS		1.47	53.9	—	59.4	$I4$	1.85	—	—	Collyer *et al* (1990)
A. niger α-amylase	SRS	Film	0.9	81.1	98.3	138.0	$C222_1$	2.1	3.0	16.9	Boel *et al* (1990)
A. oryzae α-amylase	SRS	Film	1.488	50.9	67.2	132.7	$P2_12_12_1$	2.1	—	19.5	Boel *et al* (1990)
Thermitase-Eglin-c	DORIS	Film	1.15 (1.488)	49.3	67.3	90.5	$P2_12_12_1$	1.8	—	16.5	Gros *et al* (1989)

Notes:
(1) The **SR** source acronyms are as given in table 4.1.
(2) FAST is a tradename of Enraf-Nonius, Delft for their television area detector.

ular weight of 97 000 Da and consequently a fairly large unit cell (space
group P4₃2₁2 with $a=b=128.5$ Å, $c=115.9$ Å). The crystal structure had
been solved at 3 Å resolution using conventional X-ray source data but
data were needed to extend this to 2 Å, adequate for model refinement.
On a conventional source a 0.5° oscillation photograph to 2 Å required a
13-hour exposure. With this exposure time with a conventional source
the crystal suffered severe radiation damage which led to a loss of intensity in the high angle data.

A 0.5° oscillation photograph at LURE required an exposure time of 17
minutes. During this time radiation damage was much smaller than in
the conventional source case. In fact, on average a further four usable
photographs could be collected from one crystal sample representing a
five-fold improvement in effective crystal lifetime with SR. Also, partial
reflections could be summed with these contiguous photographs, so that
the much faster rate of data collection with SR was highly effective.

10.2.1.2 Case study: High resolution data collection from 6-phosphogluconate dehydrogenase (6-PGDH) crystals

The enzyme 6-PGDH is a dimeric protein of total molecular weight
100 000 Da. The structure has been reported at a resolution of 2.5 Å using
conventional X-ray sources and the method of multiple isomorphous
replacement (Adams *et al* 1991). Using LURE, it was shown that these
crystals diffracted to at least 2.3 Å with SR (Helliwell 1979). Native data
to 2 Å were collected on film using synchrotron X-radiation at
Daresbury. These data have been processed with a merging *R*-factor on
intensity of 6% (Adams, M. J., pers. comm.), the data at 2 Å could not
have been collected on a conventional source. This is due to the fact that
this weak data would demand very long exposures on a conventional
source such that radiation damage is too great. This is an example of
time-dependent radiation damage.

10.2.1.3 Case study: Refinement of a partially oxygenated T (tense) state human haemoglobin (Hb) at 1.5 Å resolution (Waller and Liddington 1990)

An outline description of Hb was given in section 3.1.1. The essential
features of Hb are as follows. It has a molecular weight of 64 500 Da. It is
a tetramer consisting of two α-globin chains, each containing 141 amino
acid residues, and two β-globin chains, each containing 146 amino acids.
The α chains contain seven and the β chains eight helical segments
separated by non-helical 'corners'. The helices of each subunit are folded

to form a pocket for a haem group, photoporphyrin IX, which consists of an iron atom coordinated to the four pyrrole nitrogens of a porphyrin ring. One axial coordination position of the iron is occupied by a histidine side chain from the globin, forming the only covalent link between haem and globin, the other forms the ligand binding site. This is as described by Perutz (1970).

Several refined Hb model structures are now available; for the R (relaxed) state, a 2.1 Å resolution limit oxy-Hb structure (Shaanan 1983) and a 1.9 Å cross-linked deoxy-Hb structure (Luisi 1986); for the T (tense) state, a 1.7 Å structure of deoxy-Hb grown from ammonium sulphate (Fermi *et al* 1984) and 2.1 Å structures of deoxy-, semioxy- and aquomet Hbs grown from poly(ethylene glycol) (PEG) 8000 (Liddington *et al* 1988) and, finally, the 1.5 Å study to be described here of T state human Hb with the bound allosteric effector inositol hexaphosphate (IHP). These latter crystals diffracted to higher resolution and were more stable in an X-ray beam than crystals grown without IHP. This study therefore represents the most accurately refined Hb structure yet reported and serves to illustrate in detail what can be achieved in terms of accuracy of the molecular model at very high resolution using an intense short wavelength (1.00 Å) SR beam.

Data were collected on station 9.6 of the Daresbury SRS (section 5.6.5.2) on film. The space group of the crystals is $P2_12_12$ with cell parameters $a=95.8(3)$ Å, $b=97.8(3)$ Å, $c=65.5(3)$ Å. It was necessary to collect two data sets because the exposure times required for the high resolution data led to the saturation of lower angle reflections on film. The first data set extended from ∞ to 1.9 Å and the second data set contained useful data between 3.0 Å and 1.5 Å. Two crystals were used for the former and six for the latter data sets. On average each unique reflection in the 1.9 Å data set was measured five times and each unique reflection in the 1.5 Å data set was measured four times. Overall 90% of the unique data to 1.5 Å resolution were observed, namely 87 000 unique reflections from an excess of 300 000 measurements. The model contained 4560 atoms (i.e. approximately 18 000 parameters based on (x,y,z,B) per atom).

The structure was refined using a version of the restrained-least-squares refinement program PROLSQ (Hendrickson 1985). The starting point for the refinement was the model for deoxy Hb (including 89 water molecules). The course of the refinement is charted in figure 10.2(a) showing the fluctuation of the R-factor during the refinement. The final R-factor was 19.6% for all data between 10 Å and 1.5 Å. The quality of

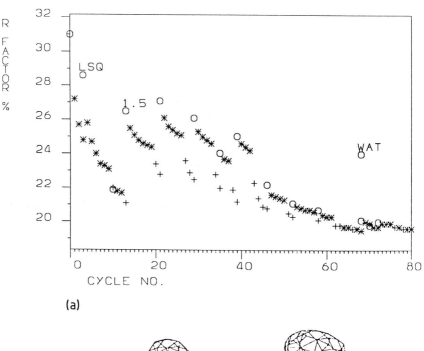

(a)

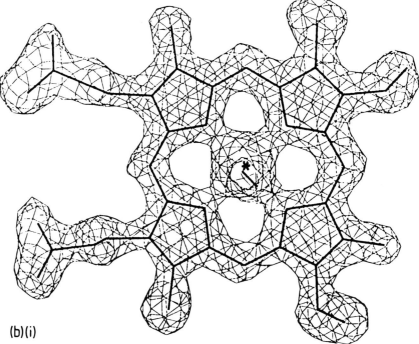

(b)(i)

Figure 10.2 High resolution (1.5 Å) human Hb study (Waller and Liddington 1990). (a) Reliability index R plotted against
refinement cycle number. Temperature factor refinement,
+; positional-parameter refinement, *; and restart after
manual intervention, o. (b) Two parts of the final

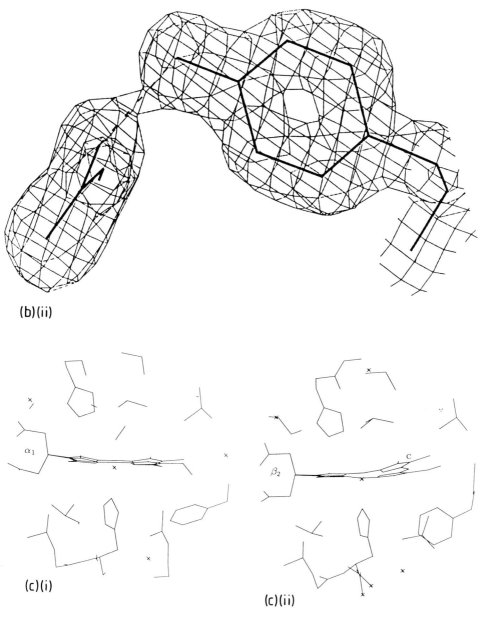

(b)(ii)

(c)(i)

(c)(ii)

$(2|F_0|-|F_c|)$ electron density map calculated at 1.5 Å resolution showing (i) α_1 haem and (ii) α_1 Tyr 42 to β_2 Asp 99
short hydrogen bond (2.43 Å) both contoured at $0.23\,e^-\,\text{Å}^{-3}$.
(c) The α_1 (i) and β_2 (ii) haem groups and their environments, illustrating the distortion of the haem group in the β
subunits. A dioxygen ligand is included in the α_1 haem and
pyrrole ring C shows the greatest displacement. These
figures kindly provided by Dr D. Waller and reproduced
with permission.

the final model is indicated in figure 10.2 (b). This shows two parts of the final $2|F_0|-|F_c|$ electron density map calculated at 1.5 Å resolution, at the α_1 haem and, in another part of the molecule, side chain density. A quantitative estimation of coordinate errors is also possible. Non-crystallographically related identical parts of the structure were overlapped by least squares and indicated that in the ordered helical regions the error in coordinates was 0.15 Å. Because of the greater scattering power of the iron atoms, the error in their positions was estimated to be no more than 0.05 Å.

The functional significance of this study is illustrated in figure 10.2 (c). It shows the haems in the α_1 and β_2 subunits where for the latter there is a significant distortion of the haem, which is, of course, precisely defined. These subtle distortions and the precise position of the iron atom during oxygenation are at the heart of the cooperative mechanism of Hb. The controversy surrounding Hb in the 1980's (see section 3.1.1) was resolved using these accurate crystallographic studies.

10.2.1.4 Case study: High resolution data from DPI (des-pentapeptide insulin) on the FAST

More sensitive and efficient detectors than film are needed for collecting the vast quantities of weak high resolution data recorded for model refinements. Data from beef DPI have been collected to 1.3 Å resolution on the Enraf-Nonius FAST TV detector system (Holden P., pers. comm.) on the Daresbury wiggler protein crystallography station (section 5.6.5.2). Several crystals would have been required to collect data to this resolution using film since the crystals were small ($0.15\times0.12\times0.3\,mm^3$) and required long exposure times leading to severe radiation damage. Data were collected from a single crystal (C2, $a=52.7$ Å, $b=26.2$ Å, $c=51.7$ Å, $\beta=93.4$) as 0.2° rotation images to give a total of 180° about the $\mathbf{b}^*$-axis, using a detector 2θ tilt of 11° and a crystal to detector distance of 44 mm. The wavelength was 0.9 Å. The data were processed using software developed by Messerschmidt and Pflugrath (1987) giving 14 828 unique data at 1.3 Å resolution and a merging R-factor of 11%.

Figure 10.3 compares the electron density observed at Tyr19 in the amino acid sequence calculated using data from a conventional source X-ray diffractometer (at 1.5 Å resolution) with the SRS FAST data to 1.3 Å (Glover et al 1988). The structure factor phases were obtained from a preliminary refined structure and the density obtained from the SRS FAST data shows a clear improvement over that from the conventional source, as evidenced by the 'hole' in the centre of the aromatic ring density.

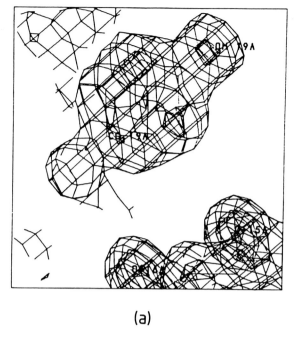

(a)

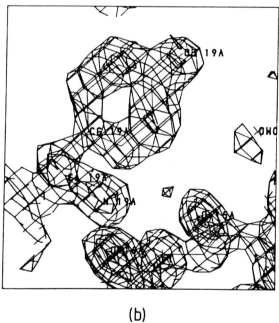

(b)

Figure 10.3 High resolution data measured with SR for protein model refinement. Electron density at tyr 19 in DPI: (a) 1.5 Å resolution diffractometer data at Cu Kα; (b) 1.3 Å resolution SRS FAST data. From P. Holden (reported in Glover *et al* (1988)) and reproduced with permission.

Modest though the resolution increase from 1.5 Å to 1.3 Å might seem, it represents an increase in the number of reflections for model refinement of $\sim (1.5/1.3)^3 = 1.54$ (see table 2.5). In addition, the intensity statistics up to 1.5 Å resolution were better with the SRS data.

High resolution data collection has been given a strong emphasis because of the central role that correct, accurate structural models play in all uses of protein crystal structure results. The trend is clearly to true atomic resolution, the use of shorter wavelengths to get rid of systematic (absorption) errors in the data and measurement of many equivalents each with many counts to cut down random errors. The IP detector developments (section 5.4.2.6) are a key step forward (figure 10.4 shows an example). The X-ray undulator should provide new impetus to this area (section 6.7).

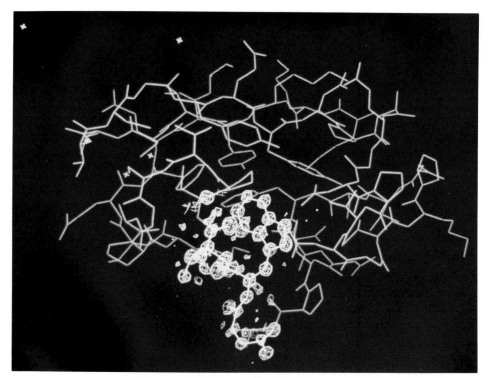

Figure 10.4 Rubredoxin studied at 1.0 Å resolution using data collected at EMBL, Hamburg ($\lambda = 0.9$ Å) on an IP. Electron density for individual atoms is shown. Figure kindly supplied by K. S. Wilson and reproduced with permission.

10.2.2 Weak diffraction

In some cases the scattering strength of a crystal falls off dramatically due to a high disorder (B)-factor and also dramatic radiation damage. Examples in this category are given in table 10.2.

10.2.2.1 Case study: Large solvent content: crystals of human erythrocyte PNP (Ealick et al 1990)

This enzyme is of considerable biological, medicinal and commercial interest. It plays a fundamental role in purine metabolism, it degrades anticancer drugs being targeted to specific cancers and its absence is associated with severe T-cell deficiency. Crystals of the enzyme on a conventional source diffractometer did not diffract beyond 4 Å resolution, a resolution which is not adequate to study the structural interactions with various substrates and inhibitors. With intense synchrotron X-radiation at Daresbury, data could be collected to between 3.2 Å and 2.8 Å resolution; the data to 3.2 Å (table 10.3) were sufficient to solve the structure. This is another example of time-dependent radiation damage. Very fast data collection methods at the SRS involved collection of 100 reflections per second, where one crystal was used for four minutes of exposure time. The structure has been described in Ealick et al (1990) (figures 10.5 and 2.1).

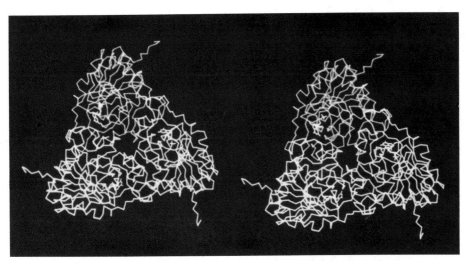

Figure 10.5 Stereo view of the human PNP trimer (C_α atoms) with bound guanine and phosphate. From Ealick et al (1990) with permission. Conventional X-ray source diffractometer data measured to 4 Å, SR data to 2.8 Å resolution.

Table 10.2. *Weak diffraction examples utilising SR.*

Sample	SR source	Detector	λ (Å)	Unit cell parameters a / α	b / β	c (Å) / γ (°)	Space group	Crystal resolution limit (Å) SR	Cu Kα	Reason for using SR	Reference
Human erythrocytic purine nucleoside phosphorylase	SRS	Film	1.488 1.0	99.2 92.3	— —	— —	R32	2.7	4.0	High solvent content (78%), high 'temperature factor'	Greenhough and Helliwell (1983) Ealick et al (1990)
E coli matrix porin	DORIS	Film	1.488	154.7	—	172.9	$P4_222$	2.9		Weak	Garavito et al (1983)
Human serum transferrin	SRS	Film	1.488	127.5	—	144.8	$P4_32_12$	3.3		High 'temperature factor'	Bailey et al (1988)
Nucleosome core particle with defined sequence DNA	SRS	Film	0.88	111	193	111	$P2_12_12_1$	3–5		High 'temperature factor'	Richmond et al (1988)
Tumour necrosis factor	SRS	Film		166	—	93	$P3_121$	2.9	3.4–5.0	High solvent content (65%)	Jones, Stuart and Walker (1989)
R-state glycogen phosphorylase b	SRS	Film	0.88	119.0	190.0 109.35	88.2	$P2_1$	2.8		Tetramer in the asymmetric unit	Barford and Johnson (1989)

Note
(1) The SR source acronyms are as given in table 4.1.

Table 10.3. *PNP from human erythrocytes: summary of data collection at the SRS, Daresbury, England and the intensity statistics for the merging of film packs within each data set. From Ealick et al (1990) with permission.*

Date set	Conditions	No. of crystals	No. of film packs	Exposure (s/°)	Beam GeV	Beam mA	No. of reflections	No. of unique	R_{sym}
Native 1		6	30	600	1.8	101	49 196	10 100	0.111
Native 2		4	52	25	2.0	158	53 669	11 424	0.117
Iodoformycin B	1 mM for 3 days	4	39	25	2.0	156	56 787	10 495	0.120
Iodoformycin B	Cocrystallised	4	51	20	2.0	165	66 661	11 414	0.122
8-Iodoguanine	50% saturated for 3 days	4	34	20	2.0	164	44 722	11 341	0.117
pCMBS	2.5 μM for 3 days	4	40	25	2.0	132	51 529	11 370	0.131
EMTS	100 μM for 3 days	4	34	250	1.8	46	27 182	10 208	0.119

Note:

(1) Crystals are space group R32 with hexagonal cell parameters $a=142.9$ (1) Å and $c=165.2$ (1) Å.

(2) The data were collected on station 7.2 (section 5.6.5.1) at ambient temperature, using a wavelength of 1.488 Å. The exposure times and the milliamperage are average values for each set of crystals.

R_{sym} is defined as

$$\sum_{hk\ell}^{N}\sum_{i} |\bar{I} - I_i| / \sum N\bar{I}$$

where $\bar{I}$ is the mean intensity of the N reflections with intensities I_i and common indices h, k, ℓ. Resolution limit of data, 3.16 Å.

10.2.3 Routine (rapid) data collection use of SR

The high intensity of SR is also used to allow the rapid collection of large numbers of data sets. Examples of these studies are given in table 10.4. Nucleotide crystal problems can yield to SR techniques. However, to date there have been relatively few examples (see table 10.5).

10.2.4 Crystal assessment

SR is now used routinely in the assessment of the diffraction quality of newly grown crystals, especially the resolution limit. Table 10.6 lists in chronological order all the examples, that it has been possible to trace, of such published tests. Occasionally, some strange diffraction effects can be unravelled using SR to the extent that it becomes absolutely clear that it is not possible to determine a structure from a given crystal.

10.2.4.1 Case study: One-dimensional disorder in spinach ribulose bisphosphate carboxylase crystals (Pickersgill 1987)

X-ray diffraction photographs of the crystalline ternary complex of spinach ribulose 1,5-bisphosphate carboxylase/oxygenase (rubisco) diffracted to high angle while showing diffraction features symptomatic of disorder. Oscillation photos about the $\mathbf{c}^*$-axis taken with a conventional source and collimation show streaks of intensity along $\mathbf{c}^*$. These intensity streaks were resolved into satellite reflections, with underlying diffuse scattering, by employing the fine collimation of SR (station 9.6 at Daresbury, section 5.6.5.2) and with slits limiting the horizontal beam convergence from the monochromator (section 6.2.3), figure 10.6(a) and (b). Individual spots on the Cu Kα photo are resolved, on the equivalent SR photograph, into a collection of satellites. The cell parameters estimated originally were $a=158.6\,\text{Å}$, $b=158.6\,\text{Å}$ and $c=203.4\,\text{Å}$. The satellite peaks suggest a super-period of order of 2000 Å.

The implication of this result was that it would not be possible to determine the structure of the enzyme from these crystals. The most probable cause of the disorder was heterogeneity in the amino termini of the large subunit of the enzyme.

This study (and table 10.6) illustrates that a great deal can be learnt from the preliminary assessment stage of a given protein crystal on the synchrotron. In addition, it illustrates the ease with which exceedingly large unit cell spacings (2000 Å here) could be resolved.

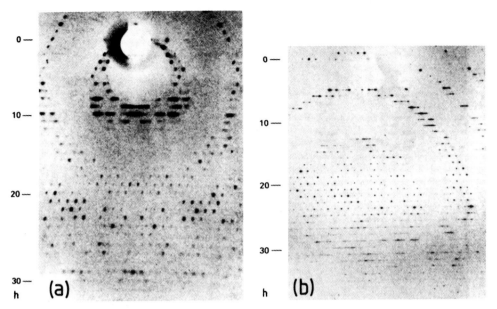

Figure 10.6 Assessment of crystal diffraction with SR usually involves a resolution test. Strange diffraction effects can also be unravelled as in the case of one-dimensional disorder in spinach rubisco crystals. (*a*) Enlargement of part of an oscillation photograph taken around the c^* axis showing broad diffuse reflections in the region of *h* equal to 10. This photograph was taken with a conventional rotating-anode source (Cu Kα, 40 kV, 40 mA). Crystal to film distance 150 mm. A 0.6 mm collimator, an oscillation range of 3.0° and exposure time of 250 min were used. (*b*) Enlargement of part of an oscillation photograph taken around the c^* axis showing satellite reflections and diffuse scattering in the regions of $h=10$ and 30. This photograph was taken using the Daresbury SRS station 9.6 ($\lambda=0.88$ Å, 1.8 GeV, 235 mA). In order to resolve the satellite reflections a 0.2 mm collimator was used and the horizontal beam convergence was limited to 0.6 mrad. Crystal to film distance was 144 mm. An oscillation range of 1.5° and exposure time of 75 min were used. Extracted from Pickersgill (1987). Photographs kindly provided by the author and reproduced with permission.

Table 10.4. *Routine (rapid) data collection use of SR.*

Sample	SR source	Detector	λ (Å)	Unit cell parameters			Space group	Resolution limit (Å)	Reference
				a / α	b / β	c (Å) / γ (°)			
Glycogen phosphorylase b (substrate studies)	SRS	Film	1.488 0.88	128.5 —	— —	116.3 —	$P4_32_12$	2.4	McLaughlin et al (1984) Johnson et al (1987) Oikonomakos et al (1987)
Tyrosyl-tRNA synthetase mutants	SRS	Film	1.488 0.88	64.6 —	— —	238.8 —	$P3_121$	2.5	Brown, Vrielink and Blow (1986), Brown, Brick and Blow (1987)
Antibody/influenza virus neuraminidase	SSRL	Film	1.54	167 —	— —	124 —	$P42_12$	3.0	Colman et al (1987)
R. sphaeroides strain R-26 reaction centre	NSLS	Film		138.0 —	77.5 —	141.8 —	$P2_12_12_1$	2.8	Allen et al (1987)
Trp repressor/operator complex	SSRL CHESS	Film Film	1.54 1.57	44.4 —	74.0 95.7	106.4 —	$P2_1$	2.4	Otwinowski et al (1988)
Human Ha-ras P21-GDP complex	SSRL	Film	1.5418 1.08	83.2 —	— —	105.1 —	$P6_522$	2.25	de Vos et al (1988) Tong et al (1989)
γIVa-crystallin	SRS	Film	1.488	35.1 —	46.2 —	186.2 —	$C222_1$	2.3	White et al (1988)
Arthrobacter Xylose isomerase/inhibitor complexes	SRS	Film	1.488 0.88	105.8 —	— —	153.4 —	$P3_121$	2.3	Henrick, Collyer and Blow (1989)

	SR source	Detector	λ (Å)	a / α	b / β	c / γ	Space group	Resolution	Reference
Spinach ribulose bisphosphate carboxylase (rubisco)	SRS DORIS	Film Film	0.87 1.49 1.69	157.2 –	– –	201.3 –	C222$_1$	2.4 2.6 2.8	Andersson et al (1989)
Unliganded phosphofructo-kinase	SRS	Film	1.488	177.0 –	66.4 118.8	154.0 –	C2	2.4	Rypniewski and Evans (1989)
Recombinant pig myoglobin	SRS	Film	1.488	156.9 –	42.6 127.9	92.2 –	C2	2.5	Smerdon et al (1990)

Note:
(1) The SR source acronyms are as given in table 4.1.

Table 10.5. *Oligonucleotide crystal studies with SR.*

Sample	SR source	Detector	λ (Å)	Unit cell parameters			Space group	Resolution	Reference
				a / α	b / β	c (Å) / γ (°)			
A-DNA octamer	LURE	Film	1.42	45.05 –	– –	41.72 –	P6$_1$	1.7	Kennard et al (1986)
d (CCGCGG) hexamer	SRS	FAST	0.88	40.9 –	– –	35.6 –	P422	2.5	Urpi et al (1989)
A and B form DNA single crystal	LURE	Film		(study of diffuse scattering)					Doucet et al (1989)

Note:
(1) SR source acronyms are as given in table 4.1.

Table 10.6. *Preliminary communications/diffraction quality assessment using SR (entries given in chronological order).*

Sample	Synchrotron source	λ (Å)	Unit cell parameters a / α	b / β	c (Å) / γ (°)	Space group	Crystal resolution limit (Å) SR	Cu Kα	Reason for using SR (see footnote for abbreviation given)	Reference
Azurin	SSRL	1.376	58.9	79.0	108.5	P2₁2₁2₁	2.7	—	RT	Phillips et al (1976)
Tyrosyl t-RNA synthetase	LURE	1.57	64.6	—	238.8	P3₁21	1.9	—	RT	Monteilhet et al (1978) (in Fourme (1978))
6-phospho gluconate dehydrogenase	LURE SRS	1.4 1.608	72.7	148.2	102.9	C222₁	2.0 1.8	2.3	RT RET	Helliwell (1979) Helliwell et al (1982a)
Glycogen phosphorylase b	LURE DORIS	1.4	128.5	—	115.9	P4₃2₁2	2.0	2.5	RT	Wilson et al (1983)
Rhinovirus	SRS DORIS	1.488	≈370	400	—	Monocl Orthor	2.7	3.5	LUC	Rossmann and Erickson (1983)
Δ⁵-3-keto-steroid isomerase	LURE	1.4	65.4	—	504	P6₁22	2.5	—	RT, LUC	Westbrook, Piro and Sigler (1984)
Cow pea mosaic virus	LURE	1.4	451	—	1038	P6₁22 or P6₅22	4.3	6.0	LUC	Usha et al (1984)
2α, 20β-hydroxy-steroid dehydrogenase	CHESS	1.6	127.3	—	112.2	P6₄22 or P6₂22	4.0	—	RT, SC	Fitzgerald, Duax, Punzi and Orr (1984)
Monoclonal antibody Fab'/hen egg white lysozyme	CHESS	—	58.3	119.3	137.9	P2₁2₁2₁	2.8	3.0	RT	Silverton et al (1984)
E. coli PNP	SRS	1.38	106.5	—	241.3	P6₁22 or P6₅22	(2.3)	2.7	RT, LUC	Cook et al (1985)
Chicken gizzard G-actin DNase I	SRS PF	1.488 1.38	42.0	225.3	77.4	P2₁2₁2₁	2.4 3.5	2.9	RT, LUC	Sakabe, Kamiya, Sakabe and Kondo (1984)
L. casei thymidylate synthase	CHESS	—	various			P6₁22 or P6₅22	2.7	—	RT, LUC	Tykarska, Lebioda, Bradshaw and Dunlop (1986)

Protein	Source	λ (Å)	a	b	c	Space group	Resolution	Resolution 2	Method	Reference
Anti progesterone monoclonal antibody Fab' and steroid –Fab' complexes	SSRL	1.54	135.2 / –	– / –	124.2 / –	$P6_222$ or $P6_422$	2.8	–	RT	Stura, Feinstein and Wilson (1987)
Malic enzyme	SRS	1.488	263	264	166	$F222$	2.5	–	RT, LUC	Baker et al (1987)
R. spheroides reaction centre	LURE	1.4	142.5	141.5	80.0	$P2_12_12_1$	3.5	3.5	RET	Ducruix and Reiss-Husson (1987)
Desulfovibrio gigas hydrogenase	LURE	1.4	257.0 / –	184.7 / 101.3	148.3 / –	$C2$	2.4	–	LUC	Nivière, Hatchikian, Cambillau and Frey (1987)
Human C-reactive protein	SRS	1.488	103.0 / – / 103.1 / –	–	308.5 / – / 312.7 / –	$P4_122$ or $P4_322$ / $P4_222$	3.5 / 2.9	–	LUC	DeLucas et al (1987)
Tobacco necrosis virus	PF		338 / –	–	–	$P4_332$	2.5	–	LUC	Fukuyama, Hirota and Tsukihara (1987)
Glycosomal glyceraldehyde phosphate dehydrogenase	SRS	1.488	135 / –	255 / –	115	$P2_12_12$	2.3	–	LUC, RET	Read et al (1987)
Lumbricus erythrocruorin	CHESS, SSRL		266.6 / 97.19 / 799.4 / 100.13 / 266.5 / 100.13 / 502.1	445.2 / 91.91 / 535.2 / 103.15 / 267.6 / 103.15 / 297.8	184.3 / 89.83 / 185.5 / 115.25 / 185.5 / 115.25 / 350.1	$P1$ / $P1$ / $P1$ / $C222_1$	3.3	–	LUC	Royer, Hendrickson and Love (1987)
L. ochrus LOL1 lectin	LURE		56.4 / –	137.8 / 91.0	62.9 / –	$P2_1$	1.8	1.8	RT	Bourne, Rouge and Cambilloau (1988)
Tumour necrosis factor	LURE	1.4	167.5 / –	–	94.8 / –	$P3_121$ or $P3_221$	–	–	RT	Lewit-Bentley et al (1988)
Oncomodulin	CHESS		39.59 / –	64.28 / –	33.07 / –	$P2_12_12_1$	2.0	1.85	RT	Przybylska, Ahmed, Birnbaum and Rose (1988)
E. coli met repressor	SRS		35.6 / –	62.6 / 102.4	44.5 / –	$P2_1$	1.5	–	RT	Rafferty et al (1988)

Table 10.6. (*cont.*)

Sample	Synchrotron source	λ (Å)	Unit cell parameters *a* α	 *b* β	 *c* (Å) γ (°)	Space group	Crystal resolution limit (Å) SR	 Cu Kα	Reason for using SR (see footnote for abbreviation given)	Reference
A. cycloclastes nitrite reductase	PF		98.4 —	— —	— —	$P2_13$	1.7	3	RT	Turley *et al* (1988)
Duck ovotransferrin 18kD fragment	SRS		41.3 —	— —	81.2 —	$P3_1$ or $P3_2$	2.3	—	RT	Jhoti *et al* (1988)
Haloalkane dehalogenase	DORIS		94.1 —	72.8 —	41.4 —	$P2_12_12$	2.4	—	RT	Rozeboom, Kingma, Janssen and Dijkstra (1988)
O^6 methyl guanine-DNA methyl transferase	SRS	1.488	46.3 —	45.8 113.3	46.9 —	$P2_1$	2.5	—	RT	Moody and Demple (1988)
Yeast $tRNA^{Asp}$/aspartyl tRNA synthetase complex	LURE CHESS		210.4 —	145.3 —	86.0 —	$P2_12_12_1$	2.7	—	RT	Ruff *et al* (1988)
Human serum amyloid P component	SRS		69.0 —	99.3 96.1	96.8 —	$P2_1$	<2.0	—	RT	Wood *et al* (1988)
Anthronilate phosphoribosyl transferase	SSRL	1.54	189.0 —	— —	— —	I432	3.0	—	RT	Edwards *et al* (1988)
B. lentus alkaline protease	DORIS		75.3	53.4	61.5 —	$P2_12_12_1$	1.8	—	RT	Betzel *et al* (1988)
Lactate dehydrogenase ternary complex (type IV)	SRS		85 —	118 96	136 —	$P2_1$	2	—	RT	Wigley, Muirhead, Gamblin and Holbrook (1988)

Protein	Source	λ (Å)	a	b	c	Space group	Resolution		Method	Reference
Horse pancreatic lipase	LURE	1.4	79.8	97.2	145.3	$P2_12_12_1$	1.8	—	RET	Lombardo, Chapus, Bourne and Cambillau (1989)
E. coli bacterioferritin	SRS	0.86	118.7	211.6	123.3	$P2_1$	2.0	—	RT	Smith, et al (1989a)
			128.7	119.1	202.8	$C222_1$				
		1.003	210.6	197.1	145.0	$P4_22_12$	1.5	—		
			146.9	—	—	$I432$				
Malate dehydrogenase T. acidophilam	DORIS		63	135	85	$P2_1$	4.0	—	RT	Stezowski et al (1989)
S. acidocaldarius			151	—	248	$P3_121$	5.0	—		
			129	—	—	$I23$ or $I2_13$	4.5	—		
S. typhimurium His J periplasmic protein	SSRL		39.3	66.2	88.3	$P2_12_12_1$	2.0	2.3	RT	Kang et al (1989)
Murine interferon-β	PF	1.04	71.4	—	79.6	$P6_1$	2.2	2.6	RT	Matsuda et al (1989)
E. coli AMP nucleosidase	SSRL	1.2	120	—	244	$P4_12_12$	3.0	—	LUC	Giranda, Berman and Schramm (1989)
β-bungarotoxin	SSRL	1.54	176.5	39.3 / 114.8	92.7	$C2$	2.3	—	RT	Kwong, Hendrickson and Sigler (1989)
T. thermophilus 70S ribosome	LURE		510	—	378	$P4_12_12$ or $P4_32_12$	20	—	LUC	Trakhanov et al (1989)
B800–850 light harvesting complex	SRS	0.86	75.8	—	97.5	$P4$	12	—	RET	Papiz et al (1989)
R. acidophila strain 10050	SRS	0.91	121.1	—	296.7	$R32$	3.5	—		
Duck di-ferric ovo-transferrin and apo-ovotransferrin	SRS	—	49.6	85.6	178.7	$P2_12_12_1$	2.1	—	RT	Rawas, Moreton, Muirhead and Williams (1989)
			77.6	98.8	127.0	$P2_12_12_1$	3.2	—	RT	

Table 10.6. (cont.)

Sample	Synchrotron source	λ (Å)	a	α	b	β	c (Å)	γ (°)	Space group	Crystal resolution limit (Å) SR	Crystal resolution limit (Å) Cu Kα	Reason for using SR (see footnote for abbreviation given)	Reference
Human serum albumin	NSLS	1.22	186.5	—	—		81.0	—	$P42_12$	3.0	2.9	RT	Carter et al (1989)
Erythrina trypsin inhibitor	SRS	—	73.4	—	—		143.0	—	$P6_122$ or $P6_522$	2.5	—		Onesti, Lloyd, Brick and Blow (1989)
Human vascular anticoagulant protein	LURE	—	83.9		80.9	108.70	71.4	—	$P2_1$	2.2	—	RT	Lewit-Bentley et al (1989)
Bovine anti-thrombin III	LURE CHESS	—	91.4	—	—	—	383.1	—	$P4_12_12$ or $P4_32_12$	3.0	—	RT, LUC	Samama et al (1989)
5-carboxymethyl-2-hydroxy muconate isomerase	SRS	1.488	88	88	89		121	—	$I222$ or $I2_12_12_1$	3.0	—	RT	Wigley, Roper and Cooper (1989)
Limulus C-reactive protein	SRS	1.488	173.3	—	—		98.8	—	$I422$ or $I4_22$	2.8	—	RT	Myles et al (1990)
Rat CD4/FAB complex	DORIS	0.98	317	—	161		41.8	—	$P2_12_12$ or $P2_12_12_1$	3.5	—	RT, LUC	Davies et al (1990)
B. subtilis AhrC	SRS	—	229.8	—	72.8		137.7	—	$C222_1$	2.6	3.0	RT	Boys et al (1990)
Octopus dofleini hemocyanin (functional fragment)	PF	—	92.6	—	167.4		59.2	—	$P2_12_12_1$	1.9	—	RT	Cuff et al (1990)
Glutathione S-transferase	DORIS	—	60.1	—	—		244.0	—	$P4_12_12$	2.7	—	RT	Parker et al (1990)

Protein	SR source	λ	a	b	c	Space group	res.	res.	Test	Reference
Blood factor XIII	DORIS	0.965	101.1	72.3 106.6	133.2	P2₁	2.7	2.8	RT	Hilgenfeld *et al* (1990)
		—	95.4	101.0	181.4	P2₁2₁2₁	3.0	3.5	RT	
Lac repressor	SSRL	—	164.7	75.6 125.5	161.2	C2	3.5	—	RT	Pace, Lu and Lewis (1990)
Histone H5 globular domain	NSLS	—	79.1	61.3	36.8	P2₁2₁2 or P2₁2₁2₁	2.5	—	RT	Graziano *et al* (1990)
Phage φX174	DORIS	1.468	306.0	361.1	299.7	P2₁	2.7		LUC	Willingmann *et al* (1990)
	SRS	0.911	—	92.91	—					
	SSRL	1.5405								
	NSLS	1.22								
N. gonorrhoeae pilin	SSRL	1.54	126.4	121.2	26.7	C222	2.4	—	RT	Parge *et al* (1990)
		1.08	—	—	—					

Notes:

(1) The SR source acronyms are as given in table 4.1.

(2) RT=diffraction resolution test; RET=reduced exposure time; LUC=large unit cell test; SC=small crystal diffraction evaluation.

(3) This table has been assembled not only from personal knowledge, but also from a literature scan encompassing the following journals, over a period from 1977 to mid 1990: *Nature, Science, Proc Nat Acad Sci USA, Acta Cryst, J Appl Cryst, J Mol Biol, J Biol Chem, EMBO J, Protein Engineering and Proteins: Structure, Function and Genetics*. I apologise for any omission; please let me know of any missing entries and/or any errors. Thank you.

10.3 SMALL CRYSTALS

Small sample volume affects the strength of the diffraction pattern as seen from Darwin's formula (equation (6.1)). High intensity overcomes this problem in part. Unfortunately, such small macromolecule crystals tend to suffer radiation damage. This problem can be overcome by freezing of the sample. Freezing of regular-sized crystals of ribosomes is now used routinely (Bartels *et al* 1988) employing the method of Hope (1988). There is also the problem of obtaining a reasonable signal-to-noise ratio of the diffraction data. To achieve this the beam cross section at the sample is set equal to the crystal cross section using motorised slits and on-line monitoring of the diffraction pattern; in this way the background can be reduced until the diffraction peak intensity just begins to be clipped, at which point one knows that the beam cross section exactly matches the crystal size. This then minimises the beam size and also any air scatter is minimised.

Many proteins currently selected for structural study crystallise only with small sample volumes. Sample volumes as small as $(20\,\mu m)^3$ have yielded very strong diffraction patterns (figure 10.7) with the white SR beam (Hedman *et al* 1985). This was a test experiment on crystals of gramicidin that were fished from their mother liquor before they grew any bigger. It was essential to use a helium-filled cone between the sample and the film in this study.

A variety of studies on small organic, inorganic and macromolecule crystals has been done. In order to compare these it is useful to define a scattering strength for a given crystal which takes account of its volume and composition. Table 10.7 compares various studies in terms of the scattering strength of the sample that was examined in each case.

10.3.1 Case study: A small molecule (piperazine silicate)

Andrews *et al* (1988) have exploited the advantages of SR using the wiggler protein crystallography workstation (section 5.6.5.2) at the SRS to solve the structure of a poorly ordered silicate crystal with monochromatic data collected from a crystal of just $18 \times 8 \times 175\,\mu m^3$. The crystal was mounted dry, in air. The data were collected using a beam of 0.2 mm diameter and a wavelength of 0.88 Å to improve signal to noise. The Enraf–Nonius FAST TV detector was used. The refined mosaic spread was high (2–3°) and the processed data of relatively low quality, due to its weakness. The structure, however, was solved using direct methods and refined to a crystallographic *R*-value of 10.3%. Although a small

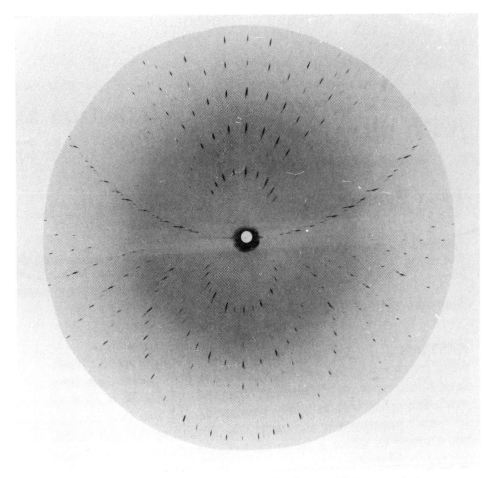

Figure 10.7 Diffraction from small crystal volumes. This example is an
SR Laue diffraction pattern from a $(20\,\mu m)^3$ crystal of gramicidin. From Hedman *et al* (1985) with permission.

molecule, the result is an encouraging example of what can be achieved
in the field of data collection from small volumes of a poorly ordered
sample.

10.3.2 Case study: Bovine growth hormone

Bell *et al* (1985) reported the preliminary characterisation of crystals
of this molecule in terms of the diffraction strength. The crystals were of
space group $P2_12_12_1$ or $P2_12_12$ with $a=219.0\,\text{Å}$, $b=51.9\,\text{Å}$, $c=68.9\,\text{Å}$ and
diffracted to 3.2 Å resolution on CHESS. The largest crystal available was
$250\times120\times10\,\mu m^3$ $(3\times10^5\,\mu m^3)$ but they were typically $120\times120\times15$

Table 10.7. *Use of SR to collect data from small crystals: macromolecule and small molecule examples.*

				Unit cell parameters		
Sample	SR source	Detector	λ (Å)	a α	b β	c (Å) γ (°)
Part A: Macromolecules						
Gramicidin A	SRS	Film	Laue	15.20 —	26.63 92.1	32.18 —
Bovine growth hormone	CHESS	Film	1.57	219.0 —	51.9 —	68.9 —
β-lactamase I	LURE	Film	1.4	143.0	35.8	57.2
	SRS	Film	1.488	—	97.8	—
FMDV	SRS	Film	0.9	345 —	— —	— —
Human class	DORIS	Film	—	60.4	80.4	56.6
histocompatibility	SSRL		1.54	—	120.4	—
antigen,	CHESS		1.57			
HLA-A2	CHESS		1.57	60.2 —	80.4 —	112.2 —
Part B: Small molecules						
CaF$_2$	DORIS	SC	0.91	5.262 —	—	—
CaF$_2$	CHESS	SC MWPC	1.56			
Zeolite	SSRL	SC	1.74	12.69 —	— —	5.16 —
Proflavine hemisulphate	SRS	Film	Laue			
Piperazine silicate	SRS	FAST	0.9	13.57 —	4.90 —	22.46 —
Rh$_2$I$_6$Cℓ (CO)$_2$ $-$ (CH$_3$OH)$^-$ · Ph$_3$PNPPh$_3^+$	SRS	FAST	0.9	17.53 —	13.34 100.23	21.41 —
Low temperature polymorph of chenodeoxycholic acid	SRS	FAST	0.895	22.25 —	— —	10.26 —

Notes:
(1) The SR source acronyms are as given in table 4.1.
(2) SC = single counter diffractometer; MWPC = multi-wire proportional chamber.

Space group	Resolution limit (Å)	Typical crystal volume (μm^3)	Unit cell scattering efficiency ($e^2/Å^6$)	Unit cell scattering power ($e^2/Å^3$) $\times 10^{12}$	Ratio of scattering powers	Reference
2_1	—	1.05×10^4	2.1×10^{-4}	2.2	110	Hedman et al (1985)
$2_12_12_1$	2.8	2.2×10^5	2.0×10^{-6}	0.44	22.5	Bell et al (1985)
2	2.5	Thin plates	—	—	—	Samraoui et al (1986) Phillips et al (1987)
3	2.6– 2.9	8.64×10^5	2.3×10^{-8}	0.02	1	Fox et al (1987) Acharya et al (1989)
2_1	2.7	5×10^6	6.1×10^{-6}	30.6	1530	Bjorkman et al (1987)
$2_12_12_1$	2.8	2.5×10^7	2.6×10^{-6}	66	3300	Bjorkman et al (1987)
m3m	1.28	200	0.084	17	850	Bachmann et al (1983, 1985)
		2.2	0.084	0.19	9.5	Rieck, Euler and Schulz (1988)
6_3	—	800	0.057	45.8	2290	Eisenberger et al (1984)
		9.6×10^4	5.4×10^{-4}	52	2600	Andrews et al (1987), Hails et al (1984)
2/c	1.0	2.5×10^4	18×10^{-3}	200	10^4	Andrews et al (1988)
2/a	1.0	7.5×10^4	5×10^{-3}	377	1.9×10^4	Rizkallah, Maginn and Harding (1990a)
6_5	1.5	1.2×10^6	3.6×10^{-4}	432	2.16×10^4	Rizkallah et al (1990b)

$(2.2\times10^5\,\mu m^3)$. These investigators stated then that 'these are the most weakly diffracting single crystals of any compound (macromolecule, small organic or inorganic molecule) to have been successfully examined by X-ray diffraction techniques'.

10.4 TIME RESOLVED MACROMOLECULAR CRYSTALLOGRAPHY

Reviews of this topic have been given by Bartunik (1983), Moffat (1989a), Moffat and Helliwell (1989) and Hajdu and Johnson (1990).

A complete molecular mechanism of action of an enzyme, for example, would consist of a picture of the structures of the free enzyme and of all intermediate complexes and transition states, as well as identification of the changes in electronic and atomic structure in the enzyme, substrate and cofactors in each intermediate. This has, however, not yet been achieved, partly because of the long timescales needed for crystallographic data collection compared with biochemical timescales. Intermediates in enzyme-catalysed reactions or in ligand-binding reactions, such as those with Mb or Hb, typically have lifetimes in the nanosecond to second time range under physiological conditions. The transition states separating these intermediates have even shorter lifetimes, in the subpicosecond range. Crystallographers have sought to prepare a series of stable structures designed to mimic these short-lived intermediates, e.g. enzyme–inhibitor or enzyme–product complexes, or enzyme–substrate complexes under conditions, e.g. pH, where turnover is greatly reduced. However, because these structures are stable they are not identical to the intermediates they try to mimic. Hence, although static structures can be determined the understanding of mechanism at the atomic level is not based on direct visualisation of intermediate structural states.

An approach to reaching such direct visualisation is the method of cryocrystallography (Alber, Petsko and Tsernoglou 1976; Makinen and Fink 1977; Fink and Petsko 1981; Douzou and Petsko 1984). Reduction of temperature greatly slows the rate of a reaction; e.g. a reaction requiring an enthalpy of activation of $10\,kcal\,mol^{-1}$ is slowed by a factor of 10^6 on cooling from $25\,°C$ to $-100\,°C$. The intrinsic difficulty of cryocrystallography as applied to studies of catalysis in the crystal is to preserve both the crystallinity and liquid nature of the solvent channels permeating the crystal. To achieve this the crystal is sequentially transferred through several cryosolvents that remain liquid at all temperatures of interest. Hence, when a substrate is diffused into the cooled crystal, catalysis will proceed at a slow enough rate to allow X-ray crystallographic data collection even on conventional X-ray sources.

Unfortunately, there are difficulties which have severely limited the method to a very few examples. The cryosolvents used often modify the kinetic behaviour of the molecule of interest. In addition, crystallinity is often lost.

SR dramatically reduces the crystallographic data collection time and thus gets away from the need to cool the crystal to such an extent. Exposure times for a full crystallographic data set are of the order of minutes to hours with the focussed, monochromatised beam and seconds or less with Laue geometry. Indeed, in Laue geometry an exposure of 120 ps for lysozyme has been realised, albeit over a somewhat limited wavelength bandpass, at Cornell, figure 10.8 (Szebenyi et al 1989).

A time resolved crystallographic experiment has three key components: reaction initiation, reaction monitoring and X-ray data acquisition and analysis. The X-ray aspects of such experiments are feasible now. However, the real challenge lies in devising suitable ways of reaction initiation and monitoring in the crystal. Progress to date will now be covered. Table 10.8 provides details of examples of time resolved crystallography.

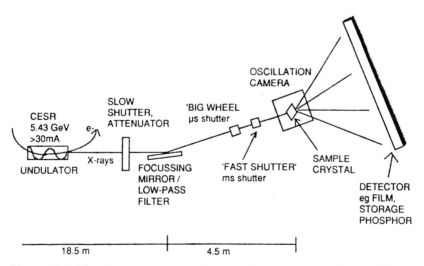

Figure 10.8 The shortest exposure time realised to date for X-ray diffraction has been at CHESS. This figure shows a schematic of the experimental set-up used to select X-rays emitted from a single bunch of the circulating electron beam. From Szebenyi et al (1989) with permission.

Table 10.8. *Time resolved crystallographic studies using (A) monochromatic SR and (B) Laue.*

	Sample	Synchrotron source	λ (Å)	Unit cell parameters			Space group	Resolution limit (Å)	Time resolution	Reference
				a α	b β	c (Å) γ (°)				
Part A: Monochromatic	CO myoglobin (feasibility test)	DORIS						30 reflections with d spacings between 2 and 5Å	500 μs	Bartunik et al (1981) Bartunik (1983)
	Glycogen phosphorylase	SRS	1.488 1.38 1.003 0.88	128.5 –	– –	116.3 –	P4$_3$2$_1$2	2.4– 3.0	25 min– 2.5 hours	Hajdu et al (1987a)
Part B: Laue	H-ras p21 (GTP hydrolysis study)	DORIS	Laue 0.6<λ<2.2	41.0 41.1	– –	164.8 164.0	P3$_2$21	2.8	3×15 s	Schlichting et al (1989, 1990)

Note
(1) The SR source acronyms are as given in table 4.1.

10.4.1 Reaction initiation

The single crystal sample must obviously be of a size and quality suitable for high resolution X-ray diffraction. Hence, a structural reaction must be initiated uniformly and promptly through the whole of the crystal. Also, in such a process the crystal must not be damaged. It is essential that spatial uniformity be achieved in less time than the lifetime of all subsequent molecular intermediates whose imaging is required.

A reaction can be initiated chemically, e.g. by altering the concentration of a reactant or physically by changing the temperature or pressure. The simplest method of reaction initiation is to allow a reactant to diffuse into a crystal mounted in a flow cell (Wyckoff *et al* 1967) (see also figure 5.32). However, diffusion is slow taking many seconds for a crystal of a few hundred microns. Diffusion is only appropriate for initiating a relatively slow reaction. However, it has been used in the monochromatic SR study of glycogen phosphorylase b as detailed in the case study below.

An alternative method of reaction initiation is photoactivation of a stable precursor. For example, caged ATP can be photoactivated to yield ATP (Kaplan, Forbush and Hoffman 1978; McCray, Herbette, Kohara and Trentham 1980) or caged GTP to yield GTP (Schlichting *et al* (1989, 1990) and case study below). The photoactivation technique eliminates macroscopic diffusion problems and the timescale of the reaction initiation is now determined by the light pulse and the diffusion time over molecular distances. For this method to work there are several requirements. Firstly, it is essential that most of the light is transmitted through the sample to avoid any spatial gradient through the crystal (Gruner 1987). Secondly, a high quantum yield for photolysis is essential to minimise heating problems. Thirdly, the stimulating pulse must deliver sufficient photons to allow complete photolysis. Finally, photochemical intermediates and waste products must not interfere chemically with the desired reactions; for example, the cage released on photolysis of caged ATP is a nitrosoketone, which is highly reactive towards sulphydryl groups. Current developments in photoenzymology have been reviewed (McCray and Trentham 1989).

A classic example of photoactivation is photodissociation of carboxy-myoglobin and carboxyhaemoglobin; a brief light pulse ruptures the haem iron–CO bond and is followed by recombination (Parkhurst and Gibson (1967), figure 10.9). Bartunik *et al* (1981) and Bartunik (1983)

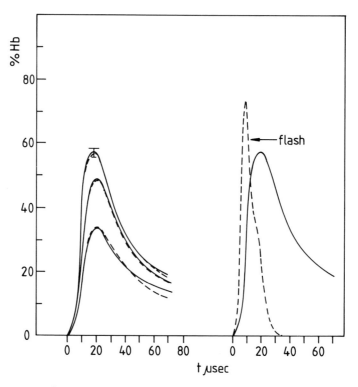

Figure 10.9 The kinetics of recombination of carbon monoxide with
crystalline horse Hb after flash photolysis (flash energies
2025 J, 1500 J and 900 J). The ordinate shows reduced Hb
expressed as a percentage of the total Hb in the sample. The
right-hand graph shows the profile of the flash in the 2025 J
case, in arbitrary units: full line, calculated; dashed line,
observed. From Parkhurst and Gibson (1967) with
permission.

used a laser for photolysis of large single crystals of carboxymyoglobin;
the process is reversible and so multiple repetitions could be used to
enhance the signal-to-noise ratio of the X-ray diffraction data; some X-
ray diffraction reflection intensities were measured on a millisecond
timescale (figure 10.10). There were changes in the intensities with time.
Unfortunately, it was not clear whether these changes arose from tertiary structural changes in the myoglobin or from artefacts such as laser-
induced heating of the crystals, mechanical jitter (to which monochromatic experiments are particularly sensitive) or incomplete, non-
uniform photolysis. The principles of the technique used in this example
followed the philosophy adopted in time resolved kinetic studies of mus-

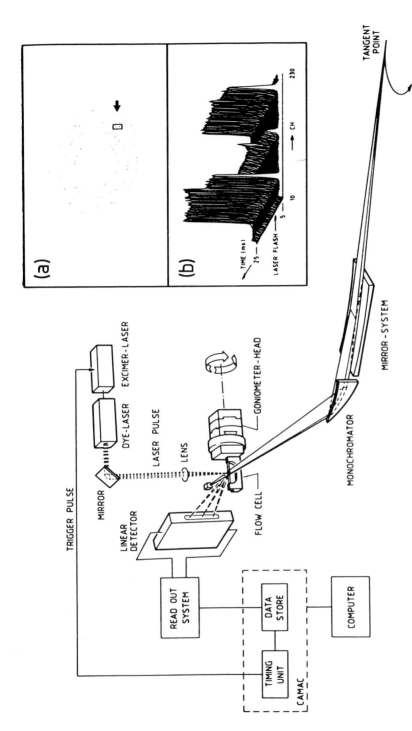

Figure 10.10 The instrumental set-up at X11/DORIS for time resolved data collection with a linear detector for carboxy myoglobin following laser photolysis of the ligand. A section of the diffraction pattern with stationary crystal, stationary detector (shown in (a)) is recorded with a linear detector. In (b) the time course of three reflections is shown before and after the laser flash. From Bartunik *et al* (1982) with permission.

cle (Huxley *et al* 1980, 1982) and of microtubule polymerisation (Mandelkow, Harmsen, Mandelkow and Bordas 1980) with a data acquisition system as described by Bordas *et al* (1980) and Boulin *et al* (1982).

10.4.2 Reaction monitoring in the crystal

Following the reaction initiation it is necessary to monitor the diffraction pattern as a function of time. To measure, say, 50 000 reflection intensities continuously as a function of time is certainly an exacting requirement. To achieve a continuous measurement photographic and IP 'streak' cameras are used (see chapter 5 and figure 10.11). The Laue method has the advantage that it gives full sampling of the RLP intensities whilst maintaining a stationary crystal. However, a Laue

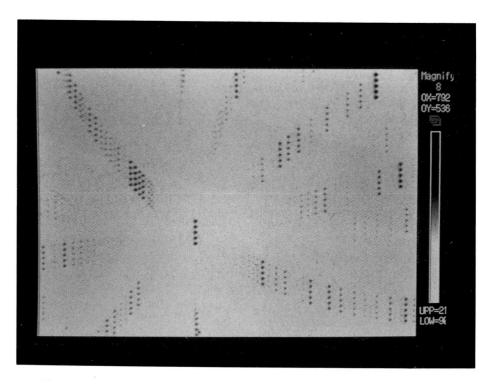

Figure 10.11 Reaction monitoring in the crystal. Possible schemes include mechanical 'streak' methods (figure kindly supplied by N. Sakabe).

pattern is already dense with spots leaving little room for spots to be drawn out into streaks. Hence, in streak methods the pattern is limited in some way, for example the bandpass is reduced to decrease the number of stimulated spots. Alternatively, only a small part of the pattern is recorded using a slit in front of a moving film or IP or an EAD (electronic area detector) such as a CCD detector (see chapter 5). In this approach the time or times at which there are significant variations in the intensities are established. The whole experiment is then redone but only at the interesting time slot is the full Laue diffraction pattern recorded. In those cases where there is a chromophore in the catalytic region of the protein, or the substrate or coenzyme itself, it should be possible to monitor the reaction by a microspectrophotometer. Again, the interesting time in the reaction in the crystal can be identified for diffraction recording to be made in a subsequent experiment. Obviously this 'piecemeal' approach requires that the processes are reproducible in a given study.

10.4.3 Kinetic crystallographic studies of the phosphorylase enzyme: a case study

These monochromatic experiments involved the substrates heptenitol and phosphate, where a long incubation time (50 hr) in the crystal gave rise to a product, heptulose-2-phosphate, bound at the active site. The group were able to trap the actual enzyme–substrate complex (McLaughlin *et al* 1984). A 1 hr soak followed by a 2 hr data collection on the SRS (over 100 000 reflections) showed a mixture of substrate and product. A 15 min soak followed by 45 min data collection at room temperature resulted in an electron density map which clearly showed substrate (phosphate and heptenitol) and no product bound. Thus, for the first time, by taking advantage of the rapid methods of data collection (especially on the SRS wiggler station), both the enzyme–substrate complex and enzyme–product complex were trapped under conditions in which the enzyme was fully competent.

Further experiments with the substrate maltoheptose and inorganic phosphate showed that this reaction was faster than with heptenitol. A 15 min soak followed by data collection at room temperature resulted in the formation of the product, glucose-1-phosphate at the active site. At lower temperature (3 °C) with only a 5 min soak of phosphate with crystals already pre-equilibrated in maltoheptose and a 35 min data collection the reaction was found to be stopped. Hence, with only a small

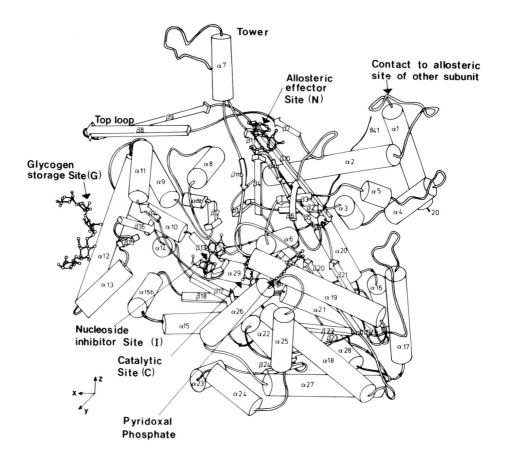

Figure 10.12 A schematic diagram of the subunit structure of phosphorylase b. α-Helices are shown as cylinders and β sheets
as arrows. The essential cofactor, pyridoxal phosphate, is
buried in the centre of the molecule. Binding studies in the
crystal have shown the existence of four major binding
sites. The catalytic site (C) is located at the centre of the
molecule, close to the pyridoxal phosphate and accessible
to the solvent only through a narrow channel some 5 Å in
diameter and some 15 Å long. The allosteric effector site
(N) is situated at the subunit–subunit interface of the
physiologically active dimer and is some 32 Å from the
catalytic site. The activators, AMP and phosphate, and the
inhibitor, glucose-6-phosphate, bind here. The glycogen
storage site (G) is located on the surface of the enzyme,
some 30 Å from the catalytic site and 39 Å from the
allosteric site. Oligosaccharides ranging in size from
maltose to maltoheptaose bind strongly to this site. The
glycogen storage site is probably the site through which

adjustment in temperature and in a range where the crystals were stable it was feasible to explore the reaction pathway of this enzyme by crystallographic methods (Hajdu *et al* 1987a). With the focussed monocrhomatic beam used at that time on the wiggler line of the SRS the intensity level was 10^3 higher than a rotating anode source. The 35 min data collection time referred to above involved about 10 min exposure time and 25 min of film cassette changes and/or hutch searches and represented a reflection measuring rate from the crystals during exposure of approximately 200 per second.

Table 10.9 summarises the conditions for data collection during all these experiments, figure 10.12 shows a schematic of the phosphorylase enzyme and figure 10.13 a sequence of difference Fourier maps corresponding to the data sets in table 10.9.

10.4.4 Application of the synchrotron Laue method to time resolved crystallography

10.4.4.1 Case study: Time resolved study of H-ras p21: identification of the conformational change induced by GTP hydrolysis (Schlichting et al 1990)

Crystals ($P3_221$; $a=b=41.0$ Å, $c=164.8$ Å, size about $0.4 \times 0.4 \times 0.2 \, mm^3$) of Ha-ras 21 protein (Pai *et al* 1989) with caged GTP have been used as the starting point for investigating the structural states occurring on a pathway of GTP hydrolysis. The structure of a fairly short-lived complex (p21·GTP) was determined by Laue diffraction methods.

Figure 10.12 (*cont.*)
> phosphorylase is attached to glycogen particles *in vivo* and there is a substantial body of evidence which supports the notion that this site is an additional control site whereby occupation by glycogen or oligosaccharide results in an increase in the rate of catalysis. The nucleotide inhibitor site (I) is also on the surface of the enzyme near the entrance to the catalytic site. The site binds purines, nucleosides or nucleotides at high concentrations and occupancy of this site results in enzyme inhibition and a stabilisation of the T state. From Hajdu *et al* (1987a) and references cited therein. This figure kindly provided by Prof. L. N. Johnson and reproduced with the permission of the authors and Oxford University Press.

Table 10.9. *Conditions for data collection experiments in the kinetic study of glycogen phosphorylase. From Hajdu* et al *(1987a), with permission of the author and Oxford University Press.*

	Exp.	Addition Substance	(mM)	Soaking time	Resting time	Temperature (°C)
Heptenitol series	1	heptenitol	100	18 hr		25
	2	heptenitol	100	15 min	33 min	22
		arsenate	50			
		AMP	2.5			
	3	heptenitol	100	10 min		13
		phosphate	50			
		AMP	2.5			
	4	heptenitol	100	15 min		20
		phosphate	50			
		AMP	2.5			
		maltoheptaose	50			
	5	heptenitol	100	60 min	29 min	23
		phosphate	50			
		AMP	2.5			
		maltoheptaose	50			
	6	heptenitol	100	10 min	8 hr	13
		phosphate	50			
		AMP	2.5			
	7	heptenitol	100	50 hr		23
		phosphate	50			
		AMP	2.5			
		maltoheptaose	50			
Phosphorolysis of oligosaccharide	8	1st soak:		15 min		
		G7	50			
		IMP	2			
		2nd soak:				
		G7	50	5 min		3
		IMP	2			
		P_i	500			
	9	G7	50	15 min		20
		IMP	2			
		P_1	500			
	10	G7	50	15 min	36 hr	22
		IMP	2			
		P_i	500			

able 10.9 (*cont.*)

me of ta llection	X-ray source	Wavelength (Å)	Resolution (Å)	No. of reflections measured	No. of unique reflections	R_m	$\Delta F/F$
veek	Rotating anode (Cu)	1.54	3.0	69 505	16 384	0.065	0.151
min	SRS PX9.6	0.88	2.4	145 035	34 323	0.058	0.104
ır	SRS PX7.2	1.488	2.8	100 163	18 662	0.097	0.142
min	SRS PX 9.6	0.88	2.5	101 926	29 417	0.112	0.136
5 hr	SRS PX7.2	1.488	3.0	94 539	19 733	0.083	0.158
ır	SRS PX7.2	1.488	2.8	54 489	13 780 (incomplete)	0.076	0.152
5 hr	SRS PX7.2	1.488	3.0	94 193	19 299	0.063	0.154
min	SRS PX9.6	1.38	2.7	108 296	21 242	0.061	0.123
min	SRS PX9.6	0.88	2.6	111 076	26 697	0.092	0.168
ır	SRS PX9.6	1.003	2.7	109 363	24 800	0.084	0.135

Table 10.9. (*cont.*)

	Exp.	Addition Substance	(mM)	Soaking time	Resting time	Temperature (°C)
Synthesis of oligosaccharide	11	G3	50	15 min		0
		AMP	5			
		Glc-1-P	100			
	12	G3	50	15 min	24 hr	22
		AMP	5			
		Glc-1-P	100			

Notes:
(1) Abbreviations: G7, maltoheptaose; G3, maltotriose; Glc-1-P, glucose-1-phosphate; P_1, sodium phosphate.

(2) $R_m = \sum_i \sum_h |\bar{I}(h) - I_i(h)| / \sum_i \sum_h I_i(h)$

where $I_i(h)$ is the ith measurement of the intensity of reflection, h, and $\bar{I}(h)$ is the mean of these measurements.

The use of caged compounds in muscle diffraction had been pioneered by the Heidelberg group. These methods were then applied by these workers to the crystalline state. After photolytic removal (figure 10.14) of the 2-nitrophenyl ethyl group from caged GTP at the active site of p21, GTP hydrolysis occurs with a half-life of about 40 min at room temperature. This is the same rate as encountered in solution. The synchrotron Laue data were recorded on photographic film at EMBL, Hamburg (X31 beam line). The data were processed using the Daresbury Laue software where the λ-curve was explicitly determined from symmetry equivalent reflections (Campbell *et al* 1986); Helliwell *et al* 1989b).

A Laue data set was recorded during the first 5 min after photolysis of the crystalline p21·caged GTP complex. The major species present at the active site was GTP as confirmed by the electron density map which showed clearly discernible phosphate peaks (figure 10.15(*b*)). Around the essential Mg^{2+} ion an expected water molecule was not apparent.

A second Laue data set was recorded 14 min after photolysis and revealed the GTP/GDP state. In the electron density map corresponding to these data there was much less electron density at the γ-phosphate position (figures 10.15(*c*) and 10.16). It is also of interest to note that the

me of ta llection	X-ray source	Wavelength (Å)	Resolution (Å)	No. of reflections measured	No. of unique reflections	R_m	$\Delta F/F$
min	SRS PX9.6	1.003	2.8	108 027	18 763	0.108	0.120
hr	SRS PX7.2	1.488	2.8	91 815	18 483	0.118	0.119

Source RA: Elliott GX6 Rotating Anode 40 kv 40 mA; exposure 75×10^2 s deg^{-1}. SRS: Daresbury Synchrotron Radiation Source. PX7.2: Protein Crystallography Station 7.2 (section 5.6.5.1). Bent quartz mirror vertical focus; germanium curved triangular crystal monochromator. Typical parameters $E = 2$ GeV, $I = 300–100$ mA ($t_{\frac{1}{2}}$ beam 6–10 hr). Exposure times 40–75 s deg^{-1}. PX9.6: Protein Crystallography Station 9.6 on Wiggler beam line (section 5.6.5.2). Platinum-coated fused quartz mirror vertical focus; horizontally focussing triangular oblique cut monochromator (Si(111) or Ge(111)) in helium. Typical parameters: $E = 2$ GeV; wiggler magnet 5 T or $E = 1.8$ GeV: wiggler magnet $= 4.5$ T, $I = 300–100$ mA ($t_{\frac{1}{2}}$ beam 6–20 hr). Exposure times 5–40 s deg^{-1} depending on crystal size, beam current and resolution.

binding of caged GTP was slightly different from that of GTP: the caged GTP was pushed out of the nucleotide binding pocket presumably due to steric hindrance arising from the presence of the cage. Also, the protecting 2-nitro phenyl ethyl group pointed towards a solvent channel (about 15×20 Å). This explained how the nitro phenyl ethyl group diffused out of the crystal after photolysis.

This study demonstrates that photolytic generation of substrates at the active site of a protein in the crystal is a viable approach to rapid initiation of the enzymatic reaction, which is needed to take advantage of the potential of SR for time resolved crystallography studies.

10.4.5 A challenge for the future: analysis of several simultaneously present conformers in a crystal

After reaction initiation, X-ray diffraction data can be recorded. If the crystal at any time t after initiation contains several conformers, can these structural states be extracted? Moffat (1989a) outlines an approach to this problem. A series of experiments will yield several X-ray data sets along a time pathway. Each data set will correspond to several conformers of different 'occupancy' and these occupancies will vary with time.

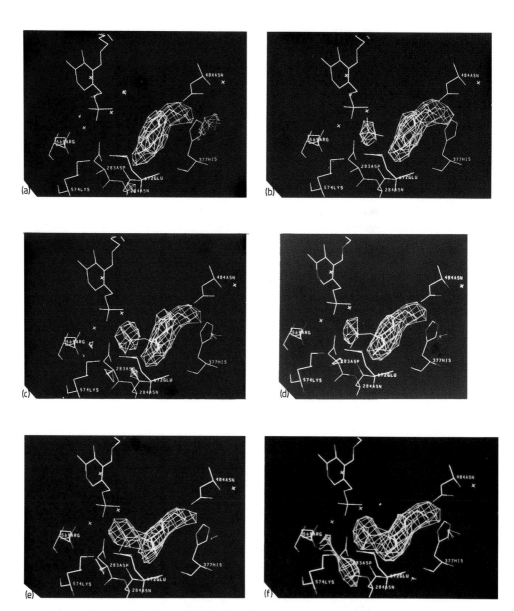

Figure 10.13 Difference Fourier syntheses in the vicinity of the catalytic
site for the heptenitol to heptulose-2-phosphate conver-
sion. A single positive contour level is shown (corresponding to 300 arbitrary units). Selected amino acids in the
native conformation are shown. The pyridoxal phosphate
is in upper left. Water molecules in the native structure are
shown as crosses. (a) Experiment 2 (100 mM heptenitol,
50 mM arsenate, 2.5 mM AMP). Only heptenitol is
observed bound. The edge-on view of the glucopyranose

ring has been chosen to emphasise the addition of phosphate in subsequent maps. In the face-on view to the ring there is considerable detail. The positive density to the right of his 377 indicates a small shift away from the substrate of this residue in order to optimise contacts. (*b*) Experiment 3 (100 mM heptenitol, 50 mM phosphate, 2.5 mM AMP; short soak, rapid data). Heptenitol and a weak phosphate peak are apparent. (*c*) Experiment 4 (100 mM heptenitol, 50 mM phosphate, 2.5 mM AMP, 50 mM maltoheptaose; short soak, rapid data). The reaction goes faster in the presence of oligosaccharide. Some product, heptulose-2-phosphate is formed. (*d*) Experiment 5 (100 mM heptenitol, 50 mM phosphate, 2.5 mM AMP, 50 mM maltoheptaose; 1 hr soak, 2.5 hr data). Some heptulose-2-phosphate formed. (*e*) Experiment 6 (100 mM heptenitol, 50 mM phosphate, 2.5 mM AMP, short soak, 8 hr rest, rapid data). Despite the lack of oligosaccharide, the reaction has proceeded to produce a significant concentration of heptulose-2-phosphate. (*f*) Experiment 7 (100 mM heptenitol, 50 mM phosphate, 2.5 mM AMP, 50 mM oligosaccharide, 50 hr soak, 2.5 hr data). The reaction has gone to completion with strong heptulose-2-phosphate binding. The positive contours indicating the movement of arg569 are apparent at bottom left. From Hajdu *et al* (1987a). This figure kindly provided by Prof. L. N. Johnson and reproduced with the permission of the authors and Oxford University Press.

Figure 10.14 Reaction and time resolved scheme of conversion of p21·caged GTP to p21·GTP to p21·GDP+phosphate (P_i). From Schlichting *et al* (1990) with permission, copyright © 1990 Macmillan Magazines Ltd.

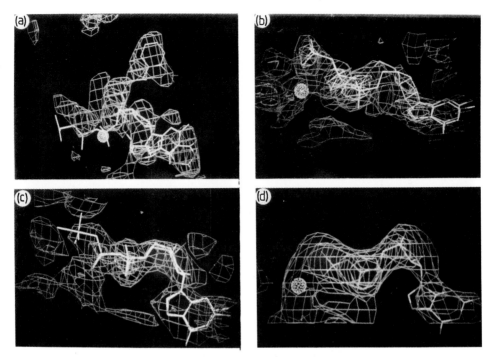

Figure l0.15 Electron density maps of the bound nucleotides at the active site of p2lc: (a) caged GTP, (b) GTP, (c) GTP/GDP, (d) GDP. From Schlichting et al (1990) with permission, copyright © 1990 Macmillan Magazines Ltd.

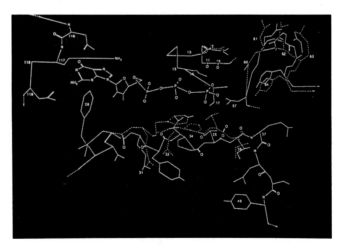

Figure l0.16 Schematic drawing of the binding region of the p21·GTP complex, with the main changes in the p21·GDP complex shown by dashed lines. From Schlichting et al (1990) with permission, copyright © 1990 Macmillan Magazines Ltd.

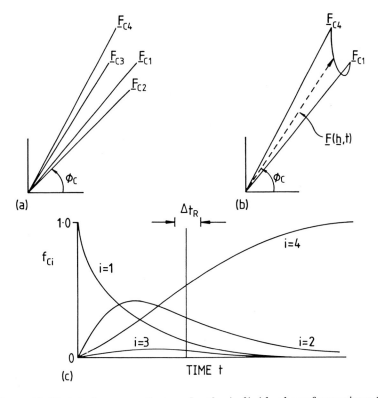

Figure 10.17 (a) Structure factors for the individual conformations C1,
C2, C3 and C4, for the reflection **h**. (b) The trace of the
structure factor $\mathbf{F}(\mathbf{h},t)$ as it evolves from $\mathbf{F}_{C1}(\mathbf{h})$ at $t=0$ to
$\mathbf{F}_{C4}(\mathbf{h})$ at $t=\infty$. At all times t, only $|\mathbf{F}(\mathbf{h},t)|$ is measurable.
(c) Schematic representation of the variation of the frac-
tional occupancies, f_{C1} etc with time. Δt_R is the exposure
time. From Moffat (1989a) with permission.

Moffat proposes that the data sets be combined and these occupancies
and their associated structures be extracted as a function of time (figure
10.17). This looks like an attractive procedure compared with trying to
derive, from one data set alone, details of all conformers present in the
crystal at that time.

10.5 LARGE UNIT CELLS (VIRUS AND RIBOSOME STUDIES)

The combination of all the advantages of SR is especially needed in
virus and ribosome crystallography where the unit cells are very large.
Data collection from very large unit cell constant crystals benefits from

the natural narrow beam collimation of SR to resolve adjacent spots. Even with SR, exposure times can still be fairly long and often only one data photograph can be collected per sample crystal due to radiation damage problems.

10.5.1 Virus crystallography

The X-ray crystallographic study of orthorhombic human rhinovirus (HRV14), the causative agent of the common cold, was the first example of extremely radiation sensitive virus crystals benefiting in a critical way from SR (Rossmann and Erickson 1983). The deterioration of these crystals commenced with the relatively short X-ray exposure required for the setting operations and continued even in the absence of radiation. The damage was even visible as an opaque volume within the crystal, initially confined to the irradiated volume but spreading fairly rapidly throughout the crystal. The use of intense radiation from a synchrotron extended the observable pattern from 3.5 Å to beyond 2.7 Å resolution because of the damage being time-, as well as dosage-, dependent, as mentioned earlier in this chapter with the proteins phosphorylase b and 6-phosphogluconate dehydrogenase (for example). Rossmann and Erickson (1983) found that the rhinovirus radiation damage was so extreme that it was not possible to perform setting of the crystal alignment *and* allow the collection of quality data. On the other hand, a fresh crystal, which had been only optically aligned, could yield two high resolution oscillation data photographs. As a result the accurate crystal setting had to be determined from 'still' photographs taken after the data collection period. This pioneering study gave the breakthrough required for tackling this most important virus and the impetus for other projects. Rossmann and Erickson termed their new procedure the 'American method' because it demanded that the shooting be done first and questions asked subsequently! 'Tradition had it that this was the normal procedure of law enforcement in the American Wild West and that this was well suited for dealing with delinquent crystals in the New World of synchrotrons'. This work was done on DORIS, Hamburg, on X11 (section 5.6.4.1) and at the SRS in Daresbury on station 7.2 (section 5.6.5.1).

The extreme radiation sensitivity of rhinovirus crystals can be contrasted with poliovirus. Poliovirus was solved using data collected on a conventional X-ray source (Hogle, Chow and Filman 1985) as were various plant viruses much earlier (see Harrison (1978) for a review).

FMDV, solved somewhat later than the rhinovirus, also depended totally on SR because of the radiation sensitivity of the crystals and their small size (Acharya *et al* 1989).

The collimation of the SR beam in these studies is as important, to allow the spot size to be kept small, as the intensity is to yield short exposure times and to reduce (or avoid) radiation damage effects. The ease with which the beam cross-fire can be controlled is important here. With more modest unit cells (e.g. $\leqslant$150 Å) the horizontal convergence angle in a focussing instrument could be up to 4 mrad in the horizontal. In the virus crystal case the use of the pre-monochromator slits to reduce this angle to $\approx$1 mrad is typical on existing machines. In the vertical, of course, the convergence angle will typically be $\approx$0.2 mrad (the natural opening angle of the SR beam).

In the early days of virus crystallography at the synchrotron, the largest lattice constant dealt with was that of cow pea mosaic virus (Usha *et al* 1984). This virus crystallised in space group $P6_122$ (or its enantiomorph) with $a=451$ Å and $c=1038$ Å. Data collected at LURE (figure 10.18) were analysed and the results showed that quality data

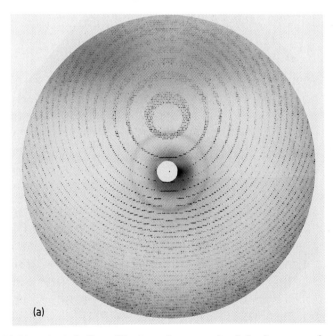

(a)

Figure 10.18 (*a*) A 0.4° oscillation photograph of the hexagonal crystal form of cow pea mosaic virus obtained at LURE. At the start of the exposure the crystal had been rotated 6.4°

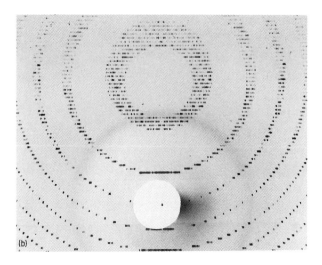

(b)

Figure 10.18 (*cont.*)

about the spindle axis (the horizontal direction in the figure) from the defined zero position in the data collection scheme (zero is defined as the **c*** axis coincident with the spindle and the **b** axis parallel to the X-ray beam). The direct beam is 0.18×0.22 mm and the crystal to film distance is 175 mm. The maximum resolution of the photograph is 4.3 Å. (*b*) An enlargement of the central region of (*a*) showing the well-resolved reflections along the **c*** direction. The maximum resolution in this print is 9.0 Å. The centre-to-centre spacing of reflections in the **c*** direction is 0.26 mm. From Usha *et al* (1984). Further details kindly provided by Prof. J. Johnson and reproduced with permission: crystal space group P6$_1$22 or P6$_5$22, $a=b=450$ Å, $c=1038$ Å (Johnson and Hollingshead 1981). Exposure time: 3 hr 20 min. Beam line and camera configuration: the small direct beam was obtained using the slit system positioned just before the germanium monochromating crystal (Lemonnier *et al* 1978), and slits in the collimator (Love *et al* 1965). A helium path was employed and found to dramatically improve the signal-to-noise ratio at this specimen to film distance.

Film processing: resolution of processing: 4.3 Å; total number of theoretically measurable reflections: 30 866; whole reflections: 11 388; partial reflections: 19 416; total number of reflections measured: 22 126; whole reflections: 9134; partial reflections: 12 992; symmetry equivalent scaling (6 Å resolution): $R=6.41\%$ based on 1456 whole reflections measured twice on the film.

could be obtained from crystals with such very large cells. Usha *et al* (1984) acknowledged Sigler and Westbrook for their unpublished work on crystals of the enzyme Δ_5-ketosteroid isomerase (Westbrook *et al* 1984) which explored the control of the collimation of the SR at LURE for larger unit cells.

These preliminary studies culminated in a wealth of activity at SR sources devoted to virus crystal data collection both for structure solving and also drug binding-studies. Table 10.10 gathers the various references to this type of work up to 1990. We now give further details on the rhinovirus work which was done primarily on CHESS after the initial work at Hamburg and Daresbury. This case study will then be followed by a description of the FMDV work done at Daresbury using the SRS.

10.5.1.1 *Case study: Rhinovirus*

Rossmann *et al* (1985) solved the structure of human rhinovirus (HRV) (a picornavirus responsible for the common cold) using nearly a million reflections collected on the protein crystallography facility at the Cornell Synchrotron source in a matter of days. This conveyed an enormous speed advantage over data collection on a conventional source and also ameliorated an otherwise impossible problem of radiation damage when long exposure times were used. The far greater rate of radiation damage in the X-ray beam in relation to plant viruses is symptomatic of an inherently less stable protein capsid. The structure was reported by Rossmann *et al* (1985) and the full details of the structure determination by Arnold *et al* (1987). Table 10.11 summarises the details of the HRV14 data sets and figure 10.19 shows a 0.3° oscillation photograph of an HRV14 crystal taken on CHESS. One data exposure was recorded from each crystal. Most crystals measured 0.3 mm from apex to apex through the centre although many were larger. The exposure time for an oscillation data photograph was 13 min in May 1984 but, by November 1984, improvements in the X-ray optics on CHESS had been made so that the exposure time had reduced to 4 min. By December 1985, the X-ray optics had been further improved so that exposure times had reduced to 2 min with smaller crystals (0.2 mm in diameter).

The processing of the SR data by Rossmann and coworkers was specifically adapted to the treatment of SR data. The prediction of reflections and their rocking width took account of the asymmetric cross-fire of the SR beam. The SR wavelength was determined by post-refinement of the apparent cell dimensions relative to the reference values established from Cu Kα radiation.

Table 10.10. *Large unit cell structure determinations (including viruses and progress on ribosome studies) based on SR data. (See also table 10.6.)*

Sample	SR source	Detector	λ (Å)	a / α	b / β	c / γ	Space group	Resolution limit (Å) SR	Resolution limit (Å) Cu Kα	Reference
Rhinovirus (HRV14)	SRS DORIS CHESS	Film Film Film	1.488 1.4 1.57	445.1 —	— —	— —	$P2_13$	2.6	3.5	Rossmann and Erickson (1983) Rossmann et al (1985) Arnold et al (1987)
HRV14 drug complexes	CHESS NSLS	Film Film		— 	— 		$P2_13$	2.6		Badger et al (1988)
HRV14 drug resistant mutants	CHESS	Film		—	—		$P2_13$	2.6		Badger et al (1989)
Cow pea mosaic virus	LURE	Film	1.4	451 —	— —	1038 —	$P6_122$	3.5		Usha et al (1984)
	LURE	Film	1.4	317.3 —	— —	— —	$I23$	3.5		Stauffacher et al (1987)
Mengo virus	CHESS	Film		441.4 —	427.3 —	421.9 —	$P2_12_12_1$	2.6		Luo et al (1987) Luo, Vriend, Kamer and Rossmann (1989)
FMDV	SRS	Film	0.9	345 —	— —	— —	$I23$	2.9		Fox et al (1987) Acharya et al (1989)
Black beetle virus	LURE CHESS	Film Film		362 —	— —	— —	$P4_232$	3.0		Hosur et al (1987)
B. subtilis Heavy riboflavin synthase	DORIS	Film	1.480	156.4 —	— —	298.5 —	$P6_322$	3.3		Ladenstein et al (1986, 1988)
Bean pod mottle virus	CHESS	Film	1.566	311.2	284.2	350.5	$P22_12_1$	3.0		Chen et al (1989)
Rhinovirus Serotype 1A (HRV1A)	CHESS	Film	1.565Å	341.3 120°	—	465.9 —	$P6_322$	3.0		Kim et al (1989)

Bacteriophage MS2	SRS	Film	0.88	465.9	287.8 121.7	274.0	$C2$	2.6	Valegård, Liljas, Fridborg and Unge (1990)
B. stearothermophilus 50S ribosome subunit	DORIS	Film	0.9	350	670	910	$P2_12_12_1$	10–13	Yonath *et al* (1984), (1986)
	SRS	Film	—	—	—	—			
	CHESS	Film	1.56	300	546	377	$C2$	11	Bartels *et al* (1988) Müssig *et al* (1989)
H. marismuorti 50S ribosome subunit	DORIS	Film	1.51	182	584	186	$P2_1$	6	Shoham, Wittmann and Yonath (1987)
	SSRL	Film	1.54	215	300	590	$C222_1$	5	Bartels *et al* (1988)
T. thermophilus 30S ribosome subunit	CHESS	Film	1.588	407	—	171	$P42_12$	9.9	Yonath *et al* (1988)

Note:
(1) The SR source acronyms are as given in table 4.1.

Table 10.11. *Data used in the structure determination of rhinovirus. From Rossmann et al (1985), with permission; copyright © 1985 Macmillan Magazines Ltd.*

| Resolution range (Å) | No. of unique observations ($F^2 > 1\sigma (F^2)$) for each data set and percentage of possible total | | | | | |
| | Native | | 1 mM KAu(CN)$_2$ | | 5 mM KAu(CN)$_2$ | |
	No.	%	No.	%	No.	%
∞–30	358	63	331	57	231	40
30–15	3 445	86	2 847	70	1 850	46
15–10	9 109	84	7 757	70	5 029	46
10–7.5	17 532	83	14,641	68	9,486	45
7.5–5.0	70 043	81	57 577	65	34 046	39
5.0–3.5	184 864	78	138 477	58		
3.5–3.0	133 101	63	83 852	39		
3.0–2.75	60 629	36	25 271	15		
2.75–2.6	26 834	11	8 175	3		
Total	509 915	58	338 928	39	50 641	41
No. of film packs	83		48		11	
R-factor (%)	11.0		12.9		12.0	

Note
(1)
$$R = \frac{\sum_h \sum_i |(I_h - I_{hi})|}{\sum_h \sum_i I_h} \times 100$$

where I_h is the mean of the I_{hi} observations of reflection h.

The intrinsic symmetry of the virus was used to provide phasing information in the final stages of the analysis; low resolution phasing was provided via two heavy atom derivative data sets, again collected using SR. The high resolution phasing was carried out using real space molecular replacement whilst extending the resolution in small steps. The final skew-averaged map allowed the location of 811 out of 855 residues in four distinct polypeptide chains.

The tertiary fold of three of the rhinovirus proteins (VP1, VP2 and VP3) and their quaternary organisation within the HRV14 capsid are similar to the structures of the plant viruses TBSV (tomato bushy stunt virus) and SBMV (southern bean mosaic virus), agreeing to within approximately 3 Å. In HRV14 VP4 is an internal structural protein in contact with VP1 and VP2. Protrusions on VP1, VP2 and VP3 together create a deep cleft or 'canyon' on the viral surface which is 25 Å deep and

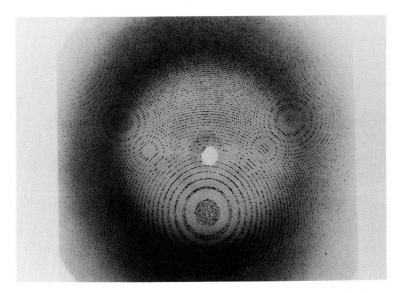

Figure 10.19 A 0.3° oscillation photograph of an HRV14 crystal taken at CHESS. The highest resolution reflections extend to 2.5 Å at the corners and the resolution cut-off at the edge, along the spindle axis, is 3.0 Å. From Arnold *et al* (1987) with permission.

12–30 Å wide. Figure 10.20 shows the layout of the icosahedron capsid and figure 10.21 shows diagrams of the viral proteins.

The cell receptor binding site is probably located in the 25 Å deep canyon, which circulates around each of 12 pentamer vertices. An antibody molecule, whose Fab fragment (figure 3.16) has a diameter of the order of 35 Å, would have difficulty in reaching the canyon floor, its entrance being blocked by the canyon rim. Thus, the residues in the deeper recesses of the canyon would not be under immune selection and could remain constant, permitting the virus to retain its ability to seek out the same cell receptor. It was expected that the canyon would be present for all picornaviruses (e.g. poliomyelitis, foot and mouth disease and hepatitis viruses but see section 10.5.1.2).

Assembly of picornaviruses proceeds via protomers of VP1, VP3 and VP0 (the latter is an initial composite of VP2 and VP4) and then via pentamers of these protomers to mature viruses. The final step involves inclusion of the RNA into empty capsids or partially assembled shells with simulatenous cleavage of VP0 into VP2 and VP4. Conversely, *in vitro* disassembly, produced by mild destruction, proceeds via the expul-

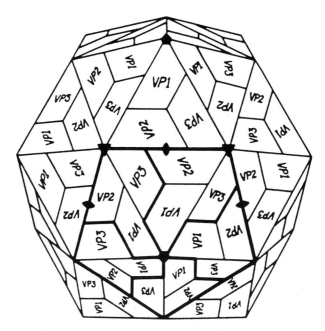

Figure 10.20 Illustration of the arrangement of the four distinct
            polypeptide chains VP1, VP2, VP3 and VP4 forming the
            icosahedral protein shell surrounding an RNA core, the
            notation refers to the unique proteins of the virus shell.
            From Rossmann *et al* (1985) with permission, copyright ©
            1985 Macmillan Magazines Ltd.

sion of VP4 followed by the RNA. Disruption of pentamer–pentamer
contacts, mediated by a slight reorientation of VP2 or its complete
removal, could provide a port by which VP4 and RNA could exit.

Evolutionary relationships can be assessed by comparison of three-
dimensional structures. Indeed, structural comparisons can be used to
trace divergent evolution over longer time spans than is possible by
amino acid sequence comparisons. The considerable similarity of VP1,
VP2 and VP3 structures to those of the plant viruses leaves little doubt as
to their divergence from a common ancestor.

There is also a similarity in the folding topology of the plant lectin
concanavalin-A and VP1 (Argos, Tsukihara and Rossmann 1980). Hence,
it is possible that the corresponding gene for a protein such as conA
(figure 3.8) provided a useful tool for the formation of biological assemblies that could attach themselves to animal cell receptors.

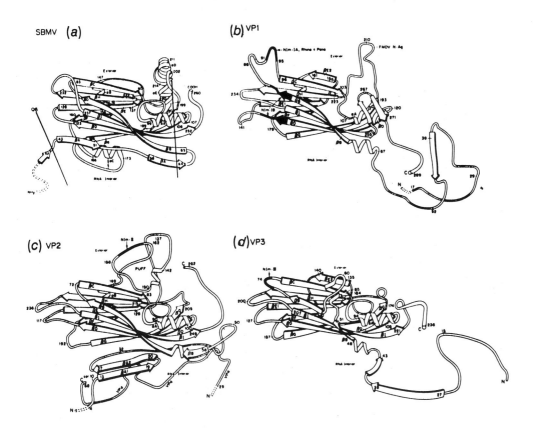

Figure 10.21 Diagrams showing the polypeptide fold of SBMV and of
each of the three larger capsid proteins of HRV14. The
nomenclature of the secondary structural elements is
derived from that of SBMV. (*a*) SBMV; (*b*) VP1 of HRV14;
(*c*) VP2 and VP4 of HRV14; (*d*) VP3 of HRV14. There are
four excursions of the polypeptide chains towards the
wedge-shaped end. Each excursion makes a sharp bend or
'corner'. The most exterior (top left) corner is formed
between the β sheets βB and βC, the second corner down is
formed between βH and βI, the third corner is between βD
and βE1 and the most internal corner connects βF and βG.
The βF–βG corner is the site of a 25-residue insertion in
SBMV, including the α-helix αC, that is not present in any
of the viral proteins of picornaviruses, nor in TBSV or satellite tobacco necrosis virus (STNV). From Rossmann *et al*
(1985) with permission, copyright © Macmillan
Magazines Ltd.

10.5.1.2 Case study: FMDV

Foot and mouth disease is a major scourge of livestock in several continents. Killed vaccines are available but despite their effectiveness they are difficult to administer in remote regions and severe economic losses due to the disease still occur. Knowledge of the structure should assist in the development of improved and novel vaccines and may lead to the design of antiviral drugs. Hence, a programme of research was initiated to undertake the X-ray crystallographic analysis of FMDV. Fox *et al* (1987) described the growth, purification, crystallisation and preliminary X-ray diffraction analysis of this virus of a particular serotype and strain. The crystals (see plate I(xi) in chapter 2) belonged to space group I23 with $a=345$ Å, with one-twelfth of the virus particle in the crystallographic asymmetric unit which therefore contained five copies of the unique protein in the structure.

The small size of the crystals (average dimensions for those used were $0.12\times0.12\times0.06$ mm^3) coupled with the large number of observable reflections, led to very weak diffraction. In common with many other picornavirus crystals (such as rhinovirus discussed earlier) the FMDV crystals were also extremely radiation sensitive. Each crystal only yielded one (0.4–0.5°) oscillation photograph. Data collection was made feasible by the characteristics of the radiation, namely, the collimation, the intensity and the relatively short wavelength of the beam (0.9 Å) on station 9.6 of the SRS (see section 5.6.5.2). All these benefits contributed to an increase in yield of data from each crystal. Many crystals (≈500) were examined using the American method (see start of this section on virus crystallography). There were 135 usable crystal exposures of which 106 were analysed. An exposed diffraction pattern is shown in figure 10.22 (from Fox *et al* (1987)).

The containment and security handling of the FMDV virus crystals was a particularly important aspect of this project. Transport of the virus crystals to the synchrotron from the south of England was approved months in advance by the Chief Veterinary Officer in the United Kingdom. Each trip could only be 24 hr in duration and a Safety Officer had to accompany the scientists on their trips. Perhaps the most important aspect of safety and security was that FMDV is unstable below pH 7. Hence, citric acid swabs were placed on the oscillation camera under the glass capillary holding an FMDV crystal; hence, in case of accidental breakage of a capillary the FMDV crystal would be neutralised by the citric acid (actually the breakage of a capillary never did happen). The lability of the virus to acid was explained qualitatively once the struc-

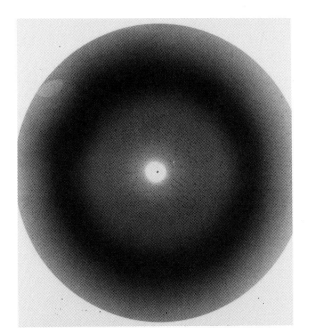

Figure 10.22 Oscillation photograph of a crystal of FMDV, strain O_1BFS. The synchrotron parameters were: wiggler field, 5 T; current, 280 mA; energy 2 GeV; wavelength, 0.876 Å. Crystal to film distance, 170 mm. Collimator diameter, 300 μm. Resolution at the edge of the film was 2.6 Å. From Fox *et al* (1987) with permission.

ture was determined. It was found that a high density of histidine residues line the pentamer interfaces in FMDV. As the pK value of this amino acid correlates with the pH at which disassembly of the virus occurs, this might explain how the virus capsid falls apart on treatment with acid. Certainly this property was a critical factor in allowing the project to be conducted, in safety, at all.

The details of the processed data used in the structure determination and the synchrotron conditions used are given in table 10.12. Phase determination proceeded in several steps. Firstly, the virion orientation in the I23 unit cell was determined using a self-rotation function. Secondly, low resolution (∞–8 Å) phases were calculated; a composite virus, 'Rhengo', was constructed by summing the electron density (calculated from the atomic coordinates) for HRV14 and Mengo virus and structure amplitudes and phases calculated. Thirdly, these phases were refined and extended to higher resolution by using five-fold molecular averaging and solvent flattening.

Table 10.12. *Data used in the structure determination of FMDV. From Acharya* et al *(1989) with permission; copyright © 1989 Macmillan Magazines Ltd. Nature* **337**.

Resolution range (Å)	No. unique observed reflections	Percentage of possible observations
10 000–15	43	18.86
25–15	888	92.50
15–10	2 800	95.96
10–7.5	4 758	95.79
7.5–5.0	20 274	96.05
5.0–3.5	51 750	95.39
3.5–3.0	43 101	85.45
3.0–2.9	5 488	39.81
2.9–2.8	1 836	10.73
Total (to 2.9)	129 100	86.98

Note:

(1) *R*-factor (%), 13.9 (all data included). The Daresbury synchrotron parameters were: Wiggler field strength, 5 T; current, 150–280 mA; energy, 2 GeV; wavelength, 0.85–1.0 Å. Crystal to film distance, 170–190 mm; collimator diameter, 300 μm; oscillation range 0.4–0.5°. Number of crystals examined, 500; number of usable filmpacks (1 filmpack per crystal), 135; number processed, 106.

The assignment of the amino acid sequence and model building began at a resolution of 3.5 Å and was completed at 2.9 Å. Over 90% of the residues of the capsid proteins were clearly visible in the electron density map. The model was refined using all the data from 5 Å to 2.9 Å to an *R*-factor (on *F*) of 17%. The accuracy of the model may be estimated from the agreement of the five crystallographically independent proteins when refined without the restraint of icosahedral symmetry; the main chain atoms deviated from the mean positions by an average of 0.17 Å. Figure 10.23 shows the intact virion structure viewed down the five-fold axis. The overall shape of the well-ordered electron density is that of a spherical shell starting at an inner radius of 100 Å and extending to an outer radius of 150 Å. There is a close structural relationship of FMDV to HRV14 and Mengo virus as shown in figure 3.21.

The outer surface of FMDV is relatively smooth. Unlike HRV14 or Mengo virus there were not any major 'canyons' or pits in the surface (figure 10.24). Rossmann *et al* (1985) proposed that a canyon on the surface of HRV14 was the site of cell attachment and which was inaccessible to antibody. This shielded HRV14 from immune surveillance.

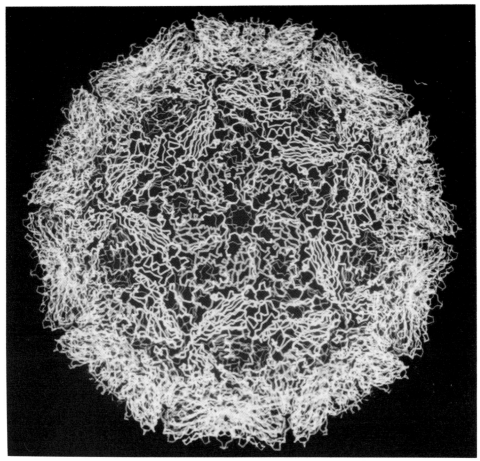

Figure 10.23 A view of the intact virion of the FMDV. Figure kindly supplied by D. Stuart with permission.

The FMDV results suggested that the 'canyon hypothesis' was not as general as once thought. As a result of the FMDV structure, Archarya *et al* (1989) proposed a more general hypothesis in which 'concealment' of the constant receptor site is achieved by 'camouflaging' a small constant region of amino acid sequence within a sea of considerable variability of amino acid sequence.

A comparison of the FMDV proteins with the other viral proteins places Mengo virus roughly midway between HRV14 and FMDV in terms of evolution.

10.5.2 Ribosome X-ray crystal structure analysis: a progress report

This has been reviewed by Bartels *et al* (1988) who open their article with the statement 'If ever a desperate project benefitted from the avail-

(a) **FMDV**   (b) **Mengo virus**

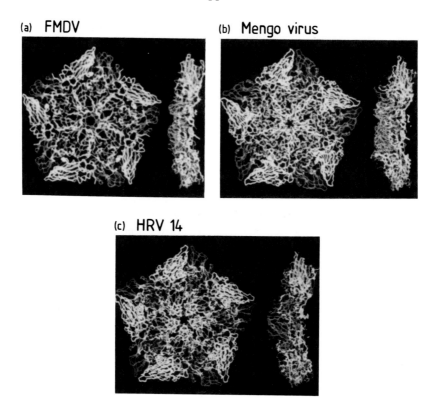

(c) **HRV 14**

Figure 10.24 Pentameric association of protomers. Orthogonal views of
             the virtual bonds joining α-carbon atoms for the pentamers
             of (a) FMDV, (b) Mengo virus and (c) HRV14 respect-
             ively. From Acharya *et al* (1989) with permission,
             copyright © 1989 Macmillan Magazines Ltd.

ability of synchrotron radiation, that project is the crystallographic
study of ribosomal particles'.

As explained in chapter 3, ribosomes are unique assemblies of several
strands of RNA and a large number of different proteins, representing the
living cell's protein factory. Upon initiation of the biosynthetic process,
one larger and one smaller subunit associate to form the active cell
organelle that reads genetic information from the messenger RNA and
translates it into a specific polypeptide chain. To this end, several binding sites are provided: one for the messenger RNA, three sites for the
t-RNAs carrying the amino acids and several for a variety of factor
proteins.

The large subunit (50S) of a typical eubacterial ribosome (MW *ca*
1 600 000) is composed of 2 RNA chains and about 33 different proteins,

the small subunit (30S, MW *ca* 700 000) comprises 1 RNA chain and about 21 proteins.

The chemical and physical properties of ribosomes are well characterised (for reviews see Chambliss *et al* (1980); Liljas (1982); Wittmann (1982, 1983); Hardesty and Kramer (1986)). The exact understanding of their function, however, still lacks a detailed molecular model. Appropriate methods such as image reconstruction from electron micrographs of two-dimensional sheets, or X-ray structure analysis, all depend on crystallisation of the material.

Experiments towards growing ribosomal crystals *in vitro* were stimulated by the observations that ribosomes may self-organise into ordered aggregates in the living cell; two-dimensional arrays have been found under special conditions, such as hibernation or lack of oxygen (e.g. Byers (1967); Unwin (1977); Milligan and Unwin (1986)). However, the complex structure, the enormous size and the flexibility of ribosomal particles render their crystallisation *in vitro* extremely difficult. Therefore, only a few successful efforts to produce three-dimensional crystals have been reported (e.g. Yonath *et al* (1980); Trakhanov *et al* (1987); Yonath and Wittmann (1988)).

10.5.2.1 Crystal growth

The first microcrystals were of *Bacillus stearothermophilus* 50S subunits (Yonath *et al* 1980) and of 70S ribosomes from *Escherichia coli* (Wittmann *et al* 1982); they were mainly grown either from lower alcohols or toluene or chloroform. Each ribosomal preparation required slightly different crystallisation conditions, and often the preparation had almost been exhausted by the time conditions were optimised. It was also found that crystals grew from active particles only.

It was found for spontaneous crystal growth of ribosomal particles that the lower the Mg^{2+} concentration, the thicker the crystals (Arad, Leonard, Wittmann and Yonath 1984). Consequently, crystals from 50S subunits from *Halobacterium marismortui*, grown spontaneously under the lowest Mg^{2+} concentration possible, were transferred as seed crystals to solutions with even a lower Mg^{2+} concentration. As a result, after about two weeks, well-ordered and relatively thick crystals of about $0.6 \times 0.6 \times 0.2$ mm were formed (figure 10.25) that diffracted to about 6 Å (Makowski *et al* 1987).

Figure 10.25 Thin plates of *H. marismortui* 50S subunits of the ribosome. From Bartels *et al* (1988) with permission.

10.5.2.2 Data collection

Only from 1985 onwards was crystallographic data collection seriously
considered. The crystals of the 50S subunits of *B. stearothermophilus*
reach a length of 2.0 mm and a cross section of 0.4 mm, and are loosely
packed in an orthorhombic unit cell of 350×670×910 Å (Yonath *et al*
1986). Fresh crystals diffracted to 10–13 Å resolution. However, the
higher resolution reflections decay within 5–10 min. Seeded crystals from
H. marismortui grow in the orthorhombic space group $C222_1$ and diffract
to a resolution of up to 6 Å. They have relatively small, densely packed
unit cells of 215×300×590 Å, in contrast to the 'open' structure and the
large   unit   cells   of   the   crystals   from   50S   subunits   of   *B.
stearothermophilus*. Up to ten photographs could be taken from an
individual crystal between −2 °C and 19 °C but the high resolution reflections appear only on the first 1–3 X-ray photographs. Hence, over 260
crystals had to be irradiated in order to obtain a (supposedly) complete
set of film data.

Crystals were aligned only visually to avoid the loss of precious reflections in the higher resolution range in the course of setting photographs.

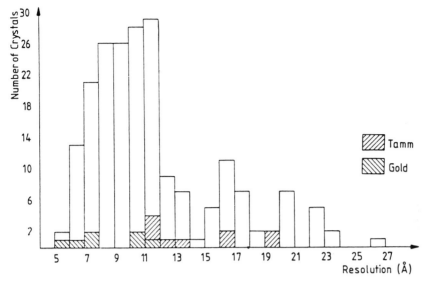

Figure 10.26 Approximate Bragg resolution for the first exposure of each
of about 200 crystals from *H. marismortui* 50S subunits
that were investigated at X11, EMBL/DORIS, Hamburg
(FRG), in August 1986 at −4°C to 19°C. Shading indicates
heavy atom derivative test crystals (undecagold cluster
and tetrakis(acetoxymercuri)-methane (TAMM)). From
Bartels *et al* (1988) with permission.

None of the data films happened to show the orthorhombic symmetry,
thus a preliminary report specified the approximate cell constants for a
primitive monoclinic cell P2$_1$ (Makowski *et al* 1987). Figure 10.26 shows
the statistical distribution of approximate Bragg resolution found for the
first film from each crystal, figure 10.27 displays the quick loss in resolution as a function of the exposure time/film number. In order to average
out the resolution decay during each exposure, the camera rotation axis
was oscillated ten times per photograph.

However, most crystals have a very large mosaic spread of up to 3°
even in the first picture. Hence, there is a severe problem with overlap, in
particular when the long 590 Å axis is in the direction of the X-ray beam.
What is worse, for many crystals there are no fully recorded reflections to
scale the partial intensities ('post-refinement', Rossmann *et al* (1979)).
These problems, taken together with the short lifetime, render even a
joint refinement of all the three axes of individual crystals nearly
impossible.

All this led to the question whether much lower temperatures might

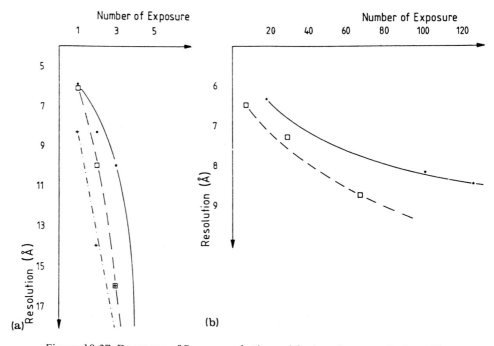

Figure 10.27 Decrease of Bragg resolution with time for crystals from *H.* 50S subunits when irradiated at −4°C to 19°C (*a*) and at cryotemperature (*b*). (Different symbols represent different crystals.) From Bartels *et al* (1988) with permission.

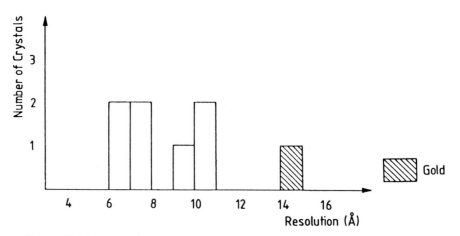

Figure 10.28 Approximate Bragg resolution for the first exposure of each of eight crystals (including one gold derivative) from *H. marismortui* 50S subunits that were investigated at SSRL beam line 7-1 (Stanford University, Ca, USA) at cryotemperature. From Bartels *et al* (1988) with permission.

Figure 10.29 A 1° rotation photograph of *M. marismortui* 50S crystals recorded with SR. From Bartels *et al* (1988) with permission.

help to increase the lifetime of the crystals in the X-ray beam. With this technique (Hope (1988), section 5.5.2) at least some crystals with Bragg resolution of about 6 Å could be transferred to liquid nitrogen temperature and would 'live' in the synchrotron beam for many hours or even days (Bartels *et al* 1988) (figure 10.27b). Figure 10.28 displays the initial Bragg resolution of those crystals that survived the procedure. As might be expected, the mosaic spread did not become smaller in the course of all this treatment, but did not become much worse either.

It was hoped that with cryotemperature methods there would be less dependence on high intensity synchrotron beams. However, neither a relatively weak synchrotron beam nor a rotating anode ever yielded a diffraction pattern comparable in Bragg resolution to figure 10.29.

10.5.2.3 Phase determination

The large size of ribosomal particles is an obstacle for crystallographic studies, but permits direct investigation by electron microscopy. A model (figure 3.20(b)) obtained by three-dimensional image reconstruction of two-dimensional sheets (e.g. Yonath, Leonard and Wittmann (1987)) may be used for gradual phasing of low resolution crystallographic data.

Heavy atom derivitisation of an object as large as a ribosomal particle requires the use of extremely dense and ultra-heavy compounds. Examples of such compounds are (a) tetrakis(acetoxy-mercuri)methane (TAMM) which was the key heavy atom derivative in the structure determination of nucleosomes (Richmond et al 1984) and the membrane reaction centre (Deisenhofer et al 1984), and (b) an undecagold cluster in which the gold core has a diameter of 8.2 Å (figure 10.30 and in Bellon, Manassero and Sansoni (1972) and Wall, Hainfeld, Barlett and Singer (1982)). The cluster compounds, in which all the moieties R are amine

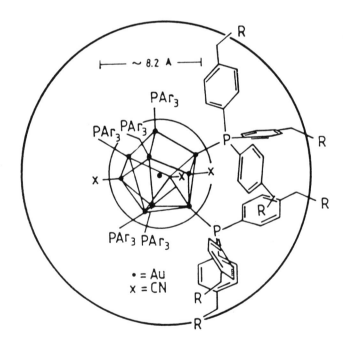

Figure 10.30 Semi-schematic representation of the undecagold cluster depicting the gold core of 8.2 Å diameter, and the arrangement of ligands around it. From Bartels et al (1988) with permission.

or alcohol, are soluble in the crystallisation solution of 50S subunits from *H. marismortui*. Thus, they could be used for soaking. Crystallographic data (to 18 Å resolution) show isomorphous unit cell constants with observable differences in the intensities.

Because surfaces of ribosomal particles have a variety of potential binding sites for such clusters, attempts are in progress to bind heavy atoms covalently to a few specific sites on the ribosomal particles prior to crystallisation. This may be achieved either by direct interaction of a heavy atom cluster with chemically active groups such as –SH or the ends of rRNA (Odom *et al* 1980) on the intact particles, or by covalent attachment of a cluster to natural or tailor-made carriers that bind specifically to ribosomes.

10.5.2.5 Resumé of the ribosome X-ray crystallographic work

Out of 15 forms of three-dimensional crystals from ribosomal particles, some appear suitable for crystallographic data collection when using SR at temperatures between 19 °C and −180 °C (Bartels *et al* 1988): 50S subunits from *H. marismortui*, and from *B. stearothermophilus*, including the –BL11 mutant, and the new crystal forms from *B. stearothermophilus* 50S and *Thermus thermophilus* 30S subunits which have only recently been grown in non-volatile precipitants (Glotz *et al* 1987). All this should eventually lead to a three-dimensional model which (if not at the atomic level) should show molecular details that may assist in the understanding of the interaction of the ribosome with the variety of other components which cooperate in the biosynthetic process.

This is a very important project in structural molecular biology. As was stated at the beginning of this section, ribosome crystallography has been made possible by SR, used in conjunction with freezing of the crystals to preserve their lifetime in the beam at high resolution.

CHAPTER 11

Conclusions and future possibilities

The pace of technological change in the field of macromolecular crystal-lography is quite breathtaking. The pursuits of particle physics have led to particle accelerators tailored for the production of synchrotron X-radiation of incredible intensity, geometric quality and wide tunable range.

The exploitation of this radiation, particularly the brilliance and use of short λ's, has made virus crystal data collection routine from difficult samples; although it is at present necessary to use hundreds of crystals in the gathering of just one data set. Maybe the use of ultra-short wavelength beams (≈ 0.33 Å) from a harmonic of an undulator could be harnessed to improve the lifetime of one such sample sufficiently to give a complete data set. Much larger macromolecular assemblies are currently under study, such as the ribosome, which possess little or no symmetry (unlike viruses) and are therefore more difficult to solve.

The revival of the Laue technique has been made possible by the polychromatic nature of the emitted spectrum and the associated high brightness. The study of time resolved phenomena in biological structures (as well as chemistry and solid state physics) with this method is a major research and development effort that is under way.

The interpretation of the diffuse scattering from macromolecular crystals will provide increasing information on the mobility of macro-molecules.

The more automatic solution of crystal structures using variable or multiple wavelength anomalous scattering measurements is an important development especially when conventional, isomorphous, heavy atom derivatives cannot be made. The phasing power of the anomalous scattering component of specific atoms can be fully optimised with SR.

The technology of SR should not be seen in isolation from other developments. Of particular note are developments in detectors, computing and crystallisation methods. A wide variety of detectors are now available or under development and are supported by increasingly sophisticated computers. The science of crystallisation has been slow to develop but is gaining momentum. Technical developments in this latter area include automatic crystallisation machines and the use of microgravity.

As well as the above there are some further, perhaps more intriguing possibilities for the future. The huge photon fluxes available with the next generation of synchrotron source could be used to compensate for very weak scattering effects. For example, magnetic scattering might be of interest for the study of the role of certain metal ions, such as iron, in biological function. The magnetic scattering cross section is 10^{-5} times or so weaker than the cross section for electric scattering. However, a 50-fold resonant enhancement of the magnetic scattering was discovered in the case of holmium as the incident X-ray energy was tuned through the L_{III} edge of holmium (Gibbs *et al* 1988). Hence, reasonable counting rates might be obtained even with a biological crystal structure, provided that the sample does not suffer marked radiation damage. Freezing of crystal samples is now being increasingly used, although with some increase of mosaicity of the sample, to prolong sample lifetime.

Another area of interest is the use of very fine spectral resolution for the exploitation of nuclear anomalous dispersion (Mossbauer) effects. The very narrow $\delta\lambda/\lambda$ has to be compensated for by a very high incident intensity. The use of the relatively huge iron nuclear anomalous effect ($f''=470e^-$) has been suggested by Mossbauer (1975) as a possible solution to the crystallographic phase problem. For larger macromolecular assemblies this suggestion might become more than just a curiosity by using SR.

The high brilliance at ultra-short wavelengths (0.15 Å) of the higher harmonics from X-ray undulators would allow X-ray diffraction from macromolecular crystals totally free of absorption errors whilst stimulating the K edge anomalous dispersion of high atomic number elements (e.g. mercury, platinum, etc).

Alternatively, native protein crystal data sets measured at wavelengths as short as 0.33 Å (i.e. utilising the K edge of the barium in an image plate as detector) would also be free of absorption errors and with greatly reduced random errors. This latter arises due to the ability to have prolonged exposure times and repeated measurements before the protein crystal suffers radiation damage. Hence, unprecedented data quality and improvements to atomic resolution of the diffraction data for protein crystals might be combined with statistical (direct) methods of the kind used routinely with small molecule crystals. Such possibilities would allow a greatly increased rate of solving protein crystal structures.

One basic problem of utilising SR is that the facility is a centralised one which is always less convenient than operating in the home laboratory. However, there is a steady increase in magnetic field strengths available. Hence, more modest machine energies coupled with high field wigglers might allow synchrotrons to become more widespread, even at the home university. The so-called table-top machine is now available to industry for soft X-ray lithography. A polychromatic hard X-ray, compact, machine for the home laboratory would be attractive for the more routine uses of SR as well as the exploitation of anomalous scattering or time resolved Laue techniques in macromolecular crystallography.

SR is now an essential tool of macromolecular crystal structure analysis. There are many fascinating possibilities for future technical development. The methods and procedures that have already been developed to exploit SR are being harnessed in a huge number of crystallographic analyses. The use of the synchrotron is now a regular feature of the way of life of the macromolecular crystallographer, from academia or industry.

Summary of various monochromatic diffraction geometries

This appendix is based, with permission, on my sections in the new *International Tables for Crystallography*, Volume C (editor A. J. C. Wilson, 1991).

There are text-books which concentrate on almost every diffraction geometry. References to these books are given in the respective sections in the following pages. However, in addition, there are several books which contain details of diffraction geometry. Blundell and Johnson (1976) described the use of the various diffraction geometries with the examples taken from protein crystallography. There is an extensive discussion and many practical details to be found in the text-books of Stout and Jensen (1968, 1989), Woolfson (1970), Glusker and Trueblood (1971, 1985), Vainshtein (1981) and McKie and McKie (1986), for example. A collection of early papers on the diffraction of X-rays by crystals involving, *inter alia*, experimental techniques and diffraction geometry, can be found in Bijvoet, Burgers and Hägg (1969, 1972). A collection of recent papers on primarily protein and virus crystal data collection via the rotation film method and diffractometry can be found in Wycoff, Hirs and Timasheff (1985); detailed references are also made to this volume later.

In this appendix which deals with monochromatic methods, the convention is adopted that the Ewald sphere takes a radius of unity and the magnitude of the reciprocal lattice vector is λ/d. This is not the convention used in chapter 7 on the Laue method. Example photographs for each monochromatic geometry are shown in plate A1. These geometries are now discussed in detail.

Plate AI Diffraction photographs recorded in different geometries.

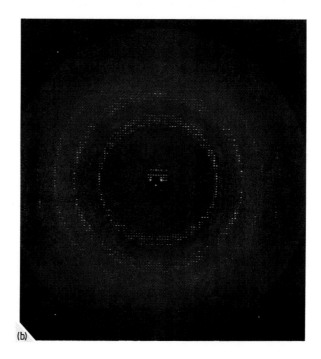

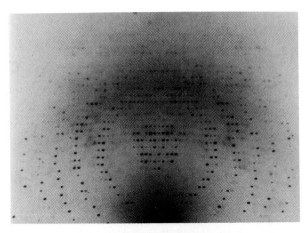

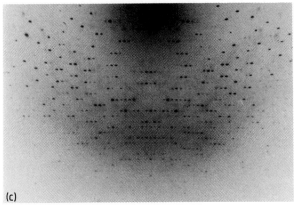

(c)

(a) Monochromatic still exposure, flat film perpendicular to X-ray beam; in this example X-ray beam parallel to a zone axis, **c**. Sample: concanavalin-A with lyxose. Space group $P2_12_12_1$, cell parameters a=121.8 Å, b=129.0 Å, c=67.3 Å. Daresbury SRS station 9.6, wavelength=0.9 Å, crystal to (flat) film distance=139.9 mm, radius of film=59 mm.

(b) Rotation/oscillation geometry for flat film, with conditions as in (a) except rotation interval, 1.25°, and also there is a change in the absolute angle between exposures for the crystal of approximately 6°. Spots visible to a resolution limit, at the edges of the pattern, of 2.4 Å.

(c) Rotation/oscillation geometry for vee cassette (α=60° in equations (A1.17) and (A1.18)). Sample: 6-phospho-gluconate dehydrogenase. Space group $C222_1$, cell parameters a=72.7 Å, b=148.2 Å, c=102.9 Å. Daresbury SRS station 7.2, λ=1.488 Å. Horizontal axis of rotation **c**. Rotation interval 2°. Resolution limit at top and bottom edges of film 1.9 Å.

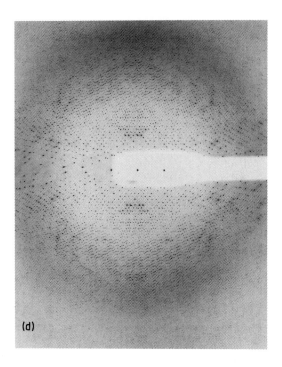

(d) Weissenberg photograph of ω-amino acid: pyruvate
aminotransferase from *Pseudomonas* sp. F-126 on IP. A
0.3×0.3 mm square collimator was used, and the radius
of the cylindrical cassette used was 287 mm. The
oscillation axis was the **c** axis ($4°$ mm^{-1}), and the
oscillation angle was $18°$. μ was set at $0°$. It took 70 s for
each oscillation, and the exposure time was 285 s. The
wavelength was 1.022 Å. The Photon Factory was
operated at 2.5 GeV and 118.5 mA. Space group I222.
$a=124.67$ Å, $b=137.9$ Å, $c=61.45$ Å, 172 000/a tetramer.
43 000/an asymmetric unit. This photograph and caption
kindly supplied by Professor N. Sakabe, with permission.

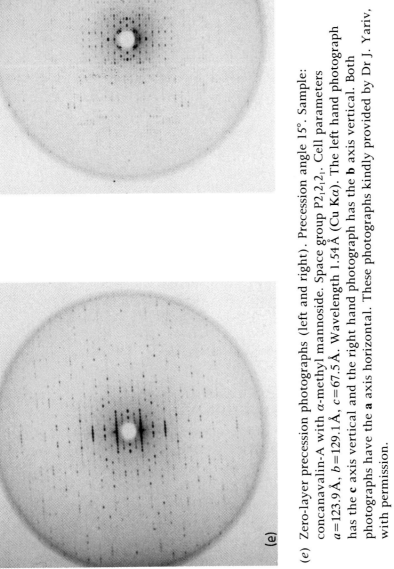

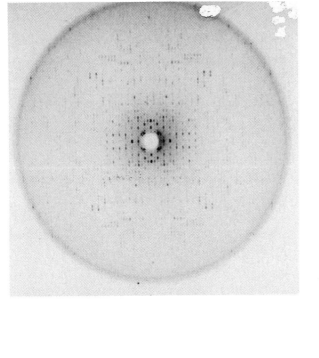

(e) Zero-layer precession photographs (left and right). Precession angle 15°. Sample: concanavalin-A with α-methyl mannoside. Space group P2₁2₁2₁. Cell parameters $a=123.9\,\text{Å}$, $b=129.1\,\text{Å}$, $c=67.5\,\text{Å}$. Wavelength 1.54 Å (Cu Kα). The left hand photograph has the **c** axis vertical and the right hand photograph has the **b** axis vertical. Both photographs have the **a** axis horizontal. These photographs kindly provided by Dr J. Yariv, with permission.

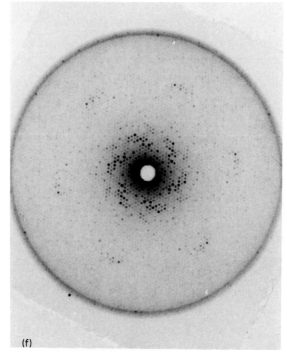

(*f*) Zero-layer precession photograph. Precession angle 15°.
Sample: concanavalin-A with α-methyl glucoside. Space
group I2₁3, cubic cell parameters $a=167.8$ Å. View down
the (111) direction – isn't it pretty!? This photograph
kindly provided by Dr J. Yariv, with permission.

A1.1 MONOCHROMATIC STILL EXPOSURE

In a monochromatic still exposure the crystal is held stationary and a
near-zero wavelength bandpass (e.g. $\delta\lambda/\lambda=0.001$) beam impinges on it.
For a small molecule crystal there are few diffraction spots. For a protein
crystal there are many (several hundred) which is due to the much more
dense reciprocal lattice. The actual number of stimulated RLPs depends
on the reciprocal cell parameters, the size of the mosaic spread of the
crystal, the angular beam divergence as well as the small, but finite,
spectral spread, $\delta\lambda/\lambda$. Diffraction spots are only partially stimulated
instead of fully integrated over wavelength as for the Laue method or
over an angular rotation (the rocking width) in moving crystal monochromatic methods.

The diffraction spots lie on curved arcs where each curve corresponds
to the intersection with a film of a cone. With a flat film the intersections

are conic sections. Monochromatic stills are used to search for a zone axis following the procedure given later in section A1.4.1. When viewed down a zone axis the pattern on a flat film or EAD has the appearance of a series of concentric circles. For example, with the beam parallel to [$00\bar{1}$] the first circle corresponds to $\ell=1$, the second to $\ell=2$, etc. The radius of the first circle R is related to the interplanar spacing between the ($hk0$) and ($hk1$) planes, i.e. λ/c (in this example), through θ, by the formulae

$$\tan 2\theta = R/D; \quad \cos 2\theta = (1 - \lambda/c) \tag{A1.1}$$

A1.2 ROTATION/OSCILLATION GEOMETRY

The main book dealing with the rotation method is that of Arndt and Wonacott (1977). The method was developed primarily by Bernal (see figures 8.2 (p. 249) and 8.8 (p. 256) of McKie and McKie (1986)) and revived for macromolecular crystallography by Arndt, Champness, Phizackerley and Wonacott (1973). The purpose of the rotation method is to stimulate a reflection fully over its rocking width via an angular rotation. Different RLPs are brought successively into the reflecting position by a larger angular rotation. The method, therefore, involves rotation of the sample about a single axis and is used in conjunction with an area detector of some sort, e.g. film, EAD or IP. The use of a repeated rotation, i.e. an oscillation, for a given exposure, is simply to average out any time-dependent changes in incident intensity or sample decay. The overall crystal rotation required to record the total accessible region of reciprocal space for a single crystal setting, and a detector picking up all the diffraction spots, is $180°+2\theta_{max}$. If the crystal has additional symmetry, then a complete asymmetric unit of reciprocal space can be recorded within a smaller angle. There is a small blind region close to the rotation axis, this is detailed in section A1.2.4.

As far as the geometry is concerned we build on the monochromatic still geometry as outlined in the previous section.

A1.2.1 Diffraction coordinates

Figures A1.1(*a*)–(*d*) are taken from *International Tables*, Volume II, p. 176. They neatly summarise the geometrical principles of reflection of a monochromatic beam in the reciprocal lattice for the general case of an incident beam inclined at an angle (μ) to the equatorial plane. The diagrams are based on an Ewald sphere of unit radius.

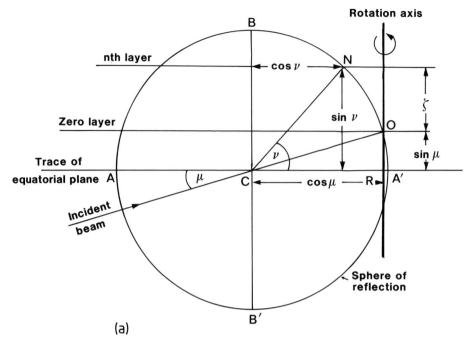

(a)

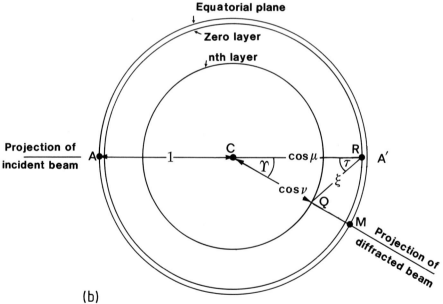

(b)

Figure A1.1 (a) Elevation of the sphere of reflection. O is the origin of reciprocal lattice. C is the centre of the Ewald sphere. The incident beam is shown in the plane. (b) Plan of the sphere of reflection. R is the projection of the rotation axis on the equatorial plane. (c) Perspective diagram. BB' is the direction of the rotation axis through the crystal sample. P is the

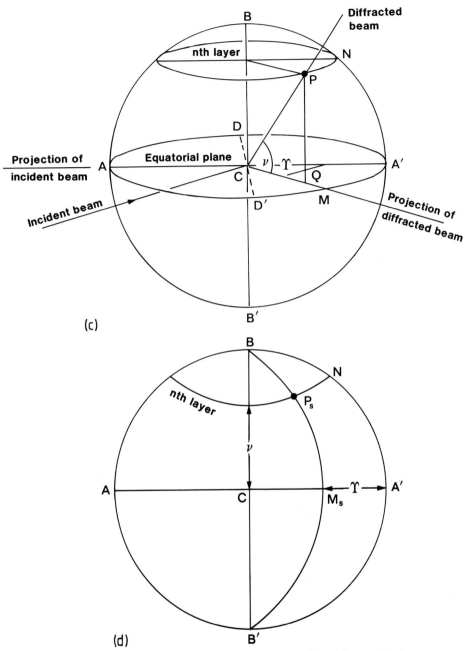

(c)

(d)

reciprocal lattice point in the reflection position with the cylindrical coordinates ζ, ξ, τ. The angular coordinates of the diffracted beam are ν, υ. (d) Stereogram to show the direction of the diffracted beam, ν, Y with DD' normal to the incident beam and in the equatorial plane, as the projection diameter. M_s and P_s are the projection points for M and P from (c). From Evans and Lonsdale (1959).

Using the nomenclature of table A1.1, figure A1.1(a) gives

$$\sin v = \sin \mu + \zeta \tag{A1.2}$$

figure A1.1(b) gives by the cosine rule

$$\cos \Upsilon = \frac{\cos^2 v + \cos^2 \mu - \xi^2}{2 \cos v \cos \mu} \tag{A1.3}$$

Table A1.1. *Glossary of symbols used to specify quantities on diffraction patterns and in reciprocal space.*

θ	Bragg angle
2θ	Angle of deviation of the reflected beam with respect to the incident beam
$\hat{\mathbf{S}}_0$	Unit vector lying along the direction of the incident beam
$\hat{\mathbf{S}}$	Unit vector lying along the direction of the reflected beam
$\mathbf{s} = \hat{\mathbf{S}} - \hat{\mathbf{S}}_0$	The scattering vector $\mathbf{s}$ of magnitude $2 \sin \theta$ is perpendicular to the bisector of the angle between $\hat{\mathbf{S}}_0$ and $\hat{\mathbf{S}}$. Also $\mathbf{s}$ is identical to the reciprocal lattice vector $\mathbf{d}^*$ of magnitude λ/d, where d is the interplanar spacing, when $\mathbf{d}^*$ is in the diffraction condition. In this notation the radius of the Ewald sphere is unity. This convention is adopted because it then follows the same as in Volume II of *International Tables* p. 175. Note that in chapter 7 'Laue' the alternative convention ($|\mathbf{d}^*| = 1/d$) is adopted whereby the radius of each Ewald sphere is $1/\lambda$. This allows a nest of Ewald spheres between $1/\lambda_{max}$ and $1/\lambda_{min}$ to be drawn.
ζ	Coordinate of a point P in reciprocal space parallel to a rotation axis as the axis of cylindrical coordinates relative to the origin of reciprocal space.
ξ	Radial coordinate of a point P in reciprocal space; that is, the radius of a cylinder having the rotation axis as axis.
τ	The angular coordinate of P, measured as the angle between ξ and $\hat{\mathbf{S}}_0$ (see figure A1.1(b)).
ϕ	Is the angle or rotation from a defined datum orientation to bring an RLP onto the Ewald sphere in the rotation method (see figure A1.3).
μ	The angle of inclination of $\hat{\mathbf{S}}_0$ to the equatorial plane
Υ	The angle between the projections of $\hat{\mathbf{S}}_0$ and $\hat{\mathbf{S}}$ onto the equatorial plane
v	The angle of inclination of $\hat{\mathbf{S}}$ to the equatorial plane
ω, χ, ϕ	These are the crystal setting angles on the four-circle diffractometer (see figure A1.6). The ϕ used here is not the same as that in the rotation method (figure A1.3). This clash in using the same symbol twice is inevitable because of the widespread use of the rotation camera and four-circle diffractometer.

Note:

(1) The equatorial plane is the plane normal to the rotation axis.

and

$$\cos \tau = \frac{\cos^2 \mu + \xi^2 - \cos^2 v}{2\xi \cos \mu} \tag{A1.4}$$

and figures A1.1(a) and (b) give

$$\xi^2 + \zeta^2 = d^{*2} = 4\sin^2 \theta \tag{A1.5}$$

The following special cases commonly occur
(a) $\mu=0$, normal-beam rotation method, then

$$\sin v = \zeta \tag{A1.6}$$

and

$$\cos \Upsilon = (2 - \xi^2 - \zeta^2)/2 (1-\zeta^2)^{\frac{1}{2}} \tag{A1.7}$$

(b) $\mu=-v$, equi-inclination (relevant to Weissenberg upper layer photography) then,

$$\zeta = -2\sin \mu = 2\sin v \tag{A1.8}$$

$$\cos \Upsilon = 1 - \xi^2 / 2\cos^2 v \tag{A1.9}$$

(c) $\mu=+v$, anti-equi-inclination

$$\zeta = 0 \tag{A1.10}$$

$$\cos \Upsilon = 1 - \xi^2 / 2\cos^2 v \tag{A1.11}$$

(d) $v=0$, flat cone

$$\zeta = -\sin \mu \tag{A1.12}$$

$$\cos \Upsilon = \frac{2 - \xi^2 - \zeta^2}{2(1-\zeta^2)^{\frac{1}{2}}} \tag{A1.13}$$

In this section we will now concentrate on case (a), the normal-beam rotation method ($\mu=0$). Firstly, the case of a plane film is considered. The notation now follows that of Arndt and Wonacott (1977) for the coordinates of a spot on the film or detector. Z_F is parallel to the rotation axis and ζ; Y_F is perpendicular to the rotation axis and the beam. *International Tables*, Volume II (1959), p. 177, follows the convention of y being parallel and x perpendicular to the rotation axis direction, i.e. $(Y_F, Z_F) \equiv (x, y)$. The advantage of the (Y_F, Z_F) notation is that the x-axis direction is then the same as the X-ray beam direction.

The coordinates of a reflection on a flat film (Y_F, Z_F) are related to the cylindrical coordinates of an RLP (ξ, ζ) (figure A1.2(a)), by

$$Y_F = D \tan \Upsilon \qquad (A1.14)$$

$$Z_F = D \sec \Upsilon \tan v \qquad (A1.15)$$

which becomes

$$Z_F = 2D\zeta / (2 - \xi^2 - \zeta^2) \qquad (A1.16)$$

where D is the crystal to film distance.

For the case of a vee-shaped cassette with the vee axis parallel to the rotation axis and the film making an angle α to the beam direction (figure A1.2(b)). Then

$$Y_F = D \tan \Upsilon / (\sin \alpha + \cos \alpha \tan \Upsilon) \qquad (A1.17)$$

$$Z_F = (D - Y_F \cos \alpha) \; \zeta / (1 - d^{*2} / 2) \qquad (A1.18)$$

This situation also corresponds to the case of a flat EAD inclined to the incident beam in a similar way.

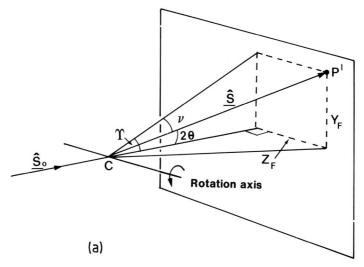

(a)

Figure A1.2 Geometrical principles of recording the pattern on (a) plane detector, (b) vee-shaped detector, (c) cylindrical

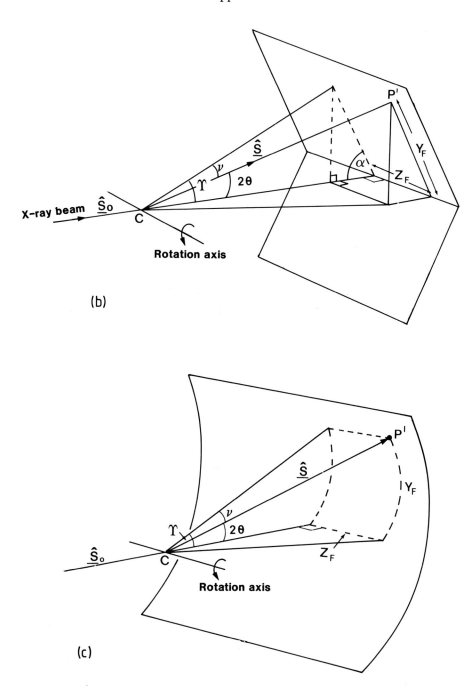

(b)

(c)

detector. P' is the position of a spot on the detector corresponding to an RLP in the diffracting condition.

For the case of a cylindrical film or IP where the axis of the cylinder is coincident with the rotation axis (figure A1.2(c)) then, for Υ in degrees,

$$Y_F = \frac{2\pi}{360} D \Upsilon \tag{A1.19}$$

$$Z_F = D \tan v \tag{A1.20}$$

which becomes

$$Z_F = \frac{D\zeta}{(1 - \zeta^2)^{\frac{1}{2}}} \tag{A1.21}$$

Here, D is then the radius of curvature of the cylinder assuming that the crystal is at the centre of curvature.

In the three geometries mentioned here detector misalignment errors have to be considered. These are three orthogonal angular errors, translation of the origin and error in the crystal to film distance.

The coordinates Y_F and Z_F are related to film scanner raster units via a scanner rotation matrix and translation vector. This is necessary because the film is placed arbitrarily on the scanner drum. Details can be found in Rossman (1985) or Arndt and Wonacott (1977).

A1.2.2 Relationship of diffraction coordinates to crystal system parameters

The reciprocal lattice coordinates, ζ, ξ, Υ, v, etc, used earlier refer to an axial system which rotates with the crystal, i.e. X_0, Y_0, Z_0 of figure A1.3. Clearly, a given RLP needs to be brought onto the Ewald sphere by the rotation about the rotation axis. We now follow the treatment of Arndt and Wonacott (1977).

The rotation angle required, ϕ, is with respect to some reference 'zero angle' direction and is determined by the particular crystal parameters. It is necessary to define a standard orientation of the crystal (i.e. a datum) when $\phi=0°$. If we define an axial system (X_0, Y_0, Z_0) which rotates with the crystal and a laboratory axial system (X, Y, Z) with X parallel to the beam and Z coincident with the rotation axis, then $\phi=0°$ corresponds to these axial systems being coincident (figure A1.3).

The angle of the crystal at which a given RLP diffracts is

$$\tan(\phi / 2) = \frac{2y_0 \pm (4y_0^2 + 4x_0^2 - d^{*4})^{\frac{1}{2}}}{(d^{*2} - 2x_0)} \tag{A1.22}$$

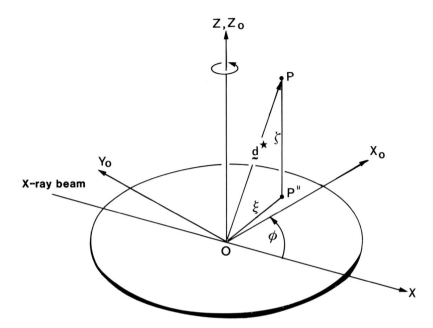

Figure A1.3 The rotation method. Definition of coordinate systems. Cylindrical coordinates of an RLP P (ξ,ζ,τ) are defined relative to axial system (X_0,Y_0,Z_0) which rotates with the crystal. The axial system (X,Y,Z) is defined such that X is parallel to the incident beam and Z is coincident with the rotation axis. From Arndt and Wonacott (1977).

The two solutions correspond to the two rotation angles at which the RLP P cuts the sphere of reflection. Note that Y_F, Z_F (section A1.2.1) are independent of ϕ.

The values of x_0 and y_0 are calculated from the particular crystal system parameters. The relationships between the coordinates x_0, y_0, z_0 and ξ and ζ are

$$\xi = (x_0^2 + y_0^2)^{\frac{1}{2}} \tag{A1.23}$$

$$\zeta = z_0 \tag{A1.24}$$

$\mathbf{X}_0$ can be related to the crystal parameters by

$$\mathbf{X}_0 = \mathbf{Ah} \tag{A1.25}$$

$\mathbf{A}$ is a crystal-orientation matrix defining the standard datum orientation of the crystal.

For example, if, by convention, $\mathbf{a}^*$ is chosen as parallel to the X-ray beam at $\phi=0$ and $\mathbf{c}$ is chosen as the rotation axis, then for the triclinic, most general, case

$$\mathbf{A} = \begin{pmatrix} a^* & b^* \cos \gamma^* & c^* \cos \beta^* \\ 0 & b^* \sin \gamma^* & -c^* \sin \beta^* \cos \alpha \\ 0 & 0 & \lambda / c \end{pmatrix} \tag{A1.26}$$

If the crystal is mounted on the goniometer head differently from this, then $\mathbf{A}$ can be modified by another matrix, $\mathbf{M}$, say, or the terms permuted. This exercise becomes clear if the reader takes an orthogonal case. For the general case see Higashi (1989).

The crystal will probably be misaligned (slightly or grossly) from the ideal orientation. To correct for this the mis-orientation matrices Φ_x, Φ_y and Φ_z are introduced, i.e.

$$\Phi_x = \begin{pmatrix} 1 & 0 & 0 \\ 0 & \cos \Delta\phi_x & -\sin \Delta\phi_x \\ 0 & \sin \Delta\phi_x & \cos \Delta\phi_x \end{pmatrix} \tag{A1.27}$$

$$\Phi_y = \begin{pmatrix} \cos \Delta\phi_y & 0 & \sin \Delta\phi_y \\ 0 & 1 & 0 \\ -\sin \Delta\phi_y & 0 & \cos \Delta\phi_y \end{pmatrix} \tag{A1.28}$$

$$\Phi_z = \begin{pmatrix} \cos \Delta\phi_z & -\sin \Delta\phi_z & 0 \\ \sin \Delta\phi_z & \cos \Delta\phi_z & 0 \\ 0 & 0 & 1 \end{pmatrix} \tag{A1.29}$$

where $\Delta\phi_x$, $\Delta\phi_y$ and $\Delta\phi_z$ are angles around the X_0, Y_0 and Z_0 axes, respectively.

Hence, the relationship between $\mathbf{X}_0$ and $\mathbf{h}$ is

$$\mathbf{X}_0 = \Phi_z \Phi_y \Phi_x \mathbf{M} \mathbf{A} \mathbf{h} \tag{A1.30}$$

A1.2.3 Maximum oscillation angle without spot overlap

For a given oscillation photograph there is a maximum value of the oscillation range, $\Delta\phi$, which avoids overlapping of spots on a film. The overlap will occur in the region of the diffraction pattern perpendicular to the rotation axis and at the maximum Bragg angle. This is where RLPs pass through the Ewald sphere with the greatest velocity. For a separation between successive RLPs of a^*, then the maximum allowable rotation angle to avoid spatial overlap is given by

$$\Delta\phi_{max} = \left(\frac{a^*}{d^*_{max}} - \phi_R \right) \qquad (A1.31)$$

where ϕ_R is the sample reflecting range (equation (6.5)). $\Delta\phi_{max}$ is a function of ϕ, even in the case of identical cell parameters. This is because it is necessary to consider, for a given orientation, the relevant reciprocal lattice vector perpendicular to d^*_{max}. In the case where the cell dimensions are quite different in magnitude (excluding the axis parallel to the rotation axis), then $\Delta\phi_{max}$ is a marked function of the orientation.

In rotation photography as large an angle as possible is used up to $\Delta\phi_{max}$. This reduces the number of films that need to be processed and the number of partially stimulated reflections per film but at the expense of signal-to-noise ratio for individual spots, which accumulate more background since $\phi_R < \Delta\phi_{max}$. In the case of a detector system $\Delta\phi$ is chosen usually to be less than ϕ_R so as to optimise the signal to noise of the measurement and to sample the rocking width profile at several points.

The value of ϕ_R, the crystal rocking width for a given $hk\ell$, depends on the reciprocal lattice coordinates of the $hk\ell$ RLP (see chapter 6). In the region close to the rotation axis ϕ_R is large. Any reflection with $\phi_R > \Delta\phi_{max}$ is defined to lie in the blind region.

A1.2.4 Blind region

In the normal-beam geometry any RLP lying close to the rotation axis will not be stimulated at all. This volume is shown in figure A1.4. It has a radius of

$$\xi_{min} = d^*_{max} \sin\theta_{max} = \lambda^2 / 2d^2_{min} \qquad (A1.32)$$

and is therefore strongly dependent on d_{min} but can be ameliorated by use of a short λ. Shorter λ makes the Ewald sphere have a larger radius, i.e. its surface moves closer to the rotation axis. At $Cu K\alpha$ for 2 Å resolution approximately 5% of the data lie in the blind region according to this simple geometrical model. However, taking account of the rocking width, ϕ_R, a greater percentage of the data than this is not fully sampled except over very large angular ranges. The actual increase in the blind

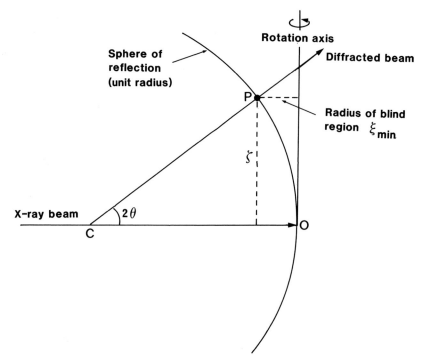

Figure A1.4 The rotation method. The blind region associated with a
single rotation axis. From Arndt and Wonacott (1977).

region volume due to this effect is minimised by use of a collimated beam
and a narrow spectral spread (i.e. finely monochromatised SR) and if the
crystal is not too mosaic.

These effects are directly related to the Lorentz factor,

$$L = 1/(\sin^2 2\theta - \zeta^2)^{\frac{1}{2}}$$
(A1.33)

It is inadvisable to measure a reflection intensity when L is large because
different parts of a spot would need a different Lorentz factor.

The blind region can be filled in by a rotation about another axis. The
total angular range that is needed to sample the blind region is $2\theta_{max}$ in
the absence of any symmetry or θ_{max} in the case of mm symmetry (for
example).

A1.3 WEISSENBERG GEOMETRY

Weissenberg geometry (Weissenberg 1924) is dealt with in the books by
Buerger (1942) and Woolfson (1970) for example. This method is not

widely used now at home. In small molecule crystallography quantitative data collection is usually done via a diffractometer. Weissenberg geometry is being revived, however, as a method for macromolecular data collection with monochromatised SR and the IP as detector (Sakabe (1983) and, e.g., Watanabe *et al* (1989)). Here the method is used without a layer-line screen where the total rotation angle used is limited to about 15°; this is a significant increase over the rotation method with a stationary film. The use of this effectively avoids the presence of partial reflections and reduces the total number of exposures required. Provided the Weissenberg camera has a large radius the X-ray background accumulated over a single spot is actually not serious. This is because the X-ray background decreases approximately according to an inverse square law of distance from the crystal to the detector.

The following sections A1.3.1 and A1.3.2 describe the standard situation where a layer-line screen is used.

A1.3.1 Recording of zero layer

Normal-beam geometry (i.e. the X-ray beam being perpendicular to the rotation axis) is used to record zero-layer photographs. The film is held in a cylindrical cassette coaxial with the rotation axis. The centre of the gap in a screen is set to coincide with the zero-layer plane. The coordinate of a spot on the film measured parallel (Z_F) and perpendicular (Y_F) to the rotation axis is given by

$$Y_F = \frac{2\pi}{360} D \Upsilon \qquad\qquad (A1.34)$$

$$Z_F = \phi / f \qquad\qquad (A1.35)$$

where ϕ is the rotation angle of the crystal from its initial setting, f is the coupling constant, which is the ratio of the crystal rotation angle divided by the film cassette translation distance, in $\deg\,mm^{-1}$, and D is the camera radius. Generally, the values of f and D are $2\,\deg\,mm^{-1}$ and $28.65\,mm$, respectively, with Mo Kα.

A1.3.2 Recording of upper layers

Upper-layer photographs are usually recorded in equi-inclination geometry (i.e. $\mu=-\nu$, see equations (A1.8) and (A1.9)). The X-ray beam direction is made coincident with the generator of the cone of the diffracted beam for the layer concerned, so that the incident and diffracted beams make equal angles (μ) with the equatorial plane where

$$\mu = \sin^{-1}(\zeta_n / 2) \tag{A1.36}$$

The screen has to be moved by an amount

$$s \tan \mu \tag{A1.37}$$

where s is the screen radius. If the cassette is held in the same position as the zero-layer photograph, then reflections produced by the same orientation of the crystal will be displaced by

$$D \tan \mu \tag{A1.38}$$

relative to the zero-layer photograph. This effect can be eliminated by initial translation of the cassette by $D \tan \mu$.

A1.4 PRECESSION GEOMETRY

The main book dealing with the precession method is that of Buerger (1964). The precession method is used to record an undistorted representation of a single plane of RLPs and their associated intensities. In order to achieve this the crystal is carefully set so that the plane of the RLPs is perpendicular to the X-ray beam. The normal to this plane, the zone axis, is then precessed about the X-ray beam axis. A layer-line screen with a transparent annulus allows RLPs of the plane of interest to pass through to the film. The screen intercepts all other diffracted rays. The motion of the crystal, screen and film are coupled together to maintain the coplanarity of the film, the screen and the zone.

A1.4.1 Crystal setting

Setting of the crystal for one zone is done in two stages. Firstly, a Laue photograph is used for small molecules or a monochromatic still for macromolecules to stimulate enough spots to identify the required zone axis and place it parallel to the X-ray beam. This is done by adjustment to the camera spindle angle and the goniometer head arc in the horizontal plane. This procedure is usually accurate to a degree or so. Note that the vertical arc will only rotate the pattern around the X-ray beam. Secondly, a screenless precession photograph is taken using an angle of

≈7–10° for small molecules or 2–3° for macromolecules. It is better to use unfiltered radiation and then the edge of the zero-layer circle is easily visible. Let the difference of the distances from the centre of the pattern to the opposite edges of the trace in the direction of displacement be called $\delta = D\Delta$ so that for the horizontal goniometer head arc and the dial: $\delta_{arc} = x_{Rt} - x_{Lt}$ and $\delta_{dial} = y_{Up} - y_{Dn}$ (figure A1.5). The corrections ε to the arc and camera spindle are given by

$$\Delta = \frac{\delta}{D} = \frac{\sin 4\varepsilon \cos \bar{\mu}}{\cos^2 2\varepsilon - \sin^2 \bar{\mu}} \qquad (A1.39)$$

in rlu where D is the crystal to film distance and $\bar{\mu}$ is the precession angle.

It is possible to measure δ to about 0.3 mm ($\delta = 1$ mm corresponds to 14′ arc error for $D = 60$ mm and $\bar{\mu} \approx 7°$, see table A1.2). (Table based on *International Tables*, Volume II, p. 200.)

A1.4.2 Recording of zero-layer photograph

Before the zero-layer photograph is taken a niobium filter (for Mo Kα) or a nickel filter (for Cu Kα) is introduced into the X-ray beam path and a screen is placed between the crystal and the film at a distance from the crystal of

$$s = r_s \cot \bar{\mu} \qquad (A1.40)$$

where r_s is the screen radius. Typical values of $\bar{\mu}$ would be 20° for a small molecule with Mo Kα and 12–15° for a protein with Cu Kα. The annulus

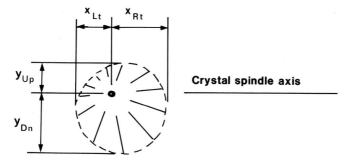

Figure A1.5 The screenless precession setting photograph (schematic) and associated mis-setting distances for a typical orientation error when the crystal has been set previously by a monochromatic still or Laue.

Table A1.2. *The distance displacement (in mm) measured on the film versus angular setting error of the crystal for a screenless precession ($\bar{\mu}=5°$) setting photograph.*

Angular correction, ε, in degrees and minutes of arc	Δ r.l.u.	Distance displacement (mm) for three crystal to film distances		
		60 mm	75 mm	100 mm
0	0	0	0	0
15'	0.0175	1.1	1.3	1.8
30'	0.035	2.1	2.6	3.5
45'	0.0526	3.2	4.0	5.3
60'	0.070	4.2	5.3	7.0
1°15'	0.087	5.2	6.5	8.7
1°30'	0.105	6.3	7.9	10.5
1°45'	0.123	7.4	9.2	12.3
2°	0.140	8.4	10.5	14.0

Notes:
(1) A value of $\bar{\mu}$ of 5° is assumed although there is a negligible variation in ε with $\bar{\mu}$ between 3° (typical for proteins) and 7° (typical for small molecules).
(2) Crystal to film distances on a precession camera are usually settable at the fixed distances $D=60$, 75, and 100 mm.
(3) This table should be used in conjunction with Fig. A1.5.
(4) Values of ε are given in intervals of 15' as this is convenient for various goniometer heads which usually have verniers in 5', 6' or 10' of arc units. The vernier on the spindle of the precession camera is often in 2' of arc units.
(5) Alternatively $\Delta=\delta/D\simeq\sin 4\varepsilon$ can be used if ε and $\bar{\mu}$ are small (from equation (A1.39)).

width in the screen is chosen usually as 2–3 mm for a small molecule and 1–2 mm for a macromolecule. A clutch slip allows the camera motor to be disengaged and the precession motion can be executed under hand control to check for fouling of the goniometer head, crystal, screen or film cassette; s and r_s need to be selected obviously to avoid this happening. The zero-layer precession photograph produced has a radius of

$$2D\sin\bar{\mu} \tag{A1.41}$$

corresponding to a resolution limit

$$d_{min} = \lambda / 2 \sin\bar{\mu} \tag{A1.42}$$

The distance between spots A is related to the reciprocal cell parameter a^* by the formula

$$a^* = A / D \tag{A1.43}$$

A1.4.3 Recording of upper-layer photographs

The recording of upper-layer photographs involves isolating the net of RLPs at a distance from the zero layer of $\zeta_n = n\lambda/b$ where b is the case of the **b**-axis being antiparallel to the X-ray beam. In order to determine ζ_n it is generally necessary to record a cone-axis photograph. If the cell parameters are known, then the camera settings for the upper-level photograph can be calculated directly without the need for a cone-axis photograph.

In the upper-layer precession photograph the film is advanced towards the crystal by a distance $D\zeta_n$ and the screen is placed at a distance

$$s_n = r_s \cot \bar{v}_n = r_s \cot \cos^{-1}(\cos \bar{\mu} - \zeta_n) \qquad (A1.44)$$

The resulting upper-layer photograph has outer radius

$$D(\sin \bar{v}_n + \sin \bar{\mu}) \qquad (A1.45)$$

and an inner blind region of radius

$$D(\sin \bar{v}_n - \sin \bar{\mu}) \qquad (A1.46)$$

A1.4.4 Recording of cone-axis photograph

A cone-axis photograph is recorded by placing a film enclosed in a light-tight envelope in the screen holder and using a small precession angle, e.g. 5° for a small molecule or 1° for a protein. The photograph has the appearance of concentric circles centred on the origin of reciprocal space, provided the crystal is perfectly aligned. The radius of each circle

$$r_n = s \tan \bar{v}_n \qquad (A1.47)$$

where

$$\cos \bar{v}_n = \cos \bar{\mu} - \zeta_n \qquad (A1.48)$$

Hence,

$$\zeta_n = \cos \bar{\mu} - \cos \tan^{-1}(r_n / s) \qquad (A1.49)$$

A1.5 DIFFRACTOMETRY

The main book dealing with single crystal diffractometry is that of Arndt and Willis (1966). For a detailed treatment of angle settings for four-circle diffractometers the reader is referred to Hamilton (1974). For up-to-date details of area detector diffractometry see Howard, Nielsen and Xuong (1985) and Hamlin (1985). In this section we will describe the following related diffractometer configurations:

(a) normal-beam equatorial geometry (ω, χ, ϕ option or ω, κ, ϕ (Kappa) option);

(b) fixed $\chi=45°$ geometry with area detector.

(a) is used with single counter detectors. The Kappa option is also used in the television area detector system of Enraf–Nonius (the FAST). (b) is used with the multiwire proportional chamber, XENTRONICS system. (FAST is a trade name of Enraf–Nonius; XENTRONICS is a trade name of Siemens/Nicolet.)

The purpose of the diffractometer goniostat is to bring a selected reflected beam into the detector aperture (see figure A1.6) or a number of reflected beams onto an area detector of limited aperture (i.e. an aperture which does not intercept all the available diffraction spots at one setting of the area detector.

The single counter diffractometer is primarily used for small molecule crystallography. In macromolecular crystallography many RLPs are simultaneously in the diffraction condition. The single counter diffractometer was extended to five separate counters (for a review see Artymiuk and Phillips (1985)) then, subsequently, a multielement linear detector (for a review see Wlodawer (1985)). EADs offer a larger aperture for simultaneous acquisition of reflections (Hamlin et al (1981) and references cited therein). Unfortunately, their aperture is still rarely adequate to measure all the stimulated RLPs in a single sweep of the sample. The increased scanning speed of modern single counter diffractometers and their use with rotating anode/high power X-ray sources or SR is leading to a revival of their use for macromolecular crystallography (unit cell edges up to 100 Å); with such a machine the unique set of RLPs can be measured specifically and very accurately; auto-indexing makes sample pre-alignment unnecessary. The rather limited aperture of EADs is a disadvantage but this is being improved by the introduction of commercial IP area detectors with a much larger aperture and greater number of true resolution elements than *electronic* area detectors.

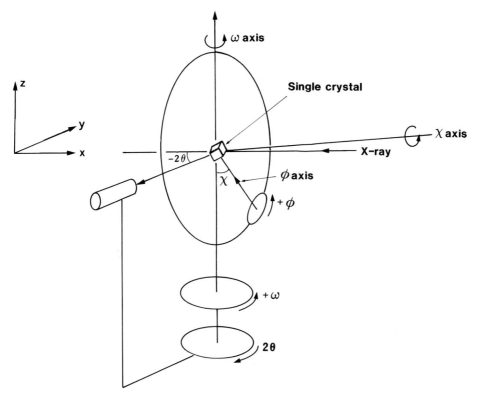

Figure A1.6 Normal-beam equatorial geometry: the angles ω, χ, ϕ, 2θ drawn in the convention of Hamilton (1974).

A1.5.1 Normal-beam equatorial geometry

In this geometry (figure A1.6) the crystal is oriented specifically so as to bring the incident and reflected beam, for a given $hk\ell$ plane, into the equatorial plane. In this way the detector is moved to intercept the reflected beam by a single rotation movement about a vertical axis (the 2θ-axis). The value of θ is given by Bragg's law as $\sin^{-1}(d^*_{hk\ell}/2)$.

In order to bring $\mathbf{d}^*$ into the equatorial plane (i.e. the Bragg plane into the meridional plane) suitable angular settings of a three-axis goniostat are necessary. The convention for the sign of the angles given in figure A1.6 is that of Hamilton (1974); his choice of sign of 2θ is adhered to despite the fact that it is left-handed but, in any case, the signs of ω, χ, ϕ are standard right-handed.

The specific reciprocal lattice vector, $\mathbf{d}^*$, shown in figure A1.7 can be rotated from the point P to Q by the ϕ rotation, from Q to R via χ and R to S via ω.

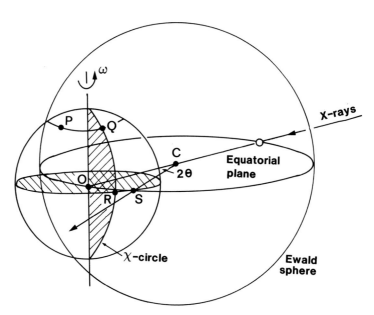

Figure A1.7 Diffractometry with normal-beam equatorial geometry and angular motions ω, χ and ϕ. The RLP at P is moved to Q via ϕ, from Q to R via χ and R to S via ω. From Arndt and Willis (1966).

In the most commonly used setting the χ plane bisects the incident and diffracted beams at the measuring position. Hence, the vector $\mathbf{d}^*$ lies in the χ plane at the measuring position. However, since it is possible for the reflection to take place for any orientation of the reflecting plane rotated about $\mathbf{d}^*$ it is feasible, therefore, that the $\mathbf{d}^*$ can make any arbitrary angle ε with the χ plane. It is conventional to refer to the azimuthal angle ψ of the reflecting plane as the angle of rotation about $\mathbf{d}^*$. It is possible with a ψ scan to keep the $hk\ell$ reflection in the diffraction condition and so to measure the sample absorption surface by monitoring the variation in intensity of this reflection. This ψ scan is achieved by adjustment of the ω, χ, ϕ angles. When $\chi=\pm90°$ the ψ scan is simply a ϕ scan and ε is 0°.

The χ circle is a relatively bulky object whose thickness can stop the measurement of diffracted beams at high θ. Also, collision of the χ circle with the collimator or X-ray tube housing has to be avoided. An alternative is the Kappa goniostat geometry. In the Kappa diffractometer (see figure A1.8) the Kappa axis is inclined at 50° to the ω-axis and can be rotated about the ω-axis; the κ-axis is an alternative to χ, therefore. The

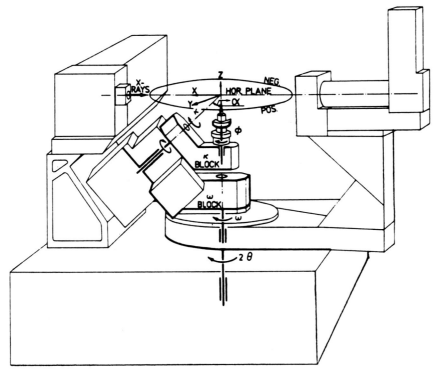

Figure A1.8 Normal-beam equatorial geometry with kappa option. From Wycoff (1985) p. 334 and with courtesy of Enraf–Nonius Ltd.

ϕ-axis is mounted on the $\varkappa$-axis and the whole κ-axis assembly rotates about the κ-axis. In this way an unobstructed view of the sample is achieved.

A1.5.2 Fixed $\chi = 45°$ geometry with area detector

This geometry was introduced by Nicolet and is now fairly common in the field. It consists of an ω-axis, a ϕ-axis and χ fixed at 45°. The rotation axis is the ω-axis. In this configuration it is possible to sample a greater number of independent reflections per degree of rotation (Xuong, Nielsen, Hamlin and Anderson 1985) because of the generally random nature of any symmetry axis.

An alternative method is to mount the crystal in a precise orientation and to use the ϕ-axis to explore the blind region of the single rotation axis. It is feasible to place the capillary containing the sample in a vertically upright position via a 135° bracket mounted on the goniometer head. The bulk of the data is collected with the ω-axis coincident with the capillary axis. This is beneficial because the effect of capillary absorption is symmetrical. At the end of this run the blind region volume whose axis is coincident with the ω-axis can be filled in by resetting around the ϕ-axis by 180°. This renders the capillary axis horizontal and a different crystal axis vertical. Hence, by rotating about this new crystal axis by $\pm 2\theta_{max}$ the blind region volume can be sampled.

Conventional X-ray sources

A2.1 THE SPECTRUM FROM A CONVENTIONAL X-RAY SOURCE

This is shown in figure A2.1.

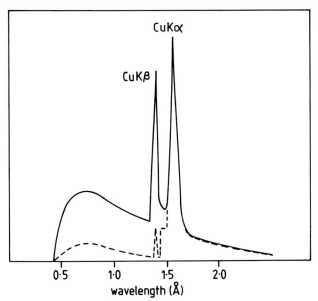

Figure A2.1 The spectrum (schematic) emitted by a conventional X-ray source using a copper anode as an example; spectral lines characteristic of the anode element superimposed on a white radiation background (the bremsstrahlung). The dashed curve represents the effect of a nickel filter on the emission spectrum (full curve). In some ways, the spectrum resembles that of an undulator! Characteristic wavelengths (Cu): $K\beta=1.3922$ Å, $K\alpha_1=1.5405$ Å, $K\alpha_2=1.5443$ Å, $K\bar{\alpha}=1.5418$ Å; weighted average.

Table A2.1. *The wavelengths of the characteristic emission lines of various anode elements used in X-ray generators (in Å).*

Element	$K\alpha_1$	$K\alpha_2$	$K\beta$	$L\alpha_1$	$L\alpha_2$	$L\beta_1$	$L\beta_2$	$L\gamma$
Chromium	2.290	2.294	2.085					
Iron	1.936	1.940	1.757					
Cobalt	1.789	1.793	1.621					
Nickel	1.6580	1.662	1.500					
Copper[a]	1.5405	1.5443	1.3922					
Molybdenum[b]	0.709	0.714	0.632					
Gold[c]				1.276	1.288	1.084	1.070	0.925

[a] $\langle K\alpha_1\, K\alpha_2 \rangle = 1.5418$ Å; weighted average.
[b] $\langle K\alpha_1\, K\alpha_2 \rangle = 0.7107$ Å: weighted average.
[c] See Ashford *et al* (1990).

A2.2 DATA COLLECTION ON A CONVENTIONAL X-RAY SOURCE WITH AN AREA DETECTOR (INCLUDING TABULATED CASES) AND RELATIONSHIP TO SYNCHROTRON RADIATION

The synchrotron should not be seen as a replacement for facilities in the home laboratory but as a means for meeting technically challenging data collection problems. Of course, in the absence of any home X-ray facilities, the central facility can be used but this is not terribly efficient because of the necessity of long-term scheduling of many users. Hence, characterisation of samples (e.g. of heavy atom derivatives) should be done at home unless the project is entirely reliant on the SR source; in this category are many virus studies and ribosome crystallography as well as small crystal projects.

The revolution in SR source technology has been paralleled by a massive expansion and investment in improved facilities in the home laboratory, particularly in the provision of EADs on rotating anode (Cu Kα) generators. In this appendix (table A2.2) a survey is presented of the data collection details for those structures solved using area detectors and Cu Kα radiation. This has been as thorough as possible in terms of searching the literature. If there are any omissions (up to mid-1990) please communicate the details to me. Not included in the survey are those cases based on use of Cu Kα rotation photography or Cu Kα single counter diffractometry. For the former the EAD has been replacing film increasingly. For the latter there will continue to be a very useful role; the single counter diffractometer is extremely efficient in terms of specifically measuring the unique portion of reciprocal space and of measuring the $hk\ell$ and $\overline{hk\ell}$ reflections for the anomalous differences. Of course, in the larger unit cell case this is relatively time-consuming but up to approximately a 100 Å cell parameter the rotating anode single counter diffractometer can be effective. For unit cells less than 50 Å it is the most effective; in this range the density of spots on an area detector is perhaps too sparse to be competitive with a diffractometer.

The survey of area detector, home laboratory data collection examples in table A2.2 shows several general points. Firstly, the size of the sample unit cell parameters is almost always less than 200 Å and, in fact, is rarely above 150 Å. This occurs for two main reasons. The weakness of the larger unit cell scattering produces inordinately long exposure times. Also, in order to resolve adjacent reflections/spots, the detector must be moved well away from the crystal and so only a part of the diffraction aperture is intercepted by the detector. Hence, to survey the required region of reciprocal space needs several or many detector settings and for each a large rotation of one or more specimens. Obviously, a detector with a larger number of useful resolution elements than EADs such as the MWPC or television systems will extend the upper limit of unit cells studied routinely. Such a detector is the IP with its relatively larger aperture and more resolution elements which solves the geometric aspect but not, of course, the weak scattering of larger and larger unit cells.

Secondly, the resolution limits in table A2.2 are rarely greater than approximately 2.5 Å. The high temperature factor of protein crystals and radiation damage restricts what can be done compared to the SR source. Strongly diffracting crystals provide examples in table A2.2 where better resolutions have been collected.

Finally, it can be seen that, year by year, the number of structures being reported is increasing steadily. Fundamentally it can be said that there is an almost inexhaustible number of proteins to be studied and, likewise, not a shortage of crystals.

What extrapolations can be made for the use of SR in macromolecular crystallography? The expanding number of structures solved at medium to high resolution (≈ 2.5 Å interplanar spacing) brings an equal number of projects requiring higher resolution data for protein model refinement (i.e. 2–1.5 Å or better interplanar spacing). Correlation of experimental results with theoretical modelling techniques requires the highest resolution determination of a structure. Indeed, even gross errors are revealed only at the refinement stage. To deal with the demand of this large quantity of projects at SR centres will require very efficient detectors with a large aperture and a large number of resolution elements. The IP is currently increasingly fulfilling this role although it is still off-line.

A significant percentage of all the protein projects tackled will present the technical challenges of small crystal volume or problems of isomorphous heavy atom derivative preparation for which the intensity and tunability of SR, respectively, will be harnessed.

The large unit cell cases, i.e. mammalian viruses and ribosome crystallography, have slotted in extremely well at SR centres because of the intrinsic weakness of scattering of these samples and the great difficulties in working with them in the home laboratory.

Time resolved macromolecular crystallography is virtually absent from conventional X-source experiments because of the long data collection times and the reliance on rotating (moving) crystal methods. The natural advantages of the Laue geometry of using a stationary sample and the high brilliance of the polychromatic SR emission means that time resolved work will be essentially unique to the SR centre and absent from the home laboratory.

Table A2.2. *Area detector data collection on conventional X-ray sources (Cu Kα sealed tube or rotating anode).*

Sample	Detector system	Unit cell parameters a (Å) α ($°$)	b (Å) β ($°$)	c (Å) γ ($°$)	Space group	Resolution limit (Å)	Reference
E. coli dihydrofolate reductase binary complex	San Diego[1] MWPC[2a]	93.2		73.6	$P6_1$	2.5	Matthews *et al* (1977)
L. casei dihydrofolate reductase ternary complex	San Diego MWPC	71.9		93.4	$P6_1$	2.5	Matthews *et al* (1978)
Yeast cytochrome C peroxidase	San Diego MWPC	107.4	76.7	51.5	$P2_12_12_1$	2.5	Poulos *et al* (1980)
Pyrochlorophyllide-apomyoglobin	San Diego MWPC	48.7	40.2	79.0	$P2_12_12_1$	2.5	Boxer *et al* (1982)
Thermostable ribonuclease A	Xentronics[3] MWPC	37.1	41.3	75.6	$P2_12_12_1$	2.2	Weber *et al* (1985)
Histone octamer	San Diego MWPC	118.7		102.9	$P3_221$	3.1	Burlingame *et al* (1985)
R-state aspartate carbamoyltransferase	San Diego MWPC	122.1		156.2	$P321$	2.5	Krause, Volz and Lipscomb (1985)
E. coli DNA polymerase I binary complexes	San Diego MWPC	102.9		85.8	$P4_3$	2.8	Ollis *et al* (1985)
P. aeruginosa exotoxin A	Virginia[4] MWPC	60.6	100.2 / 98.6	59.8	$P2_1$	3.0	Allured, Collier, Carroll and McKay (1986)
Tobacco rubisco	San Diego MWPC	148.5		137.5	$I422$	2.5	Chapman *et al* (1986)
Trimethylamine dehydrogenase	San Diego MWPC	147.8	72.0 / 97.7	83.8	$P2_1$	2.4	Lim *et al* (1986)
Phospholipase A_2	FAST[5] TV[2b]	47.0	64.5	38.1	$P2_12_12_1$	1.75	Renetseder *et al* (1986)
Calmodulin	Virginia MWPC	29.7 / 93.2	53.8 / 97.4	25.0 / 89.3	$P1$	1.65	Kretsinger, Rudnick and Weissman (1986)

Table A2.2. (cont.)

Sample	Detector system	Unit cell parameters a / α	b / β	c (Å) / γ (°)	Space group	Resolution limit (Å)	Reference
Pseudomonas putide cytochrome P-450							
(*a*) camphor bound	Xentronics	108.67	103.80	36.38	$P2_12_12_1$	1.61	Poulos, Finzel and Howard (1986)
(*b*) camphor free		108.46	104.21	36.11	$P2_12_12_1$	2.20	
Fab-lysozyme complex	Xentronics MWPC	54.9	65.2 102.4	78.6	$P2_1$	3.0	Sheriff *et al* (1987)
Fab-lysozyme complex	San Diego MWPC	54.8	74.8 101.8	79.0	$P2_1$	2.5	Sheriff *et al* (1987)
Asn^{102} trypsin	San Diego MWPC	40.4	92.0	127.4	$P2_12_12_1$	2.3	Sprang *et al* (1987)
L. casei thymidylate synthase	San Diego MWPC	78.8		230.1	$P6_122$	2.4	Hardy *et al* (1987)
L-3-hydroxyacylcoenzyme A dehydrogenase	San Diego MWPC	227.2	82.1	124.7	$C222_1$	2.7	Birktoft *et al* (1987)
Methionyl porcine growth hormone	San Diego MWPC	87.7		58.7	$P3_221$	2.8	Abdel-Meguid *et al* (1987)
Flavocytochrome b$_2$	San Diego MWPC	165.5		113.7	$P3_221$	3.0	Xia *et al* (1987)
Ricin	San Diego MWPC	72.8	79.6	114.7	$P2_12_12_1$	2.8	Montfort *et al* (1987)
Surface glycoprotein	Xentronics MWPC	55.1	98.9	172.9	$P2_12_12_1$	4.2–6.0	Metcalf *et al* (1987)
Repressor-operator complex of bacteriophage 434	Xentronics MWPC (Film)	166.0		139.0	$I422$	7.0 (3.2)	Anderson, Ptashne and Harrison (1987)
Bilin binding protein	FAST TV	132.7	121.9	63.9	$P2_12_12_1$	3.5	Huber *et al* (1987)
Cytochrome P450cam	Xentronics MWPC	108.7	103.9	36.4	$P2_12_12_1$	1.63	Poulos, Finzel and Howard (1987)
Anti-Lewis a Fab	Virginia MWPC	43.4 72.7	41.7 96.6	62.0 100.1	$P1$	3.0	Vitali *et al* (1987)

Elastase-inhibitor complex	Virginia MWPC	51.2	58.0	75.1	$P2_12_12_1$	2.1	Radhakrishnan, Presta, Meyer and Wildonger (1987)
Aspartate carbamoyltransferase ternary complex	Virginia MWPC	122.3		156.0	P321	2.6	Gouaux and Lipscomb (1988)
Dienelactone hydrolase	San Diego MWPC	48.9	71.5	78.2	$P2_12_12_1$	2.8	Pathak, Ngai and Ollis (1988)
Acyl-CoA dehydrogenase	San Diego MWPC	128.3	136.2	106.1	$C222_1$	2.4	Kim and Wu (1988)
Rh. sphaeroides photosynthetic reaction centre	San Diego MWPC	142.4	75.5	141.8	$P2_12_12_1$	2.8	Yeates et al (1988)
Lipovitellin-phosvitin complex	San Diego MWPC	192.9	87.1 100.9	91.0	C2	2.8	Raag et al (1988)
Lambda repressor-operator complex	Xentronics MWPC	37.2	68.7 92.2	57.0	$P2_1$	2.25	Jordan and Pabo (1988)
Phage 434 repressor-operator complex bromine derivative	Xentronics MWPC	148.3	27.7	65.0	$P2_12_12_1$	2.5	Aggarwal et al (1988)
α-chymotrypsin/proteinase inhibitor complex	FAST TV	66.9	70.2	86.1	$P2_12_12_1$	2.3	Grütter, Fendrich, Huber and Bode (1988)
Creatine amidinohydrolase	FAST TV	61.1	110.5 102.7	63.7	$P2_1$	3.0	Hoeffken et al (1988)
P2 myelin	Xentronics MWPC	91.8	99.5	56.5	$P2_12_12_1$	2.7	Jones, Bergfors, Sedzik and Unge (1988)
S. typhimurium $\alpha_2\beta_2$ tryptophan synthase	San Diego MWPC & Virginia MWPC	184.5	61.1 94.7	67.7	C2	2.2	Hyde et al (1988)
Klenow fragment+DNA	San Diego MWPC	104.7		86.0	$P4_3$	2.8–3.8	Freemont et al (1988)
DNA dodecamer d(CGCAAAAATGCG) +complement	San Diego MWPC	24.5	40.3	65.9	$P2_12_12_1$	2.6	DiGabriele, Sanderson and Steitz (1989)
Human α-thrombin	FAST	87.74	67.81	61.07	$P2_12_12_1$	1.9	Bode et al (1989)

Table A2.2. (cont.)

Sample	Detector system	Unit cell parameters			Space group	Resolution limit (Å)	Reference
		a α	b β	c (Å) γ (°)			
P21 Ha-ras binary complex	Xentronics MWPC	40.3		162.2	$P3_121$	2.6	Pai et al (1989)
Gln-tRNA synthase/tRNA[Gln] complex	San Diego MWPC	242.8	93.6	115.7	$C222_1$	2.8	Rould, Perona, Söll and Steitz (1989)
Glycogen phosphorylase a							Goldsmith et al (1989)
activated	San Diego MWPC	128.5		118.1	$P4_32_12$	2.8	
inhibited	San Diego MWPC	128.4		116.4	$P4_32_12$	2.1	
S. typhimurium glutamine synthetase	San Diego MWPC	235.5	134.5 102.8	200.1	$C2$	3.5	Yamashita et al (1989)
Recombinant interleukin-1β	Xentronics MWPC	54.9		77.0	$P4_3$	2.0	Finzel et al (1989)
Pig heart aconitase activated inactive	Xentronics MWPC	173.6	72.0	72.7	$P2_12_12_1$	2.5 2.1	Robbins and Stout (1989)
Carp parvalbumin	Virginia MWPC	28.2	61.0 95.0	54.3	$C2$	1.6	Swain, Kretsinger and Amma (1989)
A. vinelandii 7Fe ferrodoxin	Xentronics MWPC	55.2		95.2	$P4_12_12$	1.9	Stout (1989)
Rh. rubrum rubisco transition state analogue complex	Xentronics MWPC	65.5	70.6 92.11	104.1	$P2_1$	2.6	Lundqvist and Schneider (1989)
Phage 434 amino terminal domain	Xentronics MWPC	32.8	37.5	44.6	$P2_12_12_1$	1.9	Mondragón et al (1989)
S. rubiginosus D-xylose isomerase binary complexes	Xentronics MWPC	93.9	99.7	102.9	$I222$	1.9	Carrell et al (1989)
Colicin A fragment	Xentronics MWPC	73.0		171.3	$P4_32_12$	3.0	Parker et al (1989)
Retroviral protease	Xentronics MWPC	88.95		78.90	$P3_121$	3.0 (2.0)	Miller et al (1989)

Protein	Detector	a	b	c	Space group	Resolution (Å)	Reference
HIV-1 aspartyl protease	Xentronics MWPC FAST TV	50.29		106.8	$P4_12_12$	3.0 2.7	Navia et al (1989) Lapatto et al (1989)
HyHEL-10 fab/lysozyme complex	San Diego MWPC	57.5	118.7	137.7	$P2_12_12_1$	3.0	Padlan et al (1989)
Concanavalin-A mannoside complex	Xentronics MWPC	123.9	129.1	67.5	$P2_12_12_1$	2.9	Derewenda Z. et al (1989)
A. niger α-Amylase	Xentronics MWPC	81.1	98.3	138.0	$C222_1$	3.0	Derewenda and Helliwell (1989)
E. halophila photo-reactive yellow protein	Xentronics MWPC	66.9		40.8	$P6_3$	2.4	McRee et al (1989)
Tumour necrosis factor	Xentronics MWPC (SRS film)	166.0		93.0	$P3_121$	3.4–5.0 (2.9)	Jones et al (1989)
DNA dodecamer d(ACCGGGCGCCACA)	Xentronics MWPC	65.9		47.1	R3	2.8	Timsit, Westhof, Fuchs and Moras (1989)
E. coli Gln^{239} aspartate carbamoyltransferase	Virginia MWPC	122.3		147.1	P321	3.1	Gouaux, Stevens, Ke and Lipscomb (1989)
E. coli met aporepressor and corepressor	Xentronics MWPC (CAD4)	35.6	62.6 102.4	44.5	$P2_1$	1.7 (3.0)	Rafferty et al (1989)
	Xentronics MWPC	75.0		82.0	$P3_221$	1.9	Rafferty et al (1989)
Chaperone protein Pap D	Xentronics MWPC	58.2	64.0	67.0	$P2_12_12_1$	2.5	Holmgren and Branden (1989)
E. coli isocitrate hydrogenase	Xentronics MWPC	105.1		150.3	$P4_32_12$	2.5	Hurley et al (1989)
B. stearothermophilus T-state phosphofructokinase	FAST TV	131.85	115.29	96.06	$P2_12_12_1$	2.5	Schirmer and Evans (1989)
Synthetic [aba^{67,95}] HIV-1 protease	Xentronics MWPC	50.2		106.6	$P4_12_12$	2.8	Wlodawer et al (1989)
Yeast enolase	Virginia MWPC	124.6		67.1	$P4_12_12$	2.25	Lebioda, Stec and Brewer (1989)
Phospholipase A$_2$ (mutant, residues 62–66 deleted)	FAST	45.7	73.5 107.1	37.1	$P2_1$	2.5Å	Kuipers et al (1989)

Table A2.2. (cont.)

Sample	Detector system	Unit cell parameters			Space group	Resolution limit (Å)	Reference
		a / α	b / β	c (Å) / γ (°)			
Human serum albumin	Xentronics	186.5		81.0	$P42_12$	6.0 (3.0)	Carter et al (1989)
Collagenase	FAST	160.0		53.1	I4	3Å(4Å)	Lloyd et al (1989)
Renin	San Diego MWPC	143.0			$P2_13$	3.0Å	Lim et al (1989)
Maclura pomifera Agglutinin Gal β1-3GalNAc	San Diego MWPC	67.4		149.3	$P3_121$ or $P3_221$	2.7	Lee, Johnston, Rose and Young (1989)
Porcine Trypsin+E. elaterium trypsin inhibitor II	FAST	62.25	62.27	84.66	$P2_12_12_1$	1.9	Gaboriaud et al (1989)
Porcine pancreatic elastase+ peptidyl boronic acid	FAST	51.78	57.80	75.43	$P2_12_12_1$	2.0	Takahashi et al (1989)
Trypsin engineered D102N, S195C engineered S195C[a]	Xentronics	124.4			I23	2.0 2.5	McGrath et al (1989)
Wheat serine carboxypeptidase	San Diego MWPC	98.4		209.5	$P4_12_12$	3.5	Liao and Remington (1990)
Fructose-1,6-bis-phosphatase	San Diego MWPC	132.3		68.0	$P3_221$	2.8	Ke et al (1990)
E. coli thioredoxin	San Diego MWPC	89.5	51.1 113.5	60.5	C2	1.68	Katti, LeMasker and Eklund (1990)
C. freundii β-lactamase	Xentronics MWPC	98.1	84.6	89.8	$P2_12_12_1$	2.0–2.5	Oefner et al (1990)
H. mielet triglyceride lipase	Xentronics MWPC	71.6	75.0	55.0	$P2_12_12_1$	1.9–3.2	Brady et al (1990)
Human pancreatic lipase	Xentronics MWPC	47.8	112.8 99.3	91.0	$P2_1$	2.4–2.83	Winkler, D'Arcy and Hunziker (1990)
Papain-stefin B complex	FAST TV	67.0		169.3	$P3_121$	2.4	Stubbs et al (1990)

V_2 domain McPC603	FAST TV	86.5		75.6	$P6_122$	2.0	Glockshuber et al (1990)
Berenil-dodecanucleotide complex	Xentronics MWPC	24.5	40.0	66.2	$P2_12_12_1$	2.5	Brown et al (1990)
B. licheniformis 749/C β-lactamase	Virginia MWPC	66.8	90.7 104.5	43.6	$P2_1$	2.0	Moews et al (1990)
Human CO haemoglobin	Xentronics MWPC	54.5		197.0	$P4_12_12$	2.2 ⎫	Derewenda, Z. et al (1990)
Cowtown CO haemoglobin	Xentronics MWPC	54.4		195.5	$P4_12_12$	2.3 ⎭	
Plakalbumin	Virginia MWPC	100.6	101.6 111.9	79.1	C2	2.8	Wright, Qian and Huber (1990)
Yeast cytochrome c Peroxidase binary complex	Xentronics MWPC	107.5	76.6	51.7	$P2_12_12_1$	1.85	Edwards and Poulos (1990)
Fab'-peptide complex	Xentronics MWPC	142.5		101.5	$P6_322$	2.8	Stanfield, Fieser, Lerner and Wilson (1990)
Fab' native		98.0	151.7	80.8	$P2_12_12_1$	2.77	

Notes:

(1) San Diego, MWPC refers both to the regional/national area detector data collection facility there (for a description of the various systems and the developments over the years see Xuong et al (1978, 1985) and Hamlin et al (1981)) and to the identical type of system supplied commercially (from about 1986) by 'San Diego Multi-wire Systems'.

(2) (a) MWPC = multi-wire proportional chamber; (b) TV = television detector.

(3) Xentronics is a trade name of the Siemens/Nicolet (Wisconsin) MWPC area detector (for a description of the system see Durbin et al (1986); its software see Howard et al (1987) and its calibration and use see Derewenda and Helliwell (1989).

(4) Virginia, MWPC refers to the regional/national area detector data collection facility there (for a description of the system and its use see Sobottka et al (1990)).

(5) FAST is a trade name of the Enraf–Nonius (Delft) television area detector (see Arndt (1982, 1986) for example).

(6) This table has been assembled not only from personal knowledge, but also from a literature scan encompassing the following journals, over a period from 1977–mid 1990: Nature, Science, Proc Nat Acad Sci USA, Acta Cryst, J Appl Cryst, Biochemistry, J Mol Biol, J Biol Chem, EMBO J, Protein Engineering and Proteins: Struct, Function and Genetics. I apologise for any omissions; please let me know of any missing entries and/or any errors. Thank you.

(7) The table excludes details of protein structures solved using diffractometers or photographic film.

(8) Systems based on IPs are just entering the market place at the time of writing.

(9) MWPC's have been set up on Cu Kα conventional X-ray sources in the Soviet Union by Kheiker and Popov (pers. comm.) and by Mokulskaya et al (1981) and Anisimov et al (1981) and used to solve protein structures. These structures, reported in the Soviet literature (in Russian) are not included here.

APPENDIX 3

Fundamental Data

In the following pages fundamental data are provided in a variety of areas relevant to the utilisation of SR in macromolecular crystallography.

Section A3.1 provides details of elemental properties for hydrogen through uranium. The table provides for each element the respective atomic number and weight, the density at STP, the mass absorption coefficients at three selected wavelengths of interest for routine data collection at SR sources and finally the wavelengths of the K and L X-ray absorption edges.

Section A3.2 gives the anomalous dispersion corrections (f' and f", in e⁻) tabulated as a function of wavelength for a few elements, as an illustration, namely the K edges of selenium and bromine as well as the L edges of platinum and mercury.

Section A3.3 provides a table of fundamental constants and values needed in formulae used in this book.

Section A3.4 gives the cell parameters for silicon and germanium monochromator crystals as well as the respective spectral bandpasses of specific reflected beams. These monochromator crystals are the ones most extensively used at SR sources for macromolecular crystallography.

A3.1 PROPERTIES OF THE ELEMENTS

Element	Atomic number	Atomic weight	Density at STP (gm cm^{-3})	Mass absorption coefficients μ_m cm^2 gm^{-1} at wavelengths			Absorption edges (Å)			
				0.33 Å	0.90 Å	1.488 Å	K	L$_I$	L$_{II}$	L$_{III}$
H	1	1.007		0.398	0.446	0.458				
He	2	4.003		0.200	0.224	0.231				
Li	3	6.939	0.534	0.175	0.236	0.424				
Be	4	9.012	1.82	0.183	0.334	0.914				
B	5	10.81	2.30	0.197	0.537	1.91	226.62			
C	6	12.011	2.22	0.226	0.922	3.76	110.68			
N	7	14.007	0.0011	0.247	1.44	6.27	66.289			
O	8	15.999	0.0013	0.277	2.17	9.69	43.648			
F	9	19.00	0.0016	0.300	2.98	13.4	30.990			
Ne	10	20.183	0.0083	0.367	4.35	19.6	23.301			
Na	11	22.99	0.97	0.424	5.68	25.6	17.913			
Mg	12	24.31	1.741	0.523	7.72	34.6	14.183	197.39		
Aℓ	13	26.98	2.699	0.615	9.71	43.2	11.478	142.48		
Si	14	28.09	2.33	0.764	12.6	55.8	9.51220	105.05		
P	15	30.974	1.82	0.887	15.2	66.4	7.94813	81.02		
S	16	32.064	2.07	1.09	19.1	82.5	6.73800	64.228	76.049	76.519
Cℓ	17	35.453	0.0032	1.24	22.1	94.0	5.78400	52.084	61.366	61.672
A	18	39.948	0.0016	1.37	24.6	104.0	5.01850	43.192	50.390	50.803
K	19	39.102	0.86	1.72	31.1	130.0	4.39710	36.352	42.020	42.452
Ca	20	40.08	1.55	2.05	37.0	152.0	3.87090	31.068	35.417	35.827
Sc	21	44.96	2.5	2.21	39.8	162.0	3.43650	26.831	30.161	30.457
Ti	22	47.90	4.45	2.50	44.6	180.0	3.07030	23.389	26.831	27.184
V	23	50.94	6.0	2.79	49.5	197.0	2.76200	20.523	23.702	24.0699
Cr	24	52.00	7.19	3.24	56.9	224.0	2.49734	18.256	21.226	21.5958
							2.26910			
							2.07020			

Table A3.1. (cont.)

Element	Atomic number	Atomic weight	Density at STP (gm cm⁻³)	Mass absorption coefficients μ_m cm² gm⁻¹ at wavelengths			Absorption edges (Å)			
				0.33 Å	0.90 Å	1.488 Å	K	L_I	L_{II}	L_{III}
Mn	25	54.94	7.43	3.59	62.6	244.0	1.89643	16.268	18.896	19.2484
Fe	26	55.85	7.87	4.13	71.1	274.0	1.74346	14.601	17.169	17.4838
Co	27	58.93	8.90	4.54	77.1	294.0	1.60815	13.343	15.534	15.8314
Ni	28	58.71	8.90	5.27	88.1	41.4	1.48807	12.267	14.135	14.4476
Cu	29	63.54	8.96	5.58	92.1	44.2	1.38059	11.269	12.994	13.2578
Zn	30	65.37	7.13	6.19	101.0	49.4	1.28340	10.330	11.8395	12.1055
Ga	31	69.72	5.91	6.58	106.0	53.1	1.19580	9.535	10.6130	10.8546
Ge	32	72.59	5.36	7.14	113.0	58.1	1.11658	8.729	9.9646	10.2277
As	33	74.92	5.73	7.78	122.0	64.0	1.04500	8.107	9.1281	9.3767
Se	34	78.96	4.8	8.27	127.0	68.6	0.97974	7.467	8.4212	8.6624
Br	35	79.909	3.12	9.11	140.0	76.3	0.92040	6.925	7.7523	7.9871
Kr	36	83.80	0.0034	9.66	20.0	81.8	0.86552	6.456	7.1653	7.4227
Rb	37	85.47	1.53	10.5	22.0	89.6	0.81554	6.006	6.6538	6.8752
Sr	38	87.62	2.60	11.3	24.4	98.9	0.76973	5.604	6.1856	6.3996
Y	39	88.905	5.51	12.3	26.8	109.0	0.72762	5.19312	5.70981	5.91412
Zr	40	91.22	6.5	13.2	29.0	117.0	0.68883	4.89380	5.37088	5.57374
Nb	41	92.91	8.57	14.2	31.6	127.0	0.65298	4.59111	5.02472	5.22596
Mo	42	95.94	10.2	15.0	33.8	136.0	0.61978	4.32066	4.71330	4.90930
Tc	43	(98.0)		15.9	36.5	146.0	0.58906	4.06426	4.42714	4.62537
Ru	44	101.1	12.2	16.8	39.0	155.0	0.56051	3.84133	4.17654	4.36632
Rh	45	102.905	12.44	17.9	42.0	167.0	0.53395	3.64159	3.94902	4.13889
Pd	46	106.4	12.0	18.7	44.5	176.0	0.50920	3.42999	3.71360	3.89688
Ag	47	107.87	10.49	19.9	47.9	189.0	0.48589	3.25640	3.51640	3.69990
Cd	48	112.40	8.65	20.5	50.1	197.0	0.46407	3.08490	3.32570	3.50470
In	49	114.82	7.31	21.6	53.3	209.0	0.44371	2.92600	3.14730	3.32370
Sn	50	118.69	7.30	22.4	56.0	219.0	0.42467	2.77690	2.95230	3.15570
Sb	51	121.75	6.618	23.3	59.2	231.0	0.40668	2.63880	2.82940	3.00030
Te	52	127.60	6.24	23.8	61.1	238.0	0.38974	2.50990	2.68790	2.85550
I	53	126.40	4.93	25.5	66.3	257.0	0.37381	2.38800	2.55420	2.71960

Xe	54	131.30	0.0054	26.3	69.2	267.0	0.35840	2.27370	2.42920	2.59260
Cs	55	132.905	1.9	27.6	73.8	284.0	0.34451	2.16730	2.31390	2.47400
Ba	56	137.34	3.5	28.4	76.8	294.0	0.33104	2.06780	2.20480	2.36290
La	57	138.41	6.15	5.18	81.5	311.0	0.31844	1.97800	2.10530	2.26100
Ce	58	140.12	6.9	5.52	86.6	330.0	0.30648	1.89340	2.01240	2.16600
Pr	59	140.91	6.475	5.90	92.2	349.0	0.29518	1.81410	1.92550	2.07910
Nd	60	144.24	6.96	6.18	96.1	353.0	0.28453	1.73900	1.88400	1.99670
Pm	61	(147.0)		6.59	102.0	384.0	0.27431	1.66740	1.76760	1.91910
Sm	62	150.35	7.7	6.81	105.0	393.0	0.26464	1.60020	1.69530	1.84570
Eu	63	152.0	5.116	7.21	110.0	411.0	0.25553	1.53810	1.62710	1.77610
Gd	64	157.25	7.87	7.44	113.0	363.0	0.24681	1.47840	1.56320	1.71170
Tb	65	158.92	8.25	7.86	119.0	385.0	0.23841	1.42230	1.50230	1.64970
Dy	66	162.50	8.556	8.19	123.0	287.0	0.23048	1.36920	1.44450	1.59160
Ho	67	164.93	8.799	8.59	129.0	302.0	0.22291	1.31900	1.39050	1.53680
Er	68	167.26	9.058	9.01	134.0	113.0	0.21567	1.27060	1.33860	1.48350
Tm	69	168.93	9.318	9.47	140.0	119.0	0.20880	1.22500	1.28920	1.43340
Yb	70	173.04	6.959	9.82	145.0	123.0	0.20224	1.18180	1.24280	1.38620
Lu	71	174.97	9.85	10.3	151.0	129.0	0.19585	1.14020	1.19850	1.34050
Hf	72	178.49	13.09	10.7	156.0	134.0	0.18982	1.09790	1.15480	1.29720
Ta	73	180.95	16.6	11.2	162.0	139.0	0.18394	1.06130	1.11370	1.25530
W	74	183.85	19.3	11.6	167.0	145.0	0.17837	1.02467	1.07450	1.21550
Re	75	186.2	21.1	12.1	174.0	151.0	0.17302	0.98940	1.03710	1.17730
Os	76	190.2	22.5	12.5	178.0	156.0	0.16787	0.95580	1.00140	1.14080
Ir	77	192.2	22.52	13.1	185.0	163.0	0.16292	0.92360	0.96710	1.10580
Pt	78	195.09	21.4	13.6	165.0	170.0	0.15818	0.89310	0.93414	1.07230
Au	79	196.95	19.32	14.2	172.0	177.0	0.15344	0.86376	0.90259	1.04000
Hg	80	200.59	13.59	14.7	127.0	183.0	0.14923	0.83530	0.87220	1.00910
Tℓ	81	204.37	11.86	15.2	131.0	189.0	0.14470	0.80810	0.84340	0.97930
Pb	82	207.19	11.34	15.7	136.0	196.0	0.14077	0.78186	0.81538	0.95073
Bi	83	208.98	9.84	16.4	142.0	204.0	0.13691	0.75710	0.78870	0.92340
Po	84	(210.0)	9.32	17.2	58.5	215.0	0.13306	0.73219	0.76377	0.89761
At	85	(210.0)		18.0	61.2	224.0	0.12949	0.70915	0.73873	0.87234
Rn	86	(222.0)		17.9	60.8	222.0	0.12591	0.68675	0.71529	0.84845
Fr	87	(223.0)		18.6	63.5	232.0	0.12261	0.66537	0.69290	0.82529

Table A3.1. (cont.)

Element	Atomic number	Atomic weight	Density at STP (gm cm^{-3})	Mass absorption coefficients μ_m cm^2 gm^{-1} at wavelengths				Absorption edges (Å)			
				0.33 Å	0.90 Å	1.488 Å	K	L_I	L_{II}	L_{III}	
Ra	88	(226.0)		19.3	65.6	239.0	0.11931	0.64461	0.67114	0.80284	
Ac	89	(227.0)	10.07	20.1	68.4	249.0	0.11618	0.62479	0.65002	0.78158	
Th	90	232.04	11.5	20.5	70.1	255.0	0.11290	0.60610	0.63010	0.76151	
Pa	91	(231.0)	15.37	21.6	73.6	267.0	0.11028	0.58748	0.61064	0.74138	
U	92	238.03	18.7	21.9	74.6	270.0	0.10775	0.56974	0.59186	0.72225	

Notes:
(1) The transmitted X-ray intensity I, through a material is related to the incident X-ray intensity I_0 by the relation $I = I_0 \exp(-\mu t)$ where μ is the linear absorption coefficient and t the thickness of the material. To calculate μ for a compound from the mass absorption coefficients μ_m for the elements listed above we write $\mu_{compounds} = \sum \mu_m \varrho_P$ where the summation is taken over all the elements in the compound and ϱ_P is the partial density of the element within the compound. Obviously for the case of a pure element e.g. aluminium (foil) or nitrogen (gas) the densities (at STP) are already available in this table.

(2) I am very grateful to Dr. S. Sasaki, Photon Factory, Japan for the values of μ_m and the absorption edge wavelengths which have been reproduced with his permission and that of KEK, National Laboratory of High Energy Physics, Tsukuba.

A3.2 ANOMALOUS DISPERSION CORRECTIONS

The following pages give the anomalous dispersion corrections (f' and f'' in e^-) tabulated as a function of wavelength between 0.1 and 2.89 Å in steps of 0.01 Å and in steps of 0.0001 Å in the edge region for selected K and L edges. The elements chosen are selenium and bromine (K edge) as well as platinum and mercury (L edge) because these are commonly used for such experiments in macromolecular crystallography at SR sources. The data were extracted from Sasaki (1989) with permission. The calculations are based originally on Cromer and Liberman (1981).

Table A3.2. *Tabulated values of anomalous dispersion corrections for selenium, bromine, platinum and mercury.*

ATOMIC SYMBOL = SE   ATOMIC NUMBER = 34

I		0.00	0.01	0.02	0.03
0.1 I	F'	-0.136	-0.124	-0.111	-0.098
I	F"	0.055	0.067	0.079	0.093
0.2 I	F'	-0.003	0.011	0.024	0.035
I	F"	0.219	0.241	0.264	0.288
0.3 I	F'	0.117	0.127	0.137	0.146
I	F"	0.477	0.507	0.539	0.570
0.4 I	F'	0.194	0.198	0.201	0.203
I	F"	0.814	0.851	0.890	0.929
0.5 I	F'	0.195	0.190	0.185	0.178
I	F"	1.218	1.261	1.305	1.350
0.6 I	F'	0.096	0.079	0.061	0.041
I	F"	1.675	1.723	1.772	1.821
0.7 I	F'	-0.147	-0.182	-0.219	-0.258
I	F"	2.177	2.230	2.283	2.336
0.8 I	F'	-0.624	-0.691	-0.763	-0.840
I	F"	2.717	2.772	2.828	2.884
0.9 I	F'	-1.609	-1.773	-1.961	-2.182
I	F"	3.284	3.342	3.400	3.459
1.0 I	F'	-3.467	-3.099	-2.846	-2.654
I	F"	0.519	0.528	0.538	0.547
1.1 I	F'	-1.949	-1.889	-1.834	-1.783
I	F"	0.617	0.628	0.638	0.649
1.2 I	F'	-1.510	-1.480	-1.450	-1.422
I	F"	0.724	0.735	0.746	0.757
1.3 I	F'	-1.255	-1.235	-1.215	-1.196
I	F"	0.838	0.850	0.862	0.873
1.4 I	F'	-1.075	-1.059	-1.044	-1.029
I	F"	0.959	0.971	0.983	0.996
1.5 I	F'	-0.931	-0.918	-0.906	-0.893
I	F"	1.085	1.097	1.110	1.123
1.6 I	F'	-0.811	-0.800	-0.789	-0.778
I	F"	1.216	1.230	1.244	1.257
1.7 I	F'	-0.707	-0.697	-0.687	-0.679
I	F"	1.354	1.369	1.383	1.397
1.8 I	F'	-0.615	-0.607	-0.598	-0.589
I	F"	1.498	1.513	1.528	1.542
1.9 I	F'	-0.532	-0.524	-0.516	-0.509
I	F"	1.648	1.663	1.678	1.694
2.0 I	F'	-0.460	-0.453	-0.446	-0.439
I	F"	1.802	1.818	1.834	1.850
2.1 I	F'	-0.391	-0.385	-0.378	-0.372
I	F"	1.962	1.978	1.995	2.011
2.2 I	F'	-0.329	-0.323	-0.317	-0.311
I	F"	2.127	2.143	2.160	2.177
2.3 I	F'	-0.272	-0.267	-0.262	-0.256
I	F"	2.296	2.313	2.331	2.348
2.4 I	F'	-0.221	-0.216	-0.211	-0.206
I	F"	2.470	2.488	2.506	2.524
2.5 I	F'	-0.174	-0.170	-0.166	-0.161
I	F"	2.649	2.668	2.686	2.704
2.6 I	F'	-0.133	-0.129	-0.125	-0.122
I	F"	2.833	2.851	2.870	2.889
2.7 I	F'	-0.097	-0.093	-0.090	-0.087
I	F"	3.020	3.039	3.059	3.078
2.8 I	F'	-0.065	-0.063	-0.060	-0.057
I	F"	3.212	3.232	3.251	3.271

0.04	0.05	0.06	0.07	0.08	0.09
-0.085	-0.071	-0.058	-0.044	-0.030	-0.016
0.108	0.124	0.141	0.159	0.178	0.198
0.048	0.060	0.073	0.084	0.096	0.107
0.312	0.338	0.364	0.391	0.419	0.448
0.155	0.163	0.170	0.177	0.183	0.189
0.603	0.636	0.670	0.705	0.741	0.777
0.205	0.206	0.205	0.204	0.202	0.199
0.968	1.009	1.049	1.091	1.133	1.175
0.170	0.160	0.150	0.138	0.126	0.111
1.395	1.440	1.486	1.533	1.579	1.627
0.019	-0.004	-0.029	-0.055	-0.084	-0.114
1.871	1.921	1.971	2.022	2.073	2.125
-0.300	-0.346	-0.397	-0.448	-0.503	-0.562
2.390	2.444	2.498	2.552	2.607	2.661
-0.923	-1.013	-1.111	-1.218	-1.335	-1.465
2.940	2.997	3.054	3.111	3.168	3.226
-2.393	-2.739	-3.224	-4.051	-7.156	-4.111
3.528	3.605	3.684	3.765	0.500	0.509
-2.500	-2.373	-2.265	-2.171	-2.089	-2.015
0.557	0.567	0.577	0.587	0.597	0.607
-1.736	-1.692	-1.651	-1.613	-1.577	-1.543
0.659	0.670	0.680	0.691	0.702	0.713
-1.396	-1.370	-1.345	-1.322	-1.299	-1.277
0.769	0.780	0.791	0.803	0.814	0.826
-1.177	-1.159	-1.141	-1.124	-1.107	-1.091
0.885	0.898	0.910	0.922	0.934	0.947
-1.014	-0.999	-0.985	-0.971	-0.958	-0.944
1.008	1.021	1.033	1.046	1.059	1.072
-0.881	-0.869	-0.857	-0.845	-0.834	-0.822
1.137	1.150	1.163	1.176	1.190	1.203
-0.768	-0.757	-0.747	-0.737	-0.727	-0.717
1.271	1.285	1.299	1.312	1.326	1.340
-0.670	-0.660	-0.651	-0.642	-0.633	-0.624
1.411	1.426	1.440	1.455	1.469	1.484
-0.581	-0.573	-0.564	-0.556	-0.548	-0.540
1.557	1.572	1.587	1.602	1.617	1.632
-0.502	-0.494	-0.487	-0.480	-0.472	-0.465
1.709	1.724	1.740	1.755	1.771	1.787
-0.432	-0.425	-0.418	-0.411	-0.405	-0.398
1.866	1.882	1.898	1.914	1.930	1.946
-0.366	-0.360	-0.353	-0.347	-0.341	-0.335
2.027	2.044	2.060	2.077	2.093	2.110
-0.306	-0.300	-0.294	-0.289	-0.283	-0.278
2.194	2.211	2.228	2.245	2.262	2.279
-0.251	-0.246	-0.241	-0.236	-0.231	-0.226
2.365	2.383	2.400	2.418	2.435	2.453
-0.202	-0.197	-0.192	-0.188	-0.183	-0.179
2.541	2.559	2.577	2.595	2.613	2.631
-0.157	-0.153	-0.149	-0.145	-0.141	-0.137
2.722	2.740	2.759	2.777	2.796	2.814
-0.118	-0.114	-0.111	-0.107	-0.104	-0.100
2.907	2.926	2.945	2.964	2.983	3.002
-0.084	-0.080	-0.077	-0.074	-0.071	-0.068
3.097	3.116	3.135	3.154	3.174	3.193
-0.054	-0.052	-0.049	-0.046	-0.044	-0.041
3.290	3.310	3.330	3.349	3.369	3.389

Table A3.2. (*cont.*)

ATOMIC SYMBOL = SE   ATOMIC NUMBER = 34

I			0.0000	0.0001	0.0002	0.0003
0.965 I	F'		-3.567	-3.575	-3.583	-3.591
I		F''	3.724	3.725	3.726	3.727
0.966 I	F'		-3.649	-3.658	-3.666	-3.675
I		F''	3.732	3.733	3.734	3.735
0.967 I	F'		-3.738	-3.747	-3.756	-3.765
I		F''	3.741	3.741	3.742	3.743
0.968 I	F'		-3.833	-3.843	-3.853	-3.863
I		F''	3.749	3.750	3.750	3.751
0.969 I	F'		-3.937	-3.948	-3.959	-3.970
I		F''	3.757	3.758	3.759	3.759
0.970 I	F'		-4.051	-4.063	-4.075	-4.087
I		F''	3.765	3.766	3.767	3.768
0.971 I	F'		-4.177	-4.190	-4.204	-4.217
I		F''	3.773	3.774	3.775	3.776
0.972 I	F'		-4.318	-4.333	-4.349	-4.364
I		F''	3.782	3.783	3.783	3.784
0.973 I	F'		-4.479	-4.497	-4.514	-4.532
I		F''	3.790	3.791	3.792	3.793
0.974 I	F'		-4.667	-4.687	-4.708	-4.730
I		F''	3.798	3.799	3.800	3.801
0.975 I	F'		-4.891	-4.916	-4.942	-4.968
I		F''	3.807	3.808	3.808	3.809
0.976 I	F'		-5.172	-5.204	-5.237	-5.271
I		F''	3.815	3.816	3.817	3.818
0.977 I	F'		-5.546	-5.592	-5.639	-5.689
I		F''	3.824	3.824	3.825	3.826
0.978 I	F'		-6.117	-6.194	-6.277	-6.368
I		F''	3.832	3.833	3.834	3.834
0.979 I	F'		-7.381	-7.652	-8.016	-8.571
I		F''	3.840	3.841	3.842	3.843
0.980 I	F'		-7.157	-6.982	-6.832	-6.700
I		F''	0.500	0.500	0.500	0.500
0.981 I	F'		-6.072	-6.007	-5.945	-5.887
I		F''	0.501	0.501	0.501	0.501
0.982 I	F'		-5.554	-5.514	-5.476	-5.439
I		F''	0.502	0.502	0.502	0.502
0.983 I	F'		-5.212	-5.184	-5.156	-5.129
I		F''	0.502	0.503	0.503	0.503
0.984 I	F'		-4.958	-4.935	-4.914	-4.892
I		F''	0.503	0.503	0.504	0.504
0.985 I	F'		-4.755	-4.737	-4.719	-4.702
I		F''	0.504	0.504	0.505	0.505
0.986 I	F'		-4.587	-4.572	-4.557	-4.542
I		F''	0.505	0.505	0.505	0.506
0.987 I	F'		-4.444	-4.431	-4.418	-4.405
I		F''	0.506	0.506	0.506	0.507
0.988 I	F'		-4.320	-4.308	-4.297	-4.285
I		F''	0.507	0.507	0.507	0.507
0.989 I	F'		-4.210	-4.199	-4.189	-4.179
I		F''	0.508	0.508	0.508	0.508
0.990 I	F'		-4.111	-4.102	-4.093	-4.084
I		F''	0.509	0.509	0.509	0.509
0.991 I	F'		-4.022	-4.014	-4.005	-3.997
I		F''	0.510	0.510	0.510	0.510
0.992 I	F'		-3.941	-3.933	-3.925	-3.918
I		F''	0.511	0.511	0.511	0.511

K ABSORPTION EDGE (0.97974 Å; 12.6540 KEV)

0.0004	0.0005	0.0006	0.0007	0.0008	0.0009
-3.599	-3.607	-3.616	-3.624	-3.632	-3.641
3.728	3.728	3.729	3.730	3.731	3.732
-3.684	-3.693	-3.701	-3.710	-3.719	-3.728
3.736	3.737	3.737	3.738	3.739	3.740
-3.775	-3.784	-3.794	-3.803	-3.813	-3.823
3.744	3.745	3.746	3.746	3.747	3.748
-3.873	-3.884	-3.894	-3.905	-3.915	-3.926
3.752	3.753	3.754	3.755	3.755	3.756
-3.981	-3.992	-4.004	-4.015	-4.027	-4.039
3.760	3.761	3.762	3.763	3.764	3.764
-4.099	-4.112	-4.125	-4.137	-4.150	-4.163
3.769	3.769	3.770	3.771	3.772	3.773
-4.231	-4.245	-4.259	-4.274	-4.288	-4.303
3.777	3.778	3.778	3.779	3.780	3.781
-4.380	-4.396	-4.412	-4.429	-4.445	-4.462
3.785	3.786	3.787	3.788	3.788	3.789
-4.551	-4.569	-4.588	-4.607	-4.627	-4.647
3.793	3.794	3.795	3.796	3.797	3.798
-4.751	-4.774	-4.796	-4.819	-4.843	-4.867
3.802	3.803	3.803	3.804	3.805	3.806
-4.995	-5.023	-5.051	-5.080	-5.110	-5.140
3.810	3.811	3.812	3.813	3.813	3.814
-5.307	-5.343	-5.381	-5.420	-5.461	-5.503
3.818	3.819	3.820	3.821	3.822	3.823
-5.741	-5.795	-5.852	-5.913	-5.977	-6.044
3.827	3.828	3.829	3.829	3.830	3.831
-6.466	-6.574	-6.694	-6.830	-6.984	-7.164
3.835	3.836	3.837	3.838	3.839	3.840
-9.793	-9.633	-8.514	-7.982	-7.631	-7.367
3.844	0.499	0.499	0.499	0.499	0.500
-6.584	-6.479	-6.384	-6.297	-6.216	-6.141
0.500	0.500	0.500	0.500	0.500	0.500
-5.833	-5.780	-5.731	-5.684	-5.639	-5.595
0.501	0.501	0.501	0.501	0.501	0.501
-5.403	-5.369	-5.335	-5.303	-5.272	-5.242
0.502	0.502	0.502	0.502	0.502	0.502
-5.102	-5.077	-5.052	-5.027	-5.004	-4.980
0.503	0.503	0.503	0.503	0.503	0.503
-4.871	-4.851	-4.831	-4.812	-4.792	-4.774
0.504	0.504	0.504	0.504	0.504	0.504
-4.684	-4.668	-4.651	-4.635	-4.619	-4.603
0.505	0.505	0.505	0.505	0.505	0.505
-4.528	-4.513	-4.499	-4.485	-4.471	-4.458
0.506	0.506	0.506	0.506	0.506	0.506
-4.393	-4.380	-4.368	-4.355	-4.343	-4.332
0.507	0.507	0.507	0.507	0.507	0.507
-4.274	-4.263	-4.252	-4.241	-4.231	-4.220
0.508	0.508	0.508	0.508	0.508	0.508
-4.169	-4.159	-4.149	-4.140	-4.130	-4.121
0.508	0.509	0.509	0.509	0.509	0.509
-4.075	-4.066	-4.057	-4.048	-4.039	-4.031
0.509	0.510	0.510	0.510	0.510	0.510
-3.989	-3.980	-3.972	-3.964	-3.956	-3.949
0.510	0.510	0.511	0.511	0.511	0.511
-3.910	-3.903	-3.895	-3.888	-3.880	-3.873
0.511	0.511	0.512	0.512	0.512	0.512

Table A3.2. (*cont.*)

ATOMIC SYMBOL = BR   ATOMIC NUMBER = 35

I			0.00	0.01	0.02	0.03
0.1 I	F'		-0.145	-0.132	-0.118	-0.104
I		F"	0.062	0.075	0.090	0.105
0.2 I	F'		-0.003	0.012	0.024	0.037
I		F"	0.246	0.271	0.296	0.323
0.3 I	F'		0.119	0.129	0.138	0.146
I		F"	0.534	0.567	0.602	0.637
0.4 I	F'		0.186	0.188	0.190	0.191
I		F"	0.907	0.949	0.991	1.034
0.5 I	F'		0.160	0.152	0.142	0.130
I		F"	1.353	1.401	1.449	1.498
0.6 I	F'		0.009	-0.016	-0.042	-0.070
I		F"	1.856	1.909	1.962	2.016
0.7 I	F'		-0.334	-0.386	-0.439	-0.496
I		F"	2.406	2.464	2.521	2.579
0.8 I	F'		-1.033	-1.138	-1.254	-1.382
I		F"	2.993	3.053	3.114	3.175
0.9 I	F'		-3.106	-3.889	-8.540	-4.102
I		F"	3.651	3.735	3.822	0.512
1.0 I	F'		-2.219	-2.125	-2.043	-1.970
I		F"	0.585	0.596	0.607	0.617
1.1 I	F'		-1.609	-1.571	-1.535	-1.501
I		F"	0.696	0.708	0.720	0.731
1.2 I	F'		-1.304	-1.281	-1.258	-1.236
I		F"	0.816	0.829	0.841	0.854
1.3 I	F'		-1.100	-1.083	-1.066	-1.049
I		F"	0.944	0.957	0.971	0.984
1.4 I	F'		-0.945	-0.931	-0.917	-0.904
I		F"	1.080	1.094	1.108	1.122
1.5 I	F'		-0.816	-0.804	-0.792	-0.781
I		F"	1.221	1.236	1.250	1.265
1.6 I	F'		-0.704	-0.695	-0.685	-0.675
I		F"	1.369	1.385	1.400	1.415
1.7 I	F'		-0.607	-0.598	-0.589	-0.580
I		F"	1.524	1.540	1.556	1.572
1.8 I	F'		-0.520	-0.512	-0.504	-0.496
I		F"	1.685	1.702	1.718	1.735
1.9 I	F'		-0.443	-0.436	-0.428	-0.421
I		F"	1.853	1.870	1.887	1.904
2.0 I	F'		-0.372	-0.365	-0.359	-0.352
I		F"	2.025	2.043	2.061	2.078
2.1 I	F'		-0.307	-0.301	-0.295	-0.289
I		F"	2.204	2.222	2.240	2.259
2.2 I	F'		-0.249	-0.244	-0.238	-0.233
I		F"	2.388	2.407	2.426	2.445
2.3 I	F'		-0.197	-0.192	-0.187	-0.182
I		F"	2.578	2.597	2.616	2.636
2.4 I	F'		-0.150	-0.146	-0.142	-0.137
I		F"	2.772	2.792	2.812	2.832
2.5 I	F'		-0.109	-0.106	-0.102	-0.098
I		F"	2.972	2.993	3.013	3.033
2.6 I	F'		-0.074	-0.071	-0.068	-0.065
I		F"	3.177	3.198	3.219	3.240
2.7 I	F'		-0.045	-0.043	-0.040	-0.038
I		F"	3.387	3.408	3.429	3.451
2.8 I	F'		-0.022	-0.020	-0.018	-0.016
I		F"	3.601	3.623	3.644	3.666

0.04	0.05	0.06	0.07	0.08	0.09
-0.089	-0.075	-0.060	-0.046	-0.031	-0.017
0.122	0.140	0.159	0.179	0.200	0.223
0.050	0.063	0.075	0.087	0.098	0.109
0.350	0.378	0.408	0.438	0.469	0.501
0.154	0.161	0.168	0.174	0.179	0.183
0.674	0.711	0.748	0.787	0.826	0.866
0.190	0.189	0.186	0.179	0.174	0.168
1.078	1.123	1.168	1.213	1.259	1.306
0.118	0.103	0.088	0.070	0.051	0.031
1.548	1.598	1.648	1.699	1.751	1.803
-0.100	-0.133	-0.168	-0.205	-0.245	-0.288
2.071	2.125	2.181	2.237	2.293	2.349
-0.556	-0.622	-0.692	-0.767	-0.849	-0.937
2.637	2.696	2.754	2.813	2.873	2.933
-1.524	-1.686	-1.871	-2.089	-2.299	-2.636
3.236	3.298	3.360	3.422	3.489	3.569
-3.431	-3.056	-2.800	-2.607	-2.453	-2.326
0.522	0.532	0.543	0.553	0.564	0.574
-1.904	-1.845	-1.790	-1.740	-1.693	-1.649
0.628	0.640	0.651	0.662	0.673	0.685
-1.469	-1.438	-1.409	-1.381	-1.354	-1.329
0.743	0.755	0.767	0.779	0.791	0.804
-1.215	-1.194	-1.174	-1.155	-1.136	-1.118
0.866	0.879	0.892	0.905	0.918	0.931
-1.033	-1.018	-1.003	-0.988	-0.973	-0.959
0.998	1.011	1.025	1.039	1.052	1.066
-0.891	-0.878	-0.865	-0.852	-0.840	-0.828
1.136	1.150	1.164	1.178	1.193	1.207
-0.769	-0.758	-0.747	-0.736	-0.725	-0.715
1.280	1.295	1.309	1.324	1.339	1.354
-0.665	-0.655	-0.645	-0.635	-0.626	-0.617
1.431	1.446	1.462	1.477	1.493	1.508
-0.571	-0.562	-0.553	-0.545	-0.536	-0.528
1.588	1.604	1.620	1.636	1.653	1.669
-0.488	-0.481	-0.474	-0.466	-0.458	-0.451
1.752	1.768	1.785	1.802	1.819	1.836
-0.414	-0.407	-0.400	-0.393	-0.386	-0.379
1.921	1.938	1.956	1.973	1.990	2.008
-0.345	-0.339	-0.332	-0.326	-0.320	-0.314
2.096	2.114	2.132	2.150	2.168	2.186
-0.283	-0.277	-0.272	-0.266	-0.260	-0.255
2.277	2.295	2.314	2.332	2.351	2.370
-0.227	-0.222	-0.217	-0.212	-0.207	-0.202
2.463	2.482	2.501	2.520	2.539	2.559
-0.177	-0.173	-0.168	-0.164	-0.159	-0.155
2.655	2.674	2.694	2.714	2.733	2.753
-0.133	-0.129	-0.125	-0.121	-0.117	-0.113
2.852	2.872	2.892	2.912	2.932	2.952
-0.095	-0.091	-0.088	-0.084	-0.081	-0.078
3.054	3.074	3.095	3.115	3.136	3.156
-0.062	-0.059	-0.056	-0.053	-0.051	-0.048
3.260	3.281	3.302	3.323	3.344	3.366
-0.035	-0.033	-0.031	-0.028	-0.026	-0.024
3.472	3.493	3.515	3.536	3.558	3.579
-0.014	-0.013	-0.011	-0.009	-0.008	-0.006
3.688	3.710	3.732	3.754	3.776	3.798

ATOMIC SYMBOL = BR   ATOMIC NUMBER = 35

I			0.0000	0.0001	0.0002	0.0003
0.906 I	F '		-3.512	-3.521	-3.529	-3.537
I		F ʺ	3.701	3.702	3.703	3.704
0.907 I	F '		-3.596	-3.605	-3.613	-3.622
I		F ʺ	3.710	3.711	3.711	3.712
0.908 I	F '		-3.686	-3.695	-3.705	-3.714
I		F ʺ	3.718	3.719	3.720	3.721
0.909 I	F '		-3.783	-3.793	-3.804	-3.814
I		F ʺ	3.727	3.728	3.728	3.729
0.910 I	F '		-3.889	-3.901	-3.912	-3.923
I		F ʺ	3.735	3.736	3.737	3.738
0.911 I	F '		-4.006	-4.019	-4.031	-4.044
I		F ʺ	3.744	3.745	3.746	3.746
0.912 I	F '		-4.136	-4.150	-4.164	-4.178
I		F ʺ	3.752	3.753	3.754	3.755
0.913 I	F '		-4.282	-4.298	-4.314	-4.330
I		F ʺ	3.761	3.762	3.763	3.764
0.914 I	F '		-4.450	-4.468	-4.487	-4.505
I		F ʺ	3.770	3.771	3.771	3.772
0.915 I	F '		-4.646	-4.668	-4.690	-4.713
I		F ʺ	3.778	3.779	3.780	3.781
0.916 I	F '		-4.884	-4.911	-4.938	-4.967
I		F ʺ	3.787	3.788	3.789	3.790
0.917 I	F '		-5.186	-5.222	-5.258	-5.296
I		F ʺ	3.796	3.797	3.798	3.798
0.918 I	F '		-5.603	-5.654	-5.709	-5.766
I		F ʺ	3.805	3.806	3.806	3.807
0.919 I	F '		-6.282	-6.380	-6.489	-6.610
I		F ʺ	3.813	3.814	3.815	3.816
0.920 I	F '		-8.538	-9.941	-9.328	-8.335
I		F ʺ	3.822	3.823	0.502	0.502
0.921 I	F '		-6.591	-6.477	-6.374	-6.280
I		F ʺ	0.503	0.503	0.503	0.503
0.922 I	F '		-5.791	-5.737	-5.686	-5.637
I		F ʺ	0.504	0.504	0.504	0.504
0.923 I	F '		-5.348	-5.313	-5.279	-5.246
I		F ʺ	0.505	0.505	0.505	0.505
0.924 I	F '		-5.042	-5.016	-4.990	-4.966
I		F ʺ	0.506	0.506	0.506	0.506
0.925 I	F '		-4.808	-4.787	-4.767	-4.747
I		F ʺ	0.507	0.507	0.507	0.507
0.926 I	F '		-4.619	-4.602	-4.586	-4.569
I		F ʺ	0.508	0.508	0.508	0.508
0.927 I	F '		-4.461	-4.447	-4.433	-4.419
I		F ʺ	0.509	0.509	0.509	0.509
0.928 I	F '		-4.326	-4.314	-4.301	-4.289
I		F ʺ	0.510	0.510	0.510	0.510
0.929 I	F '		-4.208	-4.196	-4.186	-4.175
I		F ʺ	0.511	0.511	0.511	0.511
0.930 I	F '		-4.102	-4.092	-4.083	-4.073
I		F ʺ	0.512	0.512	0.512	0.512
0.931 I	F '		-4.008	-3.999	-3.990	-3.981
I		F ʺ	0.513	0.513	0.513	0.513
0.932 I	F '		-3.922	-3.914	-3.906	-3.898
I		F ʺ	0.514	0.514	0.514	0.514
0.933 I	F '		-3.844	-3.836	-3.829	-3.822
I		F ʺ	0.515	0.515	0.515	0.515

K ABSORPTION EDGE (0.92040 Å;   13.4698 KEV)

0.0004	0.0005	0.0006	0.0007	0.0008	0.0009
-3.545	-3.553	-3.562	-3.570	-3.579	-3.587
3.705	3.705	3.706	3.707	3.708	3.709
-3.631	-3.640	-3.649	-3.658	-3.667	-3.677
3.713	3.714	3.715	3.716	3.716	3.717
-3.724	-3.734	-3.743	-3.753	-3.763	-3.773
3.722	3.722	3.723	3.724	3.725	3.726
-3.825	-3.835	-3.846	-3.857	-3.867	-3.878
3.730	3.731	3.732	3.733	3.734	3.734
-3.935	-3.946	-3.958	-3.970	-3.982	-3.994
3.739	3.740	3.740	3.741	3.742	3.743
-4.056	-4.069	-4.082	-4.096	-4.109	-4.122
3.747	3.748	3.749	3.750	3.751	3.752
-4.192	-4.207	-4.222	-4.236	-4.251	-4.267
3.756	3.757	3.758	3.758	3.759	3.760
-4.346	-4.363	-4.380	-4.397	-4.414	-4.432
3.765	3.765	3.766	3.767	3.768	3.769
-4.524	-4.544	-4.564	-4.584	-4.604	-4.625
3.773	3.774	3.775	3.776	3.777	3.778
-4.736	-4.759	-4.783	-4.807	-4.832	-4.858
3.782	3.783	3.784	3.785	3.785	3.786
-4.995	-5.025	-5.055	-5.087	-5.119	-5.152
3.791	3.791	3.792	3.793	3.794	3.795
-5.335	-5.375	-5.417	-5.460	-5.506	-5.553
3.799	3.800	3.801	3.802	3.803	3.804
-5.826	-5.890	-5.958	-6.030	-6.108	-6.191
3.808	3.809	3.810	3.811	3.812	3.813
-6.746	-6.901	-7.084	-7.303	-7.580	-7.955
3.817	3.818	3.819	3.820	3.820	3.821
-7.834	-7.497	-7.242	-7.037	-6.866	-6.720
0.502	0.503	0.503	0.503	0.503	0.503
-6.194	-6.115	-6.042	-5.973	-5.909	-5.848
0.503	0.504	0.504	0.504	0.504	0.504
-5.590	-5.546	-5.503	-5.462	-5.423	-5.385
0.504	0.505	0.505	0.505	0.505	0.505
-5.214	-5.183	-5.153	-5.124	-5.096	-5.068
0.505	0.506	0.506	0.506	0.506	0.506
-4.941	-4.918	-4.895	-4.872	-4.850	-4.829
0.506	0.507	0.507	0.507	0.507	0.507
-4.728	-4.709	-4.690	-4.672	-4.654	-4.637
0.507	0.508	0.508	0.508	0.508	0.508
-4.553	-4.537	-4.522	-4.506	-4.491	-4.476
0.508	0.509	0.509	0.509	0.509	0.509
-4.405	-4.391	-4.378	-4.365	-4.352	-4.339
0.509	0.510	0.510	0.510	0.510	0.510
-4.277	-4.265	-4.253	-4.242	-4.230	-4.219
0.510	0.511	0.511	0.511	0.511	0.511
-4.164	-4.153	-4.143	-4.133	-4.122	-4.112
0.511	0.512	0.512	0.512	0.512	0.512
-4.063	-4.054	-4.045	-4.035	-4.026	-4.017
0.513	0.513	0.513	0.513	0.513	0.513
-3.973	-3.964	-3.955	-3.947	-3.939	-3.930
0.514	0.514	0.514	0.514	0.514	0.514
-3.890	-3.882	-3.874	-3.867	-3.859	-3.851
0.515	0.515	0.515	0.515	0.515	0.515
-3.814	-3.807	-3.800	-3.793	-3.786	-3.779
0.516	0.516	0.516	0.516	0.516	0.516

Table A3.2. (*cont.*)

ATOMIC SYMBOL = PT   ATOMIC NUMBER = 78

I		0.00	0.01	0.02	0.03
0.1 I	F'	-1.382	-1.483	-1.644	-1.898
I	F"	1.515	1.776	2.052	2.338
0.2 I	F'	-2.368	-2.243	-2.137	-2.045
I	F"	0.976	1.065	1.156	1.250
0.3 I	F'	-1.603	-1.557	-1.513	-1.472
I	F"	1.986	2.102	2.221	2.341
0.4 I	F'	-1.262	-1.242	-1.226	-1.212
I	F"	3.254	3.394	3.536	3.680
0.5 I	F'	-1.211	-1.223	-1.238	-1.257
I	F"	4.741	4.899	5.059	5.221
0.6 I	F'	-1.500	-1.552	-1.608	-1.670
I	F"	6.399	6.573	6.749	6.926
0.7 I	F'	-2.260	-2.371	-2.492	-2.620
I	F"	8.207	8.396	8.585	8.776
0.8 I	F'	-3.816	-4.046	-4.297	-4.571
I	F"	10.145	10.345	10.547	10.749
0.9 I	F'	-7.881	-7.957	-8.468	-9.799
I	F"	10.547	10.749	10.952	11.158
1.0 I	F'	-8.456	-8.658	-8.938	-9.317
I	F"	9.051	9.207	9.364	9.522
1.1 I	F'	-10.368	-9.780	-9.332	-8.971
I	F"	4.035	4.094	4.154	4.215
1.2 I	F'	-7.481	-7.341	-7.212	-7.092
I	F"	4.645	4.708	4.771	4.834
1.3 I	F'	-6.439	-6.363	-6.291	-6.222
I	F"	5.285	5.350	5.416	5.482
1.4 I	F'	-5.816	-5.766	-5.718	-5.671
I	F"	5.951	6.019	6.086	6.154
1.5 I	F'	-5.387	-5.351	-5.317	-5.283
I	F"	6.637	6.707	6.777	6.848
1.6 I	F'	-5.075	-5.049	-5.024	-4.999
I	F"	7.347	7.419	7.491	7.564
1.7 I	F'	-4.847	-4.827	-4.809	-4.791
I	F"	8.078	8.152	8.226	8.301
1.8 I	F'	-4.682	-4.668	-4.655	-4.643
I	F"	8.829	8.906	8.982	9.059
1.9 I	F'	-4.573	-4.565	-4.557	-4.550
I	F"	9.601	9.679	9.757	9.836
2.0 I	F'	-4.516	-4.512	-4.509	-4.506
I	F"	10.390	10.470	10.550	10.630
2.1 I	F'	-4.507	-4.513	-4.514	-4.516
I	F"	11.196	11.277	11.359	11.440
2.2 I	F'	-4.539	-4.544	-4.550	-4.555
I	F"	12.016	12.099	12.182	12.265
2.3 I	F'	-4.607	-4.616	-4.626	-4.636
I	F"	12.852	12.936	13.020	13.105
2.4 I	F'	-4.729	-4.742	-4.756	-4.770
I	F"	13.700	13.786	13.871	13.957
2.5 I	F'	-4.880	-4.897	-4.915	-4.933
I	F"	14.561	14.648	14.735	14.822
2.6 I	F'	-5.072	-5.094	-5.116	-5.138
I	F"	15.434	15.523	15.611	15.699
2.7 I	F'	-5.308	-5.334	-5.360	-5.387
I	F"	16.320	16.410	16.499	16.588
2.8 I	F'	-5.588	-5.619	-5.671	-5.703
I	F"	17.218	17.308	17.399	17.489

0.04	0.05	0.06	0.07	0.08	0.09
-2.294	-3.042	-4.485	-3.154	-2.763	-2.532
2.636	2.945	0.656	0.731	0.809	0.891
-1.964	-1.893	-1.826	-1.764	-1.707	-1.653
1.347	1.447	1.549	1.655	1.763	1.873
-1.434	-1.399	-1.366	-1.336	-1.308	-1.284
2.465	2.590	2.718	2.849	2.982	3.117
-1.202	-1.197	-1.193	-1.195	-1.197	-1.202
3.827	3.975	4.125	4.277	4.430	4.585
-1.281	-1.307	-1.337	-1.372	-1.410	-1.453
5.385	5.550	5.716	5.884	6.054	6.226
-1.736	-1.809	-1.886	-1.970	-2.060	-2.157
7.105	7.285	7.467	7.650	7.834	8.020
-2.756	-2.903	-3.060	-3.228	-3.409	-3.605
8.968	9.161	9.356	9.551	9.748	9.946
-4.878	-5.223	-5.630	-6.107	-6.726	-7.837
10.953	11.158	11.362	11.566	11.771	11.976
-9.563	-8.723	-8.408	-8.280	-8.261	-8.323
8.134	8.284	8.436	8.588	8.741	8.896
-9.833	-10.576	-11.805	-15.375	-12.813	-11.226
9.681	9.841	10.002	10.163	3.916	3.975
-8.669	-8.409	-8.181	-7.979	-7.797	-7.632
4.275	4.336	4.397	4.459	4.520	4.582
-6.980	-6.875	-6.776	-6.682	-6.596	-6.512
4.897	4.961	5.025	5.090	5.155	5.220
-6.156	-6.096	-6.036	-5.977	-5.922	-5.868
5.548	5.615	5.682	5.749	5.816	5.883
-5.627	-5.583	-5.541	-5.501	-5.462	-5.424
6.223	6.291	6.360	6.429	6.498	6.568
-5.251	-5.219	-5.189	-5.159	-5.130	-5.102
6.918	6.989	7.060	7.132	7.203	7.275
-4.975	-4.952	-4.930	-4.908	-4.887	-4.866
7.637	7.710	7.783	7.856	7.930	8.004
-4.774	-4.757	-4.741	-4.725	-4.710	-4.696
8.376	8.451	8.526	8.602	8.677	8.753
-4.631	-4.620	-4.609	-4.599	-4.589	-4.582
9.136	9.213	9.290	9.368	9.445	9.523
-4.543	-4.536	-4.530	-4.525	-4.520	-4.515
9.914	9.993	10.072	10.152	10.231	10.310
-4.504	-4.510	-4.508	-4.507	-4.507	-4.507
10.711	10.791	10.872	10.953	11.033	11.115
-4.518	-4.521	-4.524	-4.527	-4.531	-4.535
11.522	11.604	11.686	11.768	11.851	11.933
-4.562	-4.568	-4.575	-4.583	-4.591	-4.599
12.348	12.432	12.516	12.599	12.683	12.767
-4.646	-4.669	-4.680	-4.692	-4.704	-4.716
13.190	13.275	13.360	13.444	13.530	13.615
-4.784	-4.799	-4.815	-4.830	-4.846	-4.863
14.043	14.129	14.215	14.301	14.388	14.474
-4.952	-4.971	-4.990	-5.010	-5.030	-5.051
14.909	14.996	15.084	15.171	15.259	15.347
-5.161	-5.185	-5.208	-5.233	-5.257	-5.282
15.787	15.876	15.965	16.053	16.142	16.231
-5.414	-5.442	-5.470	-5.499	-5.528	-5.558
16.678	16.768	16.857	16.947	17.037	17.128
-5.735	-5.768	-5.801	-5.834	-5.869	-5.903
17.579	17.669	17.760	17.850	17.941	18.032

ATOMIC SYMBOL = PT   ATOMIC NUMBER = 78

I			0.0000	0.0001	0.0002	0.0003
0.879	I	F'	-6.654	-6.661	-6.668	-6.675
	I	F"	11.750	11.752	11.754	11.756
0.880	I	F'	-6.726	-6.734	-6.741	-6.749
	I	F"	11.771	11.773	11.775	11.777
0.881	I	F'	-6.802	-6.810	-6.818	-6.826
	I	F"	11.791	11.793	11.795	11.797
0.882	I	F'	-6.882	-6.890	-6.899	-6.907
	I	F"	11.812	11.814	11.816	11.818
0.883	I	F'	-6.967	-6.975	-6.984	-6.993
	I	F"	11.832	11.834	11.836	11.839
0.884	I	F'	-7.057	-7.066	-7.076	-7.085
	I	F"	11.853	11.855	11.857	11.859
0.885	I	F'	-7.154	-7.164	-7.174	-7.184
	I	F"	11.873	11.876	11.878	11.880
0.886	I	F'	-7.259	-7.270	-7.281	-7.292
	I	F"	11.894	11.896	11.898	11.900
0.887	I	F'	-7.375	-7.387	-7.400	-7.412
	I	F"	11.915	11.917	11.919	11.921
0.888	I	F'	-7.505	-7.519	-7.533	-7.548
	I	F"	11.935	11.937	11.939	11.941
0.889	I	F'	-7.655	-7.672	-7.689	-7.706
	I	F"	11.956	11.958	11.960	11.962
0.890	I	F'	-7.837	-7.857	-7.878	-7.889
	I	F"	11.976	11.978	11.981	11.983
0.891	I	F'	-8.061	-8.089	-8.118	-8.149
	I	F"	11.997	11.999	12.001	12.004
0.892	I	F'	-8.416	-8.465	-8.518	-8.577
	I	F"	12.018	12.020	12.022	12.025
0.893	I	F'	-9.373	-9.718	-11.069	-9.808
	I	F"	12.039	12.041	12.043	10.412
0.894	I	F'	-8.721	-8.663	-8.613	-8.567
	I	F"	10.426	10.428	10.430	10.432
0.895	I	F'	-8.341	-8.317	-8.295	-8.274
	I	F"	10.446	10.448	10.450	10.452
0.896	I	F'	-8.156	-8.142	-8.129	-8.116
	I	F"	10.466	10.468	10.470	10.472
0.897	I	F'	-8.042	-8.033	-8.025	-8.016
	I	F"	10.486	10.488	10.490	10.492
0.898	I	F'	-7.967	-7.961	-7.955	-7.949
	I	F"	10.506	10.508	10.510	10.512
0.899	I	F'	-7.916	-7.912	-7.908	-7.904
	I	F"	10.526	10.528	10.530	10.532
0.900	I	F'	-7.881	-7.878	-7.876	-7.873
	I	F"	10.547	10.549	10.551	10.553
0.901	I	F'	-7.859	-7.857	-7.855	-7.854
	I	F"	10.567	10.569	10.571	10.573
0.902	I	F'	-7.846	-7.845	-7.844	-7.844
	I	F"	10.587	10.589	10.591	10.593
0.903	I	F'	-7.841	-7.841	-7.841	-7.841
	I	F"	10.607	10.609	10.611	10.613
0.904	I	F'	-7.843	-7.843	-7.844	-7.845
	I	F"	10.627	10.629	10.631	10.633
0.905	I	F'	-7.850	-7.851	-7.853	-7.854
	I	F"	10.647	10.649	10.651	10.654
0.906	I	F'	-7.863	-7.865	-7.866	-7.868
	I	F"	10.668	10.670	10.672	10.674

L₁ ABSORPTION EDGE (0.89310 Å;   13.8816 KEV)

0.0004	0.0005	0.0006	0.0007	0.0008	0.0009
-6.682	-6.690	-6.697	-6.704	-6.711	-6.719
11.758	11.760	11.763	11.765	11.767	11.769
-6.756	-6.764	-6.771	-6.779	-6.787	-6.794
11.779	11.781	11.783	11.785	11.787	11.789
-6.834	-6.842	-6.850	-6.858	-6.866	-6.874
11.800	11.802	11.804	11.806	11.808	11.810
-6.915	-6.924	-6.932	-6.941	-6.949	-6.958
11.820	11.822	11.824	11.826	11.828	11.830
-7.002	-7.011	-7.020	-7.029	-7.038	-7.047
11.841	11.843	11.845	11.847	11.849	11.851
-7.095	-7.104	-7.114	-7.124	-7.134	-7.144
11.861	11.863	11.865	11.867	11.869	11.871
-7.195	-7.205	-7.216	-7.226	-7.237	-7.248
11.882	11.884	11.886	11.888	11.890	11.892
-7.304	-7.315	-7.327	-7.339	-7.351	-7.363
11.902	11.904	11.906	11.908	11.911	11.913
-7.425	-7.438	-7.451	-7.464	-7.478	-7.491
11.923	11.925	11.927	11.929	11.931	11.933
-7.562	-7.577	-7.592	-7.608	-7.623	-7.639
11.943	11.946	11.948	11.950	11.952	11.954
-7.723	-7.741	-7.759	-7.778	-7.797	-7.817
11.964	11.966	11.968	11.970	11.972	11.974
-7.911	-7.934	-7.957	-7.982	-8.007	-8.033
11.985	11.987	11.989	11.991	11.993	11.995
-8.181	-8.215	-8.250	-8.288	-8.328	-8.370
12.006	12.008	12.010	12.012	12.014	12.016
-8.642	-8.715	-8.798	-8.896	-9.014	-9.165
12.027	12.029	12.031	12.033	12.035	12.037
-9.430	-9.217	-9.068	-8.955	-8.863	-8.786
10.414	10.416	10.418	10.420	10.422	10.424
-8.526	-8.489	-8.454	-8.423	-8.394	-8.366
10.434	10.436	10.438	10.440	10.442	10.444
-8.254	-8.236	-8.218	-8.201	-8.185	-8.170
10.454	10.456	10.458	10.460	10.462	10.464
-8.104	-8.093	-8.082	-8.071	-8.061	-8.052
10.474	10.476	10.478	10.480	10.482	10.484
-8.009	-8.001	-7.994	-7.987	-7.980	-7.973
10.494	10.496	10.498	10.500	10.502	10.504
-7.944	-7.939	-7.934	-7.929	-7.924	-7.920
10.514	10.516	10.518	10.520	10.522	10.524
-7.900	-7.897	-7.893	-7.890	-7.887	-7.884
10.534	10.536	10.539	10.541	10.543	10.545
-7.871	-7.868	-7.866	-7.864	-7.862	-7.860
10.555	10.557	10.559	10.561	10.563	10.565
-7.853	-7.851	-7.850	-7.849	-7.848	-7.847
10.575	10.577	10.579	10.581	10.583	10.585
-7.843	-7.843	-7.842	-7.842	-7.842	-7.841
10.595	10.597	10.599	10.601	10.603	10.605
-7.841	-7.841	-7.841	-7.842	-7.842	-7.842
10.615	10.617	10.619	10.621	10.623	10.625
-7.845	-7.846	-7.847	-7.848	-7.848	-7.849
10.635	10.637	10.639	10.641	10.643	10.645
-7.855	-7.856	-7.857	-7.859	-7.860	-7.862
10.656	10.658	10.660	10.662	10.664	10.666
-7.869	-7.871	-7.873	-7.875	-7.876	-7.878
10.676	10.678	10.680	10.682	10.684	10.686

Table A3.2. (*cont.*)

ATOMIC SYMBOL = PT   ATOMIC NUMBER = 78

	I		0.0000	0.0001	0.0002	0.0003
0.920	I	F'	-8.468	-8.476	-8.483	-8.491
	I	F"	10.952	10.955	10.957	10.959
0.921	I	F'	-8.546	-8.554	-8.563	-8.571
	I	F"	10.973	10.975	10.977	10.979
0.922	I	F'	-8.631	-8.640	-8.649	-8.658
	I	F"	10.993	10.995	10.998	11.000
0.923	I	F'	-8.724	-8.734	-8.744	-8.754
	I	F"	11.014	11.016	11.018	11.020
0.924	I	F'	-8.826	-8.837	-8.848	-8.859
	I	F"	11.034	11.036	11.039	11.041
0.925	I	F'	-8.939	-8.951	-8.963	-8.975
	I	F"	11.055	11.057	11.059	11.061
0.926	I	F'	-9.065	-9.078	-9.092	-9.106
	I	F"	11.076	11.078	11.080	11.082
0.927	I	F'	-9.208	-9.223	-9.239	-9.254
	I	F"	11.096	11.098	11.100	11.102
0.928	I	F'	-9.372	-9.390	-9.408	-9.426
	I	F"	11.117	11.119	11.121	11.123
0.929	I	F'	-9.565	-9.586	-9.608	-9.630
	I	F"	11.137	11.139	11.141	11.143
0.930	I	F'	-9.799	-9.825	-9.852	-9.880
	I	F"	11.158	11.160	11.162	11.164
0.931	I	F'	-10.097	-10.132	-10.168	-10.205
	I	F"	11.178	11.181	11.183	11.185
0.932	I	F'	-10.509	-10.561	-10.615	-10.672
	I	F"	11.199	11.201	11.203	11.205
0.933	I	F'	-11.192	-11.292	-11.404	-11.528
	I	F"	11.220	11.222	11.224	11.226
0.934	I	F'	-13.885	-15.140	-13.415	-12.810
	I	F"	11.240	8.045	8.047	8.048
0.935	I	F'	-11.377	-11.274	-11.181	-11.096
	I	F"	8.059	8.060	8.062	8.063
0.936	I	F'	-10.646	-10.596	-10.549	-10.504
	I	F"	8.074	8.075	8.077	8.078
0.937	I	F'	-10.236	-10.203	-10.172	-10.141
	I	F"	8.089	8.090	8.092	8.093
0.938	I	F'	-9.952	-9.928	-9.904	-9.881
	I	F"	8.104	8.105	8.107	8.108
0.939	I	F'	-9.736	-9.717	-9.698	-9.680
	I	F"	8.119	8.120	8.122	8.123
0.940	I	F'	-9.563	-9.547	-9.532	-9.517
	I	F"	8.134	8.135	8.137	8.138
0.941	I	F'	-9.419	-9.406	-9.393	-9.381
	I	F"	8.149	8.150	8.152	8.153
0.942	I	F'	-9.297	-9.286	-9.275	-9.264
	I	F"	8.164	8.165	8.167	8.168
0.943	I	F'	-9.192	-9.182	-9.172	-9.163
	I	F"	8.179	8.180	8.182	8.183
0.944	I	F'	-9.099	-9.091	-9.082	-9.074
	I	F"	8.194	8.195	8.197	8.198
0.945	I	F'	-9.017	-9.010	-9.002	-9.004
	I	F"	8.209	8.210	8.212	8.213
0.946	I	F'	-8.952	-8.945	-8.938	-8.932
	I	F"	8.224	8.225	8.227	8.228
0.947	I	F'	-8.886	-8.880	-8.874	-8.868
	I	F"	8.239	8.240	8.242	8.244

L₂ ABSORPTION EDGE (0.93414 Å; 13.2717 KEV)

0.0004	0.0005	0.0006	0.0007	0.0008	0.0009
-8.499	-8.506	-8.514	-8.522	-8.530	-8.538
10.961	10.963	10.965	10.967	10.969	10.971
-8.579	-8.588	-8.596	-8.605	-8.614	-8.622
10.981	10.983	10.985	10.987	10.989	10.991
-8.667	-8.677	-8.686	-8.695	-8.705	-8.714
11.002	11.004	11.006	11.008	11.010	11.012
-8.764	-8.774	-8.784	-8.794	-8.805	-8.815
11.022	11.024	11.026	11.028	11.030	11.032
-8.870	-8.881	-8.892	-8.904	-8.915	-8.927
11.043	11.045	11.047	11.049	11.051	11.053
-8.988	-9.000	-9.013	-9.026	-9.039	-9.052
11.063	11.065	11.067	11.069	11.071	11.073
-9.120	-9.134	-9.148	-9.163	-9.178	-9.193
11.084	11.086	11.088	11.090	11.092	11.094
-9.270	-9.287	-9.303	-9.320	-9.337	-9.354
11.104	11.106	11.108	11.110	11.113	11.115
-9.445	-9.464	-9.483	-9.503	-9.523	-9.544
11.125	11.127	11.129	11.131	11.133	11.135
-9.652	-9.675	-9.699	-9.723	-9.748	-9.773
11.145	11.148	11.150	11.152	11.154	11.156
-9.908	-9.937	-9.967	-9.998	-10.030	-10.063
11.166	11.168	11.170	11.172	11.174	11.176
-10.243	-10.283	-10.325	-10.368	-10.413	-10.460
11.187	11.189	11.191	11.193	11.195	11.197
-10.733	-10.796	-10.865	-10.937	-11.015	-11.100
11.207	11.209	11.211	11.214	11.216	11.218
-11.670	-11.834	-12.029	-12.270	-12.585	-13.041
11.228	11.230	11.232	11.234	11.236	11.238
-12.435	-12.162	-11.948	-11.772	-11.622	-11.492
8.050	8.051	8.053	8.054	8.056	8.057
-11.017	-10.945	-10.877	-10.814	-10.755	-10.699
8.065	8.066	8.068	8.069	8.071	8.072
-10.460	-10.419	-10.379	-10.341	-10.305	-10.270
8.080	8.081	8.083	8.084	8.086	8.087
-10.111	-10.083	-10.055	-10.028	-10.002	-9.976
8.095	8.096	8.098	8.099	8.101	8.102
-9.859	-9.837	-9.816	-9.795	-9.775	-9.755
8.110	8.111	8.113	8.114	8.116	8.117
-9.662	-9.645	-9.628	-9.611	-9.595	-9.579
8.125	8.126	8.128	8.129	8.131	8.132
-9.502	-9.488	-9.474	-9.460	-9.446	-9.432
8.140	8.141	8.143	8.144	8.146	8.147
-9.368	-9.356	-9.344	-9.332	-9.320	-9.309
8.155	8.156	8.158	8.159	8.161	8.162
-9.253	-9.243	-9.232	-9.222	-9.212	-9.202
8.170	8.171	8.173	8.174	8.176	8.177
-9.153	-9.144	-9.135	-9.126	-9.117	-9.108
8.185	8.186	8.188	8.189	8.191	8.192
-9.065	-9.057	-9.049	-9.041	-9.033	-9.025
8.200	8.201	8.203	8.204	8.206	8.207
-8.995	-8.988	-8.980	-8.973	-8.966	-8.959
8.215	8.216	8.218	8.219	8.221	8.222
-8.925	-8.918	-8.912	-8.905	-8.899	-8.892
8.230	8.231	8.233	8.234	8.236	8.237
-8.862	-8.856	-8.850	-8.844	-8.838	-8.832
8.245	8.247	8.248	8.250	8.251	8.253

Table A3.2. (*cont.*)

ATOMIC SYMBOL = PT   ATOMIC NUMBER = 78

	I		0.0000	0.0001	0.0002	0.0003
1.058	I	F'	-11.491	-11.506	-11.520	-11.535
	I	F"	9.969	9.971	9.973	9.974
1.059	I	F'	-11.642	-11.658	-11.673	-11.689
	I	F"	9.985	9.987	9.989	9.990
1.060	I	F'	-11.805	-11.822	-11.839	-11.856
	I	F"	10.002	10.003	10.005	10.006
1.061	I	F'	-11.981	-12.000	-12.019	-12.038
	I	F"	10.018	10.019	10.021	10.023
1.062	I	F'	-12.175	-12.195	-12.216	-12.237
	I	F"	10.034	10.035	10.037	10.039
1.063	I	F'	-12.389	-12.411	-12.434	-12.457
	I	F"	10.050	10.052	10.053	10.055
1.064	I	F'	-12.627	-12.653	-12.678	-12.705
	I	F"	10.066	10.068	10.069	10.071
1.065	I	F'	-12.897	-12.926	-12.956	-12.985
	I	F"	10.082	10.084	10.085	10.087
1.066	I	F'	-13.208	-13.241	-13.276	-13.311
	I	F"	10.098	10.100	10.102	10.103
1.067	I	F'	-13.573	-13.614	-13.655	-13.697
	I	F"	10.115	10.116	10.118	10.119
1.068	I	F'	-14.018	-14.068	-14.120	-14.173
	I	F"	10.131	10.132	10.134	10.136
1.069	I	F'	-14.586	-14.653	-14.722	-14.793
	I	F"	10.147	10.149	10.150	10.152
1.070	I	F'	-15.375	-15.473	-15.576	-15.685
	I	F"	10.163	10.165	10.166	10.168
1.071	I	F'	-16.673	-16.863	-17.073	-17.307
	I	F"	10.179	10.181	10.183	10.184
1.072	I	F'	-21.232	-25.119	-21.900	-20.313
	I	F"	10.196	10.197	3.870	3.871
1.073	I	F'	-17.160	-16.945	-16.752	-16.575
	I	F"	3.875	3.876	3.876	3.877
1.074	I	F'	-15.648	-15.545	-15.447	-15.354
	I	F"	3.881	3.882	3.882	3.883
1.075	I	F'	-14.801	-14.733	-14.668	-14.604
	I	F"	3.887	3.887	3.888	3.889
1.076	I	F'	-14.210	-14.160	-14.111	-14.063
	I	F"	3.893	3.893	3.894	3.895
1.077	I	F'	-13.757	-13.717	-13.677	-13.639
	I	F"	3.899	3.899	3.900	3.900
1.078	I	F'	-13.389	-13.356	-13.323	-13.291
	I	F"	3.905	3.905	3.906	3.906
1.079	I	F'	-13.080	-13.051	-13.023	-12.996
	I	F"	3.911	3.911	3.912	3.912
1.080	I	F'	-12.813	-12.788	-12.764	-12.740
	I	F"	3.916	3.917	3.918	3.918
1.081	I	F'	-12.579	-12.557	-12.535	-12.514
	I	F"	3.922	3.923	3.924	3.924
1.082	I	F'	-12.378	-12.359	-12.339	-12.320
	I	F"	3.928	3.929	3.929	3.930
1.083	I	F'	-12.190	-12.172	-12.154	-12.137
	I	F"	3.934	3.935	3.935	3.936
1.084	I	F'	-12.021	-12.005	-11.988	-11.972
	I	F"	3.940	3.941	3.941	3.942
1.085	I	F'	-11.863	-11.848	-11.833	-11.819
	I	F"	3.946	3.947	3.947	3.948

L₃ ABSORPTION EDGE (1.07230 Å;   11.5617 KEV)

0.0004	0.0005	0.0006	0.0007	0.0008	0.0009
-11.550	-11.565	-11.580	-11.595	-11.611	-11.626
9.976	9.977	9.979	9.981	9.982	9.984
-11.705	-11.722	-11.738	-11.754	-11.771	-11.788
9.992	9.993	9.995	9.997	9.998	10.000
-11.873	-11.891	-11.909	-11.927	-11.945	-11.963
10.008	10.010	10.011	10.013	10.014	10.016
-12.057	-12.076	-12.095	-12.115	-12.135	-12.155
10.024	10.026	10.027	10.029	10.031	10.032
-12.258	-12.279	-12.300	-12.322	-12.344	-12.366
10.040	10.042	10.043	10.045	10.047	10.048
-12.481	-12.504	-12.528	-12.553	-12.577	-12.602
10.056	10.058	10.060	10.061	10.063	10.064
-12.731	-12.758	-12.785	-12.812	-12.840	-12.868
10.073	10.074	10.076	10.077	10.079	10.081
-13.016	-13.046	-13.078	-13.109	-13.142	-13.174
10.089	10.090	10.092	10.094	10.095	10.097
-13.346	-13.382	-13.419	-13.457	-13.495	-13.534
10.105	10.106	10.108	10.110	10.111	10.113
-13.740	-13.784	-13.829	-13.874	-13.921	-13.969
10.121	10.123	10.124	10.126	10.127	10.129
-14.227	-14.283	-14.340	-14.399	-14.459	-14.522
10.137	10.139	10.140	10.142	10.144	10.145
-14.866	-14.943	-15.022	-15.105	-15.191	-15.281
10.153	10.155	10.157	10.158	10.160	10.161
-15.800	-15.922	-16.052	-16.190	-16.339	-16.499
10.170	10.171	10.173	10.174	10.176	10.178
-17.571	-17.875	-18.232	-18.667	-19.220	-19.985
10.186	10.187	10.189	10.191	10.192	10.194
-19.439	-18.832	-18.367	-17.991	-17.674	-17.400
3.872	3.872	3.873	3.873	3.874	3.875
-16.413	-16.263	-16.124	-15.994	-15.872	-15.757
3.877	3.878	3.879	3.879	3.880	3.880
-15.265	-15.179	-15.098	-15.019	-14.944	-14.871
3.883	3.884	3.885	3.885	3.886	3.886
-14.543	-14.483	-14.426	-14.369	-14.315	-14.262
3.889	3.890	3.890	3.891	3.892	3.892
-14.016	-13.970	-13.926	-13.882	-13.839	-13.798
3.895	3.896	3.896	3.897	3.898	3.898
-13.601	-13.564	-13.528	-13.492	-13.457	-13.423
3.901	3.902	3.902	3.903	3.903	3.904
-13.259	-13.228	-13.198	-13.168	-13.138	-13.109
3.907	3.908	3.908	3.909	3.909	3.910
-12.969	-12.942	-12.915	-12.889	-12.864	-12.838
3.913	3.913	3.914	3.915	3.915	3.916
-12.716	-12.692	-12.669	-12.646	-12.623	-12.601
3.919	3.919	3.920	3.921	3.921	3.922
-12.492	-12.480	-12.459	-12.439	-12.418	-12.398
3.925	3.925	3.926	3.926	3.927	3.928
-12.301	-12.282	-12.263	-12.244	-12.226	-12.208
3.931	3.931	3.932	3.932	3.933	3.934
-12.119	-12.102	-12.085	-12.068	-12.054	-12.038
3.936	3.937	3.938	3.938	3.939	3.939
-11.956	-11.941	-11.925	-11.909	-11.894	-11.879
3.942	3.943	3.944	3.944	3.945	3.945
-11.804	-11.789	-11.775	-11.760	-11.746	-11.732
3.948	3.949	3.949	3.950	3.951	3.951

ATOMIC SYMBOL = HG   ATOMIC NUMBER = 80

I			0.00	0.01	0.02	0.03
0.1	I	F'				
	I	F"	1.662	1.944	2.239	2.547
0.2	I	F'	-2.301	-2.194	-2.101	-2.019
	I	F"	1.091	1.189	1.289	1.394
0.3	I	F'	-1.609	-1.565	-1.525	-1.488
	I	F"	2.208	2.336	2.467	2.600
0.4	I	F'	-1.312	-1.299	-1.294	-1.289
	I	F"	3.609	3.763	3.919	4.078
0.5	I	F'	-1.360	-1.386	-1.415	-1.449
	I	F"	5.241	5.414	5.590	5.767
0.6	I	F'	-1.827	-1.903	-1.986	-2.075
	I	F"	7.054	7.244	7.436	7.630
0.7	I	F'	-2.931	-3.094	-3.269	-3.459
	I	F"	9.027	9.232	9.439	9.646
0.8	I	F'	-5.440	-5.903	-6.488	-7.401
	I	F"	11.137	11.352	11.568	11.785
0.9	I	F'	-8.209	-8.103	-8.100	-8.175
	I	F"	8.354	8.515	8.677	8.840
1.0	I	F'	-12.138	-17.271	-12.016	-10.752
	I	F"	10.024	3.910	3.973	4.036
1.1	I	F'	-7.976	-7.783	-7.609	-7.451
	I	F"	4.489	4.555	4.622	4.688
1.2	I	F'	-6.635	-6.545	-6.460	-6.379
	I	F"	5.165	5.235	5.304	5.374
1.3	I	F'	-5.912	-5.856	-5.801	-5.749
	I	F"	5.872	5.945	6.017	6.090
1.4	I	F'	-5.435	-5.396	-5.358	-5.322
	I	F"	6.609	6.684	6.759	6.834
1.5	I	F'	-5.098	-5.070	-5.042	-5.016
	I	F"	7.368	7.445	7.522	7.600
1.6	I	F'	-4.855	-4.835	-4.815	-4.797
	I	F"	8.151	8.231	8.311	8.391
1.7	I	F'	-4.684	-4.670	-4.657	-4.645
	I	F"	8.959	9.041	9.123	9.206
1.8	I	F'	-4.579	-4.572	-4.564	-4.557
	I	F"	9.788	9.873	9.957	10.041
1.9	I	F'	-4.535	-4.532	-4.530	-4.529
	I	F"	10.638	10.724	10.810	10.897
2.0	I	F'	-4.532	-4.534	-4.537	-4.541
	I	F"	11.506	11.594	11.682	11.770
2.1	I	F'	-4.577	-4.584	-4.592	-4.600
	I	F"	12.392	12.482	12.572	12.662
2.2	I	F'	-4.680	-4.692	-4.704	-4.717
	I	F"	13.295	13.386	13.477	13.569
2.3	I	F'	-4.820	-4.837	-4.854	-4.872
	I	F"	14.213	14.305	14.398	14.491
2.4	I	F'	-5.009	-5.030	-5.052	-5.074
	I	F"	15.145	15.239	15.333	15.428
2.5	I	F'	-5.246	-5.273	-5.300	-5.327
	I	F"	16.092	16.188	16.283	16.379
2.6	I	F'	-5.553	-5.585	-5.618	-5.651
	I	F"	17.053	17.150	17.246	17.343
2.7	I	F'	-5.903	-5.959	-5.998	-6.037
	I	F"	18.023	18.120	18.217	18.314
2.8	I	F'	-6.326	-6.370	-6.414	-6.459
	I	F"	18.999	19.097	19.195	19.293

0.04	0.05	0.06	0.07	0.08	0.09
		-3.245	-2.832	-2.596	-2.429
2.863	0.656	0.735	0.819	0.906	0.997
-1.947	-1.879	-1.817	-1.759	-1.705	-1.655
1.501	1.611	1.725	1.841	1.961	2.083
-1.453	-1.422	-1.394	-1.369	-1.347	-1.328
2.736	2.875	3.017	3.161	3.308	3.457
-1.291	-1.293	-1.299	-1.309	-1.322	-1.339
4.238	4.401	4.566	4.732	4.900	5.069
-1.488	-1.531	-1.580	-1.633	-1.692	-1.756
5.945	6.126	6.308	6.492	6.678	6.865
-2.172	-2.276	-2.388	-2.509	-2.640	-2.780
7.825	8.022	8.220	8.420	8.621	8.823
-3.665	-3.890	-4.135	-4.406	-4.704	-5.042
9.856	10.066	10.278	10.491	10.705	10.921
-7.945	-7.914	-8.449	-10.123	-9.185	-8.485
10.387	10.599	10.812	11.027	8.035	8.194
-8.322	-8.541	-8.842	-9.242	-9.809	-10.645
9.005	9.170	9.337	9.506	9.677	9.850
-10.004	-9.472	-9.059	-8.722	-8.438	-8.192
4.100	4.164	4.228	4.293	4.358	4.423
-7.306	-7.172	-7.050	-6.934	-6.826	-6.725
4.755	4.823	4.891	4.959	5.027	5.096
-6.302	-6.232	-6.162	-6.096	-6.032	-5.971
5.445	5.515	5.586	5.657	5.728	5.800
-5.699	-5.651	-5.605	-5.560	-5.517	-5.475
6.164	6.237	6.311	6.385	6.460	6.534
-5.287	-5.252	-5.219	-5.187	-5.157	-5.127
6.909	6.985	7.061	7.137	7.214	7.291
-4.991	-4.966	-4.942	-4.919	-4.897	-4.875
7.678	7.756	7.835	7.914	7.993	8.072
-4.779	-4.761	-4.745	-4.728	-4.713	-4.698
8.472	8.552	8.633	8.714	8.795	8.877
-4.633	-4.627	-4.616	-4.606	-4.597	-4.588
9.288	9.371	9.454	9.537	9.621	9.705
-4.555	-4.549	-4.544	-4.539	-4.535	-4.531
10.126	10.211	10.296	10.381	10.467	10.553
-4.528	-4.527	-4.527	-4.528	-4.529	-4.530
10.983	11.070	11.157	11.244	11.331	11.419
-4.544	-4.549	-4.553	-4.559	-4.564	-4.571
11.858	11.947	12.036	12.125	12.214	12.303
-4.608	-4.617	-4.627	-4.637	-4.657	-4.668
12.752	12.842	12.932	13.023	13.114	13.204
-4.730	-4.744	-4.758	-4.773	-4.788	-4.804
13.660	13.752	13.844	13.936	14.028	14.120
-4.890	-4.908	-4.927	-4.947	-4.967	-4.987
14.584	14.677	14.770	14.864	14.957	15.051
-5.097	-5.121	-5.145	-5.169	-5.194	-5.220
15.522	15.617	15.712	15.807	15.902	15.997
-5.355	-5.384	-5.413	-5.443	-5.473	-5.504
16.475	16.571	16.667	16.763	16.860	16.956
-5.685	-5.719	-5.754	-5.789	-5.829	-5.866
17.440	17.537	17.634	17.731	17.828	17.925
-6.076	-6.116	-6.157	-6.198	-6.240	-6.283
18.412	18.509	18.607	18.705	18.803	18.901
-6.505	-6.551	-6.598	-6.646	-6.694	-6.743
19.392	19.490	19.589	19.688	19.787	19.885

Table A3.2. (*cont.*)

ATOMIC SYMBOL = HG ⁻   ATOMIC NUMBER = 80

	I		0.0000	0.0001	0.0002	0.0003
0.821	I	F'	-6.557	-6.565	-6.572	-6.579
	I	F"	11.590	11.592	11.594	11.596
0.822	I	F'	-6.630	-6.638	-6.645	-6.653
	I	F"	11.612	11.614	11.616	11.618
0.823	I	F'	-6.706	-6.714	-6.722	-6.730
	I	F"	11.633	11.635	11.638	11.640
0.824	I	F'	-6.786	-6.794	-6.803	-6.811
	I	F"	11.655	11.657	11.659	11.661
0.825	I	F'	-6.870	-6.879	-6.888	-6.896
	I	F"	11.677	11.679	11.681	11.683
0.826	I	F'	-6.960	-6.969	-6.978	-6.988
	I	F"	11.698	11.700	11.703	11.705
0.827	I	F'	-7.056	-7.066	-7.076	-7.086
	I	F"	11.720	11.722	11.724	11.727
0.828	I	F'	-7.160	-7.171	-7.182	-7.193
	I	F"	11.742	11.744	11.746	11.748
0.829	I	F'	-7.274	-7.286	-7.298	-7.311
	I	F"	11.763	11.766	11.768	11.770
0.830	I	F'	-7.401	-7.415	-7.429	-7.443
	I	F"	11.785	11.787	11.790	11.792
0.831	I	F'	-7.547	-7.562	-7.579	-7.595
	I	F"	11.807	11.809	11.811	11.813
0.832	I	F'	-7.713	-7.732	-7.752	-7.773
	I	F"	11.829	11.831	11.833	11.835
0.833	I	F'	-7.931	-7.957	-7.984	-8.011
	I	F"	11.851	11.853	11.855	11.857
0.834	I	F'	-8.243	-8.284	-8.328	-8.375
	I	F"	11.873	11.875	11.877	11.879
0.835	I	F'	-8.882	-9.012	-9.184	-9.438
	I	F"	11.895	11.897	11.899	11.901
0.836	I	F'	-8.844	-8.761	-8.692	-8.631
	I	F"	10.303	10.305	10.307	10.310
0.837	I	F'	-8.354	-8.326	-8.301	-8.277
	I	F"	10.324	10.326	10.328	10.331
0.838	I	F'	-8.145	-8.130	-8.116	-8.102
	I	F"	10.345	10.347	10.350	10.352
0.839	I	F'	-8.023	-8.014	-8.005	-7.996
	I	F"	10.366	10.368	10.371	10.373
0.840	I	F'	-7.945	-7.939	-7.933	-7.928
	I	F"	10.387	10.390	10.392	10.394
0.841	I	F'	-7.894	-7.890	-7.886	-7.882
	I	F"	10.409	10.411	10.413	10.415
0.842	I	F'	-7.860	-7.858	-7.855	-7.853
	I	F"	10.430	10.432	10.434	10.436
0.843	I	F'	-7.840	-7.839	-7.837	-7.836
	I	F"	10.451	10.453	10.455	10.457
0.844	I	F'	-7.830	-7.830	-7.829	-7.829
	I	F"	10.472	10.474	10.476	10.478
0.845	I	F'	-7.829	-7.829	-7.829	-7.830
	I	F"	10.493	10.495	10.497	10.499
0.846	I	F'	-7.835	-7.835	-7.836	-7.838
	I	F"	10.514	10.516	10.518	10.521
0.847	I	F'	-7.846	-7.848	-7.850	-7.851
	I	F"	10.535	10.537	10.540	10.542
0.848	I	F'	-7.864	-7.866	-7.868	-7.870
	I	F"	10.557	10.559	10.561	10.563

L$_1$ ABSORPTION EDGE (0.83530 Å;   14.8421 KEV)

0.0004	0.0005	0.0006	0.0007	0.0008	0.0009
-6.586	-6.593	-6.601	-6.608	-6.615	-6.623
11.599	11.601	11.603	11.605	11.607	11.609
-6.660	-6.668	-6.675	-6.683	-6.691	-6.698
11.620	11.622	11.625	11.627	11.629	11.631
-6.738	-6.746	-6.754	-6.762	-6.770	-6.778
11.642	11.644	11.646	11.648	11.651	11.653
-6.819	-6.828	-6.836	-6.845	-6.853	-6.862
11.664	11.666	11.668	11.670	11.672	11.674
-6.905	-6.914	-6.923	-6.932	-6.941	-6.951
11.685	11.687	11.690	11.692	11.694	11.696
-6.997	-7.007	-7.017	-7.026	-7.036	-7.046
11.707	11.709	11.711	11.713	11.716	11.718
-7.097	-7.107	-7.117	-7.128	-7.138	-7.149
11.729	11.731	11.733	11.735	11.737	11.740
-7.204	-7.216	-7.227	-7.239	-7.250	-7.262
11.750	11.753	11.755	11.757	11.759	11.761
-7.323	-7.336	-7.348	-7.361	-7.374	-7.388
11.772	11.774	11.776	11.779	11.781	11.783
-7.457	-7.471	-7.486	-7.501	-7.516	-7.531
11.794	11.796	11.798	11.800	11.803	11.805
-7.612	-7.629	-7.646	-7.658	-7.676	-7.694
11.816	11.818	11.820	11.822	11.824	11.827
-7.793	-7.815	-7.837	-7.859	-7.883	-7.907
11.838	11.840	11.842	11.844	11.846	11.848
-8.040	-8.070	-8.101	-8.134	-8.169	-8.205
11.859	11.862	11.864	11.866	11.868	11.870
-8.426	-8.481	-8.542	-8.610	-8.687	-8.776
11.881	11.884	11.886	11.888	11.890	11.892
-9.947	-10.152	-9.516	-9.245	-9.071	-8.944
11.903	10.293	10.295	10.297	10.299	10.301
-8.578	-8.531	-8.489	-8.451	-8.416	-8.383
10.312	10.314	10.316	10.318	10.320	10.322
-8.254	-8.233	-8.214	-8.195	-8.177	-8.161
10.333	10.335	10.337	10.339	10.341	10.343
-8.089	-8.077	-8.065	-8.054	-8.043	-8.033
10.354	10.356	10.358	10.360	10.362	10.364
-7.988	-7.980	-7.973	-7.965	-7.959	-7.952
10.375	10.377	10.379	10.381	10.383	10.385
-7.922	-7.917	-7.912	-7.907	-7.902	-7.898
10.396	10.398	10.400	10.402	10.404	10.406
-7.879	-7.875	-7.872	-7.869	-7.866	-7.863
10.417	10.419	10.421	10.423	10.425	10.428
-7.851	-7.849	-7.847	-7.845	-7.843	-7.842
10.438	10.440	10.442	10.444	10.447	10.449
-7.835	-7.834	-7.833	-7.832	-7.831	-7.831
10.459	10.461	10.463	10.466	10.468	10.470
-7.829	-7.829	-7.828	-7.828	-7.829	-7.829
10.480	10.482	10.485	10.487	10.489	10.491
-7.830	-7.831	-7.831	-7.832	-7.833	-7.834
10.501	10.504	10.506	10.508	10.510	10.512
-7.839	-7.840	-7.841	-7.842	-7.844	-7.845
10.523	10.525	10.527	10.529	10.531	10.533
-7.853	-7.855	-7.856	-7.858	-7.860	-7.862
10.544	10.546	10.548	10.550	10.552	10.554
-7.872	-7.875	-7.877	-7.879	-7.882	-7.884
10.565	10.567	10.569	10.571	10.574	10.576

ATOMIC SYMBOL = HG   ATOMIC NUMBER = 80

	I		0.0000	0.0001	0.0002	0.0003
0.858	I	F'	-8.299	-8.306	-8.313	-8.320
	I	F"	10.769	10.772	10.774	10.776
0.859	I	F'	-8.370	-8.378	-8.386	-8.393
	I	F"	10.791	10.793	10.795	10.797
0.860	I	F'	-8.449	-8.457	-8.465	-8.474
	I	F"	10.812	10.814	10.816	10.819
0.861	I	F'	-8.534	-8.543	-8.553	-8.562
	I	F"	10.834	10.836	10.838	10.840
0.862	I	F'	-8.629	-8.639	-8.649	-8.659
	I	F"	10.855	10.857	10.859	10.861
0.863	I	F'	-8.733	-8.744	-8.756	-8.767
	I	F"	10.876	10.879	10.881	10.883
0.864	I	F'	-8.850	-8.862	-8.875	-8.888
	I	F"	10.898	10.900	10.902	10.904
0.865	I	F'	-8.982	-8.996	-9.010	-9.024
	I	F"	10.919	10.921	10.924	10.926
0.866	I	F'	-9.132	-9.148	-9.165	-9.182
	I	F"	10.941	10.943	10.945	10.947
0.867	I	F'	-9.307	-9.326	-9.346	-9.366
	I	F"	10.962	10.964	10.967	10.969
0.868	I	F'	-9.517	-9.540	-9.564	-9.588
	I	F"	10.984	10.986	10.988	10.990
0.869	I	F'	-9.777	-9.807	-9.838	-9.870
	I	F"	11.005	11.008	11.010	11.012
0.870	I	F'	-10.123	-10.165	-10.208	-10.253
	I	F"	11.027	11.029	11.031	11.033
0.871	I	F'	-10.640	-10.709	-10.783	-10.862
	I	F"	11.048	11.051	11.053	11.055
0.872	I	F'	-11.712	-11.920	-12.183	-12.537
	I	F"	11.070	11.072	11.074	11.077
0.873	I	F'	-11.884	-11.688	-11.525	-11.385
	I	F"	7.924	7.925	7.927	7.929
0.874	I	F'	-10.739	-10.674	-10.613	-10.556
	I	F"	7.940	7.941	7.943	7.944
0.875	I	F'	-10.229	-10.191	-10.154	-10.118
	I	F"	7.955	7.957	7.959	7.960
0.876	I	F'	-9.900	-9.873	-9.847	-9.821
	I	F"	7.971	7.973	7.974	7.976
0.877	I	F'	-9.659	-9.638	-9.618	-9.598
	I	F"	7.987	7.989	7.990	7.992
0.878	I	F'	-9.470	-9.453	-9.437	-9.420
	I	F"	8.003	8.004	8.006	8.008
0.879	I	F'	-9.315	-9.301	-9.287	-9.274
	I	F"	8.019	8.020	8.022	8.023
0.880	I	F'	-9.185	-9.173	-9.161	-9.149
	I	F"	8.035	8.036	8.038	8.039
0.881	I	F'	-9.073	-9.062	-9.052	-9.042
	I	F"	8.050	8.052	8.054	8.055
0.882	I	F'	-8.975	-8.966	-8.957	-8.948
	I	F"	8.066	8.068	8.069	8.071
0.883	I	F'	-8.889	-8.881	-8.873	-8.865
	I	F"	8.082	8.084	8.085	8.087
0.884	I	F'	-8.812	-8.805	-8.798	-8.791
	I	F"	8.098	8.100	8.101	8.103
0.885	I	F'	-8.744	-8.737	-8.731	-8.724
	I	F"	8.114	8.115	8.117	8.119

L₂ ABSORPTION EDGE (0.87220 Å;   14.2142 KEV)

0.0004	0.0005	0.0006	0.0007	0.0008	0.0009
-8.327	-8.334	-8.341	-8.348	-8.356	-8.363
10.778	10.780	10.782	10.784	10.787	10.789
-8.401	-8.409	-8.417	-8.424	-8.432	-8.441
10.799	10.801	10.804	10.806	10.808	10.810
-8.482	-8.491	-8.499	-8.508	-8.517	-8.525
10.821	10.823	10.825	10.827	10.829	10.831
-8.571	-8.580	-8.590	-8.600	-8.609	-8.619
10.842	10.844	10.846	10.849	10.851	10.853
-8.669	-8.680	-8.690	-8.701	-8.712	-8.722
10.864	10.866	10.868	10.870	10.872	10.874
-8.778	-8.790	-8.802	-8.814	-8.826	-8.838
10.885	10.887	10.889	10.891	10.894	10.896
-8.901	-8.914	-8.927	-8.940	-8.954	-8.968
10.906	10.909	10.911	10.913	10.915	10.917
-9.039	-9.054	-9.069	-9.085	-9.100	-9.116
10.928	10.930	10.932	10.934	10.937	10.939
-9.199	-9.216	-9.234	-9.251	-9.270	-9.288
10.949	10.952	10.954	10.956	10.958	10.960
-9.386	-9.407	-9.428	-9.449	-9.471	-9.494
10.971	10.973	10.975	10.977	10.980	10.982
-9.613	-9.639	-9.665	-9.692	-9.720	-9.748
10.992	10.995	10.997	10.999	11.001	11.003
-9.903	-9.936	-9.971	-10.007	-10.044	-10.083
11.014	11.016	11.018	11.020	11.023	11.025
-10.301	-10.350	-10.402	-10.457	-10.514	-10.575
11.036	11.038	11.040	11.042	11.044	11.046
-10.949	-11.043	-11.146	-11.261	-11.390	-11.539
11.057	11.059	11.061	11.064	11.066	11.068
-13.086	-14.376	-13.896	-12.934	-12.451	-12.128
11.079	11.081	7.918	7.919	7.921	7.922
-11.262	-11.153	-11.055	-10.966	-10.884	-10.809
7.930	7.932	7.933	7.935	7.936	7.938
-10.502	-10.451	-10.402	-10.356	-10.312	-10.270
7.946	7.948	7.949	7.951	7.952	7.954
-10.084	-10.050	-10.018	-9.987	-9.958	-9.929
7.962	7.963	7.965	7.966	7.968	7.970
-9.796	-9.772	-9.748	-9.725	-9.703	-9.681
7.978	7.979	7.981	7.982	7.984	7.985
-9.579	-9.559	-9.541	-9.523	-9.505	-9.487
7.993	7.995	7.997	7.998	8.000	8.001
-9.405	-9.389	-9.374	-9.359	-9.344	-9.329
8.009	8.011	8.012	8.014	8.016	8.017
-9.260	-9.247	-9.234	-9.222	-9.209	-9.197
8.025	8.027	8.028	8.030	8.031	8.033
-9.138	-9.127	-9.116	-9.105	-9.094	-9.083
8.041	8.042	8.044	8.046	8.047	8.049
-9.032	-9.022	-9.013	-9.003	-8.994	-8.984
8.057	8.058	8.060	8.061	8.063	8.065
-8.939	-8.931	-8.922	-8.914	-8.905	-8.897
8.073	8.074	8.076	8.077	8.079	8.081
-8.857	-8.850	-8.842	-8.835	-8.827	-8.820
8.088	8.090	8.092	8.093	8.095	8.096
-8.784	-8.777	-8.770	-8.764	-8.757	-8.750
8.104	8.106	8.108	8.109	8.111	8.112
-8.718	-8.712	-8.706	-8.700	-8.694	-8.688
8.120	8.122	8.123	8.125	8.127	8.128

ATOMIC SYMBOL = HG   ATOMIC NUMBER = 80

I		0.0000	0.0001	0.0002	0.0003
0.995 I	F'	−11.256	−11.271	−11.285	−11.300
I	F″	9.936	9.938	9.940	9.942
0.996 I	F'	−11.405	−11.420	−11.436	−11.451
I	F″	9.954	9.956	9.957	9.959
0.997 I	F'	−11.565	−11.582	−11.598	−11.615
I	F″	9.971	9.973	9.975	9.976
0.998 I	F'	−11.739	−11.757	−11.775	−11.794
I	F″	9.989	9.990	9.992	9.994
0.999 I	F'	−11.929	−11.949	−11.969	−11.989
I	F″	10.006	10.008	10.010	10.011
1.000 I	F'	−12.138	−12.161	−12.183	−12.206
I	F″	10.024	10.025	10.027	10.029
1.001 I	F'	−12.372	−12.397	−12.422	−12.448
I	F″	10.041	10.043	10.044	10.046
1.002 I	F'	−12.635	−12.664	−12.693	−12.722
I	F″	10.058	10.060	10.062	10.064
1.003 I	F'	−12.938	−12.971	−13.004	−13.038
I	F″	10.076	10.078	10.079	10.081
1.004 I	F'	−13.293	−13.332	−13.372	−13.412
I	F″	10.093	10.095	10.097	10.099
1.005 I	F'	−13.730	−13.778	−13.828	−13.878
I	F″	10.111	10.113	10.114	10.116
1.006 I	F'	−14.273	−14.337	−14.402	−14.469
I	F″	10.128	10.130	10.132	10.134
1.007 I	F'	−15.016	−15.107	−15.203	−15.304
I	F″	10.146	10.148	10.149	10.151
1.008 I	F'	−16.196	−16.362	−16.544	−16.744
I	F″	10.163	10.165	10.167	10.169
1.009 I	F'	−19.370	−20.347	−22.326	−23.043
I	F″	10.181	10.183	10.184	3.905
1.010 I	F'	−17.272	−17.021	−16.798	−16.598
I	F″	3.910	3.910	3.911	3.912
1.011 I	F'	−15.579	−15.469	−15.365	−15.265
I	F″	3.916	3.917	3.917	3.918
1.012 I	F'	−14.684	−14.613	−14.545	−14.479
I	F″	3.922	3.923	3.923	3.924
1.013 I	F'	−14.072	−14.020	−13.970	−13.921
I	F″	3.929	3.929	3.930	3.930
1.014 I	F'	−13.608	−13.567	−13.527	−13.488
I	F″	3.935	3.935	3.936	3.937
1.015 I	F'	−13.233	−13.200	−13.167	−13.134
I	F″	3.941	3.942	3.942	3.943
1.016 I	F'	−12.920	−12.892	−12.863	−12.836
I	F″	3.947	3.948	3.949	3.949
1.017 I	F'	−12.651	−12.626	−12.602	−12.577
I	F″	3.954	3.954	3.955	3.956
1.018 I	F'	−12.415	−12.393	−12.371	−12.350
I	F″	3.960	3.961	3.961	3.962
1.019 I	F'	−12.205	−12.185	−12.166	−12.147
I	F″	3.966	3.967	3.968	3.968
1.020 I	F'	−12.016	−11.998	−11.981	−11.963
I	F″	3.973	3.973	3.974	3.975
1.021 I	F'	−11.844	−11.828	−11.812	−11.796
I	F″	3.979	3.980	3.980	3.981
1.022 I	F'	−11.687	−11.672	−11.657	−11.642
I	F″	3.985	3.986	3.987	3.987

L₃ ABSORPTION EDGE (1.00910 Å; 12.2858 KEV)

0.0004	0.0005	0.0006	0.0007	0.0008	0.0009
-11.314	-11.329	-11.344	-11.359	-11.374	-11.389
9.943	9.945	9.947	9.949	9.950	9.952
-11.467	-11.483	-11.499	-11.515	-11.532	-11.548
9.961	9.963	9.964	9.966	9.968	9.970
-11.633	-11.650	-11.667	-11.685	-11.703	-11.721
9.978	9.980	9.982	9.983	9.985	9.987
-11.813	-11.832	-11.851	-11.870	-11.889	-11.909
9.996	9.997	9.999	10.001	10.003	10.004
-12.010	-12.031	-12.052	-12.073	-12.095	-12.116
10.013	10.015	10.017	10.018	10.020	10.022
-12.229	-12.252	-12.275	-12.299	-12.323	-12.347
10.031	10.032	10.034	10.036	10.037	10.039
-12.473	-12.499	-12.526	-12.553	-12.580	-12.608
10.048	10.050	10.051	10.053	10.055	10.057
-12.751	-12.781	-12.812	-12.842	-12.874	-12.906
10.065	10.067	10.069	10.071	10.072	10.074
-13.073	-13.108	-13.143	-13.180	-13.217	-13.254
10.083	10.085	10.086	10.088	10.090	10.092
-13.454	-13.496	-13.539	-13.583	-13.628	-13.683
10.100	10.102	10.104	10.106	10.107	10.109
-13.930	-13.984	-14.038	-14.095	-14.153	-14.212
10.118	10.120	10.121	10.123	10.125	10.127
-14.539	-14.611	-14.686	-14.763	-14.844	-14.928
10.135	10.137	10.139	10.141	10.142	10.144
-15.410	-15.521	-15.639	-15.765	-15.899	-16.042
10.153	10.155	10.156	10.158	10.160	10.162
-16.965	-17.214	-17.498	-17.828	-18.224	-18.716
10.170	10.172	10.174	10.176	10.177	10.179
-20.575	-19.506	-18.815	-18.303	-17.897	-17.560
3.906	3.907	3.907	3.908	3.908	3.909
-16.416	-16.249	-16.096	-15.953	-15.821	-15.696
3.912	3.913	3.913	3.914	3.915	3.915
-15.171	-15.081	-14.995	-14.912	-14.833	-14.757
3.918	3.919	3.920	3.920	3.921	3.922
-14.416	-14.354	-14.294	-14.236	-14.180	-14.125
3.925	3.925	3.926	3.927	3.927	3.928
-13.873	-13.826	-13.780	-13.736	-13.692	-13.649
3.931	3.932	3.932	3.933	3.934	3.934
-13.449	-13.412	-13.375	-13.338	-13.303	-13.268
3.937	3.938	3.939	3.939	3.940	3.940
-13.102	-13.071	-13.040	-13.009	-12.979	-12.949
3.944	3.944	3.945	3.946	3.946	3.947
-12.808	-12.781	-12.754	-12.728	-12.702	-12.676
3.950	3.951	3.951	3.952	3.952	3.953
-12.553	-12.530	-12.506	-12.483	-12.460	-12.438
3.956	3.957	3.957	3.958	3.959	3.959
-12.328	-12.307	-12.286	-12.266	-12.245	-12.225
3.963	3.963	3.964	3.964	3.965	3.966
-12.127	-12.108	-12.090	-12.071	-12.053	-12.034
3.969	3.969	3.970	3.971	3.971	3.972
-11.946	-11.928	-11.911	-11.894	-11.877	-11.861
3.975	3.976	3.976	3.977	3.978	3.978
-11.780	-11.764	-11.748	-11.733	-11.717	-11.702
3.981	3.982	3.983	3.983	3.984	3.985
-11.627	-11.612	-11.598	-11.584	-11.569	-11.555
3.988	3.988	3.989	3.990	3.990	3.991

A3.3 FUNDAMENTAL CONSTANTS[a]

Quantity	Symbol	Value (SI units)
Electron charge	e	$1.60217733(49) \times 10^{-19}$ C
Speed of light in vacuo	c	2.99792458×10^{8} m s^{-1}
Electron rest mass	m	$9.1093897(54) \times 10^{-31}$ kg
Classical radius of the electron	e^2/mc^2	$2.81794092(38) \times 10^{-15}$ m
Planck's constant	h	$6.6260755(40) \times 10^{-34}$ J s
Permittivity of free space	ε_0	$8.854187817(12) \times 10^{-12}$ F m^{-1}
Boltzmann's constant	k_B	$1.380658(12) \times 10^{-23}$ JK^{-1}
Avogadro's number	N_A	$6.0221367(36) \times 10^{-23}$ mol^{-1}

[a] Fundamental constants taken from Cohen and Taylor (1987). The figure in brackets after the values gives the one sigma uncertainties in the last digits.

A3.4 CELL PARAMETERS OF SILICON AND GERMANIUM MONOCHROMATOR CRYSTALS

These are the monochromator crystal materials that are used extensively for macromolecular crystallography at SR sources. The wavelength of the reflected beam from the monochromator can be calculated from the specific lattice plane spacing and the angle of reflection.

Silicon and germanium both crystallise in the cubic diamond structure. Their absolute lattice parameters have been measured to an accuracy level of $\leqslant 1$ ppm, i.e. at 22.5 °C they are (M. Hart, pers. comm., based on Becker *et al* (1981), Baker and Hart (1975)):

$$a_0 \text{ (Si)} \quad 5.43102018 \ (\pm 34) \ \text{Å}$$

$$a_0 \text{ (Ge)} \quad 5.65779696 \ (\pm 500) \ \text{Å}$$

These values are obviously sensitive to temperature; an increase in temperature from 22.5 °C to 25 °C increases these values by a factor of 1.000011.

In routine data collection for macromolecular crystallography we are interested in knowing the wavelength to ≈ 0.001 Å, whereas in current anomalous scattering experiments an accuracy of 0.0001 Å is required. Hence, the values of $2d$ for the (111) and (220) planes of silicon and germanium are given in table A3.4 to four decimal places calculated

Table A3.4. *Lattice plane spacings (given as 2d) for perfect silicon and germanium monochromators at 22.5°C.*

	2d (Å)	
Element	(111)	(220)
Silicon	6.2712	3.8403
Germanium	6.5331	4.0007

Table A3.5. *Plane wave (intrinsic) spectral bandpass of reflected beams for perfect silicon and germanium monochromators.*

	$\delta\lambda/\lambda$	
Element	(111)	(220)
Silicon	1.33×10^{-4}	5.6×10^{-5}
Germanium	3.2×10^{-4}	1.5×10^{-4}

from the values of a_0 given above. These values are for 22.5°C. Using the factor of 1.000011 quoted above, then at 25°C Ge(111) $2d$ is 6.53313 Å and at 32.5°C $2d$ is 6.53342. For setting of the wavelength to 1 part in 10^4 the monochromator temperature *in the beam* needs to be stable to ±10°C. For setting of the wavelength to better than 1 part in 10^4, if this ever became necessary, would require better temperature control and correction for refraction effects (Compton and Allison (1935), p. 672).

The values of $\delta\lambda/\lambda$ were given in section 5.1 but are repeated in table A3.5 for convenience.

Extended X-ray absorption fine structure (EXAFS)

A4.1 INTRODUCTION

The absorption of X-rays by any material may be expressed as:

$$I = I_0 \exp(-\mu x) \tag{A4.1}$$

where I_0 is the incident X-ray intensity, I is the transmitted X-ray intensity, μ is the absorption coefficient and x is the thickness of absorber. However, the absorption cross section, μ, of an element in the X-ray region of the electromagnetic spectrum has a complex structure as a function of energy extending as far as 1000 eV above an absorption threshold. This structure is known as EXAFS. The X-ray absorption near edge (within $\approx$20 eV) structure is known as XANES. See figure A4.1.

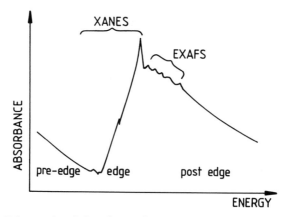

Figure A4.1 Schematic of the absorption cross section in the vicinity of an elemental absorption edge showing the EXAFS and XANES.

A4.2 PHENOMENOLOGICAL DESCRIPTION

X-rays are absorbed by matter primarily through the photoelectric effect. Photoelectric absorption occurs when a bound electron (e.g. K shell) is excited to a continuum state by an incident photon of energy $=\hbar\omega$. The final electron has energy $(\hbar\omega-E_K)$ where E_K is the K shell binding energy. For $\hbar\omega<E_K$ excitation is not possible (pre-edge). For $\hbar\omega>E_K$ a photoelectron is ejected which can be backscattered by atoms in positions near the absorbing site which, in turn, affects the photon absorption in a periodic manner. For instance, consider the following simple case of four nearest neighbours.

The outgoing electron is shown as a spherical wave, whose wavelength $\lambda_e=2\pi/k$ depends on the energy of the electron $(\hbar\omega-E_K)$ according to the formula

$$k = \left(\frac{2m(\hbar\omega - E_k)}{\hbar^2}\right)^{\frac{1}{2}} \qquad\qquad (A4.2)$$

The scattered wave from the near neighbours overlaps the outgoing wave from the absorption site, the K shell, and can interfere constructively or destructively depending on the total phase shift experienced by the electron (see figure A4.2). This phase shift depends, in turn, on the distance between atoms and λ_e (i.e. on $\hbar\omega-E_K$) and the propagation of the electron between the absorbing site and the scattering atom.

Because a given absorption energy (E_K) is produced at a specific energy it is *element specific*. Thus it is possible to discriminate between different elements in a given sample. By using not only K but also L edges and possibly M edges all elements in the periodic table are accessible using SR.

Since the effect is dependent only on the presence of an excitable atom and an ordered *local* environment of that atom, the sample can be disordered (e.g. amorphous, solution) or ordered (e.g. crystalline); the only constraint being that sufficient material be present and the concentration of the excitable atom be enough for a reasonable signal-to-noise ratio.

In summary, information concerning the electronic structure at and/or in the immediate environment about the primary absorbing atom can be obtained by the accurate measurement of:

(*a*) the position (point of maximum inflection) of the
        absorption edge;

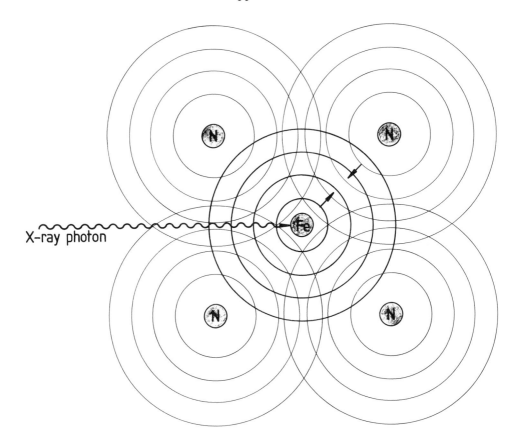

Figure A4.2 An incident X-ray photon of sufficient energy is absorbed
by an atom (e.g. iron is depicted here). The inner shell
photoelectron ejected can be considered as an outgoing
spherical wavelet which is scattered by neighbouring
atoms (e.g. nitrogen neighbours to the iron).

(b) the details of the edge structure;
(c) the intervals and amplitudes of the EXAFS ripples.

An energy-dispersive detector can be used to isolate the fluorescent radiation from the Compton and Rayleigh components.

A4.3 DATA ANALYSIS

The analysis of the data obtained aims at extracting information on the local neighbours of the stimulated atom, namely the number, distance and type of neighbours.

The EXAFS interference expressed in terms of the fractional modulation of the absorption coefficient, $\chi(k)$ is given by

$$\chi(k) = \frac{\mu - \mu_0}{\mu_0} = \frac{\Delta\mu}{\mu_0} \qquad (A4.3)$$

and μ_0 is the absorption coefficient without EXAFS and

$$\mu \ (\text{or } \mu_0) = \frac{1}{t} \log_e \frac{I_0}{I}$$

for a sample of thickness t where I_0 and I are respectively the incident and attenuated (transmitted) intensity.

A simplified expression for the EXAFS modulation can be written as

$$\chi(k) = -\sum_j \frac{N_j}{kR_j^2} |f_j(k, \pi)| \ \exp(-2\sigma_j^2 k^2) \exp(-2R_j/\lambda) \sin(2kR_j + 2\delta + \psi_j) \qquad (A4.4)$$

This is known as the plane wave approximation, i.e. the electron wave is approximated by a plane wave. The latter is only applicable when the radius of the emitting atom is small in comparison to the curvature of the outgoing wave which is the case when the photoelectron energy is high. A more accurate description, often known as exact curved wave theory has been given by Lee and Pendry (1975). For a review of the theories and procedures see Gurman (1990).

The N_j are the number of scattering atoms at distance R_j in the jth shell of atoms and are the parameters of main investigative interest.

$|f_j(k, \pi)|$	is the electron scattering amplitude in the backward (π phase change) direction for the jth shell, which is a function of k.
$1/R_j^2$	inverse square law decrease of the photoelectron wave intensity.
$\exp(-2\sigma_j^2 k^2)$	is a Debye–Waller factor term due to thermal vibrations or static disorder with root mean square fluctuation σ_j.
$\exp(-2R_j/\lambda)$	is an attenuation term to account for the loss of photoelectrons inelastically scattered and λ is a mean-free path.
$\sin(2kR_j + 2\delta + \psi_j)$	is the sinusoidal interference term, $2\delta + \psi_j$ being the phase shift function.

$\chi(k)$ is summed over all (j) contributions.

The steps involved in obtaining the N_j, R_j and phase shift data are as follows:

(1) Extraction of $\chi(k)$ from $\mu(E)$. The quantity $\mu_0(E)$ is obtained from the smooth part of $\mu(E)$ and $\Delta\mu$ is the 'wiggly' part. There is some ambiguity in the value of E_0 since there is found to be variation in a given edge position from sample to sample. This causes ambiguity in k and $\delta_j(k)$ and the problem is resolved only by making E_0 a variable in later stages of the analysis.

(2) $\chi(k)$ is multiplied by k^3 to compensate for the fall off of the EXAFS amplitude with k. The quantity k^3 is chosen since $|f_j(k,\pi)|\sim^{-1}/k^2$.

(3) Fourier transform of $k^3\chi(k)$ which shows peaks corresponding to the nearest neighbour shells, next nearest neighbour shells and so on (see figure A4.3).

(4) By using the numerical filter to exclude all contributions except from the first shell. The inverse transform contains information then of only one term of the summation over j in $\chi_j(k)$ to give information about the number of nearest neighbours (N_j) and the electron mean-free path (λ) for the first shell. By

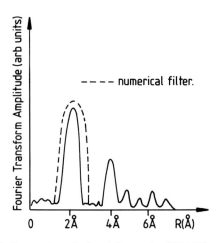

Figure A4.3 The information derived from the EXAFS measurements is a radial distribution function showing peaks corresponding to the nearest neighbour shells of atoms to the absorbing atom.

repeating this procedure for more distant shells similar information can be gained but at a deteriorating accuracy as higher shells contribute to a smaller and smaller k-range.

(5) Direct curve fitting of the EXAFS data in k-space provides the most accurate information for the whole of the coordination sphere of the photoexcited atom.

A4.4 MULTIPLE SCATTERING

Section A4.3 outlined an analysis procedure based on single scattering, i.e. the outgoing electron wave generates backscattered waves which are of sufficiently small amplitude that their rescattering can be ignored. There are situations where this assumption breaks down and multiple scattering (i.e. rescattering effects) becomes important. For metallo-proteins this is especially true in the presence of pyrrole or histidine ligands. In the case of deoxy Hb, for example, both protein crystallography and multiple scattering EXAFS analysis agree that the iron atom lies out of the plane of the porphyrin by $\sim 0.5 \, \text{Å}$ (see section 3.1.1 and Hasnain and Strange (1990)).

A4.5 CONCLUDING REMARKS

EXAFS is capable of providing precise information on the distances and type of a few nearest neighbours. These distances are, however, more precise than those derived from macromolecular crystallography. There are situations where the information gained from EXAFS is very valuable. These include cases of macromolecules that have resisted crystallisation or where, through an unfortunate choice of space group or not very careful structure analysis, errors in a structure have resulted. These and other examples are reviewed by Hasnain (1988), by Garner (1990) and by Lindley, Garratt and Hasnain (1990). There is a complementarity of the two methods.

Synchrotron X-radiation laboratories: addresses and contact names (given in alphabetical order of country)

NB: Table 4.1 describes the parameters of each of the machines listed below.

(1) Brazil	LNLS, Laboratorio Nacional de Luz Sincrotron, Cx. Postal 6192, Campinas, 13081, São Paulo, Brazil. Director: Cylon ET Goncalves de Silva
(2) China	BEPC, Institute of High Energy Physics, 19 Yucuan Rd, PO Box 918, Beijing, People's Republic of China. Director: Dingchang Xian
(3) European source	ESRF, BP 220, F-38043 Grenoble Cédex, France. Director: R. Haensel. Diffraction contact: Å. Kvick
(4) France	DCI, LURE, Batiment 209D, Centre Scientifique, 91405 Orsay Cédex, France. Director: Y. Petroff. Diffraction contact: R. Fourme
(5) Germany	EMBL, DESY, Notkestrasse 85, D-2000 Hamburg 52, Federal Republic of Germany. Director: K. S. Wilson
(6) India	INDUS-1, Centre for Advanced Technology, Rajendrangar, Indore 452012, India. Director: S. S. Ramamurthi
(7) Italy	ADONE, Instituto Nazionale di Fisica Nucleare (INFN), via E. Fermi 40, 00044 Frascati, Rome, Italy. Director: E. Burattini ELETTRA, Sincrotrone Trieste, Padriciano 99, Trieste, Italy. Head of Scientific Division: L. Fonda

(8) Japan	Photon Factory, National Laboratory for High Energy Physics (KEK), Oho 1-1, Tsukuba-shi, Ibaraki-ken, 305 Japan. Director: Jun-ichi Chikawa SPRING-8, JAERI/RIKEN PROJECT, 2–28–2 Honkomagome, Bunkyo-ku, Tokyo 113, Japan. Co-chairman T. Sasaki.
(9) Sweden	(Big)MAX-Laboratory, Lund University, PO Box 118, S-22100 Lund, Sweden, Director: M. Eriksson
(10) Taiwan	Synchrotron Radiation Research Center, 8th Floor, No. 6 Roosevelt Road, Sec 1, Taipei 10757, Taiwan, RoC. Director: E. Yen
(11) United Kingdom	SRS, SERC, Daresbury Laboratory, Daresbury, Warrington, Cheshire WA4 4AD, England. Co-directors: J. Bordas (Science) and D. J. Thompson (Machine).
(12) USA	Cornell Electron Storage Ring (CESR), CHESS, Cornell University, 200 L. Wilson Laboratory, Ithaca, New York, NY 14853, USA. Director: B. W. Batterman SPEAR and PEP, Stanford Synchrotron Radiation Laboratory, Bin 69, PO Box 4349, Stanford, California 94309–0210, USA. Director: A. Bienenstock NSLS (X-ray ring), Brookhaven National Laboratory, Upton, New York, NY 11973, USA. Director: M. Knotek APS, Argonne National Laboratory, 9700 South Cass Avenue, Argonne, Illinois, IL 60439, USA. Director: D. Moncton
(13) USSR	VEPP-3, Institute of Nuclear Physics, 630090 Novosibirsk, Siberia, USSR. Director: G. N. Kulipanov

Bibliography

The books (or parts thereof) given in this list are either text-books or volumes of a journal devoted to a topic or edited books on various aspects of crystallography or synchrotron radiation.

1. GENERAL TEXTS ON CRYSTALLOGRAPHY

Blundell, T. L. and Johnson, L. N. (1976) *Protein Crystallography*, New York: Academic Press.

Bragg, W. H. (1928) *An Introduction to Crystal Structure Analysis*, London: G. Bell and Sons Ltd.

Bragg, W. H. and Bragg, W. L. (1915) *X-rays and Crystal Structure*, London: G. Bell and Sons Ltd.

Bragg, W. L. (1949) *The Crystalline State*, Volume I, *A General Survey*, London: G. Bell and Sons Ltd

Bragg, W. L. (1975) *The Development of X-ray Analysis*, edited by D. C. Phillips and H. Lipson, London: G. Bell and Sons Ltd.

Buerger, M. J. (1942) *X-ray Crystallography*, New York: John Wiley.

Buerger, M. J. (1964) *The Precession Method*, New York: John Wiley.

Cochran, W. (1973) *The Dynamics of Atoms in Crystals*, London: Arnold.

Compton, A. H. and Allison, A. H. (1935) *X-rays in Theory and Experiment*, New York: Van Nostrand.

Glazer, A. M. (1987) *The Structures of Crystals*, Bristol: IOP Publishing Ltd.

Glusker, J. P. and Trueblood, K. (1971) *Crystal Structure Analysis*, first edition, Oxford: Oxford University Press; (1985) second edition.

James, R. W. (1954) *The Crystalline State*, Volume II, *The Optical Principles of the Diffraction of X-rays*, London: G. Bell and Sons.

Lonsdale, K. (1948) *Crystals and X-rays*, London: G. Bell and Sons Ltd.

McKie, D. and McKie, C. (1986) *Essentials of Crystallography*, Oxford: Blackwell Scientific Publications

Stout, G. H. and Jensen, L. H. (1968) *X-ray Structure Determination: A Practical Guide*, New York: Macmillan.

Stout, G. H. and Jensen, L. H. (1989) *X-ray Structure Determination: A Practical Guide*, New York, John Wiley.

Vainshtein, B. K. (1981) *Modern Crystallography*, Volumes I and II, Berlin: Springer Verlag.

Warren, B. E. (1969) *X-ray Diffraction*, Reading, Mass: Addison-Wesley.

Woolfson, M. M. (1970) *X-ray Crystallography*, Cambridge: Cambridge University Press.

Wyckoff, R. W. G. (1924) *The Structure of Crystals*, New York: The Chemical Catalog Company Inc.

2. BOOKS (OR PARTS THEREOF) ON DIFFRACTION GEOMETRY AND DATA COLLECTION METHODS

Amorós, J. L., Buerger, M. J. and Amorós, M. C. (1975) *The Laue Method*, New York: Academic Press.

Arndt, U. W. and Willis, B. T. M. (1966) *Single Crystal Diffractometry*, Cambridge: Cambridge University Press.

Arndt, U. W. and Wonacott, A. J. (1977) *The Rotation Method in Crystallography*, Amsterdam: North-Holland.

Buerger, M. J. (1964) *The Precession Method*, New York: John Wiley.

Evans, H. T. and Lonsdale, K. (1959) Diffraction geometry in *International Tables for X-ray Crystallography*, Volume II, Birmingham: Kynoch Press (Kluwer Academic Press, Dordrecht).

Helliwell, J. R. (1991) Diffraction geometry and its practical realisation in *International Tables for X-ray Crystallography*, Volume C, edited by A. J. C. Wilson, Oxford: Oxford University Press.

Henry, N. F. M., Lipson, H. and Wooster, W. A. (1951) *The Interpretation of X-ray Diffraction Photographs*. London: Macmillan.

3. BOOKS ON SPECIALISED OR ADVANCED ASPECTS OF CRYSTALLOGRAPHY

Amorós, J. L. and Amorós, M. (1968) *Molecular Crystals: Their Transforms and Diffuse Scattering*, New York: John Wiley.

Bijvoet, J. M., Burgers, W. G. and Hägg, G. (1969) *Early Papers on Diffraction of X-rays by Crystals*, Volume I, Dordrecht: Kluwer Academic Publishers; (1972) *Early Papers on Diffraction of X-rays by Crystals*, Volume II, Dordrecht: Kluwer Academic Publishers.

Dodson, G., Glusker, J. P. and Sayre, D., Eds. (1981) *Structural Studies on Molecules of Biological Interest – A Volume in Honour of Dorothy Hodgkin*, Oxford: Clarendon Press.

Giegé, R., Ducruix, A., Fontecilla-Camps, J. C., Feigelson, R. S., Kern, R. and McPherson, A. (1988) Crystal growth of biological macromolecules. *Journal of Crystal Growth* **90**, 1–374, Amsterdam: North-Holland.

Harburn, G., Taylor, C. A. and Welberry, T. R. (1975) *Atlas of Optical Transforms*, London: G. Bell and Sons Ltd.

International Tables for X-ray Crystallography (1959), Volumes I–IV. Present distributor: Kluwer Academic Publishers, Dordrecht.

Mandelkow, E., Ed. (1988, 1988, 1989) *Topics in Current Chemistry*, Volumes 145, 157, 151, *Synchrotron Radiation in Chemistry and Biology*, Heidelberg: Springer Verlag.

McPherson, A., Jr (1982) *The Preparation and Analysis of Protein Crystals*, New York: John Wiley.

Moras, D., Drenth, J., Strandberg, B., Suck, D. and Wilson, K. S., Eds. (1987) *Life Sciences*, **126A**, Crystallography in molecular biology, New York: Plenum Press.

Ramaseshan, S. and Abrahams, S. C. (1975) *Anomalous Scattering*, Copenhagen: Munksgaard.

†Schoenborn, B. P., Ed. (1984) *Neutrons in Biology*, New York: Plenum Press.

Stuhrmann, H. B., Ed. (1982) *Uses of Synchrotron Radiation in Biology*, New York: Academic Press.

Willis, B. T. M. and Pryor, A. W. (1975) *Thermal Vibrations in Crystallography*, Cambridge: Cambridge University Press.

---

† This book is mentioned because reactor and spallation neutrons are also polychromatic in nature. There are parallels therefore in the diffraction geometry and its treatment for neutrons and for synchrotron radiation.

Winick, H. and Doniach, S., Eds. (1980) *Synchrotron Radiation Research*, New York: Plenum Press.

Wooster, W. A. (1962) *Diffuse X-ray Reflections from Crystals*, Oxford: Clarendon Press.

Wyckoff, H., Ed. (1985) *Methods in Enzymology*, Volumes 114A and 114B, London: Academic Press.

4. SYNCHROTRON RADIATION (SR)

(a) *Proceedings of SR instrumentation and research meetings* (in chronological order)

Wuilleumier, F. and Farge, Y., Eds. (1978) *SR instrumentation and new developments. Proc. Int. Conf., Orsay, held in 1977. Nuclear Instruments and Methods* **152**, 1–335, Amsterdam: North-Holland Publishing Company.

Ederer, D. L. and West, J. B., Eds. (1980) SR instrumentation. Proc. National Conf., Gaithesburg, held 1979. *Nuclear Instruments and Methods* **172**, 1–401, Amsterdam: North-Holland Publishing Company.

Bordas, J., Fourme, R. and Koch, M. J. H., Eds. (1982) X-ray detectors for synchrotron radiation. *Nuclear Instruments and Methods* **201**, 1–279, Amsterdam: North-Holland Publishing Company.

Mills, D. M. and Batterman, B. W., Eds. (1982) SR instrumentation. Proc. National Conf., Ithaca, held 1981. *Nuclear Instruments and Methods* **195**, 1–423, Amsterdam: North-Holland Publishing Company.

Koch, E. E., Ed. (1983) SR instrumentation. Proc. Int. Conf., Hamburg, held 1982. *Nuclear Instruments and Methods* **208**, 1–865. Amsterdam: North-Holland Publishing Company.

Tomlinson, W. and Williams, G. P., Eds. (1984) SR instrumentation. *Nuclear Instruments and Methods* **222**, 1–409, Amsterdam: North-Holland Physics Publishing.

Brown, G. S. and Lindau, I., Eds. (1986) SR instrumentation. Proc. Int. Conf., Stanford, held 1985. *Nuclear Instruments and Methods* **246**, 1–876, Amsterdam, North-Holland Publishing Company.

Ando, M. and Miyahara, T. (1989) SR instrumentation. Proc. Int. Conf., Tsukuba, held September 1988. *Review of Scientific Instruments* **60**(7), 1373–2566, New York: American Institute of Physics.

(b) Books, edited books and handbooks involving SR

Bianconi, A. and Congiu-Castellano, A., Eds. (1987) *Biophysics and Synchrotron Radiation*, Heidelberg: Springer Verlag.

Catlow, R. and Greaves, G. N., Eds. (1989) *Applications of Synchrotron Radiation*, Glasgow: Blackie.

Chemistry in Britain (1986) Synchrotron Radiation Special Issue, **22**, No. 9, September.

Hasnain, S. S., Ed. (1990) *Synchrotron Radiation and Biophysics*, Chichester: Ellis Horwood.

Koch, E. E., Ed. (1983) *Handbook on Synchrotron Radiation*, Volumes 1a and 1b, Amsterdam: North-Holland

Mandelkow, E., Ed. (1988) *Topics in Current Chemistry* **145**, *Synchrotron Radiation in Chemistry and Biology*, Volume I, 1–239, Heidelberg: Springer Verlag.

Mandelkow, E., Ed. (1988) *Topics in Current Chemistry* **147**, *Synchrotron Radiation in Chemistry and Biology*, Volume II, 1–166, Heidelberg: Springer Verlag.

Mandeklow, E., Ed. (1989) *Topics in Current Chemistry* **151**, *Synchrotron Radiation in Chemistry and Biology*, Volume III, 1–231, Heidelberg: Springer Verlag.

Sokolov, A. A. and Ternov, I. M. (1968) *Synchrotron Radiation*, Oxford, Pergamon.

Stuhrmann, H. B., Ed. (1982) *Uses of Synchrotron Radiation in Biology*, New York: Academic Press.

Sweet, R. M. and Woodhead, A. D., Eds. (1989) *Synchrotron Radiation in Structural Biology*, Basic Life Sciences, Volume 51, New York: Plenum Press.

Tanner, B. K. and Bowen, D. K., Eds. (1980) *Characterisation of Crystal Growth Defects by X-ray Methods*, New York: Plenum Press.

Winick, H. and Doniach, S., Eds. (1980) *Synchrotron Radiation Research*, New York: Plenum Press.

5. BOOKS ON THE STRUCTURE AND FUNCTION OF BIOLOGICAL MACROMOLECULES

Dickerson, R. E. and Geis, I. (1969) *The Structure and Action of Proteins*, Menlo Park, California: Benjamin/Cummings Publishing Company.

Fersht, A. (1985) *Enzyme Structure and Mechanism*, New York: W. H. Freeman.

Oxender, D. L. and Fox, C. F., Eds. (1987) *Protein Engineering*, New York: Alan Liss Inc.

Saenger, W. (1984) *Principles of Nucleic Acid Structure*, Heidelberg, Springer Verlag.

Stryer, L. (1988) *Biochemistry*, third edition, New York: W. H. Freeman and Company.

Voet, D. and Voet, J. (1990) *Biochemistry*, New York: John Wiley and Sons Inc.

6. BOOKS AND REVIEWS ON OTHER STRUCTURE DETERMINING METHODS FOR BIOLOGICAL MACROMOLECULES

Dwek, R. A. (1973) *NMR in Biochemistry*, Oxford: Clarendon Press.

Dwek, R. A. (1977) *NMR in Biology* (Proceedings of the British Biophysical Soc. Spring Meeting, Oxford), London: Academic Press.

Garner, C. D. (1990) *X-ray Absorption Spectroscopy of Biological Molecules in Applications of Synchrotron Radiation*, edited by R. Catlow and N. Greaves, pp. 268–82, Glasgow: Blackie.

Hasnain, S. S. (1988) Application of EXAFS to biochemical systems in *Topics in Current Chemistry* **147**, 73–93, edited by E. Mandelkow.

Hukins, D. W. L. (1981) *X-ray Diffraction by Disordered and Ordered Systems*, Oxford: Pergamon.

Jardetzky, O. and Roberts, G. C. K. (1981) *NMR in Molecular Biology*, New York: Academic Press.

Wuthrick, K. (1986) *NMR of Proteins and Nucleic Acids*, New York: John Wiley.

References

Abad-Zapatero, C., Griffith, J. P., Sussmann, J. L. and Rossmann, M. G. (1987) *Journal of Molecular Biology* **198**, 445–67.

Abdel-Meguid, S. S., Shieh, H.-S., Smith, W. W., Dayringer, H. E., Violand, B. N. and Bentle, L. A. (1987) *Proceedings of the National Academy, USA* **84**, 6434–7.

Acharya, K R., Fry, E., Stuart, D., Fox, G., Rowlands, D. and Brown, F. (1989) *Nature* **337**, 709–16.

Adams, M. J., Helliwell, J. R. and Bugg, C. E. (1977) *Journal of Molecular Biology* **112**, 183–97.

Adams, M. J., Gover, S., Leaback, R., Phillips., C, and Somers, D. O'N. (1991) *Acta Cryst* B47, 817–820.

Adman, E. T., Stenkemp, R. E., Sieker, L. C. and Jensen, L. H. (1978) *Journal of Molecular Biology* **123**, 35–47.

Aggarwal, A. K., Rodgers, D. W., Drottar, M., Ptashne, M. and Harrison, S. C. (1988) *Science* **242**, 899–907.

Alber, T., Petsko, G. A. and Tsernoglou, D. (1976) *Nature* **263**, 297–300.

Alferov, D. F., Bashmakov, Yu. A., Belovintsev, E. G., Bessonov, E. G. and Cerenkov, P. A. (1979) *Particle Accelerators* **9**, 223.

Al-Hilal, D., Baker, E., Carlisle, C. H., Gorinsky, B. A., Horsburgh, R. C., Lindley, P. F., Moss, D. S., Schneider, H. and Stimpson, R. (1976) *Journal of Molecular Biology* **108**, 255–7.

Allen, J. P., Feher, G., Yeates, T. O., Komiya, H. and Rees, D. C. (1987) *Proceedings of the National Academy, USA* **84**, 5730–4.

Allinson, N. M. (1982) *Nuclear Instruments and Methods* **201**, 53–64,

Allinson, N. M. (1989) *Nuclear Instruments and Methods* A**275**, 587–96.

Allinson, N. M. and Greaves, G. N. (1988) *Nuclear Instruments and Methods* A**273**, 620–4.

Allinson, N. M., Brammer, R., Helliwell, J. R., Harrop, S., Magorrian, B. G. and Wan, T. (1989) *Journal of X-ray Science and Technology* **1**, 143–53.

Allured, V. S., Collier, R. J., Carroll, S. F. and McKay, D. B. (1986) *Proceedings of the National Academy, USA* **83**, 1320–4.

Amemiya, Y., Kishimoto, S., Matsushita, T., Satow, Y. and Ando, M. (1989) *Review of Scientific Instruments* **60**(7), 1552–6.

Amemiya, Y., Wakabayashi, K., Tanaka, H., Ueno, Y. and Miyahara, J. (1987) *Science* **237**, 164–8.

Amemiya, Y., Matsushita, T., Nakagawa, A., Satow, Y., Miyahara, J. and Chikawa, J. (1988a) *Nuclear Instruments and Methods* A**266**, 645–53.

Amemiya, Y., Satow, Y., Matsushita, T., Chikawa, J., Wakabayashi, K. and Miyahara, J. (1988b) *Topics in Current Chemistry* **147**, edited by E. Mandelkow, Heidelberg: Springer Verlag, pp. 121–44.

Amit, A. G., Mariuzza, R. A., Phillips, S. E. V. and Poljak, R. J. (1986) *Science* **233**, 747–53.

Amorós, J. L. and Amorós, M. (1968) *Molecular Crystals: Their Transforms and Diffuse Scattering*, New York: John Wiley, p. 187.

Amorós, J. L., Buerger, M. and Amorós, M. C. (1975) *The Laue Method*, New York: Academic Press.

Anderson, J. E., Ptashne, M. and Harrison, S. C. (1987) *Nature* **326**, 846–52.

Andersson, I., Knight, S., Schneider, G., Lindquist, Y., Lundquist, T., Brändén, C.-I. and Lorimer, G. H. (1989) *Nature* **337**, 229–34.

Andrews, S. J., Hails, J. E., Harding, M. M. and Cruickshank, D. W. J. (1987) *Acta Crystallographica* A**43**, 70–3.

Andrews, S. J., Papiz, M. Z., McMeeking, R., Blake, A. J., Lowe, B. M., Franklin, K. R., Helliwell, J. R. and Harding, M. M. (1988) *Acta Crystallographica* B**44**, 73–7.

Anisimov, Yu. S., Chernenko, S. P., Ivanov, A. B., Peshekhonov, V. D., Rozhnyatovskaya, S. A., Zanevsky, Yu. V., Kheiker, D. M., Malakhova, L. F. and Popov, A. N. (1981) *Nuclear Instruments and Methods* **179**, 503–7.

Arad, T., Leonard, K. R. Wittmann, H. G. and Yonath, A. (1984) *EMBO Journal* **3**, 127–31.

Argos, P., Tsukihara, T. and Rossmann, M. G. (1980) *Journal of Molecular Evolution* **15**, 169–79.

Arndt, U. W. (1977) in *The Rotation Method in Crystallography*, edited by U. W. Arndt and A. J. Wonacott, Amsterdam: North-Holland, pp. 245–61.

Arndt, U. W. (1978) *Journal of Physics* E**11**, 671–3.

Arndt, U. W. (1990) *Synchrotron Radiation News* **3**, No. 4, 17–22.

Arndt, U. W. (1984) *Journal of Applied Crystallography* **17**, 118–19.

Arndt, U. W. (1985) *Progress in Enzymology* **114**, 472–85.

Arndt, U. W. (1986) *Journal of Applied Crystallography* **19**, 145–63.

Arndt, U. W. (1990) *Synchrotron Radiation News* **3**, No. 4, 17–22.

Arndt, U. W. and Faruqi, W. (1977) in *The Rotation Method in Crystallography*, edited by U. W. Arndt and A. J. Wonacott, Amsterdam: North-Holland, pp. 219–26.

Arndt, U. W. and Gilmore, D. J. (1979) *Journal of Applied Crystallography* **12**, 1–9.

Arndt, U. W. and Sweet, R. M. (1977) in *The Rotation Method in Crystallography*, edited by U. W. Arndt and A. J. Wonacott, Amsterdam: North-Holland, pp. 43–63.

Arndt, U. W. and Thomas, D. J. (1982) *Nuclear Instruments and Methods* **201**, 21–6.

Arndt, U. W. and Wonacott, A. J. (1977) *The Rotation Method in Crystallography*, Amsterdam: North-Holland.

Arndt, U. W., Gilmore, D. J. and Wonacott, A. J. (1977) in *The Rotation Method in Crystallography*, edited by U. W. Arndt and A. J. Wonacott, Amsterdam: North-Holland, pp. 207–18.

Arndt, U. W., Champness, J. N., Phizackerley, R. P. and Wonacott, A. J. (1973) *Journal of Applied Crystallography* **6**, 457–63.

Arndt, U. W., Greenhough, T. J., Helliwell, J. R., Howard, J. A. K., Rule, S. A. and Thompson, A. W., (1982) *Nature* **298**, 835–8.

Arnold, E., Vriend, G., Luo, M., Griffith, J. P., Kamer, G., Erickson, J. W., Johnson, J. E. and Rossmann, M. G. (1987) *Acta Crystallographica* A**43**, 346–61.

Artymiuk, P. (1987) *Nature* **325**, 575.

Artymiuk, P. (1988) *Nature* **332**, 582.

Artymiuk, P. and Phillips, D. C. (1985) *Methods in Enzymology* **114A**, 397–415.

Artymiuk, P., Blake, C. C. F., Grace, D. E. P., Oatley, S. J., Phillips, D. C. and Sternberg, M. J. E. (1979) *Nature* **280**, 563–8.

Ashford, V., Dai, X., Nielsen, C., Sullivan, D. and Xuong, N.-H. (1990) *Acta Crystallographica* A**43**, C-15.

Azaroff, L. (1955) *Acta Crystallographica* **8**, 701–4.

Azaroff, L., Kaplow, R., Kato, N., Weiss, R. J., Wilson, A. J. C., Young, R. A. (1974), *X-ray Diffraction*, New York: McGraw Hill, p. 468.

Bachmann, R., Kohler, H., Schulz, H. and Weber, H.-P. (1985) *Acta Crystallographica* A**41**, 35–40.

Bachmann, R., Kohler, H., Schulz, H., Weber, H. P., Kupcik, V., Wendschuh-Josties, M., Wolf, A. and Wulf, R. (1983) *Angewandte Chemie International*, edition in English, **22**, 1011–12.

Bade, D., Parak, F., Mossbauer, R. L., Hoppe, W., Levai, N. and Charpak, G. (1982) *Nuclear Instruments and Methods* **201**, 193–6.

Badger, J., Krishnaswamy, S., Kremer, M. J., Oliveira, M. A., Rossmann, M. G., Heinz, B. A., Rueckert, R. R., Dutko, F. J. and McKinlay, M. A. (1989) *Journal of Molecular Biology* **207**, 163–74.

Badger, J., Minor, I., Kremer, M. J., Oliveira, M. A., Smith, T. J., Griffith, J. P., Guerin, D. M. A., Krishnaswamy, S., Luo, M., Rossmann, M. G., McKinlay, M. A., Diana, G. D., Dutko, F. J., Fancher, M., Rueckert, R. G. and Heinz, B. A. (1988) *Proceedings of the National Academy, USA* **85**, 3304–8.

Bailey, S., Evans, R. W., Garratt, R. C., Gorinsky, B., Hasnain, S. S., Horsburgh, C., Jhoti, H., Lindley, P. F., Mydin, A., Sarra, R. and Watson, J. L. (1988) *Biochemistry* **27**, 5804–12.

Baker, J. F. C. and Hart, M. (1975) *Acta Crystallographica* A**31**, 364–7.

Baker, P. J., Thomas, D. H., Barton, C. H., Rice, D. W. and Bailey, E. (1987) *Journal of Molecular Biology* **193**, 233–5.

Baker, P. J., Farrants, G. W., Stillman, T. J., Britton, K. L., Helliwell, J. R. and Rice, D. W. (1990) *Acta Crystallographica* A**46**, 721–5.

Banner, D. W., Bloomer, A. C., Petsko, G. A., Phillips, D. C. and Wilson, I. A. (1976) *Biochemistry and Biophysics Research Communications* **72**, 146–55.

Barford, D. and Johnson, L. N. (1989) *Nature* **340**, 609–16.

Barrington-Leigh, J. and Rosenbaum, G. (1974) *Journal of Applied Crystallography* **7**, 117–21.

Bartels, K. S. (1977) in *The Rotation Method in Crystallography*, edited by U. W. Arndt and A. J. Wonacott, Amsterdam: North-Holland, pp. 153–71.

Bartels, K. S. Weber, G., Weinstein, S., Wittmann, H.-G. and Yonath, A. (1988) *Topics in Current Chemistry* **147**, edited by E. Mandelkow, Heidelberg: Springer Verlag, pp. 57–72.

Bartunik, H. D. (1978) *Acta Crystallographica* A**34**, 747–50.

Bartunik, H. D. (1983) *Nuclear Instruments and Methods* **208**, 523–33.

Bartunik, H. D. and Borchert, T. (1989) *Acta Crystallographica* A**45**, 718–26.

Bartunik, H. D. and Fuess, H. (1975) *Proceedings of the Neutron Diffraction Conference Petten*, Reactor Centrum Conference, RCN-234, 527–34.

Bartunik, H. D. and Schubert, P. (1982) *Journal of Applied Crystallography* **15**, 227–31.

Bartunik, H. D., Clout, P. N. and Robrahn, B. (1981) *Journal of Applied Crystallography* **14**, 134–6.

Bartunik, H. D., Fourme, R. and Phillips, J. C. (1982) in *Uses of Synchrotron Radiation in Biology*, edited by H. B. Stuhrmann, London: Academic Press.

Bartunik, H. D., Summers, L. J. and Bartsch, H. H. (1989) *Journal of Molecular Biology* **210**, 813–28.

Bartunik, H. D., Jerzembek, E., Pruss, D., Huber, G. and Watson, H. C. (1981) *Acta Crystallographica* A**37**, C-51.

Baru, S. E., Proviz, G. A., Savinov, V. A., Sidorov, A. G. Khabakhpashev, A. G., Shuvalov, B. N. and Yakovlev, V. A. (1978) *Nuclear Instruments and Methods* **152**, 209–12.

Baru, S. E., Proviz, G. A., Savinov, V. A., Sidorov, V. A. Khabakhpashev, A. G., Shekhtman, L. I., Shuvalov, B. N. and Yasenev, M. V. (1983) *Nuclear Instruments and Methods* **208**, 445–7.

Beaumont, J. H. and Hart, M. (1974) *Journal of Physics* E**7**, 823–9.

Becker, P., Dorenwendt, K., Ebeling, G., Lauer, R., Lucas, W., Probst, R., Rademacher, H.-J., Reim, G., Seyfried, P. and Siegert, H. (1981) *Physical Review Letters* **46**, 1540–3.

Bell, J. A., Moffat, K., Vonderhaar, B. K. and Golde, D. W. (1985) *Journal of Biological Chemistry* **260**, 8520–5.

Bellon, P., Manassero, P. M. and Sansoni, M. (1972) *Journal of the Chemical Society, Dalton Transactions*, 1481–7.

Berghuis, J., Haanapell, I. J. M., Potters, M., Loopstra, Bo., MacGillavry, C. H. and Veenendaal, A. L. (1955) *Acta Crystallographica* **8**, 478–83.

Bernal, J. D. and Crowfoot, D. (1934) *Nature* **133**, 794–5.

Berreman, D. W. (1955) Ph.D. Thesis, Cal. Tech., USA.

Betzel, C., Pal G. P. and Saenger, W. (1988a) *Acta Crystallographica* B**44**, 163–72.

Betzel, C., Dauter, Z., Dauter, M., Ingelman, M., Papendorf, G., Wilson, K. S. and Branner, S. (1988b) *Journal of Molecular Biology* **204**, 803–4.

Bienenstock, A. (1980) *Nuclear Instruments and Methods* **172**, 13–20.

Bienenstock, A., Brown, G., Wiedemann, H. and Winick, H. (1989) *Review of Scientific Instruments* **60**(7), 1393–8.

Bijvoet, J. M. (1949) *Proceedings of the Academy of Sciences, Amsterdam* **52**, 313.

Bijvoet, J. M., Burgers, W. G. and Hägg, G. (1969) *Early Papers on Diffraction of X-rays by Crystals*, Volume I, Dordrecht: Kluwer Academic Publishers.

Bijvoet, J. M., Burgers, W. G. and Hägg, G. (1972) *Early Papers on Diffraction of X-rays by Crystals*, Volume II, Dordrecht: Kluwer Academic Publishers.

Bilderback, D. H. (1981) *Proceedings of SPIE – International Society for Optical Engineering* **315**, 90–102.

Bilderback, D. H. (1982) *Nuclear Instruments and Methods* **195**, 67–72.

Bilderback, D. H. (1986) *Nuclear Instruments and Methods* A**246**, 434–6.

Bilderback, D. H., Moffat, K. and Szebenyi, D. (1984) *Nuclear Instruments and Methods* **222**, 245–51.

Bilderback, D. H., Lairson, B. M., Barbee, T. W., Ice, G. E. and Sparks, C. J. (1983) *Nuclear Instruments and Methods* **208**, 251–61.

Bilderback, D. H., Moffat, K., Owen, J., Rubin, B., Schildkamp, W., Szebenyi, D., Temple B. S., Volz, K. and Whiting, B. (1988) *Nuclear Instruments and Methods* A**226**, 636–44.

Bilderback, D. H., Batterman, B. W., Bedzyk, M. J., Finkelstein, K., Henderson, C., Merlini, A., Schildkamp, W., Shen, Q., White, J., Blum, E., Viccaro, P. J., Mills, D. M., Kim, S., Shenoy, G. K., Robinson, K. E., James, F. E. and Slater, J. M. (1989) *Review of Scientific Instruments* **60**(7), 1419–25.

Birktoft, J. T., Holden, H. M., Hamlin, R., Xuong, N.-H. and Banaszak, L. J. (1987) *Proceedings of the National Academy, USA* **84**, 8262–6.

Bjorkman, P. J., Saper, M. A., Samraoui, B., Bennett, W. S., Strominger, J. L. and Wiley, D. C. (1987) *Nature* **329**, 506–12.

Black, P. J. (1965) *Nature* **206**, 1223–6.

Blake, C. C. F. and Phillips, D. C. (1962) in *Biological Effects of Ionising Radiation at the Molecular Level*, IAEA Symposium, Vienna, and discussed in Blundell and Johnson (1976), p. 253.

Blow, D. M. and Crick, F. H. C. (1959) *Acta Crystallographica* **12**, 794–802.

Blundell, T. L. and Johnson, L. N. (1976) *Protein Crystallography*, New York: Academic Press.

Bode, W., Chen, Z., Bartels, K., Kutzbach, C., Schmidt-Kastner, G. and Bartunik, H. D. (1983) *Journal of Molecular Biology* **164**, 237–82.

Bode, W., Mayr, I., Baumann, U., Huber, R., Stone, S. R. and Hofsteenge, J. (1989) *EMBO Journal* **8**, 3467–75.

Boel, E., Brady, L., Brzozowski, A. M., Derewenda, Z., Dodson, G. G., Jensen, V. J., Petersen, S. B., Swift, H., Thim, L. and Woldike, H. F. (1990) *Biochemistry* **29**, 6244–9.

Bœuf, A., Lagomarsino, S., Mazkedian, S., Melone, S., Puliti, P. and Rustichelli, F. (1978) *Journal of Applied Crystallography* **11**, 442–9.

Bonse, U. and Materlik, G. (1972) *Zeitschrift für Physik* **253**, 232–9.

Bonse, U. and Materlik, G. (1976) *Zeitschrift für Physik* B**24**, 189–91.

Bonse, U., Materlik, G. and Schroder, W. (1976) *Journal of Applied Crystallography* **9**, 223–30.

Bonse, U., Spicker, P., Hein, J. T. and Materlik, G. (1980) *Nuclear Instruments and Methods* **172**, 223–6.

Bordas, J. (1982) in *Uses of Synchrotron Radiation in Biology*, edited by H. B. Stuhrmann, New York: Academic Press, pp. 107–44.

Bordas, J. and Mandelkow, E. (1983) in *Fast Methods in Physical Biochemistry and Cell Biology*, edited by R. I. Sha'afi and S. M. Fernandez, Amsterdam: Elsevier, pp. 137–72.

Bordas, J., Koch, M. H. J., Clout, P. N., Dorrington, E., Boulin, C. and Gabriel, A. (1980) *Journal of Physics* E**13**, 938–44.

Boulin, C., Dainton, D., Dorrington, E., Elsner, G., Gabriel, A., Bordas, J. and Koch, M. H. J. (1982) *Nuclear Instruments and Methods* **201**, 209–20.

Bourne, Y., Rouge, P. and Cambilloau, C. (1988) *Journal of Molecular Biology* **202**, 685–7.

Boxer, S. G., Kuki, A., Wright, K. A., Katz, B. A., Xuong, N.-H. (1982) *Proceedings of the National Academy, USA* **79**, 1121–5.

Boyce, R., Brown, G., Hower, N., Hussain, Z., Pate, T., Umbach, E. and Winick, H. (1983) *Nuclear Instruments and Methods* **208**, 127–37.

Boylan, D. and Phillips, G. N. Jr (1986) *Biophysical Journal* **49**, 76–8.

Boys, C. W. G., Czaplewski, L. G., Phillips, S. E. V., Baumberg, S. and Stockley, P. G. (1990) *Journal of Molecular Biology* **231**, 227–8.

Boysen, H., Frey, F. and Jagodzinski, H. (1984) *Rigaku Journal* **1**, 3–14.

Brady, L., Brzozowski, A. M., Derewenda, Z. S., Dodson, E. J., Dodson, G. G., Tolley, S., Turkenburg, J. P., Christiansen, L., Huge-Jensen, B., Norskov, L., Thim, L. and Menge, U. (1990) *Nature* **343**, 767–70.

Bragg, W. L. (1949) 'General Survey' in *The Crystalline State* by W. H. and W. L. Bragg, p. 27.

Bragg, W. L. (1975) in *The Development of X-ray Analysis*, edited by D. C. Phillips and H. Lipson, London: G. Bell and Sons.

Brammer, R. C. (1987) Daresbury Laboratory Technical Memorandum, DL/SCI/TM56E.

Brammer, R., Helliwell, J. R., Lamb, W., Liljas, A., Moore, P. R., Thompson, A. W. and Rathbone, K. (1988) *Nuclear Instruments and Methods* A**271**, 678–87.

Bravais, M. A. (1849) 'On the systems formed by points regularly distributed on a plane or in space', Crystallographic Society of America. Translated by A. J. Shaler from J. de l'Ecole Polytechnique, *Cahier* 33, **19**, 1–128, Paris (1850).

Brown, D. G., Sanderson, M. R., Skelly, J. V., Jenkins, T. C., Brown, T., Garman, E., Stuart, D. I. and Neidle, S. (1990) *EMBO Journal* **9**, 1329–34.

Brown, G., Halbach, K., Harris, J. and Winick, H. (1983) *Nuclear Instruments and Methods* **208**, 65–77.

Brown, K. A., Brick, P. and Blow, D. M. (1987) *Nature* **326**, 416–18.

Brown, K. A., Vrielink, A. and Blow, D. M. (1986) *Biochemical Society Transactions* **14**, 1228–9.

Brown, M., Peierls, R. E. and Stern, E. A. (1977) *Physical Review* B**15**, 738–44.

Brown, N. M. D., McMonagle, J. B. and Greaves, G. N. (1984) *Journal of the Chemical Society, Faraday Transactions* **1**, **80**, 589–97.

Brünger, A. T., Kuriyan, J. and Karplus, M. (1987) *Science* **235**, 458–60.

Brzozowski, A., Derewenda, Z., Dodson, E. J., Dodson, G., Grabowski, M., Liddington, R., Skarzynski, T. and Vallely, D. (1984) *Nature* **307**, 74–6.

Buerger, J. M. (1964) *The Precession Method*, New York: John Wiley.

Buras, B. and Gerward, L. (1975) *Acta Crystallographica* A**31**, 372–4.

Buras, B., Fourme, R. and Koch, H. H. J. (1983) in *Handbook on Synchrotron Radiation*, Volume 1B, edited by E. E. Koch, Amsterdam: North-Holland, pp. 1015–90.

Burattini, E., Reale, A., Bernieri, E., Cavallo, N., Morone, A., Masullo, M. R., Rinzivillo, R., Dallsa, G., Fornasini, P. and Mencuccini, C. (1983) *Nuclear Instruments and Methods* **208**, 91–6.

Burlingame, R. W., Love, W. E., Wang, B.-C., Hamlin, R., Xuong, N.-H. and Mondrianakis, N. (1985) *Science* **228**, 546–53.

Byers, B. (1967) *Journal of Molecular Biology* **26**, 155–67.

Caffrey, M. and Bilderback, D. H. (1983) *Nuclear Instruments and Methods* **208**, 495–510.

Campbell, J. W., Habash, J., Helliwell, J. R. and Moffat, K. (1986) *Information Quarterly for Protein Crystallography*, No. 18, Daresbury Laboratory, pp. 23–31.

Carlisle, C. H., Lindley, P. F., Moss, D. S. and Slingsby, C. (1977) *Journal of Molecular Biology* **110**, 417–19.

Carlisle, C. H., Palmer, R. A., Mazumdar, S. I., Gorinsky, B. A. and Yeates, D. G. R. (1974) *Journal of Molecular Biology* **85**, 1–18.

Carrell, H. L., Glusker, J. P., Burger, V., Manfre, F., Tritsch, D. and Biellmann, J.-F. (1989) *Proceedings of the National Academy, USA* **86**, 4440–4.

Carter, D. C., He, X.-M., Munson, S. H., Twigg, P. D., Gemert, K. M., Broom, M. B. and Miller, T. Y. (1989) *Science* **244**, 1195–8.

Cascarano, G., Giacovacco, C., Peerdeman, A. F. and Kroon, J. (1982) *Acta Crystallographica* A**38**, 710–17.

Caspar, D. L. D., Clarage, J., Salunke, D. M. and Clarage, M. (1988) *Nature* **352**, 659–62.

Chambliss, G., Craven, G. R., Davies, J., Davies, K., Kahan, L., Nomura, M. (eds.) (1980) *Ribosomes: Structure, Function, and Genetics*. Baltimore: University Park Press.

Chapman, M. S., Smith, W. W., Suh, S. W., Cascio, D., Howard, A., Hamlin, R., Xuong, N.-H. and Eisenberg, D. (1986) *Philosophical Transactions of the Royal Society* B**313**, 367–78.

Charpak, G. (1982) *Nuclear Instruments and Methods* **201**, 181–92.

Charpak, G., Hajduk, Z., Jeavons, A. P., Kahn, R. and Stubbs, R. J. (1974) *Nuclear Instruments and Methods* **122**, 307–12.

Chen, Z., Stauffacher, C., Li, Y., Schmidt, T., Bomu, W., Kamer, G., Shanks, M., Lomonossoff, G. and Johnson, J. E. (1989) *Science* **245**, 154–9.

Clifton, I. J., Cruickshank, D. W. J., Diakun, G., Elder, M., Habash, J., Helliwell, J. R., Liddington, R. C., Machin, P. A. and Papiz, M. Z. (1985) *Journal of Applied Crystallography* **18**, 296–300.

Clucas, J. A., Harding, M. M. and Maginn, S. J. (1988) *Journal of the Chemical Society, Chemical Communications* No. 1481, 185–7.

Cohen, E. R. and Taylor, B. N. (1987) *Review of Modern Physics* **59**, 1121.

Coisson, R. and Winick, H. (eds.) (1987) Workshop on PEP as a synchrotron radiation source, held 1987 (Proceedings available from the Stanford Synchrotron Radiation Laboratory).

Collyer, C. A., Guss, J. M., Sugimura, Y., Yoshizaki, F. and Freeman, H. C. (1990) *Journal of Molecular Biology* **211**, 617–32.

Colman, P. M., Laver, W. G., Varghese, J. N., Baker, A. T., Tulloch, P. A., Air, G. M. and Webster, R. G. (1987) *Nature* **326**, 358–62.

Compton, A. H. and Allison, S. K. (1935) *X-rays in Theory and Experiment*, New York: Van Nostrand.

Cook, W. J., Ealick, S. E., Krenitsky, T. A., Stoeckler, J. D., Helliwell, J. R. and Bugg, C. E. (1985) *Journal of Biological Chemistry* **260**, 12968–9.

Coster, D. (1924) *Zeitschrift für Physik* **25**, 83–98.

Coster, D., Knol, K. S. and Prins, J. A. (1930) *Zeitschrift für Physik* **63**, 345–69.

Cox, M. J. and Weber, P. C. (1987) *Journal of Applied Crystallography* **20**, 366–73.

Crick, F. H. C. and Magdoff, B. S. (1956) *Acta Crystallographica* **9**, 901–8.

Cromer, D. T. (1965) *Acta Crystallographica* **18**, 17–23.

Cromer, D. T. (1983) *Journal of Applied Crystallography* **16**, 437.

Cromer, D. T. and Liberman, D. (1970) *Journal of Chemistry and Physics* **53**. 1891–8. (See also *Los Alamos Scientific Report* LA-4403 (1970).)

Cromer, D. T. and Liberman, D. (1981) *Acta Crystallographica* A**37**, 267–8.

Cruickshank, D. W. J., Helliwell, J. R. and Moffat, K. (1987) *Acta Crystallographica* A**43**, 656–74.

Cruickshank, D. W. J., Helliwell, J. R. and Moffat, K. (1991) *Acta Crystallographica* A **47**, 352–73.

Cuff, M. E., Hendrickson, W. A., Lamy, J., Lamy, J. N., Miller, K. I. and van Holde, K. E. (1990) *Journal of Molecular Biology* **213**, 11–15.

Darwin, C. G. (1914) *Philosophical Magazine* **27**, 315.

Dauben, C. H. and Templeton, D. H. (1955) *Acta Crystallographica* **8**, 841–2.

Davies, S. J., Brady, R. L., Barclay, A. N., Harlos, K., Dodson, G. G. and Williams, A. F. (1990) *Journal of Molecular Biology* **212**, 7–10.

de Vos, A. M., Tong, L., Milburn, M. V., Matias, P. M., Jancarik, J., Noguchi, S., Nishimura, S., Miura, K., Ohtsuka, E. and Kim, S.-H. (1988) *Science* **239**, 888–93.

Deisenhofer, J. and Michel, H. (1989) *EMBO Journal* **8**, 2149–69.

Deisenhofer, J., Epp, O., Miki, K., Huber, R. and Michel, H. (1984) *Journal of Molecular Biology* **180**, 385–98.

Deisenhofer, J., Epp, O., Miki, K., Huber, R. and Michel, H. (1985) *Nature* **318**, 618–24.

DeLucas, L. J., Greenhough, T. J., Rule, S. A., Myles, D. A. A., Babu, Y. S., Volanakis, J. E. and Bugg, C. E. (1987) *Journal of Molecular Biology* **196**, 741–2.

DeLucas, L. J., Smith, C. D., Smith, H. W., Vijay-Kumar, S., Senadhi, S. E., Ealick, S. E., Carter, D. C., Snyder, R. S., Weber, P. C., Salemme, F. R., Ohlendorf, D. H., Einspahr, H. M., Clancy, L. L., Navia, M. A., McKeeves, B. M., Nagabhushan, T. L., Nelson, G., McPherson, A., Koszelak, S., Taylor, G., Stammers, D., Powell, K., Darby, G. and Bugg, C. E. (1989) *Science* **246**, 651–4.

Derewenda, U., Derewenda, Z., Dodson, E. J., Dodson, G. G., Reynolds, C. D., Smith, G. D., Sparks, C. and Swenson, D. (1989) *Nature* **338**, 594–6.

Derewenda, Z. and Helliwell, J. R. (1989) *Journal of Applied Crystallography* **22**, 123–37.

Derewenda, Z., Dodson, G. G., Emsley, P., Harris, D., Nagai, K., Perutz, M. and Reynaud, J.-P. (1990) *Journal of Molecular Biology* **221**, 515–19.

Derewenda, Z., Yariv, J., Helliwell, J. R., Kalb (Gilboa), A. J., Dodson, E. J., Papiz, M. Z., Wan, T. and Campbell, J. (1989) *EMBO Journal* **8**, 2189–93.

Dewan, J. C. and Tilton, R. F. (1987) *Journal of Applied Crystallography* **20**, 130–2.

Diamond, R. (1974) *Journal of Molecular Biology* **82**, 371–91.

Diamond, R. (1990) *Acta Crystallographica* A**46**, 425–35.

Didenko, A. N., Kozhevnikov, A. V., Medvedev, A. F., Nikitin, M. M. and Epp, V. Ya. (1979) *Soviet Physics Journal of Experimental and Theoretical Physics* **49**, 973.

Dietrich, I., Dubochet, J., Fox, F., Knapek, E. and Weyl, R. (1980) in *Electron Microscopy at Molecular Dimensions*, edited by W. Baumeister and W. Vogell, Heidelberg: Springer Verlag, pp. 234–44.

DiGabriele, A. D., Sanderson, M. R. and Steitz, T. A. (1989) *Proceedings of the National Academy, USA* **86**, 1816–20.

Dobson, C. M. (1990) *Nature* **348**, 198–9.

Dodson, E. J. and Vijayan, M. (1971) *Acta Crystallographica* B**27**, 2402–11.

Dodson, E. J., Evans, P. R. and French, S. (1975) in *Anomalous Scattering*, edited by S. Ramaseshan and S. C. Abrahams, Copenhagen: Munksgaard, pp. 423–36.

Doucet, J. and Benoit, J. P. (1987) *Nature* **325**, 643–6.

Doucet, J., Benoit, J.-P., Cruse, W. B. T., Prange, T. and Kennard, O. (1989) *Nature* **337**, 190–2.

Douzou, P. and Petsko, G. A. (1984) *Advances in Protein Chemistry* **36**, 246.

Drenth, J., Helliwell, J. R. and Littke, W. (1987) in *Fluid Sciences and Materials Science in Space*, edited by H. U. Walter, Heidelberg: Springer Verlag, pp. 451–76.

Ducruix, A. and Reiss-Husson, F. (1987) *Journal of Molecular Biology* **193**, 419–21.

Du Mond, J. W. M. (1937) *Physical Review* **52**, 872–83.

Durbin, R. M., Burns, R., Moulai, J., Metcalf, P., Freymann, D., Blum, M., Anderson, J. E., Harrison, S. C. and Wiley, D. C. (1986) *Science* **232**, 1127–32.

Ealick, S. E., Rule, S. A., Carter, D. C., Greenhough, T. J., Babu, Y. S., Cook, W. J., Habash, J., Helliwell, J. R., Stoeckler, J. D., Parks, R. E. Jr, Chen, S.-F. and Bugg, C. E. (1990) *Journal of Biological Chemistry* **265**, 1812–20.

Edwards, C., Palmer, S. B., Emsley, P., Helliwell, J. R., Glover, I. D., Harris, G. W. and Moss, D. S. (1990) *Acta Crystallographica* A**46**, 315–20.

Edwards, S. C. and Poulos, T. C. (1990) *Journal of Biological Chemistry* **265**, 2588–95.

Edwards, S. L., Kraut, J., Xuong, N., Ashford, V., Halloran, T. P. and Mills, S. E. (1988) *Journal of Molecular Biology* **203**, 523–4.

Einspahr, H., Suguna, K., Suddath, F. L., Ellis, G., Helliwell, J. R. and Papiz, M. Z. (1985) *Acta Crystallographica* B**41**, 336–41.

Eisenberger, P., Newsam, J. M., Leonowicz, M. E. and Vaughan, D. E. W. (1984) *Nature* **309**, 45–7.

Elder, M. (1987) CCP4 Newsletter, No. 19, Daresbury Laboratory, pp. 32–6.

Ellaume, P. (1987a) ESRF Preprint SR/ID-87-10.

Ellaume, P. (1987b) ESRF Preprint SR-ID-87-19.

Ellaume, P. (1989a) *Review of Scientific Instruments* **60**(7), 1830.

Ellaume, P. (1989b) *Synchrotron Radiation News* **1**(4), 18–23.

Ellaume, P. (1989c) ESRF Preprint SR/ID-89-26.

Evans, H. T. and Lonsdale, K. (1959) *International Tables of Crystallography*, Volume II, Birmingham, England: Kynoch Press, pp. 161–215.

Fano, U. (1961) *Physical Review* **124**, 1866–78.

Farber, G. K., Machin, P. A., Almo, S., Petsko, G. A. and Hajdu, J. (1988) *Proceedings of the National Academy, USA* **85**, 112–15.

Farber, H. R. and Matthews, B. W. (1990) *Nature* **348**, 263–6.

Faruqi, W. (1977) in *The Rotation Method in Crystallography*, edited by U. W. Arndt and A. J. Wonacott, Amsterdam: North-Holland, pp. 227–43.

Fermi, G., Perutz, M. F. and Shulman, R. G. (1987) *Proceedings of the National Academy, USA* **84**, 6167–8.

Fermi, G., Perutz, M. F., Shaanan, B. and Fourme, R. (1984) *Journal of Molecular Biology* **175**, 159–74.

Fink, A. L. and Petsko, G. (1981) *Advances in Enzymology* **52**, 177–246.

Finzel, B. C., Clancy, L. L., Holland, D. R., Muchmore, S. W., Watenpaugh, K. D. and Einspahr, H. M. (1989) *Journal of Molecular Biology* **209**, 779–91.

Fitzgerald, P. M. D., Duax, W. L., Punzi, J. S. and Orr, J. C. (1984) *Journal of Molecular Biology* **175**, 225–7.

Flank, A. M., Fontaine, A., Lagarde,P., Lemonnier, M., Mimault, J., Raoux, D., Sadoc, A. (1981) in Daresbury Study Weekend, March 1981, Proceedings, edited by S. S. Hasnain and C. D. Garner, Daresbury Laboratory, pp. 70–5.

Flook, R. J., Freeman, H. C. and Scudder, M. L. (1977) *Acta Crystallographica* B**33**, 801–9.

Fourme, R. (1978) in *Trends in Physics, Proceedings of the 4th EPS Meeting*, York: Adam Hilger Ltd, pp. 507–14.

Fourme, R. (1979) in *Synchrotron Radiation Applied to Biophysical and Biochemical Research*, edited by A. Castellani and I. F. Quercia, New York: Plenum Press, pp. 349–59, 361–75.

Fourme, R. and Kahn, R. (1981) *Biochimie* **63**, 887–90.

Fox, G., Stuart, D. I., Acharya, K. R., Fry, E., Rowlands, D. and Brown, F. (1987) *Journal of Molecular Biology* **196**, 591–7.

Franklin, R. and Gosling, R. G. (1953) *Nature* **172**, 156–7.

Franks, A. and Breakwell, P. R. (1974) *Journal of Applied Crystallography* **7**, 122–5.

Franks, A., Gale, B., Lindsey, K., Stedman, M. and Bailey, W. P. (1983) *Nuclear Instruments and Methods* **208**, 223–6.

Frauenfelder, H., Petsko, G. A. and Tsernoglou, D. (1979) *Nature* **280**, 558–63.

Freemont, P. S., Friedman, J. M., Beese, L. S., Sanderson, M. R. and Steitz, T. A. (1988) *Proceedings of the National Academy, USA* **85**, 8924–8.

Freund, A. (1975) in *Anomalous Scattering*, edited by S. Ramaseshan and S. C. Abrahams, Copenhagen: Munksgaard, pp. 69–84.

Friedrich, W. Knipping, P. and von Laue, M. (1912) *Bayerische Akademie der Wissenschaften*, 303–22.

Fukamachi, T. and Hosoya, S. (1975) *Acta Cyrstallographica* A**31**, 215–20.

Fukamachi, T., Hosoya, S., Kawamura, T. and Okunki, M. (1977) *Acta Crystallographica* A**33**, 54–8.

Fukuyama, K., Hirota, S. and Tsukihara, T. (1987) *Journal of Molecular Biology* **196**, 961–2.

Fuller, W., Forsyth, V. T. and Mahendrasingam, A. (1990) in *Synchrotron Radiation and Biophysics*, edited S. S. Hasnain, Chichester: Ellis Horwood, pp. 201–22.

Gaboriaud, C., Vaney, M. C., Bachet, B., Le-Nguyen, D., Castro, B. and Mornon, J. P. (1989) *Journal of Molecular Biology* **210**, 883–4.

Galloway, J. W. (1985) *Nature* **318**, 602.

Gaponov, Yu. A., Lyakhov, N. Z., Tolochko, B. P., Boldyrev, V. V. and Sheromov, M. A. (1989a) *Nuclear Instruments and Methods* A**282**, 698–700.

Gaponov, Yu. A., Lyakhov, N. Z., Tolochko, B. P., Baru, S. E., Savinov, G. A. and Sheromov, M. A. (1989b) *Review of Scientific Instruments* **60**(7), 2429–31.

Gaponov, Yu. A., Lyakhov, N. Z., Tolochko, B. P., Baru, S. E., Savinov, G. A. and Sheromov, M. A. (1989c) *Nuclear Instruments and Methods* A**282**, 695–7.

Garavito, R. M. and R. Rosenbusch, J. P. (1980) *Journal of Cell Biology* **86**, 327–9.

Garavito, R. M., Jenkins, J., Jansonius, J. N., Karrlsson, R. and Rosenbusch, J. P. (1983) *Journal of Molecular Biology* **164**, 313–27.

Garner, C. D. (1990) in *Applications of Synchrotron Radiation*, edited by R. Catlow and G. N. Greaves, Glasgow: Blackie.

Gibbs, D., Harshmann, D., Isaacs, E., McWhan, D. B., Mills, D. M. and Vettier, C. (1988) *Physical Review Letters* **61**, 1241–4.

Giegé, R., Ducruix, A., Fontecilla-Camps, J. C., Feigelson, R. S., Kern, R. and McPherson, A. Jr (eds.) (1988) *Journal of Crystal Growth* **90**, 1–374.

Giranda, V. L., Berman, H. M. and Schramm, V. L. (1989) *Journal of Biological Chemistry*, **264**, 15674–80.

Glockshuber, R., Steipe, B. Huber, R. and Plückthun, A. (1990) *Journal of Molecular Biology* **213**, 613–15.

Glotz, C., Müssig, J., Gewitz, H. S., Makowski, I., Arad, T., Yonath, A. and Wittmann, H. G. (1987) *Biochemistry International* **15**, 953–60.

Glover, I. D., Helliwell, J. R. and Papiz, M. Z. (1988) *Topics in Current Chemistry* **147**, edited by E. Mandelkow, Heidelberg: Springer Verlag, pp. 31–55.

Glover, I. D., Harris, G. W., Helliwell, J. R. and Moss, D. S. (1991) *Acta Crystallographica* B In Press.

Glusker, J. P. and Trueblood, K. N. (1985) *Crystal Structure Analysis*, second edition, Oxford: Oxford University Press, pp. 42–60

Goldsmith, E. J., Sprang, S. R., Hamlin, R., Xuong, N.-H. and Fletterick, R. J. (1989) *Science* **245**, 528–32.

Gomez de Anderez, D., Helliwell, M., Habash, J., Dodson, E. J., Helliwell, J. R., Bailey, P. D. and Gammon, R. E. (1989) *Acta Crystallographica* B**45**, 482–8.

Goodfellow, J., Henrick, K. and Hubbard, R. (eds.) (1989) Proceedings of the Daresbury Study Weekend entitled 'Molecular Simulation and Protein Crystallography', DL/SCI/R27.

Gouaux, J. E. and Lipscomb, W. N. (1988) *Proceedings of the National Academy, USA* **85**, 4205–8.

Gouaux, J. E., Stevens, R. C., Ke, H. and Lipscomb, W. N. (1989) *Proceedings of the National Academy, USA* **86**, 8212–16.

Graves, B. J., Hatada, M. H., Hendrickson, W. A., Miller, J. K., Madison, V. S. and Satow, Y. (1990) *Biochemistry* **29**, 2679–84.

Graziano, V., Gerchman, S. E., Wonacott, A. J., Sweet, R. M., Wells, J. R. E., White, S. W. and Ramakrishman, V. (1990) *Journal of Molecular Biology* **212**, 253–7.

Greaves, G. N., Durham, P. J., Diakun, G. and Quinn, P. (1981) *Nature* **294**, 139–42.

Greaves, G. N., Bennett, R., Duke, P. J., Holt, R. and Suller, V. P. (1983a) *Nuclear Instruments and Methods* **208**, 139–142.

Greaves, G. N., Diakun, G. P., Quinn, P. D., Hart, M. and Siddons, D. P. (1983b) *Nuclear Instruments and Methods* **208**, 335–9.

Green, D. W., Ingram, V. M. and Perutz, M. F. (1954) *Proceedings of the Royal Society* **225**, 287–302.

Greenhough, T. J., (1983) *Information Quarterly for Protein*

Crystallography, No. 10, February, edited by P. Machin, Daresbury Laboratory.

Greenhough, T. J. and Helliwell, J. R. (1982a), *Journal of Applied Crystallography* **15**, 338–51.

Greenhough, T. J. and Helliwell, J. R. (1982b), *Journal of Applied Crystallography* **15**, 493–508.

Greenhough, T. J. and Helliwell, J. R. (1983) *Progress in Biophysics and Molecular Biology* **41**, 67–123.

Greenhough, T. J., Helliwell, J. R. and Rule, S. A. (1983), *Journal of Applied Crystallography* **16**, 242–50.

Gros, P., Betzel, Ch., Dauter, Z., Wilson, K. S. and Hol. W. G. J. (1989) *Journal of Molecular Biology* **210**, 347–67.

Gruner, S. (1987) *Science* **238**, 305.

Gruner, S. M., Milch, J. R. and Reynolds, G. T. (1978) *Institute of Electrical and Electronic Engineers Transactions in Nuclear Science* **NS-25**, No. 1, 562–5.

Grütter, M. G., Fendrich, G., Huber, R. and Bode, W. (1988) *EMBO Journal* **7**, 345–51.

Gurman, S. J. (1990) in *Synchrotron Radiation and Biophysics*, edited by S. S. Hasnain, Chichester: Ellis Horwood, pp. 9–42.

Guss, J. M., Merritt, E. A., Phizackerley, R. P., Hedman, B., Murata, M., Hodgson, K. O. and Freeman, H. C. (1988) *Science* **241**, 806–11.

Haas, D. J. and Rossmann, M. G. (1970) *Acta Crystallographica* **B26**, 998–1004.

Habash, J., Harrop, S., Helliwell, J. R., Nave, C. and Thompson, A. W. (1990), Daresbury Laboratory Technical Memorandum, DL/SCI/TM64E.

Hails, J., Harding, M. M., Helliwell, J. R., Liddington, R. and Papiz, M. Z. (1984) Daresbury Laboratory Preprint DL/SCI/P428E.

Hajdu, J. and Johnson, L. N. (1990) *Biochemistry* **29**, 1669–78.

Hajdu, J., Acharya, K. R., Stuart, D. I. and Johnson, L. N. (1988) *Trends in Biochemical Sciences* **13**, 104–9.

Hajdu, J., Greenhough, T. J., Clifton, I. J., Campbell, J. W., Schrive, A. K., Harrison, S. C. and Liddington, R. (1989) in *Synchrotron Radiation in Structural Biology*, edited by R. M. Sweet, New York: Plenum Press, pp. 331–9.

Hajdu, J., Acharya, K. R., Stuart, D. I., McLaughlin, P. J., Barford, D., Oikonomakos, N. G., Klein, H. and Johnson, L. N. (1987a) *EMBO Journal* **6**, 539–46.

Hajdu, J., Machin, P. A., Campbell, J. W. Greenhough, T. J.,

Clifton, I. J., Zurek, S., Gover, S., Johnson, L. N. and Elder, M. (1987b) *Nature* **329**, 178–81.

Hamilton, W. C. (1974) 'Angle settings for four-circle diffractometers' in *International Tables for X-ray Crystallography*, Volume IV, Birmingham, England: Kynoch Press, pp. 273–84. (Present distributor: Kluwer Academic Publishers, Dordrecht.)

Hamlin, R. (1985) *Methods in Enzymology* **114A**, 416–51.

Hamlin, R., Cork, C., Howard, A., Nielsen, C., Vernon, W., Matthews, D., Xuong, Ng-h. and Perez-Mendez, V. (1981) *Journal of Applied Crystallography* **14**, 85–93.

Harada, S., Yasui, M., Murakawa, K. and Kasai, N. (1986) *Journal of Applied Crystallography* **19**, 448–52.

Hardesty, B. and Kramer, G. (eds.) (1986) *Structure, Function, and Genetics of Ribosomes*, Heidelberg: Springer Verlag.

Harding, M. M. (1962) D.Phil. Thesis, University of Oxford.

Harding, M. M., Maginn, S. J., Campbell, J. W., Clifton, I. and Machin, P. (1988) *Acta Crystallographica* B**44**, 142–6.

Hardy, L. W., Finer-Moore, J. S., Montfort, W. R., Jones, M. O., Santi, D. V. and Stroud, R. M. (1987) *Science* **235**, 448–55.

Harmsen, A., Leberman, R. and Schultz G. E. (1976) *Journal of Molecular Biology* **104**, 311–14.

Harrison, S. C. (1968) *Journal of Applied Crystallography* **1**, 84–90.

Harrison, S. C. (1973) in *Research Applications of Synchrotron Radiation*, edited by R. W. Watson and M. L. Perlman, Brookhaven National Laboratory Report BNL 50381, pp. 105–8.

Harrison, S. C. (1978) *Trends in Biochemical Sciences* 3, 4.

Hart, M. (1971) *Reports on Progress in Physics* **34**, 435–90.

Hart, M. (1980) *Nuclear Instruments and Methods* **172**, 209–14.

Hart, M. (1990) *Nuclear Instruments and Methods* A**297**, 306–11.

Hart, M. and Rodrigues, A. R. D. (1978) Journal of Applied Crystallography **11**, 248–53.

Hart, M. and Siddons, D. P. (1981) *Proceedings of the Royal Society, London* **376**, 465–82.

Hartree, D. (1933) *Proceedings of the Royal Society* **A141**, 282–301.

Haselgrove, J. C., Faruqi, A. R., Huxley, H. E. and Arndt, U. W. (1977) *Journal of Physics* E**10**, 1035–44.

Hashizume, H., Wakabayashi, K., Amemiya, Y., Hamanaka, T., Wakabayashi, T., Matsushita, T., Ueki, T., Hiiragi, Y., Izumi, Y. and Tagawa, H. (1982) Photon Factory Kek, Internal Report, p. 81.

Hasnain, S.S. (1988) *Topics in Current Chemistry*, Volume 147, 73–93 Heidelberg, Springer Verlag.

Hasnain, S.S. and Strange, R.W. (1990) in *Synchrotron Radiation and Biophysics*, edited by S.S. Hasnain, Chichester: Ellis Horwood, pp. 104–21.

Hastings, J.B. (1977) *Journal of Applied Physics* **48**, 1576–84.

Hastings, J.B., Kincaid, B.M. and Eisenberger, P. (1978) *Nuclear Instruments and Methods* **152**, 167–71.

Hastings, J.B., Suortti, P., Thomlinson, W., Kvick, A. and Koetzle, T.F. (1983) *Nuclear Instruments and Methods* **208**, 55–8.

Heald, S. (1982) *Nuclear Instruments and Methods* **195**, 59–62.

Hedman, B., Hodgson, K.O., Helliwell, J.R., Liddington, R. and Papiz, M.Z. (1985) *Proceedings of the National Academy, USA* **82**, 7604–7.

Helliwell, J.R. (1977a) D.Phil. Thesis, University of Oxford.

Helliwell, J.R. (1977b) *New Scientist* **76**, 646–8.

Helliwell, J.R. (1979) in *Applications of Synchrotron Radiation to the Study of Large Molecules of Chemical and Biological Interest*, edited by R.B. Cundall and I.H. Munro, Proceedings of the Daresbury Study Weekend, DL/SCI/R13, pp. 1–6.

Helliwell, J.R. (1982) *Nuclear Instruments and Methods* **201**, 153–74.

Helliwell, J.R. (1984) *Reports on Progress in Physics* **47**, 1403–97.

Helliwell, J.R. (1985) *Journal of Molecular Structure* **130**, 63–91.

Helliwell, J.R. (1987) in the *ESRF Foundation Phase Report* ('Red Book'), Grenoble: ESRF, pp. 329–40.

Helliwell, J.R. (1988) *Journal of Crystal Growth* **90**, 259–72.

Helliwell, J.R. (1989) *Physics World*, January issue, pp. 29–32.

Helliwell, J.R. (1991) *Nuclear Instruments and Methods* **A308**, 260–6.

Helliwell, J.R. and Fourme, R. (1983) Design study for the proposed European Synchrotron Radiation Facility (ESRF). Report presented to the Instrumentation Sub Group of the ESF Ad-hoc Committee on Synchrotron Radiation.

Helliwell, J.R., Moore, P.R., Papiz, M.Z. and Smith, J.M.A. (1984b) *Journal of Applied Crystallography* **17**, 417–19.

Helliwell, J.R., Glover, I.D., Jones, A. Pantos, E. and Moss, D.S. (1986a) *Biochemical Society Transactions* **14**, 653–5.

Helliwell, J.R., Cruickshank, D.W.J., Ellis, G.H., Habash, J., Papiz, M.Z., Rule, S. and Smith, J.M.A. (1984a) Daresbury Laboratory Study Weekend, DL/SCI/R22, pp. 41–59.

Helliwell, J.R., Greenhough, T.J., Carr, P.D., Rule, S.A., Moore,

P. R., Thompson, A. W. and Worgan, J. S. (1982a) *Journal of Physics* E**15**, 1363–72.

Helliwell, J. R., Hughes, G., Pryzbylski, M. M., Ridley, P. A., Sumner, I., Bateman, J. E., Connolly, J. F. and Stephenson, R. (1982b) *Nuclear Instruments and Methods* **201**, 175–80.

Helliwell, J. R., Harrop, S., Habash, J., Magorrian, B. G., Allinson, N. M., Gomez de Anderez, D., Helliwell, M., Derewenda, Z. and Cruickshank, D. W. J. (1989a) *Review of Scientific Instruments* **60**, 1531–6.

Helliwell, J. R., Papiz, M. Z., Glover, I. D., Habash, J., Thompson, A. W., Moore, P. R., Harris, N., Croft, D. and Pantos, E. (1986b) *Nuclear Instruments and Methods* A**246**, 617–23.

Helliwell, J. R., Habash, J., Cruickshank, D. W. J., Harding, M. M., Greenhough, T. J., Campbell, J. W., Clifton, I. J., Elder, M., Machin, P. A., Papiz, M. Z. and Zurek, S. (1989b) *Journal of Applied Crystallography* **22**, 483–97.

Helliwell, M., Gomez de Anderez, D., Habash, J., Helliwell, J. R. and Vernon, J. (1989) *Acta Crystallographica* B**45**, 591–6.

Henderson, R. (1990) *Proceedings of the Royal Society, London B* **241**, 6–8.

Hendrickson, W. A. (1980) Proceedings of the Daresbury Study Weekend on the 'Refinement of Protein Structures', DL/SCI/R16, pp. 1–8.

Hendrickson, W. A. (1985) *Transactions American Crystallographic Association* **21**, 11–21.

Hendrickson, W. A. and Teeter, M. M. (1981) *Nature* **290**, 107–113.

Hendrickson, W. A., Smith, J. L., Phizackerley, R. P. and Merritt, E. A. (1988) *Proteins: Structure Function Genetics* **4**, 77–88.

Hendrickson, W. A., Pähler, A., Smith, J. L., Satow, Y., Merritt, E. A. and Phizackerley, R. P. (1989) *Proceedings of the National Academy, USA* **86**, 2190–4.

Hendrix, J., Koch, M. H. J. and Bordas, J. (1979) *Journal of Applied Crystallography* **12**, 467–72.

Henrick, K., Collyer, C. A. and Blow, D. M. (1989) *Journal of Molecular Biology* **208**, 129–57.

Herzenberg, A. and Lau, H. S. M. (1967) *Acta Crystallographica* **22**, 24–8.

Higashi, T. (1989) *Journal of Applied Crystallography* **22**, 9–18.

Hilgenfeld, R., Liesum, A., Storm, R., Metzner, H. J. and Karges, H. E. (1990) *FEBS Letters* **265**, 110–12.

Hingerty, B. E., Brown, R. S. and Jack, A. (1978) *Journal of Molecular Biology* **124**, 523–34.

Hoeffken, H. W., Knof, S. H., Bartlett, P. A., Huber, R.,

Moellering, H. and Schumacher, G. (1988) *Journal of Molecular Biology* **20**, 417–33.

Hogle, J. M., Chow, M. and Filman, D. J. (1985) *Science* **229**, 358.

Hol, W. (1986) *Angewandte Chemie* **25**(9), 767–78.

Holland, B. W., Pendry, J. B., Pettifer, R. F. and Bordas, J. (1978) *Journal of Physics* C**11**, 633–42.

Holmes, K. C. (1974) *Proceedings of the Fourth International Congress on VUV Radiation Physics, Hamburg*, pp. 809–22.

Holmgren, A. and Branden, C.-I. (1989) *Nature* **342**, 248–51.

Hope, H. (1988) *Acta Crystallographica* B**44**, 22–6.

Hope, H., Frolow, F., von Böhlen, K., Makowski, I., Kratky, C., Halfon, Y., Danz, H., Webster, P., Bartels, K. S., Wittmann, H. G. and Yonath, A. (1989) *Acta Crystallographica* B**45**, 190–9.

Hoppe, W. and Jakubowski, U. (1975) in *Anomalous Scattering*, edited by S. Ramaseshan and S. C. Abrahams, Copenhagen: Munksgaard pp. 437–61.

Horowitz, P. and Howell, J. A. (1972) *Science* **178**, 608–11.

Hosur, M. V., Schmidt, T., Tucker, R. C., Johnson, J. E., Gallagher, T. M., Selling, B.-H. and Rueckert, R. R. (1987) *Proteins: Structure Function Genetics* **2**, 167–76.

Hough, E., Hansen, L. K., Birknes, B., Jynge, K., Hansen, S., Hordvik, A., Little, C., Dodson, E. and Derewenda, Z. (1989) *Nature* **338**, 357–60.

Hovmoller, S. (1981) *Journal of Applied Crystallography* **14**, 75.

Howard, A., Nielsen, C. and Xuong, Ng. H. (1985) *Methods in Enzymology* **114A**, 452–72.

Howard, A. J., Gilliland, G. L., Finzel, B. C., Poulos, T. L., Ohlendorf, D. H. and Salemme, F. R. (1987) *Journal of Applied Crystallography* **20**, 383–7.

Hoyer, E. (1983) *Nuclear Instruments and Methods* **208**, 117–25.

Hsieh, H., Krinsky, S., Luccio, A., Pellegrini, C. and van Steenbergen, A. (1983) *Nuclear Instruments and Methods* **208**, 79–90.

Hubbard, D. J. and Pantos, E. (1983) *Nuclear Instruments and Methods* **208**, 319.

Huber, R., Schneider, M., Epp, O., Mayr, I., Messerschmidt, A., Pflugrath, J. and Kayser, H. (1987) *Journal of Molecular Biology* **195**, 423–34.

Hull, S. E., Karlsson, R., Main, P., Woolfson, M. M. and Dodson, E. J. (1978) *Nature* **275**, 206–7.

Hurley, J. H., Thorsness, P. E., Ramalingam, V., Helmers, N. H., Koshland, D. E. and Stroud, R. M. (1989) *Proceedings of the National Academy, USA* **86**, 8635–9.

Huxley, H. E., Faruqi, A. R., Bordas, J., Koch, M. H. J. and Milch, J. R. (1980) *Nature* **284**, 140–3.

Huxley, H. E., Faruqi, A. R., Kress, M., Bordas, J. and Koch, M. H. J. (1982) *Journal of Molecular Biology* **158**, 637–84.

Hyde, C. C., Ahmed, S. A., Padlan, E. A., Miles, E. W. and Davies, D. R. (1988) *Journal of Biological Chemistry* **263**, 17857–71.

Jackson, J. D. (1975) *Classical Electrodynamics*, New York: Wiley.

James, R. W. (1954) *The Crystalline State*, Volume II, *The Optical Principles of the Diffraction of X-rays*, London: G. Bell and Sons.

Jeffery, J. W. (1958) *Zeitschrift für Kristallographie* **110**, 321–8.

Jhoti, H., Gorinsky, B., Garratt, R. C., Lindley, P. F. and Walton, A. R. (1988) *Journal of Molecular Biology* **200**, 423–5.

Johnson, J. E. and Hollingshead, C. (1981) *Journal of Ultrastructure Research* **74**, 223–31.

Johnson, L. N., Acharya, K. R., Stuart, D. I., Barford, D., Oikonomakos, N. G., Hajdu, J. and Varvill, K. M. (1987) *Biochemical Society Transactions* **15**, 1001–5.

Jones, E. Y., Stuart, D. I. and Walker, N. P. C. (1989) *Nature* **338**, 225–8.

Jones, T. A., Bergfors, T., Sedzik, J. and Unge, T. (1988) *EMBO Journal* **7**, 1597–604.

Jordan, S. R. and Pabo, C. O. (1988) *Science* **242**, 893–9.

Kahn, R., Fourme, R., Bosshard, R. and Saintagne, V. (1986) *Nuclear Instruments and Methods* A**246**, 596–603.

Kahn, R., Fourme, R., Bosshard, R., Lewit-Bentley, A. and Prangé, T. (1989) *Review of Scientific Instruments* **60**(7), 1568 (abstract).

Kahn, R., Fourme, R., Gadet, A., Janin, J. and Andre, D. (1982a) *Journal of Applied Crystallography* **15**, 330–7.

Kahn, R., Fourme, R., Bosshard, R., Caudron, B., Santiard, J. C. and Charpak, G. (1982b) *Nuclear Instruments and Methods* **201**, 203–9.

Kahn, R., Fourme, R., Bosshard, R., Chiadmi, M., Risler, J. L., Dideberg, O. and Wery, J. P. (1985) *FEBS Letters* **179**, 133–7.

Kahn, R., Fourme, R., Caudron, B., Bosshard, R., Benoit, R., Bouclier, P., Charpak, G., Santiard, J. C. and Sauli, F. (1980) *Nuclear Instruments and Methods* **172**, 337–44.

Kalata, K. (1982) *Nuclear Instruments and Methods* **201**, 35–42.

Kalb (Gilboa), A. J., Yariv, J., Helliwell, J. A. and Papiz, M. Z. (1988) *Journal of Crystal Growth* **88**, 537–40.

Kalman, Z. H. (1979) *Acta Crystallographica* A**35**, 634–41.

Kang, C., Kim, S.-H., Nikaido, K., Gokcen, S. and Ames, G. F.-L. (1989) *Journal of Molecular Biology* **207**, 643–4.

Kaplan, J. H., Forbush, B. and Hoffman, J. F. (1978) *Biochemistry* **17**, 1929.

Karle, J. (1967) *Applied Optics* **6**, 2132–5.

Karle, J. (1980) *International Journal of Quantum Chemistry* **7**, 356–67.

Karle, J. (1989) *Physics Today* **42**, 6, 22–9.

Kartha, G. (1975) in *Anomalous Scattering*, edited by S. Ramaseshan and S. C. Abrahams, Copenhagen: Munksgaard, pp. 363–92.

Katti, S. K., LeMasker, D. M. and Eklund, H. (1990) *Journal of Molecular Biology* **212**, 167–84.

Kawata, S. and Maeda, K. (1973) *Journal of Physics F Metal Physics* **3**, 167–78.

Ke, H., Thorpe, C. M., Seaton, B. A., Lipscomb, W. N. and Marcus, F. (1990) *Journal of Molecular Biology* **212**, 513–39.

Kendrew, J. C., Bodo, G., Dintzis, H., Parrish, R. G., Wyckoff, H. and Phillips, D. C. (1958) *Nature* **181**, 662–6.

Kennard, O. and Hunter, W. N. (1989) *Quarterly Reviews of Biophysics* **22**, 3, 327–79.

Kennard, O., Cruse, W. B. T., Nachman, J., Prange, T., Shakhed, Z. and Rabinovich, D. (1986) *Journal of Biomolecular Structure and Dynamics* **3**, 623–47.

Kim, J.-J. P. and Wu, J. (1988) *Proceedings of the National Academy, USA* **85**, 6677–81.

Kim, S., Smith, T. J., Chapman, M. S., Rossmann, M. G., Pevear, D. C., Dutko, F. J., Felock, P. J., Diana, G. D. and McKinlay, M. A. (1989) *Journal of Molecular Biology* **210**, 91–111.

Kirkland, J. P., Nagel, D. J. and Cowan, P. L. (1983) *Nuclear Instruments and Methods* **208**, 49–54.

Kitigawa, Y., Tanaka, N., Hata, Y., Katsube, Y. and Satow, Y. (1987) *Acta Crystallographica* B**43**, 272–5.

Knapek, E. and Dubochet, J. (1980) *Journal of Molecular Biology* **141**, 147–61.

Koetzle, T. F. and Hamilton, W. C. (1975) in *Anomalous Scattering*, edited by S. Ramaseshan and S. C. Abrahams, Copenhagen: Munksgaard pp. 489–502.

Kohra, K., Ando, M., Matsushita, T. and Hashizume, H. (1978) *Nuclear Instruments and Methods* **152**, 161–6.

Kopfmann, G. and Huber, R. (1968) *Acta Crystallographica* A**24**, 348–51.

Korszun, Z. R. (1987) *Journal of Molecular Biology* **196**, 413–19.

Krause, K. L., Volz, K. W. and Lipscomb, W. N. (1985) *Proceedings of the National Academy, USA* **82**, 1643–7.

Kretsinger, R. H., Rudnick, S. E. and Weissman, L. J. (1986) *Journal of Inorganic Biochemistry* **28**, 289–302.

Kroes, R. L. and Reiss, D. (1984) *Journal of Crystal Growth* **69**, 414–20.

Kronig, R. de L. and Kramers, H. A. (1928) *Zeitschrift für Physik* **48**, 174–9.

Kuipers, O. P., Thunnissen, M. M. G. M., de Geus, P., Dijkstra, V. W., Drenth, J., Verheij, H. M. and de Haas, G. H. (1989) *Science* **244**, 82–5.

Kwong, P. D., Hendrickson, W. A. and Sigler, P. B. (1989) *Journal of Biological Chemistry* **264**, 19349–53.

Laclare, J.-L. (1989) *Review of Scientific Instruments* **60**(7), 1399–402.

Ladenstein, R., Schneider, M., Huber, R., Bartunik, H. D., Wilson, K. S., Schott, K. and Bacher, A. (1988) *Journal of Molecular Biology* **203**, 1045–70.

Ladenstein, R., Meyer, B., Huber, R., Labischinski, H., Bartels, K., Bartunik, H., Bachmann, L., Ludwig, H. and Bacher, A. (1986) *Journal of Molecular Biology* **187**, 87–100.

Lairson, B. M. and Bilderback, D. H. (1982) *Nuclear Instruments and Methods* **195**, 79–83.

Lamvik, M. K., Kopf, D. A. and Robertson, J. D. (1983) *Nature* **301**, 332–4.

Lapatto, R., Blundell, T. L., Hemmings, A., Overington, J., Wilderspin, A., Wood, S., Merson, J. R., Whittle, P. J., Danley, D. E., Geoghegan, K. F., Hawrylik, S. J., Lee, S. E., Scheld, K. G. and Hobart, P. M. (1989) *Nature* **342**, 299–302.

Larmor, J. (1897) *Philosophical Magazine* **44**, 503–12.

Lavender, W., Brown, G., Troxel, T. and Coisson, R. (1989) *Review of Scientific Instruments* **60**(7), 1414–18.

Lebioda, L., Stec, B. and Brewer, J. M. (1989) *Journal of Biological Chemistry* **264**, 3685–93.

Lee, P. A. and Pendry, J. B. (1979) *Physical Review* B**11**, 2795–811.

Lee, X., Johnston, R. A. Z., Rose, D. R. and Young, N. M. (1989) *Journal of Molecular Biology* **210**, 685–6.

Lemonnier, M., Fourme, R., Rousseaux, F. and Kahn, R. (1978), *Nuclear Instruments and Methods* **152**, 173–7.

Leslie, A. G. W. (1990) *Journal of Molecular Biology* **213**, 167–86.

Leslie, A. G. W., Moody, P. C. E. and Shaw, W. V. (1988) *Proceedings of the National Academy, USA* **85**, 4133–7.

Levitt, M. (1974) *Journal of Molecular Biology* **82**, 393–420.

Lewit-Bentley, A., Doublié, S., Fourme, R. and Bodo, G. (1989) *Journal of Molecular Biology* **210**, 875–6.

Lewit-Bentley, A., Fourme, R., Kahn, R., Prangé, T., Vachette, P., Tavermier, J., Hauquier, G. and Fiers, W. (1988) *Journal of Molecular Biology* **199**, 389–92.

Liao, D.-I. and Remington, S. J. (1990) *Journal of Biological Chemistry* **265**, 6528–31.

Liddington, R. C. (1985) D.Phil. Thesis, University of York.

Liddington, R. C., Derewenda, Z., Dodson, G. and Harris, D. (1988) *Nature* **331**, 725–8.

Liljas, A. (1982) *Progress in Biophysics and Molecular Biology* **40**, 161–228.

Lim, L. W., Stegeman, R. A., Leimgruber, N. K., Gierse, J. K. and Abdel-Meigud, S. S. (1989) *Journal of Molecular Biology* **210**, 239–40.

Lim, L. W., Shamala, N., Mathews, F. S., Steenkamp, D. J., Hamlin, R. and Xuong, N.-H. (1986) *Journal of Biological Chemistry* **261**, 15140–6.

Lindahl, M., Vidgren, J., Eriksson, E., Habash, J., Harrop, S., Helliwell, J. R., Liljas, A., Lindeshoz, M. and Walker, N. (1990) *Proceedings of the Symposium on Carbonic Anhydrase* (in press).

Lindley, P. F., Garratt, R. C. and Hasnain, S. S. (1990) in *Synchrotron Radiation and Biophysics*, edited by S. S. Hasnain, Chichester: Ellis Horwood, pp. 176–200.

Lindqvist, Y. and Brändén, C.-I. (1984) in *Flavins and Flavoproteins*, edited by R. C. Bray, P. C. Engel and S. G. Mayhew, Walter de Gruyter and Co., p. 277.

Lindqvist, Y. and Brändén, C. I. (1985) *Proceedings of the National Academy, USA* **82**, 6855–9.

Lindqvist, Y. and Brändén, C. I. (1989) *Journal of Biological Chemistry* **264**, 3624–8.

Lloyd, L. F., Skarzynski, T., Wonacott, A. J., Cawston, T. E., Clark, I. M., Mannix, C. J. and Harper, G. P. (1989) *Journal of Molecular Biology* **210**, 237–8.

Lombardo, D., Chapus, C., Bourne, Y. and Cambillau, C. (1989) *Journal of Molecular Biology* **205**, 259–61.

Love, W. E., Hendrickson, W. A., Herriott, J. R., Lattman, E. E. and McCorkle, G. L. (1965) *Review of Scientific Instruments* **36**, 1655–6.

Luisi, B. F. (1986) Ph.D. Thesis, University of Cambridge, England.

Lundqvist, T. and Schneider, G. (1989) *Journal of Biological Chemistry* **264**, 7078–83.

Luo, M., Vriend, G., Kamer, G. and Rossman, M. G. (1989) *Acta Crystallographica* **B45**, 85–92.

Luo, M., Vriend, G., Kamer, G., Minor, I., Arnold, E., Rossmann, M. G., Boege, U., Scraba, D. G., Duke, G. M. and Palmenberg, A. C. (1987) *Science* **235**, 182–91.

Lye, R. C., Phillips, J. C., Kaplan, D., Doniach, S. and Hodgson, K. O. (1980) *Proceedings of the National Academy, USA* **77**, 5884–8.

MacDonald, A. C. and Sikka, S. K. (1969) *Acta Crystallographica* B**25**, 1804–11.

Machin, P. A. (1987) in 'Computational aspects of protein crystal data analysis', Proceedings of the Daresbury Study Weekend, DL/SCI/R25, edited by J. R. Helliwell, P. A. Machin and M. Z. Papiz, pp. 75–83.

Machin, P. A. and Harding, M. M. (1985) *Information Quarterly for Protein Crystallography*, No. 15, Daresbury Laboratory.

Machin, P. A., Campbell, J. W. and Elder, M. (eds.) (1980) Proceedings of the Daresbury Study Weekend entitled 'Refinement of Protein Structures', DL/SCI/R16, 107 pages.

Magorrian, B. G. and Allinson, N. M. (1988) *Nuclear Instruments and Methods* A**273**, 599–604.

Mahan, G. D. (1967) *Physical Review Letters* **18**, 448–50.

Mahendrasingam, A., Forsyth, V. T., Hussain, R., Greenall, R. J., Pigram, W. J. and Fuller, W. (1986) *Science* **233**, 195.

Makinen, M. W. and Fink, A. L. (1977) *Annual Review of Biophysics and Bioengineering* **6**, 301–42.

Makowski, I., Frolow, F., Saper, M. A., Shoham, M., Wittmann, H. G. and Yonath, A. (1987) *Journal of Molecular Biology* **193**, 819–22.

Mandelkow, E. M., Harmsen, A., Mandelkow, E. and Bordas, J. (1980) *Nature* **287**, 595–9.

Marks, N., Greaves, G. N., Poole, M. W., Suller, V. P. and Walker, R. P. (1983) *Nuclear Instruments and Methods* **208**, 97–103.

Mathews, F. S., Argos, P. and Levine, M. (1972) *Cold Spring Harbor Symposium on Quantitative Biology* **36**, 387–97.

Matsuda, S., Senda, T., Itoh, S., Kawano, G., Mizuro, H. and Mitsui, Y. (1989) *Journal of Biological Chemistry* **264**, 13381–2.

Matsushita, T. (1980) *AIP Conference Proceedings 1980-64*, 109–10.

Matsushita, T. and Kaminaga, U. (1980a) *Journal of Applied Crystallography* **13**, 465–71.

Matsushita, T. and Kaminaga, U. (1980b) *Journal of Applied Crystallography* **13**, 472–8.

Matsushita, T. and Phizackerley, R. P. (1981) *Journal of Applied Physics* **20**, 2223–8.

Matthews, B. W. (1966) *Acta Crystallographica* **20**, 82–6.

Matthews, D. A., Alden, J. T., Bolin, J. T., Freer, S. T., Hamlin, R., Xuong, N., Kraut, J., Poe, M., Williams, M. and Hoogsteen, K. V. (1977) *Science* **197**, 452–5.

Matthews, D. A., Alden, R. A., Bolin, J. T., Filman, D. J., Freer, S. T., Hamlin, R., Hol, W. G. J., Kisliuk, R. L., Pastore, E. JU., Plante, L. T., Xuong, N.-H. and Kraut, J. (1978) *Journal of Biological Chemistry* **253**, 6946–54.

Mauguen, Y., Hartley, R. W., Dodson, E. J., Dodson, G. G., Bricogne, G., Chothia, C. and Jack, A. (1982) *Nature* **297**, 162–4.

McCammon, J. A. (1984) *Reports on Progress in Physics* **47**, 1–46.

McCray, J. A. and Trentham, D. R. (1989) *Annual Review of Biophysics and Biophysical Chemistry* **18**, 239–70.

McCray, J. A., Herbette, L., Kohara, T. and Trentham, D. R. (1980) *Proceedings of the National Academy, USA* **77**, 7237.

McGrath, M. W., Wilke, M. E., Higaki, J. N., Craik, C. S. and Fletterick, R. J. (1989) *Biochemistry* **28**, 9264–70.

McKie, D. and McKie, C. (1986) *Essentials of Crystallography*, Oxford: Blackwell Scientific Publications, Oxford.

McLaughlin, P. J., Stuart, D. I., Klein, H. W., Oikonomakos, N. G. and Johnson, L. N. (1984) *Biochemistry* **23**, 5862–73.

McPherson, A. Jr (1982) *The Preparation and Analysis of Protein Crystals*, New York: Wiley.

McRee, D. E., Tainer, J. A., Meyer, T. E., Van Beeuman, J., Casanovich, M. A. and Getzoff, E. D. (1989) *Proceedings of the National Academy, USA* **86**, 6533–7.

Messerschmidt, A. and Pflugrath, J. W. (1987) *Journal of Applied Crystallography* **20**, 306–15.

Metcalf, P., Blum, M., Freymann, D., Turner, M. and Wiley, D. C. (1987) *Nature* **325**, 84–6.

Michel, H. (1982) *Journal of Molecular Biology* **158**, 567–72.

Milch, J. R., Gruner, S. M. and Reynolds, G. T. (1982) *Nuclear Instruments and Methods* **201**, 43–52.

Miller, A. (1990) *ESRF Newsletter* No. 6, 2–4.

Miller, M., Jaskólski, M., Mohana Rao, J. K., Leis, J. and Wlodawer, A. (1989) *Nature* **337**, 576–9.

Milligan, R. A. and Unwin, P. N. T. (1986) *Nature* **319**, 693–5.

Mitchell, C. M. (1957) *Acta Crystallographica* **10**, 475–6.

Miyahara, J., Takahashi, K., Amemiya, Y., Kamiya, N. and Satow, Y. (1986) *Nuclear Instruments and Methods* A**246**, 572–8.

Moews, P. C., Knox, J. R., Dideberg, O., Charlier, P. and Frère,

J.-M. (1990) *Proteins: Structure, Function, and Genetics* **7**, 156–71.

Moffat, K. (1987) *Proceedings of a Workshop held at Stanford (SSRL) October 1987*, Stanford: SSRL.

Moffat, K. (1989a) *Annual Review of Biophysics and Biophysical Chemistry* **18**, 309–32.

Moffat, K. (1989b) *Nature* **336**, 422–3.

Moffat, K. and Helliwell, J. R. (1989) in *Uses of Synchrotron Radiation in Chemistry and Biology. Topics in Current Chemistry* **151**, edited by E. Mandelkow, Heidelberg: Springer Verlag, pp. 62–74.

Moffat, K., Szebenyi, D. and Bilderback, D. W. (1984) *Science* **223**, 1423–5.

Moffat, K., Bilderback, D., Schildkamp, W. and Volz, K. (1986a) *Nuclear Instruments and Methods* A**246**, 627–35.

Moffat, K., Bilderback, D., Schildkamp, W., Szebenyi, D. and Loane, R. (1986b) in *New Methods in X-ray Absorption, Scattering and Diffraction*, edited by B. Chance and H. Bartunik, New York: Academic Press, pp. 125–33.

Mokulskaya, T. (1981) Unpublished.

Mokulskaya, T. D., Kuzev, S. V., Myshko, G. E., Khrenov, A. A., Mokulskii, M. A., Dobrokhotova, Z. D., Volodenkov, A. Ya., Rubanov, V. P., Ryanzina, N. A., Shitikov, B. I. Baru, S. E., Khabakhpashev, A. G. and Sidorov, V. A. (1981) *Journal of Applied Crystallography* **14**, 33–7.

Moncton, D. E., Crosbie, E. and Shenoy, G. K. (1989) *Review of Scientific Instruments* **60**(7), 1403–5.

Mondragón, A., Subbiah, S., Almo, S. C., Drottar, M. and Harrison, S. C. (1989) *Journal of Molecular Biology* **205**, 189–200.

Monteilhet, C., Fourme, R. and Blow, D. M., (1978), in *Trends in Physics, Proceedings of the 4th EPS Meeting, York*, Bristol: Adam Hilger Ltd (reported by R. Fourme), p. 510.

Montfort, W., Villafranca, J. E., Monzingo, A. F., Ernst, S. R., Katzin, B., Rutenber, E., Xuong, N.-H., Hamlin, R. and Robertus, J. D. (1987) *Journal of Biological Chemistry* **262**, 5398–403.

Moody, P. C. E. and Demple, B. (1988) *Journal of Molecular Biology* **200**, 751–2.

Moon, P. B. (1961) *Nature* **185**, 427–9.

Morimoto, H. and Uyeda, R. (1963) *Acta Crystallographica* **16**, 1107–19.

Mossbauer, R. L. (1975) in *Anomalous Scattering*, edited by S.

Ramaseshan and S. C. Abrahams, Copenhagen: Munksgaard, pp. 463–83.

Mukherjee, A. K., Helliwell, J. R. and Main, P. (1989) *Acta Crystallographica* A**45**, 715–18.

Murthy, H. M. K., Hendrickson, W. A., Orme-Johnson, W. H., Merritt, E. A. and Phizackerley, R. P. (1988) *Journal of Biological Chemistry* **263**, 18430–6.

Müssig, J., Makowski, I., von Böhlen, K., Hansen, H., Bartels, K. S., Wittmann, H. G. and Yonath, A. (1989) *Journal of Molecular Biology* **205**, 619–21.

Myles, D. A. A., Bailey, S., Rule, S. A., Jones, G. R. and Greenhough, T. J. (1990) *Journal of Molecular Biology* **231**, 223–5.

Naday, I., Strauss, M. G., Sherman, I. S., Kraimer, M. R. and Westbrook, E. M. (1987) *Optical Engineering* **26**, 788.

Nagel, D. J., Gilfrich, J. V. and Barbee, T. W. (1982) *Nuclear Instruments and Methods* **195**, 63–5.

Narayan, R. and Ramaseshan, S. (1981) *Acta Crystallographica* A**37**, 636–41.

Nave, C., Helliwell, J. R., Moore, P. R., Thompson, A. W., Worgan, J. S., Greenall, R. J., Miller, A., Burley, S. K., Bradshaw, J., Pigram, W. J., Fuller, W., Siddons, D. P., Deutsch, M. and Tregear, R. T. (1985) *Journal of Applied Crystallography* **18**, 396–403.

Navia, M. A., Fitzgerald, P. M. D., McKeever, B. M., Leu, C.-T., Heimbach, J. C., Herber, W. K., Sigal, I. S., Darke, P. L. and Springer, J. P. (1989) *Nature* **337**, 615–20.

Nivière, V., Hatchikian, C., Cambillau, C. and Frey, M. (1987) *Journal of Molecular Biology* **195**, 969–71.

North, A. C. T. (1965) *Acta Crystallographica* **18**, 212–16.

North, A. C. T., Phillips, D. C. and Matthews, F. S. (1968) *Acta Crystallographica* A**24**, 351–9.

Nozieres, P. and De Dominicis, C. T., (1969) *Physical Review* **178**, 1097–107.

Odom, O. W. Jr, Robbins, D. R., Lynch, J., Dottavio-Martin, D., Kramer, G. and Hardesty, B. (1980) *Biochemistry* **19**, 5947–54.

Oefner, C., D'Arcy, A., Daly, J. J., Gubernator, K., Charnas, R. L., Heinze, I., Hubschwerlen, C. and Winkler, F. K. (1990) *Nature* **343**, 284–8.

Ohshima, K.-I. and Tanaka, M. (1981) *Journal of Applied Crystallography* **14**, 75.

Oikonomakos, N. G., Johnson, L. N., Acharya, K. R., Stuart, D. I., Barford, D., Hajdu, J., Varvill, K. M., Melpidou, A. E.,

Papageorgiou, T., Graves, D. J. and Palm, D. (1987) *Biochemistry* **26**, 8381–9.

Okaya, Y. and Pepinsky, R. (1956) *Physical Review* **103**, 1645.

Ollis, D. L., Brick, P., Hamlin, R., Xuong, N. G. and Steitz, T. A. (1985) *Nature* **313**, 762–6.

Onesti, S., Lloyd, L. F., Brick, P. and Blow, D. M. (1989) *Journal of Molecular Biology* **210**, 241–2.

Otwinowski, Z., Schevitz, R. W., Zhang, R.-G., Lawson, C. L., Jonchimiak, A., Marmorstein, R. A., Luisi, B. F. and Sigler, P. B. (1988) *Nature* **335**, 321–9.

Oxender, D. L. and Fox, C. F. (eds.) (1987) *Protein Engineering*, New York: Alan Liss Inc.

Pace, H. C., Lu, P. and Lewis, M. (1990) *Proceedings of the National Academy, USA* **87**, 1870–3.

Padlan, E. A., Silverton, E. W., Sheriff, S., Cohen, G. H., Smith-Gill, S. J. and Davies, D. R. (1989) *Proceedings of the National Academy, USA* **86**, 5938–42.

Padmaja, N., Ramakumar, S. and Viswamitra, M. A. (1990) *Acta Crystallographica* A**46**, 725–30.

Pai, E. F., Kabsch, W., Krengel, U., Holmes, K. C., John, J. and Wittinghofer, A. (1989) *Nature* **341**, 209–14.

Papiz, M. Z. and Helliwell, J. R. (1984) Daresbury Laboratory Technical Memorandum, DL/SCI/TM42E.

Papiz, M. Z., Hawthornthwaite, A. M., Cogdell, R. J., Woolley, K. J., Wightman, P. A., Ferguson, L. A. and Lindsay, J. G. (1989) *Journal of Molecular Biology* **209**, 833–5.

Parak, F., Mossbauer, R. L., Hoppe, W., Thomanek, U. F. and Bade, D. (1976) *Journal de Physique* **37** C6, 703–6.

Parge, H. E., Bernstein, S. L., Deal, C. D., McRee, D. E., Christensen, D., Capozza, M. A., Kays, B. W., Fieser, T. R., Draper, D., So, M., Getzoff, E. D. and Tainer, J. A. (1990) *Journal of Biological Chemistry* **265**, 2278–85.

Parker, M. W., Lo Bello, M. and Federici, G. (1990) *Journal of Molecular Biology* **213**, 221–2.

Parker, M. W., Pattus, F., Tucker, A. D. and Tsernoglou, D. (1989) *Nature* **337**, 93–6.

Parkhurst, L. J. and Gibson, Q. H. (1967) *Journal of Biological Chemistry* **242**, 5762–70.

Parratt, L. G. (1954) *Physical Review* **95**, 359–69.

Parratt, L. G. and Hampstead, C. F. (1954) *Physical Review* **94**, 1593–600.

Pathak, D., Ngai, K. L. and Ollis, D. (1988) *Journal of Molecular Biology* **204**, 435–45.

Perutz, M. F. (1970) *Nature* **228**, 726–34.

Perutz, M. F., Hasnain, S. S., Duke, P. J., Sessler, J. L. and Hahn, J. E. (1982) *Nature* **295**, 535–8.

Perutz, M. F., Rossmann, M. G., Cullis, A. F., Muirhead, H., Will, G. and North, A. C. T. (1960) *Nature, London* **185**, 416–22.

Petratos, K., Dauter, Z. and Wilson, K. S. (1988) *Acta Crystallographica* B**44**, 628–36.

Phillips, D. C. (1966) *Scientific American* **215**(5), 78–90.

Phillips, D. C., Cordero-Borboa, A., Sutton, B. J. and Todd, R. J. (1987) *Pure and Applied Chemistry* **59**, 279–86.

Phillips, J. C. (1978) Ph.D. Thesis, Stanford University.

Phillips, J. C. and Hodgson, K. O. (1980) in *Synchrotron Radiation Research*, edited by H. Winick and S. Doniach, New York: Plenum Press, pp. 565–605.

Phillips, J. C., Cerino, J. A. and Hodgson, K. O. (1979) *Journal of Applied Crystallography* **12**, 592–600.

Phillips, J. C., Templeton, L. K., Templeton, D. H. and Hodgson, K. O. (1978) *Science* **201**, 257–9.

Phillips, J. C., Wlodawer, A., Yevitz, M. M. and Hodgson, K. O. (1976) *Proceedings of the National Academy, USA* **73**, 128–32.

Phillips, J. C., Wlodawer, A., Goodfellow, J., Watenpaugh, K. D., Sieker, L. C., Jensen, L. H. and Hodgson, K. O. (1977) *Acta Crystallographica* A**33**, 445–55.

Phizackerley, R. P., Cork, C. W. and Merritt, E. A. (1986) *Nuclear Instruments and Methods* A**246**, 579–95.

Phizackerley, R. P., Cork, C. W., Hamlin, R. C., Nielsen, C. P., Vernon, W., Xuong, N. H. and Perez-Mendez, V. (1980) *Nuclear Instruments and Methods* **172**, 393–5.

Phizackerley, R. P., Rek, Z. U., Stephenson, G. B., Conradson, S. D., Hodgson, K. O., Matsushita, T. and Oyanagi, H. (1983) *Journal of Applied Crystallography* **16**, 220–32.

Pickersgill, R. W. (1987) *Acta Crystallographica* A**43**, 502–6.

Poole, M. W., Munro, I. H., Taylor, D. G., Walker, R. P. and Marr, G. V. (1983) *Nuclear Instruments and Methods* **208**, 143–8.

Popov, A. N., Antson, A. A., Bondarenko, K. P., Belyaev, E. G., Harutunyan, E. G., Kheiker, D. M., Sheromov, M. A. and Mytnytchenko, S. V. (1989) *Review of Scientific Instruments* **60**(7), 2440 (abstract).

Poulos, T. L., Finzel, B. C. and Howard, A. J. (1986) *Biochemistry* **25**, 5314–22.

Poulos, T. L., Finzel, B. C. and Howard, A. J. (1987) *Journal of Molecular Biology* **195**, 687–700.

Poulos, T. L., Freer, S. T., Alden, R. A., Edwards, S. L., Skogland,

U., Koji, T., Eriksson, B., Xuong, N.-H., Yonetani, T. and Kraut, J. (1980) *Journal of Biological Chemistry* **255**, 575–80.

Priestle, J. P. (1988) *Journal of Applied Crystallography* **21**, 572–6.

Priestle, J. P., Schär, H.-P. and Grütter, M. G. (1989) *Proceedings of the National Academy, USA* **86**, 9667–71.

Przybylska, M., Ahmed, F. R., Birnbaum, G. I. and Rose, D. R. (1988) *Journal of Molecular Biology* **199**, 393–4.

Pusey, M., Witherow, W. and Naumann, R. (1988) *Journal of Crystal Growth* **90**, 105–11.

Quinn, P. D., Corlett, J. N., Poole, M. W. and Thomson, S. L. (1990) Daresbury Preprint DL/SCI/P699E.

Raag, R., Appelt, K., Xuong, N.-H. and Banaszak, L. (1988) *Journal of Molecular Biology* **200**, 553–69.

Rabinovitch, D. and Lourie, B. (1987) *Acta Crystallographica* A**43**, 774–80.

Radhakrishnan, R., Presta, L. G., Meyer, E. F. and Wildonger, R. (1987) *Journal of Molecular Biology* **198**, 417–24.

Rafferty, J. B., Somers, W. S., Saint-Girons, I. and Phillips, S. E. V. (1989) *Nature* **341**, 705–10.

Rafferty, J. B., Phillips, S. E. V., Rojas, C., Boulot, G., Saint-Girons, I., Guillou, Y. and Cohen, G. N. (1988) *Journal of Molecular Biology* **200**, 217–19.

Raghavan, R. S. (1961) *Proceedings of the Indian Academy of Science* A**53**, 265–71.

Ramachandran, G. N. and Raman, S. (1956) *Current Science (India)* **25**, 348–51.

Ramaseshan, S. (1964) in *Advanced Methods in X-ray Crystallography*, edited by G. N. Ramachandran, New York: Academic Press.

Ramaseshan, S. and Abrahams, S. C. (eds.) (1975) *Anomalous Scattering*, Copenhagen: Munskgaard.

Ramaseshan, S. and Narayan, R. (1981) Unpublished report.

Rawas, A., Moreton, K., Muirhead, H. and Williams, J. (1989) *Journal of Molecular Biology* **208**, 213–14.

Read, R. J., Wierenga, R. K., Groendijk, H., Hol, W. G. J., Lambeir, A. and Opperdoes, F. R. (1987) *Journal of Molecular Biology* **194**, 573–7.

Reeke, G. N. Jr, Becker, J. W. and Edelman, G. M. (1975) *Journal of Biological Chemistry* **250**, 1525–47.

Renetseder, R., Dijkstra, B. W., Kalk, K. H., Verpoote, J. and Drenth, J. (1986) *Acta Crystallographica* B**42**, 602–5.

Reynolds, C. D., Stowell, B., Joshi, K. K., Harding, M. M. and Maginn, S. J. (1988) *Acta Crystallographica* B**44**, 512–15.

Rice, D. W., Hornby, D. P. and Engel, P. C. (1985) *Journal of Molecular Biology* **181**, 147–9.

Richardson, J. S. (1981) *Advances in Protein Chemistry* **34**, 168–339.

Richardson, J. S. (1985) *Methods in Enzymology* **115**, 359–80.

Richmond, T. J., Searles, M. A. and Simpson, R. T. (1988) *Journal of Molecular Biology* **199**, 161–70.

Richmond, T. J., Finch, J. T., Rushton, B., Rhodes, D. and Klug, A. (1984) *Nature* **311**, 532–7.

Rieck, W., Euler, H. and Schulz, H. (1988) *Acta Crystallographica* A**44**, 1099–101.

Rizkallah, P. J., Maginn, S. J. and Harding, M. M. (1990a) *Acta Crystallographica* B**46**, 193–5.

Rizkallah, P. J., Harding, M. M., Lindley, P. F., Aigner, A. and Bauer, A. (1990b) *Acta Crystallographica* B**46**, 262–6.

Robbins, A. H. and Stout, C. D. (1989) *Proceedings of the National Academy, USA* **86**, 3639–43.

Rosenbaum, G., (1980) Ph.D. Thesis, Rupprecht-Karl-University, Heidelberg.

Rosenbaum, G. and Holmes, K. C. (1980) in *Synchrotron Radiation Research*, edited by H. Winick and S. Doniach, New York: Plenum Press, pp. 533–64.

Rosenbaum, G., Holmes, K. C. and Witz, J. (1971) *Nature* **230**, 129–31.

Rossmann, M. G., (1961) *Acta Crystallographica* **14**, 383–8.

Rossmann, M. G. (1972) *Molecular Replacement Method*, New York: Gordon and Breach.

Rossmann, M. G., (1985) *Methods of Enzymology* **114A**, 237–80.

Rossmann, M. G. and Erickson, J. W. (1983) *Journal of Applied Crystallography* **16**, 629–36.

Rossmann, M. G., Leslie, A. G. W., Abdel-Meguid, S. S. and Tsukihara, T. (1979) *Journal of Applied Crystallography* **12**, 570–81.

Rossmann, M. G., Arnold, E., Erickson, J. W., Frankenberger, E. A., Griffith, J. P., Hecht, H.-J., Johnson, J. E., Kamer, G., Luo, M., Mosser, A. G., Rueckert, R. R., Sherry, B. and Vriend, G. (1985) *Nature* **317**, 145–53.

Rould, M. A., Perona, J. T., Söll, D. and Steitz, T. A. (1989) *Science* **246**, 1135–42.

Royer, W. E., Hendrickson, W. A. and Love W. E. (1987) *Journal of Molecular Biology* **197**, 149–53.

Rozeboom, H. J., Kingma, J., Janssen, D. B. and Dijkstra, B. W. (1988) *Journal of Molecular Biology* **200**, 611–12.

Ruff, M., Cavarelli, J., Mikol, V., Lorber, B., Mitschler, A., Giegé, R., Thierry, J. C. and Moras, D. (1988) *Journal of Molecular Biology* **201**, 235–6.

Rypniewski, W. R. and Evans, P. R. (1989) *Journal of Molecular Biology* **207**, 805–21.

Sakabe, N. (1983) *Journal of Applied Crystallography* **16**, 542–7.

Sakabe, N., Kamiya, N., Sakabe, K. and Kondo, H. (1984) *Journal of Biochemistry* **95**, 887–90.

Sakabe, N., Nakagawa, A., Sasaki, K., Sakabe, K., Watanabe, N., Kondo, H. and Shimomura, M. (1989) *Review of Scientific Instruments* **60**(7), 2440–1 (abstract).

Salem, S. I., Chang, C. N., Lee, P. L. and Severson, V. (1978) *Journal of Physics C: Solid State Physics* **11**, 4085–93.

Samama, J. P., Delarue, M., Mourey, L., Choay, J. and Moras, D. (1989) *Journal of Molecular Biology* **210**, 877–9.

Samraoui, B., Sutton, B. J., Todd, R. J., Artymuik, P. J., Waley, S. G. and Phillips, D. C. (1986) *Nature* **320**, 378–80.

Sasaki, S. (1989) KEK Report 88-14, National Laboratory for High Energy Physics, Tsukuba, Japan.

Satow, Y. and Iitaka, Y. (1989) *Review of Scientific Instruments* **60**(7), 2390–3.

Satow, Y., Mikuni, A., Kamiya, N. and Ando, M. (1989) *Review of Scientific Instruments* **60**(7), 2394–7.

Sauvage, M. (1980) in *Characterisation of Crystal Growth Defects by X-ray Methods*, edited by B. K. Tanner and D. K. Bowen, New York: Plenum Press.

Saxena, A. M. and Schoenborn, B. P. (1977) *Acta Crystallographica* A**33**, 805–13.

Schierbeek, A. J., Swarte, M. B. A., Dijkstra, B. W., Vriend, G., Read, R. J., Hol, W. G. J., Drenth, J. and Betzel, C. (1989) *Journal of Molecular Biology* **206**, 365–79.

Schildkamp, W., Bilderback, D. and Moffat, K. (1989) *Review of Scientific Instruments* **60**(7), 2439 (abstract).

Schirmer, T. and Evans, P. R. (1989) *Nature* **343**, 140–5.

Schirmer, T., Bode, W. and Huber, R. (1987) *Journal of Molecular Biology* **196**, 677–95.

Schlichting, I., Rapp, G., John, J., Wittinghofer, A., Pai, E. F. and Goody, R. S. (1989) *Proceedings of the National Academy, USA* **86**, 7687–90.

Schlichting, I., Almo, S. C., Rapp, G., Wilson, K. S., Petratos, K., Lentfer, A., Wittinghofer, A., Kabsch, W., Pai, E. F., Petsko, G. A. and Goody, R. S. (1990) *Nature* **345**, 309–15.

Schneider, G., Lindqvist, Y. and Lundqvist, T. (1990) *Journal of Molecular Biology* **211**, 989–1008.

Schoenborn, B. P. (1975) in *Anomalous Scattering*, edited by S. Ramaseshan and S. C. Abrahams, Copenhagen: Munksgaard, pp. 407–16.

Schoenborn, B. P. (1983), *Acta Crystallographica* A**39**, 315–21.

Schreuder, H. A., Prick, P. A. J., Wierenga, R. K., Vriend, G., Wilson, K. S., Hol, W. G. J. and Drenth, J. (1989) *Journal of Molecular Biology* **208**, 679–96.

Schwinger, J. (1949) *Physical Review* **75**, 1912–25.

Sevcik, J., Dodson, E. J. and Zelinka, J. (1987) in *Metabolism of Nucleic Acids including Gene Manipulation*, Bratislava: Slovak Academy of Science, pp. 33–46.

Shaanan, B. (1983) *Journal of Molecular Biology* **171**, 31–59.

Sheldrick, G. M. (1976) *SHELX: A Program for Crystal Structure Determination*, available from Sheldrick at the University of Gottingen.

Sheriff, S., Silverton, G. W., Padlan, E. A., Cohen, G. H., Smith-Gill, S. J., Finzel, B. C. and Davies, D. R. (1987) *Proceedings of the National Academy, USA* **84**, 8075–9.

Shoham, M., Wittmann, H. G. and Yonath, A. (1987) *Journal of Molecular Biology* **193**, 819–22.

Shrive, A. K., Clifton, I. J., Hajdu, J. and Greenhough, T. J. (1990) *Journal of Applied Crystallography* **23**, 169–74.

Siddons, D. P. (1979) Ph.D. Thesis, King's College, London.

Sikka, S. K. (1969a) *Acta Crystallographica* A**25**, 396–7.

Sikka, S. K. (1969b) *Acta Crystallographica* A**25**, 539–43.

Sikka, S. K. and Rajagopal, H. (1975) in *Anomalous Scattering*, edited by S. Ramaseshan and S. C. Abrahams, Copenhagen: Munksgaard, pp. 503–14.

Silverton, E. W., Navia, M. A. and Davies, D. R. (1977) *Proceedings of the National Academy, USA* **74**, 5142.

Silverton, E. W., Padlan, E. A., Davies, D. R., Smith-Gill, S. and Potter, M. (1984) *Journal of Molecular Biology* **180**, 761–5.

Skarzynski, T., Moody, P. C. E. and Wonacott, A. J. (1987) *Journal of Molecular Biology* **193**, 171–87.

Skuratovskii, I. Ya., Kapitonova, K. A. and Volkova, L. I. (1978) *Journal of Applied Crystallography* **11**, 238–42.

Skuratovskii, I. Ya., Volkova, L. I., Kapitonova, K. A. and Bartenev, V. N. (1979) *Journal of Molecular Biology* **134**, 369–74.

Smerdon, S. J., Oldfield, T. J., Dodson, E. J., Dodson, G. G.,

Hubbard, R. E. and Wilkinson, A. J. (1990) *Acta Crystallographica* B**46**, 370–7.

Smith, J. L., Hendrickson, W. A. and Addison, A. W. (1983) *Nature* **303**, 86–8.

Smith, J. M. A., Helliwell, J. R. and Papiz, M. Z. (1985) *Inorganica Chimica Acta* **106**, 193–6.

Smith, J. M. A., Ford, G. C., Harrison, P. M., Yariv, J. and Kalb (Gilboa), A. J. (1989a) *Journal of Molecular Biology* **205**, 465–7.

Smith, J. M. A., Rice, D. W., White, J. L., Ford, G. C. and Harrison, P. M. (1989b) *Journal of Applied Crystallography* **22**, 284–6.

Smither, R. K., Forster, G. A., Bilderback, D. H., Bedzyk, M., Finkelstein, K., Henderson, C., White, J., Berman, L. E., Stefan, P. and Oversluizen, T. (1989) *Review of Scientific Instruments* **60**(7), 1486–92.

Sobottka, S. E., Chandross, R. J., Cornick, G. G., Kretsinger, R. H. and Rains, R. G. (1990) *Journal of Applied Crystallography* **23**, 199–208.

Sokolov, A. A. and Ternov, I. M. (1968) *Synchrotron Radiation*, Oxford: Pergamon Press.

Spencer, J. and Winick, H. (1979) in *Synchrotron Radiation Research*, edited by H. Winick and S. Doniach, New York: Plenum Press.

Sprang, S., Standing, T., Fletterick, R. J., Stroud, R. M., Finer-Moore, J. S., Xuong, N.-H., Hamlin, R., Rutter, W. J. and Craik, C. S. (1987) *Science* **237**, 905–9.

Sprang, S. R., Archarya, K. R., Goldsmith, E. J., Stuart, D. I., Varvill, K., Fletterick, R. J., Madsen, N. B. and Johnson, L. N. (1988) *Nature* **336**, 215–21.

Stanfield, R. L., Fieser, T. M., Lerner, R. A. and Wilson, I. A. (1990) *Science* **248**, 712–19.

Staudenmann, J.-L., Hendrickson, W. A. and Abramowitz, R. (1989) *Review of Scientific Instruments* **60**(7), 1939–42.

Stauffacher, C. V., Usha, R., Harrington, M., Schmidt, T., Hosur, M. V. and Johnson, J. E. (1987) in *Crystallography in Molecular Biology*, edited by D. Moras, J. Drenth, B. Strandberg, D. Suck and K. Wilson, New York: Plenum Press, pp. 293–308.

Steigemann, W. (1980) Proceedings of the Daresbury Study Weekend on the 'Refinement of Protein Structures', DL/SCI/R16, pp. 40–6.

Steinberger, I. T., Bordas, J. and Kalman, Z. H. (1977) *Philosophical Magazine* **35**, No. 5, 1257–67.

Steitz, T. (1990) *Quarterly Reviews of Biophysics* **23**, 3, 205–80.

Stezowski, J. J., Englmaier, R., Galdiga, C., Hartl, T., Rommel, I.,

Dauter, Z., Görisch, H., Grossebüter, W., Wilson, K. S. and Musil, D. (1989) *Journal of Molecular Biology* **208**, 507–8.

Stout, C. D. (1989) *Journal of Molecular Biology* **205**, 545–55.

Stout, G. H. and Jensen, L. H. (1989) *X-ray Structure Determination: A Practical Guide*, New York: Macmillan, pp. 93–153.

Strauss, M. G., Naday, I., Sherman, J. S., Kraimer, M. R. and Westbrook, E. M. (1987) *Institute of Electrical and Electronic Engineers Transactions on Nuclear Science* **34**, 389–95.

Stubbs, M. T., Laber, B., Bode, W., Huber, R., Jerala, R., Lenarčič, B. and Turk, V. (1990) *EMBO Journal* **9**, 1939–47.

Stuhrmann, H. B. (1978) *Quarterly Reviews of Biophysics* **11**, 1, 71–98.

Stura, E. A., Feinstein, A. and Wilson, I. A. (1987) *Journal of Molecular Biology* **193**, 229–31.

Suller, V. (1989) Daresbury Accelerator Physics Note.

Svensson, S. and Nyholm, R. (1985) Uppsala University Institute of Physics Report UUIP-1139, July 1985.

Swain, A. L., Kretsinger, R. H. and Amma, E. L. (1989) *Journal of Biological Chemistry* **264**, 16620–8.

Szebenyi, D. M. E., Bilderback, D., LeGrand, A., Moffat, K., Schildkamp, W. and Teng, T.-Y. (1989) *Transactions of the American Crystallographic Association* **24**, 167–72.

Takahashi, L. H., Radhakrishan, R., Rosenfield, R. E. Jr and Meyer, E. F. Jr (1989) *Biochemistry* **28**, 7610–17.

Temple, B. and Moffat, K. (1987) in 'Computational Aspects of Protein Crystal Data Analysis', Proceedings of the Daresbury Study Weekend, DL/SCI/R25, edited by J. R. Helliwell, P. A. Machin and M. Z. Papiz, pp. 84–9.

Templeton, D. H. (1955) *Acta Crystallographica* **8**, 842.

Templeton, D. H. and Templeton, L. K. (1980) *Acta Crystallographica* A**36**, 237–41.

Templeton, D. H. and Templeton, L. K. (1982) *Acta Crystallographica* A**38**, 62–7.

Templeton, D. H. and Templeton, L. K. (1985a) *Acta Crystallographica* A**41**, 133–42.

Templeton, D. H. and Templeton, L. K. (1985b) *Acta Crystallographica* A**41**, 365–71.

Templeton, D. H. and Templeton, L. K. (1989) *Acta Crystallographica* A**45**, 39–42.

Templeton, D. H., Templeton, L. K., Phillips, J. C. and Hodgson, K. O. (1980b) *Acta Crystallographica* A**36**, 436–42.

Templeton, L. K., Templeton, D. H. and Phizackerley, R. P.

(1980a) *Journal of the American Chemical Society* **102**, 1185–6.

Templeton, L. K., Templeton, D. H., Phizackerley, R. P. and Hodgson, K. O. (1982a) *Acta Crystallographica* A**38**, 74–8.

Templeton, L. K., Templeton, D. H., Zalkin, A. and Ruben, H. W. (1982b) *Acta Crystallographica* B**38**, 2155–9.

Thomas, D. J. (1982a) *Nuclear Instruments and Methods* **201**, 27–30.

Thomas, D. J. (1982b) *Nuclear Instruments and Methods* **201**, 31–4.

Thompson, A. W. and Helliwell, J. R. (1986) Daresbury Laboratory Technical Memorandum, DL/SCI/TM51E.

Thong, N. and Schwarzenbach, D. (1979) *Acta Crystallographica* A**35**, 658–64.

Timsit, Y., Westhof, E., Fuchs, R. P. P. and Moras, D. (1989) *Nature* **341**, 459–62.

Tonegawa, S. (1985) *Scientific American* **253**, 104–13.

Tong, L., de Vos, A. M., Milburn, M. V., Jancarik, J., Noguchi, S., Nishimura, S., Miura, K., Ohtsuka, E. and Kim, S.-H. (1989) *Nature* **337**, 90–3.

Trakhanov, S. D., Yusupov, M. M., Agalarov, S. C., Garber, M. B., Ryazantsev, S. N., Tischenko, S. V. and Shirovok, V. A. (1987) *FEBS Letters* **220**, 319–22.

Trakhanov, S. D., Yusupov, M. M., Shirovok, V. A., Garber, M. B., Mitschler, A., Ruff, M., Thierry, J.-C. and Moras, D. (1989) *Journal of Molecular Biology* **209**, 327–8.

Turley, S., Adman, E. T., Sieker, L. C., Liu, M.-Y., Payne, W. J. and LeGall, J. (1988) *Journal of Molecular Biology* **200**, 417–19.

Tykarska, E., Lebioda, L., Bradshaw, T. P. and Dunlop, R. B. (1986) *Journal of Molecular Biology* **191**, 147–50.

Unangst, D., Muller, E., Muller, J. and Keinert, B. (1967) *Acta Crystallographica* **23**, 898–901.

Underwood, J. H. and Turner, D. (1977) *Society of Photo-optical and Instrumentation Engineers* **106**, 125.

Unwin, P. N. T. (1977) *Nature* **269**, 118–22.

Urpi, L., Ridoux, J. P., Verdaguer, N., Fita, I., Subirana, J. A., Iglesias, F., Huynh-Dinh, T., Igolen, J. and Taillandier, E. (1989) *Nucleic Acids Research* **17**, 6669–80.

Usha, R., Johnson, J. E., Moras, D., Thierry, J. C., Fourme, R. and Kahn, R. (1984) *Journal of Applied Crystallography* **17**, 147–53.

Vainshtein, B. K. (1981) *Modern Crystallography* I, Berlin: Springer Verlag, pp. 297–300.

Valegård, K., Liljas, L., Fridborg, K. and Unge, T. (1990) *Nature* **345**, 36–41.

Varghese, J. N., Laver, W. G. and Colman, P. M. (1983) *Nature* **303**, 35–40.

Veenendaal, A. L., MacGillavry, C. H., Stam, B., Potters, M. L. and Romgens, M. J. H. (1959) *Acta Crystallographica* **12**, 242–6.

Vellieux, F. M. D., Huitema, F., Groendijk, H., Kalk, K. H., Frank, J., Jongejan, J. A., Duire, J. A., Petratos, K., Drenth, J. and Hol, W. G. J. (1989) *EMBO Journal* **8**, 2171–8.

Verma, L. P. and Agarwal, B. K. (1968) *Journal of Physics C: Solid State Physics* **1**, 1658–61.

Vitali, J., Young, W. W., Schatz, V. B., Sobottka, S. E. and Kretsinger, R. H. (1987) *Journal of Molecular Biology* **198**, 351–5.

Walker, R. P. (1986) Daresbury Laboratory Preprint DL/SCI/P513A.

Wall, J. S., Hainfeld, J. F., Barlett, P. A. and Singer, S. J. (1982) *Ultramicroscopy* **8**, 397–402.

Waller, D. A. and Liddington, R. C. (1990) *Acta Crystallographica* B**46**, 409–18.

Walter, J., Steigemann, W., Singh, T. P., Bartunik, H. D., Bode, W. and Huber, R. (1982) *Acta Crystallographica* B**38**, 1462–72.

Watanabe, N., Sakabe, K., Sakabe, N., Higashi, T., Sasaki, K., Aibara, S., Morita, Y., Yonaha, K., Toyama, S. and Fukutani, H. (1989) *Journal of Biochemistry* **105**, 1–3.

Watson, J. D. and Crick, F. H. C. (1953) *Nature* **171**, 737–8.

Webb, N. G. (1976) *Review of Scientific Instruments* **47**, 545–7.

Webb, N. G., Samson, S., Stroud, R. M., Gamble, R. C. and Baldeschwieler, J. D. (1976) *Review of Scientific Instruments* **47**, 836–9.

Webb, N. G., Samson, S., Stroud, R. M., Gamble, R. C. and Baldeschwieler, J. D. (1977) *Journal of Applied Crystallography* **10**, 104–10.

Weber, P. C., Sheriff, S., Ohlendorf, D. H., Finzel, B. C. and Salemme, F. R. (1985) *Proceedings of the National Academy, USA* **82**, 8473–7.

Weissenberg, K. (1924) *Zeitschrift für Physik* **23**, 229–38.

Wendin, G., (1980) *Physica Scripta* **21**, 535–42.

Westbrook, E. M. (1988) Conceptual Design Report Document No. J9001-2001-SA-01, Argonne National Laboratory.

Westbrook, E. M., Piro, O. E. and Sigler, P. B. (1984) *Journal of Biological Chemistry* **259**, 9096–103.

White, H. E., Driessen, H. P. C., Slingsby, C., Moss, D. S., Turnell,

W. G. and Lindley, P. F. (1988) *Acta Crystallographica* B**44**, 172–8.

Wigley, D. B., Roper, D. I. and Cooper, R. A. (1989) *Journal of Molecular Biology* **210**, 881–2.

Wigley, D. B., Muirhead, H., Gamblin, S. J. and Holbrook, J. J. (1988) *Journal of Molecular Biology* **204**, 1041–3.

Willingmann, P., Krishnaswamy, S., McKenna, R., Smith, T. J., Olson, N. H., Rossmann, M. G., Stow, P. L. and Incardona, N. L. (1990) *Journal of Molecular Biology* **212**, 345–50.

Willis, B. T. M. and Pryor, A. W. (1975) *Thermal Vibrations in Crystallography*, Cambridge: Cambridge University Press.

Wilson, I. A., Skehel, J. J. and Wiley, D. C. (1981) *Nature* **289**, 366–73.

Wilson, K. S. (1989) 'Synchrotron Radiation Beam Lines for Protein Crystallography at the EMBL Outstation Hamburg', EMBL Preprint.

Wilson, K. S., Stura, E. A., Wild, D. L., Todd, R. J., Stuart, D. I., Babu, Y. S., Jenkins, J. A., Standing, T. S., Johnson, L. N., Fourme, R., Kahn, R., Gadet, A., Bartels, K. S. and Bartunik, H. D. (1983) *Journal of Applied Crystallography* **16**, 28–41.

Winick, H. (1980) in *Synchrotron Radiation Research*, edited by H. Winick and S. Doniach, New York: Plenum Press, pp. 11–60.

Winick, H. (1989) *Synchrotron Radiation News* **2**, 2, 25.

Winick, H. and Knight, T. (eds.) (1977) 'Wiggler Magnets' Workshop held at SLAC, SSRP Report No. 77/05.

Winick, H. and Spencer, J. (1980) *Nuclear Instruments and Methods* **172**, 45–53.

Winick, H., Brown, G., Halbach, K. and Harris J. (1981) *Physics Today*, May issue, 50–63.

Winick, H., Boyce, R., Brown, G., Hower, N., Hussain, Z., Pate, T. and Umbach, E. (1983) *Nuclear Instruments and Methods* **208**, 127–37.

Winkler, F. K., D'Arcy, A. and Hunziker, W. (1990) *Nature* **343**, 771–4.

Wise, D. S. and Schoenborn, B. P. (1982) 'Design notes of the Biology/NSLS small angle X-ray scattering station'.

Wittmann, H. G. (1982) *Annual Review of Biochemistry* **51**, 155–83.

Wittmann, H. G. (1983) *Annual Review of Biochemistry* **52**, 35–65.

Wittmann, H. G., Müssig, J., Piefke, J., Gewitz, H. S., Rheinberger, H. J. and Yonath, A. (1982) *FEBS Letters* **146**, 217–20.

Witz, J. (1969) *Acta Crystallographica* A**25**, 30–41.

Wlodawer, A. (1980) *Acta Crystallographica* B**36**, 1826–31.

Wlodawer, A. (1985) *Methods in Enzymology* **114A**, 551–64.

Wlodawer, A., Miller, M., Jaskólski, M., Sathyanarayana, B. K., Baldwin, E., Weber, I. T., Selk, L. M., Clawson, L., Schneider, J. and Kent, D. B. H. (1989) *Science* **245**, 616–21.

Wood, I. G., Thompson, P. and Mathewman, J. C. (1983) *Acta Crystallographica* B**39**, 543–7.

Wood, S. P., Pitts, J. E., Blundell, T. L. and Jenkins, J. A. (1977) *European Journal of Biochemistry* **78**, 119–26.

Wood, S. P., Oliva, G., O'Hara, B. P., White, H. E., Blundell, T. L., Perkins, S. J., Sardharwalla, I. and Pepys, M. B. (1988) *Journal of Molecular Biology* **202**, 169–73.

Woolfson, M. M. (1970) *X-Ray Crystallagraphy*, Cambridge: Cambridge University Press.

Wooster, W. A. (1962) *Diffuse X-ray Reflections from Crystals*, Oxford: Clarendon Press.

Wooster, W. A. (1964) *Acta Crystallographica* **17**, 878–82.

Wright, H. T., Qian, H. X. and Huber, R. (1990) *Journal of Molecular Biology* **213**, 513–28.

Wuthrich, K. (1986) *NMR of Proteins and Nucleic Acids*, New York: Wiley.

Wyckoff, H. W. (1973) in *Research Applications of Synchrotron Radiation*, edited by R. W. Watson and M. L. Perlman, Brookhaven National Laboratory Report, BNL 50381, pp. 133–8.

Wyckoff, H. W., Hirs, C. H. W. and Timasheff, S. N. (1985) *Methods in Enzymology* **114A**, 199–588.

Wyckoff, H. W., Doscher, M., Tsernoglou, D., Inagami, T., Johnson, L. N., Hardman, K. D., Allewell, N. M., Kelly, D. M. and Richards, F. M. (1967) *Journal of Molecular Biology* **27**, 563–78.

Wyckoff, R. W. G. (1924) *The Structure of Crystals*, New York: Chemical Catalog Co., pp. 142–3.

Xia, Z.-X., Shamala, N., Bethge, P. H., Lim, L. W., Bellamy, H. D., Xuong, N.-H., Lederer, F. and Mathews, F. S. (1987) *Proceedings of the National Academy, USA* **84**, 2629–33.

Xuong, Ng. H., Nielsen, C., Hamlin, R. and Anderson, D. (1985) *Journal of Applied Crystallography* **18**, 342–50.

Xuong, N. H., Freer, S. T., Hamlin, R., Nielsen, C. and Vernon, W. (1978) *Acta Crystallographica* A**34**, 289–96.

Yamashita, M. M., Almassy, R. J., Janson, C. A., Cascio, D. and Eisenberg, D. (1989) *Journal of Biological Chemistry* **264**, 17681–90.

Yeates, T. O., Komiya, H., Chirino, A., Rees, D. C., Allen, J. P.

and Feher, G. (1988) *Proceedings of the National Academy, USA* **85**, 7993–7.

Yonath, A. and Wittmann, H. G. (1988) *Biophysical Chemistry* **29**, 17–29.

Yonath, A., Leonard, K. R. and Wittmann, H. G. (1987) *Science* **236**, 813.

Yonath, A. E., Bartunik, H. D., Bartels, K. S. and Wittmann, H. G. (1984) *Journal of Molecular Biology* **177**, 201–6.

Yonath, A. E., Müssig, J., Tesche, B., Lorenz, S., Erdmann, V. A. and Wittmann, H. G. (1980) *Biochemistry International* **1**, 428–35.

Yonath, A., Glotz, C., Gewitz, M. S., Bartels, K. S., von Böhlen, K., Makowski, I. and Wittmann, H. G. (1988) *Journal of Molecular Biology* **203**, 831–4.

Yonath, A. E., Saper, M. A., Makowski, I., Müssig, J., Piefke, J., Bartunik, H. D., Bartels, K. S. and Wittmann, H. G. (1986) *Journal of Molecular Biology* **187**, 633–6.

Zachariasen, W. H. (1945) *Theory of X-ray Diffraction in Crystals*, New York: John Wiley.

Zurek, S., Papiz, M. Z., Machin, P. A. and Helliwell, J. R. (1985) *Information Quarterly for Protein Crystallography* No. 16, Daresbury Laboratory, pp. 37–40.

Glossary of commonly used symbols in the text (see also table A1.1)

λ Wavelength of incident beam

a, b, c Base vectors of the unit cell (magnitudes a, b, c)

α, β, γ Angles between **b** and **c**; **a** and **c**; and **a** and **b**, respectively

a*, **b***, **c*** Reciprocal lattice base vectors

d Interplanar spacing of a set of planes in a crystal

d_{min} Minimum value of d (resolution limit)

d_{max}^* Resolution limit in reciprocal space

θ or θ_B Bragg angle of diffraction; 2θ is the angle between incident and reflected beams

$(hk\ell)$ Miller indices defining a unique plane of reflection

F$(hk\ell)$ Structure factor

F(h) Structure factor (**h**$\equiv hk\ell$)

F$_P$, F$_{PH}$ Structure factor for protein (P) and protein+heavy atom (PH)

$|\mathbf{F}(hk\ell)|$ Structure amplitude

$|\mathbf{F(h)}|$ Structure amplitude

F Structure amplitude

α_P Phase of structure factor for protein crystal

α_{PH} Phase of structure factor for protein+heavy atom crystal

Φ_T Phase of structure factor for the wavelength-independent parts of the totality of scattering atoms

Φ_A Phase of the structure factor for the wavelength-dependent parts of the anomalously scattering atoms

$f=f_0+f'+if''$ Atomic scattering factor including the normal scattering component (f_0) and anomalous dispersion corrections (f' and f'')

$\Delta_{ANO}(hk\ell)$ Anomalous difference between structure amplitudes

$\Delta_{ISO}(hk\ell)$ Isomorphous difference between structure amplitudes

$P(u,v,w)$ Patterson synthesis

$\varrho(x,y,z)$ Electron density as a function of the coordinate in the unit cell (x,y,z)

$\varrho(\mathbf{r})$ Ditto ($\mathbf{r} \equiv x,y,z$)

μ Linear absorption coefficient

μ_m Mass absorption coefficient

$\eta_{hk\ell}$ Perfect crystal sample rocking width for reflecting plane ($hk\ell$)

η Mosaic spread of an ideally imperfect crystal sample

γ_H Horizontal beam convergence or divergence angle

γ_V Vertical beam convergence or divergence angle

ϕ_R Angular reflecting range for a reflecting plane ($hk\ell$)

ω Angular velocity of sample rotation or frequency of light beam (as per context)

V_x Crystal sample volume

dV Volume element of crystal

V_0 Unit cell volume

γ^{-1} Opening angle of synchrotron radiation

E Synchrotron machine energy or photon energy (as per context)

ϱ Radius of curvature of electron or positron in orbit in synchrotron

I_b Circulating current in storage ring

I_0 Incident photon intensity at the specimen (photons s^{-1} mm^{-2})

$\lambda_c(\varepsilon_c)$ Critical wavelength (energy) of synchrotron radiation spectrum

σ Electron source size

σ' Divergence angle of synchrotron radiation emitted (including electron divergence and opening angle, γ^{-1}, contribution)

$\varepsilon\ (\approx\sigma\sigma')$ Machine emittance

$\beta\approx\sigma/\sigma')$ Machine beta-function (in context)

λ_u Undulator magnet period

K Angular deflection of electron moving in a periodic magnet (as a multiple of γ^{-1})

θ Angle parameter with respect to magnet axis (in context)

$\delta\lambda/\lambda$ Spectral bandpass

p Monochromator (or mirror) to synchrotron tangent point distance

p' Monochromator (or mirror) to focus distance

α Monochromator oblique-cut angle

θ_c Critical angle of mirror reflection

λ_{min} Minimum wavelength in the polychromatic beam used for Laue diffraction

λ_{max} Maximum wavelength in the polychromatic beam used for Laue diffraction

$E(hk\ell)$ Energy in the diffracted beam from reflecting plane $hk\ell$

L Lorentz factor or monochromator illuminated length (as per context)

P Polarisation correction

A Absorption correction

e Charge on the electron

m Electron rest mass

c Speed of light

v Speed of electron

ε_0 Permittivity of free space

k Boltzmann constant or wave number (latter in context of EXAFS data analysis)

COMMONLY USED ABBREVIATIONS

SR   Synchrotron radiation
RLP   Reciprocal lattice point
DQE   Detective quantum efficiency
CCD   Charge coupled device
IP   Image plate
SP   Storage phosphor
MWPC   Multiwire proportional chamber
TV   Television detector
MAD   Multiwavelength anomalous dispersion
SIROAS   Single isomorphous replacement with optimised anomalous scattering
MIROAS   Multiple isomorphous replacement with optimised anomalous scattering
OAD   One or single wavelength anomalous dispersion experiment
RAE   Reduced absorption error
LUC   Large unit cell
TDS   Thermal diffuse scattering
XANES   X-ray absorption near edge structure
EXAFS   Extended X-ray absorption fine structure

SYNCHROTRON X-RADIATION LABORATORIES AND MACHINE ABBREVIATIONS

ESRF   European Synchrotron Radiation Facility, Grenoble

EMBL	European Molecular Biology Laboratory, Hamburg and Grenoble
SSRL	Stanford Synchrotron Radiation Laboratory
SRS	Synchrotron Radiation Source, Daresbury
APS	Advanced Photon Source, Argonne
NSLS	National Synchrotron Light Source, Brookhaven
PF	Photon Factory, Tsukuba
SPRING-8	Super Photon Ring, Harima
LURE	Laboratoire Utilisation Rayonnement Electromagnetique, Orsay

A full list of laboratories and addresses is given in Appendix 5.

UNITS AND DEFINITIONS

Wherever possible SI. Major exception – use of the Å preferred over the nm.

$$[\text{Flux}] = \text{photons}\,s^{-1}\,(0.1\%\ \delta\lambda/\lambda)^{-1}$$
$$[\text{Brightness}] = \text{photons}\,s^{-1}\,(0.1\%\ \delta\lambda/\lambda)^{-1}\,\text{mrad}^{-2}$$
$$[\text{Brilliance}] = \text{photons}\,s^{-1}\,(0.1\%\ \delta\lambda/\lambda)^{-1}\,\text{mrad}^{-2}\,\text{mm}^{-2}$$

Index